CAMBRIDGE STUDIES IN ADVANCED MATHEMATICS 215

ABELIAN MODEL CATEGORY THEORY

Offering a unique resource for advanced graduate students and researchers, this book treats the fundamentals of Quillen model structures on abelian and exact categories. Building the subject from the ground up using cotorsion pairs, it develops the special properties enjoyed by the homotopy category of such abelian model structures. A central result is that the homotopy category of any abelian model structure is triangulated and characterized by a suitable universal property – it is the triangulated localization with respect to the class of trivial objects. The book also treats derived functors and monoidal model categories from this perspective, showing how to construct tensor triangulated categories from cotorsion pairs.

For researchers and graduate students in algebra, topology, representation theory, and category theory, this book offers clear explanations of difficult model category methods that are increasingly being used in contemporary research.

James Gillespie is Professor of Mathematics at Ramapo College of New Jersey. His research interests are homological algebra and abstract homotopy theory and he is the author of 35 well-cited articles in the area, particularly on topics such as rings and modules, chain complexes, and sheaves.

CAMBRIDGE STUDIES IN ADVANCED MATHEMATICS

All the titles listed below can be obtained from good booksellers or from Cambridge University Press. For a complete series listing, visit www.cambridge.org/mathematics.

Already Published

175 G. Navarro *Character Theory and the McKay Conjecture*
176 P. Fleig, H. P. A. Gustafsson, A. Kleinschmidt & D. Persson *Eisenstein Series and Automorphic Representations*
177 E. Peterson *Formal Geometry and Bordism Operators*
178 A. Ogus *Lectures on Logarithmic Algebraic Geometry*
179 N. Nikolski *Hardy Spaces*
180 D.-C. Cisinski *Higher Categories and Homotopical Algebra*
181 A. Agrachev, D. Barilari & U. Boscain *A Comprehensive Introduction to Sub-Riemannian Geometry*
182 N. Nikolski *Toeplitz Matrices and Operators*
183 A. Yekutieli *Derived Categories*
184 C. Demeter *Fourier Restriction, Decoupling and Applications*
185 D. Barnes & C. Roitzheim *Foundations of Stable Homotopy Theory*
186 V. Vasyunin & A. Volberg *The Bellman Function Technique in Harmonic Analysis*
187 M. Geck & G. Malle *The Character Theory of Finite Groups of Lie Type*
188 B. Richter *Category Theory for Homotopy Theory*
189 R. Willett & G. Yu *Higher Index Theory*
190 A. Bobrowski *Generators of Markov Chains*
191 D. Cao, S. Peng & S. Yan *Singularly Perturbed Methods for Nonlinear Elliptic Problems*
192 E. Kowalski *An Introduction to Probabilistic Number Theory*
193 V. Gorin *Lectures on Random Lozenge Tilings*
194 E. Riehl & D. Verity *Elements of ∞-Category Theory*
195 H. Krause *Homological Theory of Representations*
196 F. Durand & D. Perrin *Dimension Groups and Dynamical Systems*
197 A. Sheffer *Polynomial Methods and Incidence Theory*
198 T. Dobson, A. Malnič & D. Marušič *Symmetry in Graphs*
199 K. S. Kedlaya *p-adic Differential Equations*
200 R. L. Frank, A. Laptev & T. Weidl *Schrödinger Operators:Eigenvalues and Lieb–Thirring Inequalities*
201 J. van Neerven *Functional Analysis*
202 A. Schmeding *An Introduction to Infinite-Dimensional Differential Geometry*
203 F. Cabello Sánchez & J. M. F. Castillo *Homological Methods in Banach Space Theory*
204 G. P. Paternain, M. Salo & G. Uhlmann *Geometric Inverse Problems*
205 V. Platonov, A. Rapinchuk & I. Rapinchuk *Algebraic Groups and Number Theory, I (2nd Edition)*
206 D. Huybrechts *The Geometry of Cubic Hypersurfaces*
207 F. Maggi *Optimal Mass Transport on Euclidean Spaces*
208 R. P. Stanley *Enumerative Combinatorics, II (2nd Edition)*
209 M. Kawakita *Complex Algebraic Threefolds*
210 D. Anderson & W. Fulton *Equivariant Cohomology in Algebraic Geometry*
211 G. Pineda *Villavicencio Polytopes and Graphs*
212 R. Pemantle, M. C. Wilson & S. Melczer *Analytic Combinatorics in Several Variables (2nd Edition)*
213 A. Yadin *Harmonic Functions and Random Walks on Groups*
214 Y. Kawamata *Algebraic Varieties: Minimal Models and Finite Generation*

Abelian Model Category Theory

JAMES GILLESPIE
Ramapo College of New Jersey

CAMBRIDGE
UNIVERSITY PRESS

CAMBRIDGE
UNIVERSITY PRESS

Shaftesbury Road, Cambridge CB2 8EA, United Kingdom

One Liberty Plaza, 20th Floor, New York, NY 10006, USA

477 Williamstown Road, Port Melbourne, VIC 3207, Australia

314–321, 3rd Floor, Plot 3, Splendor Forum, Jasola District Centre, New Delhi – 110025, India

103 Penang Road, #05–06/07, Visioncrest Commercial, Singapore 238467

Cambridge University Press is part of Cambridge University Press & Assessment, a department of the University of Cambridge.

We share the University's mission to contribute to society through the pursuit of education, learning and research at the highest international levels of excellence.

www.cambridge.org
Information on this title: www.cambridge.org/9781009449465

DOI: 10.1017/9781009449489

When citing this work, please include a reference to the DOI 10.1017/9781009449489

First published 2025

A catalogue record for this publication is available from the British Library

A Cataloging-in-Publication data record for this book is available from the Library of Congress

ISBN 978-1-009-44946-5 Hardback

To my wife, Tennille,
and my son, David

Contents

Preface

Quillen model categories provide a language to develop abstract homotopy theory and the tools to gain control over localizations of categories. This book focuses on the case of when the ground category is an abelian category, or more generally a Quillen exact category, and the model structure is compatible with the short exact sequences. In this case, Quillen's homotopy theory encompasses a vast generalization of traditional homological algebra. However, the model category perspective puts the emphasis on the homotopy category, and one focus of this book is on developing properties of the homotopy category of an abelian model structure. Examples of these include traditional derived categories of rings and schemes, as well as the more recently studied coderived and contraderived categories. Another source of examples are the stable module categories arising in representation theory and Gorenstein homological algebra.

In fact, examples and applications of abelian model structures are appearing more frequently in contemporary research in fields adjacent to homological algebra. As a result, many algebraists now have a general understanding of an abelian model structure as being equivalent to the interlacing of two complete cotorsion pairs. Yet there is a disconnect between traditional methods of homological algebra and the homotopy-theoretic approach of model categories. Among the reasons for this is that model categories have a reputation of being difficult, and naturally, the standard model category references have topology in mind. A related reason is that the categorical machinery needed for algebraic topology goes beyond what is actually needed for additive categories. Indeed abelian model categories are easier than general model categories, and we have learned that they possess quite special properties that general model categories do not. Here are a few points to support this idea.

- The definition of an abelian model category moves almost all of the focus from morphisms to (the simpler) objects. Indeed the fundamental result known as Hovey's Correspondence allows all the data to be packaged as a triple $\mathfrak{M} = (\mathcal{Q}, \mathcal{W}, \mathcal{R})$ of classes of objects rather than a triple of classes of morphisms. The morphisms known as cofibrations, fibrations, and weak equivalences are completely determined by these three classes of objects.
- In the abelian case, the left and right homotopy relations identify to a single homotopy relation, and it is *always* an equivalence relation. Indeed two morphisms f, g in an abelian model structure $\mathfrak{M} = (\mathcal{Q}, \mathcal{W}, \mathcal{R})$ are left homotopic if and only if they are right homotopic and this happens if and only if $g - f$ factors through some trivial object, that is, an object in $\mathcal{W}$. This actually stems from the fact that finite coproducts identify with finite products in additive categories. Moreover, the classical notions of *very good left (and right) homotopy* also determine equivalence relations. These play an interesting role in defining the triangulated structure on the homotopy category.
- The homotopy category of an abelian model category is *always* a triangulated category. This follows from work of Nakaoka and Palu. This book gives a detailed treament of the triangulated structure, and in a way that relates very closely to Happel's construction of a triangulated structure on the stable category associated to a Frobenius category. In fact, an important case is that of an *hereditary* abelian model structure where the full subcategory of cofibrant–fibrant objects actually is a Frobenius category.
- In practice, abelian (and exact) categories typically possess a set of generators that can "detect" the short exact sequences. It is also typically the case that every object is some sort of directed colimit of objects from some set. These properties lead to further special properties of the homotopy category. In particular, they are often well generated in the sense of Neeman.

It is valuable for these ideas and the methods of abelian model categories to be accessible to more mathematicians and their graduate students. So in the same way that one may reach for a book on additive or abelian categories, this book is intended to be a resource on the theory of *abelian* model categories. We develop the theory from the ground up, in terms of cotorsion pairs.

This book is well suited for researchers and graduate students who are familiar with the basic language of categories, functors, and natural transformations, and who have had some exposure to introductory homological algebra. Ideally the reader will be familiar with projective, injective, and flat modules over a ring R, chain complexes of R-modules, and the cohomology groups $\mathrm{Ext}^n_R(M, N)$. In a few places we will describe applications and examples in the setting of Grothendieck categories, and discuss Grothendieck's (AB3)–(AB5) axioms in order to motivate some of our definitions for Quillen exact categories. Having said this, the exposition here is very elementary and most homological notions are defined in this book, and in the general setting of exact categories. It is not unreasonable for a student just learning homological algebra to study from this book simultaneously. However, this book focuses on an elementary development of the *theory*, rather than examples and applications of abelian model structures. One reason for this is that the existing literature already contains quite many readable constructions of examples and applications of abelian model structures. A focused and updated presentation of the basic theoretical consequences of abelian model structures is what is now needed. Another reason is that the axioms for model categories are very strong. So any particular area of application (e.g. to sheaves and schemes, Gorenstein homological algebra, functor categories, or categories relevant to functional analysis) requires a significant amount of field-specific knowledge in order to build and apply model structures. We instead will be content to periodically point the reader to the literature for further details on certain examples. The overall goal of this book is to provide the average mathematician who uses some form of homological algebra, the theoretical foundations to understand and apply model category methods.

This book is a result of the inspiration I have received from the work of many mathematicians. It starts with the direct influence of Mark Hovey, but now there are many others whose work and support have made this book possible. In addition to my friends and colleagues across the United States and Europe I am thankful to many mathematicians throughout our international community. Rather than attempt an exhaustive list of names, I will just say thank you to my friends and colleagues in some particular hubs of the community: Murcia, Almería, Aarhus, Copenhagen, Padova, Verona, Prague, Nanjing, Lanzhou, Athens, Montevideo, and Valdivia. This book would not be possible without the contributions of everyone in the field. I also thank my colleagues at Ramapo College. Their support for pure academic research led to a sabbatical leave in which much of this book was written. Most of all, I thank my family for their lasting support.

Introduction and Main Examples

Since the introduction of the derived category of an abelian category, triangulated structures have become an integral part of homological algebra. Abelian model structures provide a convenient method for implementing and studying triangulated structures arising as a type of localization of an abelian (or exact) category. Indeed the homotopy category associated to any abelian model structure is a triangulated category. This section is meant to give the reader a broad overview of this idea along with a survey of the most fundamental examples appearing on R-Mod, the category of (say left) R-modules over a ring R, and Ch(R), the associated category of chain complexes of R-modules. These categories are the simplest ones with meaningful applications and they serve as a common ground for anyone that might be interested in learning some of the theory of abelian model categories. These examples will also be referenced throughout the book, to illustrate the general theory as it is being developed.

Cotorsion pairs are the cornerstone of the theory of abelian model categories. Let $(\mathcal{X}, \mathcal{Y})$ be a pair of classes of objects in an abelian category $\mathcal{A}$, such as R-Mod or Ch(R), and consider short exact sequences

$$0 \to Y \to Z \to X \to 0 \tag{1}$$

in $\mathcal{A}$. We say that $(\mathcal{X}, \mathcal{Y})$ is a *complete cotorsion pair* if it satisfies the following.

- $X \in \mathcal{X}$ if and only if every such short exact sequence (1) with $Y \in \mathcal{Y}$ splits, thus inducing a direct sum decomposition $Z \cong Y \bigoplus X$.
- Dually, $Y \in \mathcal{Y}$ if and only if every such short exact sequence (1) with $X \in \mathcal{X}$ splits.
- Given any $A \in \mathcal{A}$, there exists a short exact sequence $0 \to Y \to X \to A \to 0$ with $X \in \mathcal{X}$ and $Y \in \mathcal{Y}$. We say $(\mathcal{X}, \mathcal{Y})$ *has enough projectives.*
- Dually, there exists a short exact sequence $0 \to A \to Y' \to X' \to 0$ with $Y' \in \mathcal{Y}$ and $X' \in \mathcal{X}$. We say $(\mathcal{X}, \mathcal{Y})$ *has enough injectives.*

The first two conditions can be expressed more succinctly by writing $\mathcal{X}^{\perp} = \mathcal{Y}$, and ${}^{\perp}\mathcal{Y} = \mathcal{X}$. The usual definition of this orthogonality is given in terms of the Yoneda Ext functor, $\mathrm{Ext}^1_{\mathcal{A}}(-,-)$, discussed in Section 1.6.

Note that the concept of a complete cotorsion pair generalizes the fundamental idea from homological algebra that $\mathcal{A} = R$-Mod has enough projectives and enough injectives. Indeed this can be summarized in the language of cotorsion pairs by saying that the pair $(\mathcal{A}, \mathcal{I})$, where $\mathcal{I}$ is the class of all injective R-modules, is a complete cotorsion pair. Dually, $(\mathcal{P}, \mathcal{A})$ is a complete cotorsion pair, where $\mathcal{P}$ is the class of projective R-modules. We call these, respectively, the *canonical injective* and *canonical projective* cotorsion pairs in R-Mod. In Chapter 2 we define and study cotorsion pairs in the quite general setting of Quillen exact categories, in terms of Yoneda's Ext functor.

A result known as Hovey's Correspondence reduces an abelian model structure (whatever that means precisely) on $\mathcal{A}$ to a triple of classes of objects, $\mathfrak{M} = (\mathcal{Q}, \mathcal{W}, \mathcal{R})$, where (i) $\mathcal{W}$ satisfies the property that if two out of three terms in a short exact sequence are in $\mathcal{W}$ then so must be the third, and (ii) $(\mathcal{Q}, \mathcal{W} \cap \mathcal{R})$ and $(\mathcal{Q} \cap \mathcal{W}, \mathcal{R})$ are each complete cotorsion pairs. Objects in $\mathcal{Q}$ (resp. $\mathcal{R}$) are called *cofibrant* (resp. *fibrant*), and objects in $\mathcal{W}$ are called *trivial*. Hovey's Correspondence, Theorem 4.25, shows that such a triple $\mathfrak{M}$ is equivalent to the seemingly more complicated notion of an abelian model structure. Thus one could even define an abelian model structure to be such a *Hovey triple*, and in fact this is the philosophy and approach taken in this book. Since a Hovey triple packages a great amount of data in a very simple way, this perspective has proven to be beneficial. It also makes the subject more accessible.

Before proceeding, let us now give what are perhaps the two most fundamental examples of abelian model categories. Here we take $\mathcal{A} = \mathrm{Ch}(R)$, the category of chain complexes of R-modules. By a chain complex X, we mean a $\mathbb{Z}$-indexed sequence of R-module homomorphisms

$$\cdots \to X_{n+1} \xrightarrow{d_{n+1}} X_n \xrightarrow{d_n} X_{n-1} \to \cdots$$

satisfying $d_n d_{n+1} = 0$ for all $n \in \mathbb{Z}$. A morphism $f \colon X \to Y$ of chain complexes is a *chain map*, that is, a collection of R-module homomorphisms $f_n \colon X_n \to Y_n$ making all squares commute with the d_n.

- The *standard projective model structure* on $\mathrm{Ch}(R)$ is most easily described by a triple

$$\mathrm{Ch}(R)_{proj} = \left(dg\widetilde{\mathcal{P}}, \widetilde{\mathcal{E}}, All\right) \tag{2}$$

  of classes of chain complexes. Here, $\mathcal{W} = \widetilde{\mathcal{E}}$ is the class of all acyclic (i.e. exact) chain complexes of R-modules; those for which all homology groups are 0. The chain complexes in the class $\mathcal{Q} = dg\widetilde{\mathcal{P}}$ of cofibrant objects are usually called *DG-projective*, (or sometimes *semiprojective*, or *homotopically projective*). More details on this model structure are given shortly in Example 3. But we note now that the homotopy category of $\text{Ch}(R)_{proj}$ is $\mathcal{D}(R)$, the derived category of R. A fundamental idea, discussed a bit more shortly and made precise in Section 6.7, is that $\mathcal{D}(R)$ is the triangulated localization of $\text{Ch}(R)$ with respect to the class $\widetilde{\mathcal{E}}$ of acyclic complexes. Moreover, each chain complex X is isomorphic in the homotopy category, $\mathcal{D}(R)$, to a DG-projective complex, QX.

- Dually, there exists the *standard injective model structure* on $\text{Ch}(R)$, given by the triple

$$\text{Ch}(R)_{inj} = \left(All, \widetilde{\mathcal{E}}, dg\widetilde{\mathcal{I}}\right) \tag{3}$$

of classes of chain complexes. Again, $\mathcal{W} = \widetilde{\mathcal{E}}$ denotes the class of all acyclic chain complexes. But this time every chain complex is cofibrant while the fibrant objects form the class $dg\widetilde{\mathcal{I}}$ of *DG-injective* chain complexes. This gives an injective model for the derived category $\mathcal{D}(R)$.

We will explain how to establish the existence of these two model structures later, in Examples 3 and 4. We will also relate them to the standard fact that $\text{Ext}^n_R(M, N)$ may be computed by way of either a projective resolution of M, or an injective coresolution of N.

The first big idea behind model categories is the construction of $\text{Ho}(\mathfrak{M})$, the homotopy category of $\mathfrak{M}$. In general, we want it to be a category with the same objects as $\mathcal{A}$, and in the abelian case it should, at the very least, be an additive category for which the trivial objects have been "killed". That is, each $W \in \mathcal{W}$ should identify as a 0 object in $\text{Ho}(\mathfrak{M})$. But the additive category obtained by setting the objects of $\mathcal{W}$ to 0 is merely $\text{St}(\mathcal{A})$, a category we will call the *stable category* of $\mathfrak{M}$. By definition, the category $\text{St}(\mathcal{A})$ has the same objects as $\mathcal{A}$, but its morphism sets are defined by

$$\underline{\text{Hom}}(A, B) := \text{Hom}_{\mathcal{A}}(A, B)/\sim,$$

where $\sim$ is the equivalence relation defined by $f \sim g$ if and only if $g - f$ factors through some trivial object $W \in \mathcal{W}$. The stable category $\text{St}(\mathcal{A})$ is not the homotopy category $\text{Ho}(\mathfrak{M})$, but it is an important first step. While the stable category $\text{St}(\mathcal{A})$ may be thought of as the additive localization with respect to $\mathcal{W}$, the homotopy category, $\text{Ho}(\mathfrak{M})$, may be thought of as the triangulated localization with respect to $\mathcal{W}$. This is made precise in Section 6.7, but let us now note this: We also want any short exact sequence $0 \to A \to B \to C \to 0$ in $\mathcal{A}$ to identify as a *distinguished triangle* in $\text{Ho}(\mathfrak{M})$, with the idea being that distinguished triangles hold onto much of the homological data encoded within the class of short exact sequences. This condition implies in particular that any monomorphism $i\colon A \to B$, with trivial cokernel, $\operatorname{Cok} i \in \mathcal{W}$, shall become an isomorphism in $\operatorname{Ho}(\mathfrak{M})$. We call such a morphism i a trivial monic. Similarly, the dual statement will be true: Each trivial epic, that is, epimorphism $p\colon B \to C$ with $\operatorname{Ker} p \in \mathcal{W}$, will become an isomorphism in $\operatorname{Ho}(\mathfrak{M})$. This leads to the idea that we wish to invert the class of all morphisms f which factor as $f = pi$, where i is a trivial monic and p is a trivial epic. Such morphisms make up the class of **weak equivalences**, which we denote by We.

Following a purely categorical approach that does not involve the classes $\mathcal{Q}$ or $\mathcal{R}$, one may formally invert the morphisms of We. With this approach one defines the homotopy category $\operatorname{Ho}(\mathfrak{M}) := \mathcal{A}\left[We^{-1}\right]$, by keeping the objects the same as $\mathcal{A}$, but defining the morphisms to be finite "zig-zags" of $\mathcal{A}$-morphisms where we allow the reversal of any arrow in We; see Exercise 5.4.1 for more details. This forces all weak equivalences to become isomorphisms and one obtains a canonical functor $\gamma\colon \mathcal{A} \to \mathcal{A}\left[We^{-1}\right]$ which is universally initial with respect to the property of inverting the morphisms of We. The standard result about Quillen model categories is that this construction does indeed produce a category in the sense that we still have small hom-sets, but more importantly that there is a more useful way to construct the homotopy category, $\operatorname{Ho}(\mathfrak{M})$. For our case of an abelian model category, $\mathfrak{M} = (\mathcal{Q}, \mathcal{W}, \mathcal{R})$, we get the following elegant theorem. To describe it, we need the notion of (co)fibrant approximations. Their existence follows immediately from the definition of a complete cotorsion pair, as given above. Indeed since $(\mathcal{Q}, \mathcal{W} \cap \mathcal{R})$ is a complete cotorsion pair, there exists for each object A, a short exact sequence

$$0 \to R_A \xrightarrow{i_A} QA \xrightarrow{p_A} A \to 0 \tag{4}$$

with $QA \in \mathcal{Q}$ and $R_A \in \mathcal{W} \cap \mathcal{R}$. Although such a short exact sequence is not unique, any such morphism p_A is unique in $\operatorname{St}(\mathcal{A})$, up to a canonical isomorphism. We call any such QA a *cofibrant approximation*, or a *cofibrant replacement*, of A. Note that p_A is a trivial epic, so it will provide an isomorphism $QA \cong A$ in $\operatorname{Ho}(\mathfrak{M})$. On the other hand, the dual notion is that of a *fibrant approximation* (or *fibrant replacement*), RA, obtained by taking a short exact sequence

$$0 \to A \xrightarrow{j_A} RA \xrightarrow{q_A} Q_A \to 0 \tag{5}$$

with $RA \in \mathcal{R}$ and $Q_A \in \mathcal{Q} \cap \mathcal{W}$. Objects in $\mathcal{Q} \cap \mathcal{R}$ are called *bifibrant* and by a *bifibrant approximation* of A we mean RQA, that is, a fibrant approximation of a cofibrant approximation of A.

Theorem 1 (The Fundamental Theorem of Abelian Model Categories) *Let $\mathfrak{M} = (Q, \mathcal{W}, \mathcal{R})$ be an abelian model category (i.e. Hovey triple) on an abelian category $\mathcal{A}$. Then there is an additive category,* $\mathrm{Ho}(\mathfrak{M})$, *called the* **homotopy category of** $\mathfrak{M}$, *whose objects are the same as those of $\mathcal{A}$ but whose morphisms are given by*

$$\mathrm{Ho}(\mathfrak{M})(A, B) := \underline{\mathrm{Hom}}(RQA, RQB),$$

where these are morphism sets in the stable category, $\mathrm{St}(\mathcal{A})$, *between any choice of bifibrant approximations. Moreover, we have the following.*

(1) *The inclusion $Q \cap \mathcal{R} \hookrightarrow \mathcal{A}$ induces an equivalence of categories*

$$\mathrm{St}(Q \cap \mathcal{R}) \xrightarrow{\cong} \mathrm{Ho}(\mathfrak{M}),$$

whose inverse is given by any bifibrant approximation assignment RQ.

(2) *There is a functor $\gamma\colon \mathcal{A} \to \mathrm{Ho}(\mathfrak{M})$ which is the identity on objects but bifibrant approximation on morphisms. It is a localization of $\mathcal{A}$ with respect to the class We (of all weak equivalences), and hence there is a canonical isomorphism of categories*

$$\mathrm{Ho}(\mathfrak{M}) \cong \mathcal{A}\left[We^{-1}\right].$$

(3) *For any choice of cofibrant approximation QA and fibrant approximation RB, there is a natural isomorphism* $\mathrm{Ho}(\mathfrak{M})(A, B) \cong \underline{\mathrm{Hom}}(QA, RB)$.

These results are all proved in Chapter 5, where our initital study of the homotopy category takes place. Chapter 6 then studies the triangulated structure that exists on the homotopy category. Our proofs are in the more general setting of any Quillen exact category for which every split monomorphism admits a cokernel. These are the so-called *weakly idompotent complete* exact categories which we find to be the most natural abstract setting to develop the theory of abelian model categories.

Examples: Model Structures for the Derived Category of R

As promised, the remainder of this introductory section will illustrate some of the most well-known examples of abelian model structures for the cases $\mathcal{A} = R$-Mod, and $\mathcal{A} = \mathrm{Ch}(R)$. We start by describing the construction of the standard projective and injective model structures on $\mathrm{Ch}(R)$.

Through Hovey's Correspondence, the problem of showing a cotorsion pair to be complete corresponds to proving the Factorization Axiom for model categories. Quillen's small-object argument is typically the tool used to construct such factorizations. Translating back to the abelian case, this corresponds to the notion of a cotorsion pair that is cogenerated by a set (as opposed to a proper class). Using the notation alluded to above, after the definition of a cotorsion pair, a set $\mathcal{S}$ is said to **cogenerate** a cotorsion pair $(\mathcal{X}, \mathcal{Y})$ if $\mathcal{S}^{\perp} = \mathcal{Y}$. It just means $Y \in \mathcal{Y}$ if and only if every short exact sequence $0 \to Y \to Z \to S \to 0$ in $\mathcal{A}$, with $S \in \mathcal{S}$, splits. Said another way, $Y \in \mathcal{Y}$ if and only if $\mathrm{Ext}^1_{\mathcal{A}}(Y, S) = 0$ for all $S \in \mathcal{S}$. Chapter 9 details a version of Quillen's small object argument that is useful for constructing complete cotorsion pairs in very general exact categories. Our approach is inspired by the *efficient exact categories* introduced in Saorín and Šťovíček [2011]. The following is a special case of the powerful Theorem 9.34; see also Corollary 9.40 and Corollary 12.4.

Theorem 2 (Eklof and Trlifaj [2001]) *Let $\mathcal{A} = R$-Mod, or $\mathrm{Ch}(R)$. Then any set $\mathcal{S}$ (not a proper class) of objects in $\mathcal{A}$ cogenerates a complete cotorsion pair $({}^{\perp}(\mathcal{S}^{\perp}), \mathcal{S}^{\perp})$.*

With this, we can easily construct the standard projective model structure $\mathrm{Ch}(R)_{proj} = \left(dg\widetilde{\mathcal{P}}, \widetilde{\mathcal{E}}, All\right)$, described earlier in (2). A comment on notation: Given any R-module M, we denote by $S^n(M)$ the chain complex consisting only of M in degree n and 0 elsewhere. We call $S^n(M)$ the *n-sphere on M*.

Example 3 (The Standard Projective Model Structure for $\mathcal{D}(R)$) Again, $\widetilde{\mathcal{E}}$ denotes the class of all acyclic (i.e. exact) chain complexes of R-modules. Let $\mathcal{S} = \{S^n(R)\}$ be the set of all n-spheres on R, where here R is considered as a (left) R-module over itself. There is an isomorphism $\mathrm{Ext}^1_{\mathrm{Ch}(R)}\left(S^{n+1}(R), X\right) \cong H_nX$, from which it follows that $\mathcal{S}^{\perp} = \widetilde{\mathcal{E}}$. So by Theorem 2 we have a complete cotorsion pair, $\left({}^{\perp}\widetilde{\mathcal{E}}, \widetilde{\mathcal{E}}\right)$. This is the key to the existence of the Hovey triple

$$\mathrm{Ch}(R)_{proj} = \left(dg\widetilde{\mathcal{P}}, \widetilde{\mathcal{E}}, All\right),$$

where $dg\widetilde{\mathcal{P}} := {}^{\perp}\widetilde{\mathcal{E}}$ is the class of *DG-projective* chain complexes. The DG-projective complexes are characterized as those complexes P such that each P_n is a projective R-module and any chain map $P \to E$ is null homotopic whenever E is an exact chain complex. The latter condition is automatic if P is a bounded below complex of projectives.

The chain homotopy category of R, denoted $K(R)$, is the category whose objects are chain complexes but whose morphisms are chain homotopy classes of chain maps. Results of Sections 10.5 and 10.6 relate abelian model structures on $\mathrm{Ch}(R)$ to the classical notion of Verdier quotients of $K(R)$. In particular, the formalities associated to the triangulated structure on $\mathrm{Ho}(\mathrm{Ch}(R))$ (such as the suspension functor, Σ, the mapping cone, $\mathrm{Cone}(f)$, etc.) will typically coincide with the classical notions in $K(R)$. In the current case, $\mathrm{Ho}\left(\mathrm{Ch}(R)_{proj}\right)$ identifies as the Verdier quotient $\mathcal{D}(R) := K(R)/\widetilde{\mathcal{E}}$, and shows the category to be equivalent to the isomorphic closure of $dg\widetilde{\mathcal{P}}$ in $K(R)$. In the language of Spaltenstein [1988], these are the precisely the K-projective complexes. □

Example 3 is a special case of Corollary 10.42 where the projective model structure is constructed in a far more general setting.

It is often stated that model categories encompass homological algebra. Let us give an example, beyond the above construction of $\mathcal{D}(R)$, supporting this sentiment. A well-known fact in algebra is that for two given R-modules M and N, the cohomology groups $\mathrm{Ext}^n_R(M,N)$ may be computed by taking a projective resolution

$$\cdots \to P_2 \xrightarrow{d_2} P_1 \xrightarrow{d_1} P_0 \xrightarrow{\epsilon} M \to 0 \tag{6}$$

of M, and then taking the nth-cohomology group of the cochain complex obtained by applying $\mathrm{Hom}_R(-,N)$ to the projective resolution

$$\mathcal{P}_\circ \equiv \cdots \to P_2 \xrightarrow{d_2} P_1 \xrightarrow{d_1} P_0 \to 0.$$

That is,

$$\mathrm{Ext}^n_R(M,N) \cong H^n[\mathrm{Hom}_R(\mathcal{P}_\circ, N)] \cong K(R)(\mathcal{P}_\circ, S^n(N)).$$

From the abelian model category perspective, the projective resolution $\mathcal{P}_\circ \xrightarrow{\epsilon} M \to 0$ of (6) is a cofibrant approximation of M in the projective model structure $\mathrm{Ch}(R)_{proj} = \left(dg\widetilde{\mathcal{P}}, \widetilde{\mathcal{E}}, All\right)$. To see this, we identify M with the 0-sphere complex $S^0(M)$. Note that the projective resolution of (6) may then be construed as a short exact sequence of chain complexes

$$0 \to E \to \mathcal{P}_\circ \xrightarrow{\epsilon} S^0(M) \to 0.$$

The projective resolution $\mathcal{P}_\circ$ is indeed a DG-projective complex while the kernel E is an exact complex. Referring to the very definition of a cofibrant approximation, given previously in Equation (4), this means that the projective resolution $\mathcal{P}_\circ$ is a cofibrant approximation of $S^0(M)$ in the projective model structure $\mathrm{Ch}(R)_{proj}$. So by part (3) of the Fundamental Theorem 1,

$$\mathrm{Ho}\left(\mathrm{Ch}(R)_{proj}\right)\left(S^0(M), S^n(N)\right) \cong \underline{\mathrm{Hom}}(\mathcal{P}_\circ, S^n(N)).$$

But in this case we have $\underline{\mathrm{Hom}}(\mathcal{P}_\circ, S^n(N)) = K(R)(\mathcal{P}_\circ, S^n(N))$; in the stable category $\mathrm{St}(\mathrm{Ch}(R)_{proj})$, morphisms with cofibrant domain are precisely chain homotopy classes of chain maps. Putting all this together we have

$$\mathrm{Ho}\left(\mathrm{Ch}(R)_{proj}\right)\left(S^0(M), S^n(N)\right) \cong \mathrm{Ext}^n_R(M,N).$$

Note that this identifies $\mathrm{Ext}^n_R(M,N)$ with a morphism set in the homotopy category.

On the other hand, recall that $\mathrm{Ext}^n_R(M, N)$ may be computed by taking an injective coresolution in the second variable N. This corresponds to the existence of the dual standard injective model structure described earlier in (3). However, the dual of Theorem 2 doesn't hold, simply because R-modules don't satisfy the dual properties needed to carry out the constructions. The following example indicates the set-theoretic flavor of arguments that are typically used to construct (cofibrantly generated) abelian model structures.

Example 4 (The Standard Injective Model Structure for $\mathcal{D}(R)$) Again, let $\widetilde{\mathcal{E}}$ denote the class of all exact chain complexes. Let κ be a regular cardinal number satisfying $\kappa \geq \max\{|R|, \omega\}$. Up to isomorphism, there exists a set (that is not a proper class) of exact chain complexes E with each $|E_n| \leq \kappa$. Let $\widetilde{\mathcal{E}}_\kappa$ denote a set of isomorphism representatives for all the exact chain complexes with this property. It is not too hard to argue that for each exact chain complex $E \in \widetilde{\mathcal{E}}$, there exists an exact subcomplex $E' \subseteq E$ with each $|E'_n| \leq \kappa$. It follows from this that the set $\widetilde{\mathcal{E}}_\kappa$ cogenerates a complete cotorsion pair $\left(\widetilde{\mathcal{E}}, \widetilde{\mathcal{E}}^\perp\right)$; see Exercise 10.9.1. This provides the abelian model structure,

$$\mathrm{Ch}(R)_{inj} = \left(All, \widetilde{\mathcal{E}}, dg\widetilde{\mathcal{I}}\right),$$

where this time, the class $dg\widetilde{\mathcal{I}} := \widetilde{\mathcal{E}}^\perp$ is the class of *DG-injective* chain complexes. Such complexes are characterized as chain complexes I of injective R-modules such that any chain map $E \to I$ is null homotopic whenever E is an exact chain complex. In other words, these are Spaltenstein's K-injective complexes but with injective components. □

Example: Modules over Iwanaga–Gorenstein Rings

The category of R-modules over an Iwanaga–Gorenstein ring possesses a beautiful homotopy theory which can nicely be described in terms of abelian model structures. This corresponds to part of the subject known as Gorenstein homological algebra, as presented in the book by Enochs and Jenda [2000].

Let's first consider some immediate consequences of having an abelian model structure $\mathfrak{M} = (\mathcal{Q}, \mathcal{W}, \mathcal{R})$ on R-Mod, regardless of the ring R we are considering. If we are to have such an $\mathfrak{M}$, then since $(\mathcal{Q} \cap \mathcal{W}, \mathcal{R})$ is a cotorsion pair, it is immediate that $\mathcal{Q} \cap \mathcal{W}$ must contain all projective R-modules. On the other hand, $\mathcal{W} \cap \mathcal{R}$ must contain all injective modules, and so the class $\mathcal{W}$ must contain all projective and injective modules. But then by the 2 out of 3 property on short exact sequences, $\mathcal{W}$ must contain any module of finite injective dimension, or of finite projective dimension. Consequently, the smallest class of trivial objects possible is the class of all modules having either finite injective dimension, or finite projective dimension.

In the case of an Iwanaga–Gorenstein ring R, there exist abelian model structures on R-Mod for which $\mathcal{W}$ is nothing more than this minimal class of modules. In fact, *Iwanaga–Gorenstein* rings are characterized as the two-sided Noetherian rings for which the class of all (left and right) modules of finite injective dimension coincides with the class of all (left, resp. right) modules of finite projective dimension. This class of Noetherian rings includes the simplest case of quasi-Frobenius rings which are characterized by the property that a module is injective if and only if it is projective. A key example is the group ring $R = k[G]$ where k is a field and G is a finite group. In representation theory, the stable module category $\mathrm{St}(k[G])$ naturally arises from $k[G]$-Mod by killing the projective–injective modules. Formally, the objects of $\mathrm{St}(k[G])$ are just $k[G]$-modules, but morphisms are identified by the relation $f \sim g$ if and only if $g - f$ factors through a projective–injective $k[G]$-module. The Tate cohomology groups of G reside as morphism sets in $\mathrm{St}(k[G])$. Beyond quasi-Frobenius rings, Iwanaga–Gorenstein rings include, for instance, integral group rings $\mathbb{Z}[G]$ and p-adic group rings $\mathbb{Z}_p[G]$ over finite groups G.

Example 5 (Hovey [2002]) Let R be an Iwanaga–Gorenstein ring and let $\mathcal{W}$ denote the class of all (say left) R-modules of finite projective dimension, equivalently, finite injective dimension. Then we have the following.

- There is an abelian model structure $\mathfrak{M}_{inj} = (All, \mathcal{W}, \mathcal{GI})$ on R-Mod in which every module is cofibrant. The modules in $\mathcal{GI}$ are called *Gorenstein injective*, and they are precisely the modules appearing as a cycle in some (possibly unbounded) exact chain complex of injective modules.
- There is an abelian model structure $\mathfrak{M}_{proj} = (\mathcal{GP}, \mathcal{W}, All)$ on R-Mod in which every module is fibrant. The modules in $\mathcal{GP}$ are called *Gorenstein projective*, and they are precisely the modules appearing as a cycle in some exact chain complex of projective modules.

Since these two models share the same class of trivial objects they each model the same homotopy category which is a generalization of $\mathrm{St}(R)$, the stable module category of a quasi-Frobenius ring R. Referring to the Fundamental Theorem 1, the homotopy category is equivalent to each of the two stable module categories, $\mathrm{St}(\mathcal{GI})$, and $\mathrm{St}(\mathcal{GP})$. □

Examples: Frobenius Categories and Chain Homotopy Categories

Let R be a quasi-Frobenius ring, such as $R = k[G]$. Since the injective modules and projective modules coincide we get that the model structures $\mathfrak{M}_{inj}$ and $\mathfrak{M}_{proj}$ of Example 5 each coincide and become the simple Hovey triple

$\mathfrak{M} = (All, \mathcal{W}, All)$ on R-Mod. So the associated homotopy category, $\mathrm{Ho}(\mathfrak{M})$, is exactly the stable module category $\mathrm{St}(k[G])$. These properties are reflecting that R-Mod is a *Frobenius category* whenever R is a quasi-Frobenius ring. Such structures will appear throughout this book. Let us give another standard example of a Frobenius category, one arising in the context of chain complexes of modules over a general ring R. Since model categories capture the idea of homotopy it is not surprising that $K(R)$, the chain homotopy category of R, is itself the homotopy category of an abelian model structure on $\mathrm{Ch}(R)$.

Example 6 (Frobenius Model Structure for $K(R)$) Let $\mathrm{Ch}(R)_{dw}$ denote the category of chain complexes along with the class of all short exact sequences

$$0 \to W \xrightarrow{f} X \xrightarrow{g} Y \to 0$$

that are *degreewise* split exact. That is, each $0 \to W_n \xrightarrow{f_n} X_n \xrightarrow{g_n} Y_n \to 0$ is a split exact sequence of R-modules. Then $\mathrm{Ch}(R)_{dw}$ is an example of a Quillen exact category in the sense studied in Chapter 1. We get a model structure on $\mathrm{Ch}(R)$ which is abelian relative to $\mathrm{Ch}(R)_{dw}$, and it may be described quite succinctly by a Hovey triple

$$\mathfrak{M}_{K(R)} = (\mathrm{Ch}(R), \mathcal{W}, \mathrm{Ch}(R)). \tag{7}$$

This time $\mathcal{W}$ denotes the class of all contractible chain complexes; they are precisely the projective–injective objects of $\mathrm{Ch}(R)_{dw}$. Since two chain maps are chain homotopic if and only if their difference factors through a contractible complex, it follows again from the above Fundamental Theorem 1 that $\mathrm{Ho}(\mathfrak{M}_{K(R)}) = K(R)$. □

Chain complexes over general additive categories are studied in Chapter 10. The above example is a special case of part of Theorem 10.20.

The relevance of Frobenius categories to the theory of abelian model categories goes far beyond providing easy examples of such model structures. The vast majority of the abelian model structures that have arisen in applications have been *hereditary* model structures. These are studied in Chapter 8. A notable feature is that their homotopy categories are triangle equivalent to the stable category of a Frobenius category with its classical triangulation from Happel [1988]. See Theorem 8.6.

Examples: Flat Model Structures on Modules and Complexes

Note that Example 3 lifts the canonical projective cotorsion pair, $(\mathcal{P}, \mathcal{A})$ on R-Mod, to the standard projective model structure on $\mathrm{Ch}(R)$. On the other hand, Example 4 lifts the canonical injective cotorsion pair, $(\mathcal{A}, \mathcal{I})$, to the standard injective model structure on Ch(R). In general, there are infinitely many intermediate cotorsion pairs that will lift to an abelian model structure on Ch(R). The following is a special case of Theorem 10.49 which is stated for Grothendieck categories.

Example 7 (General Model Structures for $\mathcal{D}(R)$) Let $(\mathcal{X}, \mathcal{Y})$ be an hereditary cotorsion pair on R-Mod, that is cogenerated by a set. Then it lifts to an hereditary model structure on chain complexes,

$$\mathrm{Ch}(R)_{(\mathcal{X},\mathcal{Y})} = \left(dg\widetilde{\mathcal{X}}, \widetilde{\mathcal{E}}, dg\widetilde{\mathcal{Y}}\right), \tag{8}$$

whose homotopy category again is equivalent to the derived category, $\mathcal{D}(R)$. The complexes in $dg\widetilde{\mathcal{X}}$ (resp. $dg\widetilde{\mathcal{Y}}$) are built up from R-modules in $\mathcal{X}$ (resp. $\mathcal{Y}$). □

An important example comes from Enochs' flat cotorsion pair, $(\mathcal{F}, C)$, in R-Mod. Here, $\mathcal{F}$ is the class of all flat R-modules and $C := \mathcal{F}^{\perp}$ is the class of all *cotorsion* R-modules. Using theory developed in Chapter 9, the reader will be asked to prove the well-known fact that $(\mathcal{F}, C)$ is a complete hereditary cotorsion pair, cogenerated by a set $\mathcal{S}$; see Exercise 9.9.4. As a special case of Example 7, we have a flat model structure $\mathrm{Ch}(R)_{flat} = \left(dg\widetilde{\mathcal{F}}, \widetilde{\mathcal{E}}, dg\widetilde{C}\right)$ for the derived category. Complexes in $dg\widetilde{\mathcal{F}}$ are called *DG-flat* and these include all DG-projective complexes. However, we develop technical tools in Section 9.10 to simplify the description of the fibrant objects. It turns out that, unlike the situation for the classes $dg\widetilde{\mathcal{I}}$, $dg\widetilde{\mathcal{P}}$, and $dg\widetilde{\mathcal{F}}$, every (even unbounded) chain complex of cotorsion modules is in the class $dg\widetilde{C}$ of fibrant objects. This result was proved by Bazzoni, Cortés-Izurdiaga, and Estrada [2020].

Example 8 (The Flat Model Structure for $\mathcal{D}(R)$) Let $dw\widetilde{C}$ denote the class of all chain complexes that are *degreewise* cotorsion, meaning $C \in dw\widetilde{C}$ if and only if each C_n is a cotorsion R-module. Then Enochs' flat cotorsion pair, $(\mathcal{F}, C)$, lifts to an hereditary model structure on chain complexes,

$$\mathrm{Ch}(R)_{flat} = \left(dg\widetilde{\mathcal{F}}, \widetilde{\mathcal{E}}, dw\widetilde{C}\right), \tag{9}$$

for the derived category, $\mathcal{D}(R)$. In particular, $dw\widetilde{C} = dg\widetilde{C}$ is the class of fibrant objects. The interested reader will be able to prove this result from the tools we develop in Section 9.10; see Exercises 10.9.4 and 10.9.6. □

Although sheaves and schemes are beyond the scope of this book, it should be pointed out that the significance of the flat model structure is that it generalizes to quasi-coherent sheaves over quite general schemes. Its existence on the category of complexes of quasi-coherent sheaves over such a scheme provides an abelian *monoidal* model structure which is a nice way to put the derived tensor product functor on solid theoretical ground. We study abelian monoidal model structures in Chapter 7, showing in particular that their homotopy categories are always tensor triangulated in the sense of Balmer [2005]. See Example 10.51 for further comments on the flat model structure for quasi-coherent sheaves.

Returning to the example of modules over an Iwanaga–Gorenstein ring R, there is another interesting point. We call any abelian model structure such as $\mathfrak{M}_{inj} = (All, \mathcal{W}, \mathcal{GI})$ from Example 5 an *injective model structure*. Here, every object is cofibrant, equivalently, the trivially fibrant objects are precisely the injectives. It follows from a technical result, Corollary 9.57, that if there is to exist an abelian model structure of the form $\mathfrak{M} = (All, \mathcal{W}, \mathcal{R})$ on R-Mod, then the class $\mathcal{W}$ must be closed under direct limits. So it then follows from the Govorov–Lazard Theorem that $\mathcal{W}$ must even contain all flat R-modules. Then again, the 2 out of 3 property for $\mathcal{W}$ implies that it must contain all modules of finite flat dimension. In particular, for Iwanaga–Gorenstein rings, any flat module has finite projective/injective dimension, and the class $\mathcal{W}$ of trivial objects is precisely the class of modules of finite flat dimension.

Example 9 Let R be an Iwanaga–Gorenstein ring and let $\mathcal{W}$ denote the class of all R-modules of finite projective (equivalently finite injective, or finite flat) dimension. There is an abelian model structure $\mathfrak{M}_{flat} = (\mathcal{GF}, \mathcal{W}, C)$ on R-Mod in which the fibrant objects form the class C of cotorsion R-modules. The modules in $\mathcal{GF}$ are called *Gorenstein flat*, and they are precisely the modules appearing as a cycle in some exact chain complex of flat modules. Its homotopy category coincides with those of Example 5 and shows them to also be equivalent to $\mathrm{St}(\mathcal{GF} \cap C)$.

For which rings R can we define such nice stable module categories? Such questions have motivated much work in Gorenstein homological algebra. For instance, the results of Examples 5 and 9 generalize to the larger class of Ding–Chen rings [Gillespie, 2017b]. More amazing is that it was shown in Šaroch and Šťovíček [2020] that for a general ring R, there exists a complete cotorsion pair $(\mathcal{GF}, \mathcal{GF}^{\perp})$ where $\mathcal{GF}$ is the general class of all Gorenstein flat modules, as in Enochs and Jenda [2000]. On the other hand, they show that we always have a complete cotorsion pair $({}^{\perp}\mathcal{GI}, \mathcal{GI})$, where $\mathcal{GI}$ is the general class of all Gorenstein injective modules. Both of these cotorsion pairs do indeed correspond to abelian model structures on R-Mod although their trivial objects may not coincide for general rings. As of the writing of this book, it is an open question whether or not the dual statement about Gorenstein projectives holds.

1

Additive and Exact Categories

This chapter covers the foundations of additive and exact category theory that will be the basis for the rest of the book. After reviewing the basics of additive categories, we develop the most fundamental and crucial consequences of the axioms for Quillen exact categories. These facts about short exact sequences will be used constantly throughout the text. The category of abelian groups will be denoted by **Ab**.

1.1 Additive Categories

Additive categories are truly fundamental structures in homological algebra. They include all abelian categories, such as abelian groups, vector spaces, and modules over a general ring R, not to mention categories of (quasi-coherent) sheaves and their associated categories of chain complexes. Moreover, *exact categories* and *triangulated categories*, both of which play a central role in this book, are special classes of additive categories.

A category $\mathcal{A}$ is called a **preadditive category** if every hom-set, which we will denote by $\mathrm{Hom}_{\mathcal{A}}(A, B)$, is an abelian group and composition of morphisms is bilinear. That is, $f(g + h) = fg + fh$ and $(f + g)h = fh + gh$ whenever these compositions make sense. Alternatively, $\mathcal{A}$ is often called an **Ab**-*category*, for in a technical sense it is a category "enriched" over the category **Ab**.

A morphism of preadditive categories is called an **additive functor**. It is a functor $F\colon \mathcal{A} \to \mathcal{B}$ between additive categories $\mathcal{A}$ and $\mathcal{B}$ for which each function

$$F_{A,B}\colon \mathrm{Hom}_{\mathcal{A}}(A, B) \to \mathrm{Hom}_{\mathcal{B}}(FA, FB)$$

is an abelian group homomorphism.

It is sometimes a helpful perspective to think of a preadditive category as a "gigantic ring with unities", where the morphisms are the ring elements and composition is the multiplication. Of course not all arrows can be added and/or multiplied, but whenever they can we have properties similar to a ring with unity. Here are some examples of what we have in mind.

- Bilinearity corresponds to the left and right distribute laws.
- The maps 1_A and 0_A act like "unities" and "zeros". For example, composing any morphism with a 0 yields another 0. This follows from bilinearity in the same way that one uses the distributive law to show that a ring element r satisfies the properties $0 \cdot r = r \cdot 0 = 0$.
- Also similar to the usual ring theory proofs, we have the compatibilities

$$(-g)f = -1(gf) = -(gf) = g(-f).$$

- Additive functors are analogous to ring homomorphisms. In fact, for any object $A \in \mathcal{A}$, $\mathrm{Hom}_{\mathcal{A}}(A, A)$ is a ring, called the endomorphism ring of A. If $F\colon \mathcal{A} \to \mathcal{B}$ is an additive functor, then each function

$$F_{A,A}\colon \mathrm{Hom}_{\mathcal{A}}(A, A) \to \mathrm{Hom}_{\mathcal{B}}(FA, FA)$$

  is a ring homomorphism.
- In the same way that a monoid is equivalent to a category with one object, a ring with unity is equivalent to a preadditive category with one object.

The simplicity of preadditive categories stems from the fact that finite products and coproducts coincide, if they exist. The first instance of this is the following lemma. Recall that, by definition, an *initial* object I is an object of a category C for which there is exactly one morphism $I \to C$ for each object $C \in C$. The dual notion defines a *terminal* object.

Lemma 1.1 *If an object of a preadditive category $\mathcal{A}$ is either an initial or a terminal object, then that object must be both initial and terminal.*

Before proving the lemma, we note that such an object is called a **zero object**, or a **null object**, and typically denoted by 0. Clearly any such object is unique up to a unique isomorphism. Our proof of the lemma will show that $X = 0$ if and only if $1_X = 0_X$.

Proof If I is initial, then $\mathrm{Hom}_{\mathcal{A}}(I, I)$ must have exactly one element. Hence $1_I = 0_I$. We then can see that I must also be terminal as follows. First, there does exist at least one morphism $A \to I$, namely the zero morphism $A \xrightarrow{0} I$. But given any morphism $f\colon A \to I$, we have $f = 1_I f = 0_I f = 0$. This proves I is terminal. By duality, any terminal object must also be initial. □

A preadditive category need not have a 0 object. For example, any (nonzero) ring R with unity, considered as a preadditive category with one object, doesn't have a 0 object. Another example is the category $\{R^n \mid n \geq 1\}$ of free R-modules of finite rank $n \geq 1$. Note that the hom-sets, $\mathrm{Hom}(R^n, R^m)$, are equivalent to $M_{m\times n}$, the additive group of all $m \times n$ matrices over R.

Exercise 1.1.1 Let $\mathcal{A}$ be a preadditive category with a zero object 0. Show that the zero morphism $(A \xrightarrow{0} B) \in \mathrm{Hom}_{\mathcal{A}}(A, B)$ coincides with the unique composition $A \to 0 \to B$, through the zero object.

A zero object may be viewed as an "empty biproduct". In fact, the amalgamation of initial and terminal objects in preadditive categories generalizes to the fact that finite coproducts coincide with finite products. Here we will only explicitly define biproducts for a given collection of two objects ($n = 2$), but it can be generalized in a straightforward way to any $n \geq 0$ (with $n = 0$ being a null object).

Definition 1.2 Let A and C be objects of a preadditive category $\mathcal{A}$. A **biproduct diagram** for A and C is a diagram of the form

$$A \underset{i_A}{\overset{p_A}{\leftrightarrows}} B \underset{p_C}{\overset{i_C}{\leftrightarrows}} C$$

satisfying (i) $p_A i_A = 1_A$, (ii) $p_C i_C = 1_C$, and (iii) $i_A p_A + i_C p_C = 1_B$. In this case we write $B = A \bigoplus C$, and say that B is the **direct sum**, or **biproduct**, of A and C. We also say that A and C are **direct summands** of B.

An **additive category** is simply a preadditive category $\mathcal{A}$ in which all finite biproducts exist. In particular, this includes the existence of a zero object.

The categorical notion of the kernel of a morphism makes sense in any preadditive category. If the kernel of $f \colon A \to B$ exists, we denote it by $\ker f$. Note that $\ker f$ is exactly the same thing as an equalizer of the pair

$$A \underset{0}{\overset{f}{\rightrightarrows}} B.$$

The dual notion of cokernel, denoted $\mathrm{cok}\, f$ when it exists, is the coequalizer of f and 0. The following exercise is fundamental as it states that (i_A, p_C) and (i_C, p_A) form what we call kernel–cokernel pairs.

Exercise 1.1.2 Show $i_A = \ker p_C$, $p_C = \mathrm{cok}\, i_A$, $i_C = \ker p_A$, and $p_A = \mathrm{cok}\, i_C$ in any biproduct diagram as in Definition 1.2.

The following exercise asks the reader to prove the fact alluded to previously: Finite coproducts coincide with finite products in preadditive categories, and determine biproducts.

Exercise 1.1.3 Let A and C be objects of a preadditive category $\mathcal{A}$. Then, if they exist, their product $A \prod C$, coproduct $A \coprod C$, and biproduct $A \bigoplus C$, all coincide. More precisely, suppose we have a biproduct diagram as in Definition 1.2. Then $\left(A \bigoplus C, p_A, p_C\right)$ is the product of A and C, and $\left(A \bigoplus C, i_A, i_C\right)$ is the coproduct of A and C. Conversely, any product diagram or coproduct diagram determines a biproduct diagram.

A convenient feature of additive categories is that we may use matrix notation to denote morphisms in and out of a biproduct. Being a product, a morphism $X \to A \bigoplus C$ is equivalent to a matrix $\left[\begin{smallmatrix} f \\ g \end{smallmatrix}\right]$ for some $X \xrightarrow{f} A$ and $X \xrightarrow{g} C$. On the other hand, being a coproduct, a morphism $A \bigoplus C \to Y$ is equivalent to a matrix $[\, f \; g \,]$ for some $A \xrightarrow{f} Y$ and $C \xrightarrow{g} Y$. Note that if we are given compositions $X \xrightarrow{f} A \xrightarrow{g} Y$ and $X \xrightarrow{\alpha} C \xrightarrow{\beta} Y$, then we have the morphism $X \xrightarrow{gf+\beta\alpha} Y$ which factors as

$$X \xrightarrow{\left[\begin{smallmatrix} f \\ \alpha \end{smallmatrix}\right]} A \bigoplus C \xrightarrow{[\, g \; \beta \,]} Y,$$

where we use the usual matrix multiplication $[\, g \; \beta \,]\left[\begin{smallmatrix} f \\ \alpha \end{smallmatrix}\right] = gf + \beta\alpha$.

We perhaps should call an object B as in Definition 1.2 the *internal direct sum* of A and C. For conversely, given any two objects A and C, assuming their coproduct (equivalently product) exists we may denote it by $A \bigoplus C$. We have the **canonical split exact sequence**

$$A \underset{\left[\begin{smallmatrix} 1_A \\ 0 \end{smallmatrix}\right]}{\overset{[\, 1_A \; 0 \,]}{\leftrightarrows}} A \bigoplus C \underset{[\, 0 \; 1_C \,]}{\overset{\left[\begin{smallmatrix} 0 \\ 1_C \end{smallmatrix}\right]}{\leftrightarrows}} C \tag{1.1}$$

for A and C, and we call $A \bigoplus C$ the *external direct sum* of A and C.

Lemma 1.3 *Given any biproduct B as in Definition 1.2, the morphism $\left[\begin{smallmatrix} p_A \\ p_C \end{smallmatrix}\right]$: $B \to A \bigoplus C$ is an isomorphism to the external direct sum, with inverse $[\, i_A \; i_C \,]$: $A \bigoplus C \to B$. The isomorphisms are compatible in the sense that they make the entire biproduct diagram for the internal direct sum commute with the biproduct diagram for the canonical split exact sequence of Diagram* (1.1).

Proof We have $[\, i_A \; i_C \,]\left[\begin{smallmatrix} p_A \\ p_C \end{smallmatrix}\right] = i_A p_A + i_C p_C = 1_B$. On the other hand,

$$\left[\begin{smallmatrix} p_A \\ p_C \end{smallmatrix}\right][\, i_A \; i_C \,] = \left[\begin{smallmatrix} p_A i_A & p_A i_C \\ p_C i_A & p_C i_C \end{smallmatrix}\right] = \left[\begin{smallmatrix} 1_A & 0 \\ 0 & 1_C \end{smallmatrix}\right] = 1_{A \bigoplus C}.$$

Moreover, the isomorphisms make all diagrams in sight commute. □

We end this section by clarifying the connection between split exact sequences (i.e. biproduct diagrams) and the notion of *split monics* and *split epics*. We will return to this important idea is Section 1.8.

In any category, a morphism $f\colon A \to B$ is called a **split monomorphism** if it admits a left inverse $r\colon B \to A$, called a **retraction**. That is, we have $rf = 1_A$. We then also say that A is a **retract** of B. The retraction r then satisfies the dual notion of being a split epimorphism. Let us state formally that a morphism $g\colon B \to C$ is called a **split epimorphism** if it admits a right inverse $s\colon C \to B$, called a **section**, and so satisfying $gs = 1_C$. Certainly s is a split monomorphism, and so C is a retract of B. Note that a split monomorphism is indeed a monomorphism since it is easily seen to be left cancellable, while a split epimorphism is an actual epimorphism. In most familiar additive categories split monomorphisms (and split epimorphisms) give rise to biproduct diagrams. The next statement clarifies when this actually happens.

Proposition 1.4 *Let $\mathcal{A}$ be an additive category and let $A \xrightarrow{f} B$ and $B \xrightarrow{g} C$ be morphisms. The following are equivalent.*

(1) *We have a biproduct diagram* $A \underset{f}{\overset{r}{\leftrightarrows}} B \underset{g}{\overset{s}{\leftrightarrows}} C$. *In particular, A and C are direct summands of $B = A \bigoplus C$.*
(2) *The morphism g is a split epimorphism and admits a kernel. In this case, let $f = \ker g$ denote a kernel, and $C \xrightarrow{s} B$ denote a section $gs = 1_C$.*
(3) *The morphism f is a split monomorphism and admits a cokernel. In this case, let $g = \operatorname{cok} f$ denote a cokernel, and $B \xrightarrow{r} A$ denote a retraction $rf = 1_A$.*

Proof In light of Exercise 1.1.2, we clearly have (1) implies (2) and (3). We only prove (2) implies (1) as (3) implies (1) is dual. So suppose $f = \ker g$ and $gs = 1_C$. Then $gsg = g$, which implies $g(1_B - sg) = 0$. By the universal property of f, there exists a unique morphism $r\colon B \to A$ such that $fr = 1_B - sg$. Hence $1_B = fr + sg$, and it only remains to show $rf = 1_A$. For this, we note $frf = (1_B - sg)f = f - s(0) = f = f1_A$. Since f is left cancellable we conclude $rf = 1_A$. □

Exercise 1.1.4 Show that an additive functor between additive categories necessarily preserves null objects (zero objects) and biproducts.

Exercise 1.1.5 Consider a sequence of morphisms $A \xrightarrow{f} B \xrightarrow{g} C$ in an additive category $\mathcal{A}$. Verify that the following statements are equivalent.

(1) $f = \ker g$.

(2) $\left[\begin{smallmatrix} f \\ 0 \end{smallmatrix}\right]$ is the kernel of $\left[\begin{smallmatrix} g & 0 \\ 0 & 1_D \end{smallmatrix}\right]$: $B \bigoplus D \to C \bigoplus D$ for any object D.

(3) $\left[\begin{smallmatrix} f & 0 \\ 0 & 1_D \end{smallmatrix}\right]$: $A \bigoplus D \to B \bigoplus D$ is a kernel of $[\, g \; 0 \,]$ for any object D.

Dually we have $g = \operatorname{cok} f$ if and only if $[\, g \; 0 \,]$ is the cokernel of $\left[\begin{smallmatrix} f & 0 \\ 0 & 1_D \end{smallmatrix}\right]$: $A \bigoplus D \to B \bigoplus D$ for any object D, if and only if $\left[\begin{smallmatrix} g & 0 \\ 0 & 1_D \end{smallmatrix}\right]$: $B \bigoplus D \to C \bigoplus D$ is a cokernel of $\left[\begin{smallmatrix} f \\ 0 \end{smallmatrix}\right]$ for any object D.

1.2 Pushouts and Pullbacks in Additive Categories

The point of this brief section is simply to record some useful facts about pushouts and pullbacks in additive categories. These are standard results and are straightforward to prove, so much of the proofs will be left as exercises. Let $\mathcal{A}$ be an additive category throughout.

A key observation is that pushouts and pullbacks are essentially equivalent to cokernels and kernels, respectively. Indeed a square as shown is equivalent to a sequence of morphisms as shown:

$$\begin{array}{ccc} A & \xrightarrow{g} & C \\ {\scriptstyle f}\downarrow & & \downarrow{\scriptstyle f'} \\ B & \xrightarrow{g'} & D \end{array} \qquad\qquad A \xrightarrow{\left[\begin{smallmatrix} f \\ -g \end{smallmatrix}\right]} B \bigoplus C \xrightarrow{[\, g' \; f' \,]} D. \tag{1.2}$$

Moreover, the square commutes if and only if the sequence is a null sequence. By a **null sequence** we mean any pair of composable morphisms, $A \xrightarrow{f} B \xrightarrow{g} C$, such that $gf = 0$. It is easy to prove the following lemma by translating between the universal properties involved.

Lemma 1.5 *Consider the square and corresponding sequence given in Diagrams* (1.2).

(1) *The square is a pullback if and only if* $\left[\begin{smallmatrix} f \\ -g \end{smallmatrix}\right]$ *is the kernel of* $[\, g' \; f' \,]$.

(2) *The square is a pushout if and only if* $[\, g' \; f' \,]$ *is the cokernel of* $\left[\begin{smallmatrix} f \\ -g \end{smallmatrix}\right]$.

The single negative sign, appearing as $-g$ *in Diagram* (1.2), *may be placed in any of the four possibilities within* $\left[\begin{smallmatrix} f \\ g \end{smallmatrix}\right]$ *and* $[\, g' \; f' \,]$.

We ask the reader to also prove the following lemma.

Lemma 1.6 *Suppose we have a commutative diagram in $\mathcal{A}$:*

$$\begin{array}{ccccc} A & \xrightarrow{f} & B & \xrightarrow{g} & C \\ \downarrow{\scriptstyle\alpha} & & \downarrow{\scriptstyle\beta} & & \downarrow{\scriptstyle\gamma} \\ A' & \xrightarrow{f'} & B' & \xrightarrow{g'} & C', \end{array}$$

where both rows are null sequences.

(1) *If $f = \ker g$ and γ and f' are each monomorphisms, then the left-hand square is necessarily a pullback.*
(2) *If $g' = \operatorname{cok} f'$ and α and g are each epimorphisms, then the right-hand square is necessarily a pushout.*

Next we will prove that pushout and pullback squares may be composed, or "pasted" together.

Lemma 1.7 (Composing and Factoring Pushout Squares) *Assume the following left square is a pushout and the entire diagram commutes:*

$$\begin{array}{ccccc} A & \xrightarrow{f} & B & \xrightarrow{g} & C \\ {\scriptstyle\alpha}\downarrow & & \downarrow{\scriptstyle\beta} & & \downarrow{\scriptstyle\gamma} \\ D & \xrightarrow[f']{} & E & \xrightarrow[g']{} & F. \end{array}$$

Then the right square is a pushout if and only if the outer rectangle is a pushout.

Of course there is also a dual statement concerning pullback squares.

Proof First let us assume that the outer rectangle is a pushout and prove that the right square is a pushout. So assume we are given morphisms $C \xrightarrow{s} Z$ and $E \xrightarrow{t} Z$ such that $t\beta = sg$. Then $t\beta f = sgf \implies (tf')\alpha = sgf$. So since the outer rectangle is a pushout we get a unique $F \xrightarrow{\xi} Z$ such that

$$\begin{aligned} \xi g' f' &= tf', \\ \xi\gamma &= s. \end{aligned} \qquad (*)$$

We wish to show that ξ also uniquely satisfies the set of equations:

$$\begin{aligned} \xi g' &= t, \\ \xi\gamma &= s. \end{aligned} \qquad (**)$$

But the uniqueness portion is obvious since the Equations $(**)$ imply Equations $(*)$. So we only need to show $\xi g' = t$. Note that since the left square is a pushout, and we have the "impostor" square

$$\begin{array}{ccc} A & \xrightarrow{f} & B \\ {\scriptstyle\alpha}\downarrow & & \downarrow{\scriptstyle 0} \\ D & \xrightarrow{0} & Z, \end{array}$$

it is enough to show that $(\xi g' - t)f' = 0$ and $(\xi g' - t)\beta = 0$. Clearly, $(\xi g' - t)f' = 0$ by Equations $(*)$. Now we compute:

$$\begin{aligned} (\xi g' - t)\beta &= \xi g'\beta - t\beta \\ &= \xi\gamma g - t\beta \\ &= sg - t\beta, \text{ by Equations } (**) \\ &= 0. \end{aligned}$$

Conversely, assume the right square is a pushout and that $C \xrightarrow{u} W$ and $D \xrightarrow{v} W$ are morphisms such that $u(gf) = v\alpha$. Using the pushout property of the left square, followed by the pushout property of the right square, we easily find a morphism $F \xrightarrow{\xi} W$ such that $\xi(g'f') = v$ and $\xi\gamma = u$. To prove that the outer rectangle is a pushout it is only left to show that ξ is the unique morphism with this property. So let us assume that $F \xrightarrow{\xi'} W$ also satisfies $\xi'(g'f') = v$ and $\xi'\gamma = u$ and our goal is to show that $\xi' - \xi = 0$. By the universal property of the right pushout square, it is enough to show $(\xi' - \xi)g' = 0$ and $(\xi' - \xi)\gamma = 0$. It is immediate that $\xi'\gamma - \xi\gamma = 0$, so we just need to show $\xi'g' - \xi g' = 0$. But by the universal property of the left pushout square, it is enough to show $(\xi'g' - \xi g')f' = 0$ and $(\xi'g' - \xi g')\beta = 0$. It is immediate that $(\xi'g' - \xi g')f' = 0$ and for the second equation we have $\xi'g'\beta - \xi g'\beta = \xi'\gamma g - \xi\gamma g = ug - ug = 0$. We conclude that the outer rectangle is also a pushout. □

We have one final lemma that is fundamental to pullback and pushout arguments in additive and exact categories.

Lemma 1.8 *Suppose we have a commutative diagram in $\mathcal{A}$:*

$$\begin{array}{ccc} A & \xrightarrow{g} & C \\ {\scriptstyle f}\downarrow & & \downarrow{\scriptstyle f'} \\ B & \xrightarrow{g'} & D. \end{array}$$

(1) *Assume the square is a pullback. If $k = \ker g$ exists, then $fk = \ker g'$. On the other hand, if $k' = \ker g'$, then $k' = fk$ for some unique morphism k and $k = \ker g$.*

(2) *Assume the square is a pushout. If* $c' = \operatorname{cok} g'$ *exists, then* $c'f' = \operatorname{cok} g$. *On the other hand, if* $c = \operatorname{cok} g$, *then* $c = c'f'$ *for some unique morphism* c' *and* $c' = \operatorname{cok} g'$.

Exercise 1.2.1 Prove Lemma 1.5.

Exercise 1.2.2 Prove Lemma 1.6.

Exercise 1.2.3 Prove Lemma 1.8.

1.3 Exact Categories: The Axioms

We now introduce the notion of an exact category which will be the categorical foundation for our theory of abelian model structures. Interestingly enough, exact categories were also introduced by Quillen, though independently of his introducing model categories. Loosely, an exact category is an additive category $\mathcal{A}$ along with a distinguished class of kernel–cokernel pairs satisfying several closure axioms. An abelian category $\mathcal{A}$, along with the class of all short exact sequences in $\mathcal{A}$, is the primary and motivating example of an exact category. Extension closed full subcategories of abelian categories are another prime example. Several specific examples will appear in the exercises and elsewhere throughout this book.

First, by a **kernel–cokernel pair** we mean an ordered pair (i, p) of composable morphisms $A \xrightarrow{i} B \xrightarrow{p} C$ such that $i = \ker p$ and $p = \operatorname{cok} i$. Note that such an i is necessarily monic and such a p is necessarily epic. (We will use the term *monic* interchangeably with *monomorphism*, and *epic* interchangeably with *epimorphism*.) A morphism from a kernel–cokernel pair (i, p) to another (i', p') is a triple of morphisms making the diagram commute:

$$\begin{array}{ccccc} A & \xrightarrow{i} & B & \xrightarrow{p} & C \\ \downarrow & & \downarrow & & \downarrow \\ A' & \xrightarrow{i'} & B' & \xrightarrow{p'} & C'. \end{array}$$

Such a morphism is an isomorphism from (i, p) to (i', p') when all three morphisms are isomorphisms.

To motivate the axioms for exact categories, let us recall some well-known facts about abelian categories, (which will appear in more detail in Section 1.9). A defining aspect of abelian categories is that every monic i is the kernel of its cokernel and therefore is part of a kernel–cokernel pair (i, p). The dual is true as well; each epic appears as a p in some kernel–cokernel pair. So in the abelian case, kernel–cokernel pairs are equivalent to the usual notion of *short exact sequences*, and they satisfy a number of useful properties. In particular, the composite of two monics is again a monic and hence the first component of another short exact sequence. Also, the pushout of any morphism along a monic is again a monic. The dual statements hold for epics, and it turns out that a great deal of pure homological algebra follows from these formal properties. Extracting them leads to Quillen's notion of an exact category.

Definition 1.9 Let $\mathcal{A}$ be an additive category and $\mathcal{E}$ be a class of kernel–cokernel pairs in $\mathcal{A}$. We say that $\mathcal{E}$ is an **exact structure** on $\mathcal{A}$, and that $(\mathcal{A}, \mathcal{E})$ is an **exact category**, if $\mathcal{E}$ satisfies the following axioms. The axioms utilize this standard terminology: If (i, p) is a kernel–cokernel pair in $\mathcal{E}$, then we say that (i, p) is a **short exact sequence**, i is an **admissible monic**, and p is an **admissible epic**.
(Ex1) (*Replete*) $\mathcal{E}$ is closed under isomorphisms.
(Ex2) (*Split Exact Sequences*) $\mathcal{E}$ contains the canonical split exact sequence

$$A \xrightarrow{\left[\begin{smallmatrix} 1_A \\ 0 \end{smallmatrix}\right]} A \bigoplus C \xrightarrow{[0\ 1_C]} C$$

for any given pair of objects A and C.
(Ex3) (*Pushouts and Pullbacks*) The pushout of an admissible monic along any morphism exists and is again an admissible monic. Dually, the pullback of an admissible epic along any morphism exists and is again an admissible epic.
(Ex4) (*Compositions*) The class of admissible monics is closed under composition. Dually, the class of admissible epics is closed under composition.

It is standard to use the symbol $\rightarrowtail$ to denote an admissible monic and $\twoheadrightarrow$ to denote an admissible epic. A short exact sequence is denoted by $A \rightarrowtail B \twoheadrightarrow C$. In the literature, the terms *conflation*, *inflation*, and *deflation*, are also sometimes used for *short exact sequence*, *admissible monic*, and *admissible epic*.

Note that a morphism $f : A \to B$ is an isomorphism if and only if $f : A \rightarrowtail B$ is an admissible monic with $\operatorname{cok} f = 0$, if and only if $f : A \twoheadrightarrow B$ is an admissible epic with $\ker f = 0$. This can be seen from Axiom **(Ex2)** and Proposition 1.4.

Given any additive category $\mathcal{A}$, taking $\mathcal{E}$ to be the class of all split exact sequences determines an exact category $(\mathcal{A}, \mathcal{E})$.

Exercise 1.3.1 Let $\mathcal{A}$ be any additive category. Let $\mathcal{E}$ be the class of all split exact sequences. Show that $(\mathcal{A}, \mathcal{E})$ is an exact category. Note Exercise 1.1.2.

Note that the exact category axioms **(Ex1)**–**(Ex4)** are self-dual. So as a matter of convenience, each result we prove has an equally valid dual.

In the remainder of this section, we will develop some immediate consequences of the exact category axioms. Our first lemma is a foundational result clarifying the pushouts and pullback in Axiom **(Ex3)**.

Lemma 1.10 *Let* $A \overset{i}{\rightarrowtail} B \overset{p}{\twoheadrightarrow} C$ *be a short exact sequence in an exact category* $(\mathcal{A}, \mathcal{E})$.

(1) *The pushout of* i *along any morphism* $f \colon A \to A'$ *yields a commutative diagram with each row a short exact sequence:*

$$\begin{array}{ccccc} A & \overset{i}{\rightarrowtail} & B & \overset{p}{\twoheadrightarrow} & C \\ {\scriptstyle f}\downarrow & & {\scriptstyle f'}\downarrow & & \| \\ A' & \underset{i'}{\rightarrowtail} & B' & \underset{p'}{\twoheadrightarrow} & C. \end{array}$$

Moreover, the left square is not just a pushout, but also a pullback square.

(2) *The pullback of* p *along any morphism* $g \colon C' \to C$ *yields a commutative diagram with each row a short exact sequence:*

$$\begin{array}{ccccc} A & \overset{i'}{\rightarrowtail} & B' & \overset{p'}{\twoheadrightarrow} & C' \\ \| & & \downarrow{\scriptstyle g'} & & \downarrow{\scriptstyle g} \\ A & \underset{i}{\rightarrowtail} & B & \underset{p}{\twoheadrightarrow} & C. \end{array}$$

Moreover, the right square is not just a pullback, but also a pushout square.

Proof The statements are dual to one another and we will prove (1). By Axiom **(Ex3)** the pushout exists and i' is an admissible monic. By Lemma 1.8(2) there exists a unique morphism $p' : B' \to C$, satisfying $p'f' = p$, and such that $p' = \operatorname{cok} i'$. This proves we have the indicated commutative diagram with each row a short exact sequence. The fact that the left-hand square is also a pullback follows from Lemma 1.6(1). □

In fact, Lemma 1.10 has a converse which we will see in Proposition 1.12. But first we need the Short Five Lemma. The Short Five Lemma is an easy diagram chase in concrete categories such as modules over a ring. But it is a little bit of work in our abstract setting, essentially because not all monic–epic morphisms are isomorphisms.

Lemma 1.11 (Short Five Lemma) *For any commutative diagram*

$$\begin{array}{ccccc} A & \overset{i}{\rightarrowtail} & B & \overset{p}{\twoheadrightarrow} & C \\ \| & & {\scriptstyle v}\downarrow & & \| \\ A & \underset{i'}{\rightarrowtail} & B' & \underset{p'}{\twoheadrightarrow} & C \end{array}$$

with exact rows, the morphism v is necessarily an isomorphism.

Proof Using Lemma 1.10(2), we may form the commutative diagram with exact rows, where the upper right square is a pullback:

$$\begin{array}{ccccc}
A & \overset{f}{\rightarrowtail} & P & \overset{g}{\twoheadrightarrow} & B' \\
\| & & \downarrow{\scriptstyle g'} & & \downarrow{\scriptstyle p'} \\
A & \overset{i}{\rightarrowtail} & B & \overset{p}{\twoheadrightarrow} & C \\
\| & & \downarrow{\scriptstyle v} & & \| \\
A & \overset{i'}{\rightarrowtail} & B' & \overset{p'}{\twoheadrightarrow} & C.
\end{array}$$

The idea of the proof is to show that the short exact sequence on top is split exact, and in fact there is a section s for g such that $v^{-1} = g's$. We will use that v is necessarily monic (left cancellable), so we start by proving this. It is enough to show that $vt = 0 \Longrightarrow t = 0$. So assume $vt = 0$. Then also $0 = p'vt = pt$. Since $i = \ker p$, there exists a uniqe map l such that $il = t$. Hence $0 = vt = vil = i'l$. Since i' is certainly monic, we may left cancel on $i'l = 0$ to obtain $l = 0$. Hence $t = il = 0$ too.

Now let us show that (f, g) is a split exact sequence. First, $p'(vg' - g) = p'vg' - p'g = pg' - p'g = 0$. Since $i' = \ker p'$, there exists a unique morphism $r\colon P \to A$ satisfying $i'r = vg' - g$. Then $i'rf = vg'f - gf = vg'f = i'$. Since i' is monic, we conclude $rf = 1_A$. So by Proposition 1.4 the top row is a split exact sequence. Let $s\colon B' \to P$ denote the section, so that we have (i) $rf = 1_A$, (ii) $gs = 1_{B'}$, and (iii) $fr + sg = 1_P$. Then, since $rs = 0$, we have

$$v(g's) = (vg')s = (i'r + g)s = i'rs + gs = 0 + gs = 1_{B'}.$$

On the other hand, we want to show $(g's)v = 1_B$. We have

$$v(g'sv) = (vg')sv = (i'r + g)sv = i'rsv + gsv = 0 + gsv = 1_{B'}v = v = v1_B.$$

Since we have already seen that v is left cancellable we obtain $(g's)v = 1_B$, proving that v is an isomorphism. □

Now the Short Five Lemma allows us to quickly deduce the following converse to Lemma 1.10. This statement, concerning a particular class of commutative diagrams, will be used repeatedly in our development of abelian model category theory.

Proposition 1.12 (Pushout and Pullback Diagrams) *Let $(\mathcal{A}, \mathcal{E})$ be an exact category.*

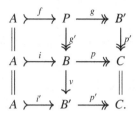

(1) *The left-hand square in any morphism of short exact sequences of the form*

$$\begin{array}{ccccc} A & \rightarrowtail & B & \twoheadrightarrow & C \\ \downarrow & & \downarrow & & \| \\ X & \rightarrowtail & Y & \twoheadrightarrow & C \end{array}$$

is necessarily both a pullback and a pushout square.

(2) *The right-hand square in any morphism of short exact sequences of the form*

$$\begin{array}{ccccc} A & \rightarrowtail & Y & \twoheadrightarrow & Z \\ \| & & \downarrow & & \downarrow \\ A & \rightarrowtail & B & \twoheadrightarrow & C \end{array}$$

is necessarily both a pullback and a pushout square.

A square that is both a pullback and a pushout is often called a **bicartesian** square.

Proof The statements are dual to one another and we will just prove (1). First, just like the statement in Lemma 1.10(1), the fact that the left square is automatically a pullback follows from Lemma 1.6(1). We just need to show that any such square is necessarily a pushout. But using Lemma 1.10(1), we may easily construct the following commutative diagram with each row a short exact sequence and the upper left square a pushout:

$$\begin{array}{ccccc} A & \rightarrowtail & B & \twoheadrightarrow & C \\ \downarrow\!\!\curlyvee & & \downarrow\!\!\curlyvee & & \| \\ X & \rightarrowtail & P & \twoheadrightarrow & C \\ \| & & {\scriptstyle \exists!}\,\downarrow{\scriptstyle v} & & \| \\ X & \rightarrowtail & Y & \twoheadrightarrow & C. \end{array}$$

The map v is induced by the universal property of the pushout P, and is the unique map making both the lower left-hand square commute, and the vertical composite equal to $B \to Y$. (The lower right-hand square then must commute too by the universal property of the pushout P, for the two maps agree upon composition with $X \rightarrowtail P$ and $B \rightarrowtail P$.) The Short Five Lemma 1.11 asserts v is an isomorphism and this completes the proof. □

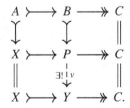

1.4 Exact Categories: First Consequences and (3×3)-Lemma

In this section we will prove some more fundamental properties of exact categories. We start by deducing several useful corollaries from Proposition 1.12. Assume throughout that $(\mathcal{A}, \mathcal{E})$ is an exact category.

Corollary 1.13 (Factoring Morphisms of Short Exact Sequences) *Any morphism from a short exact sequence $A \rightarrowtail B \twoheadrightarrow C$ to another short exact sequence $A' \rightarrowtail B' \twoheadrightarrow C'$ factors through a third short exact sequence as shown:*

$$\begin{array}{ccccc} A & \rightarrowtail & B & \twoheadrightarrow & C \\ \downarrow & & \downarrow & & \| \\ A' & \rightarrowtail & P & \twoheadrightarrow & C \\ \| & & \downarrow & & \downarrow \\ A' & \rightarrowtail & B' & \twoheadrightarrow & C'. \end{array}$$

The upper left and lower right squares are each bicartesian.

Proof Begin the construction of the commutative diagram by taking the pushout of $A' \leftarrow A \rightarrowtail B$ and using Lemma 1.10(1) to get the top two rows of the diagram. Then we may continue to construct the entire diagram by arguing exactly as in the proof of Proposition 1.12. (In particular, the universal property of the pushout is used to show the lower right-hand square commutes.) But then the two squares are bicartesian by Proposition 1.12. □

The following corollary to Proposition 1.12 is extremely useful. It describes the pullback of an admissible epic along an admissible monic, as well as the dual situation.

Corollary 1.14 (Epic–Monic Pullbacks and Pushouts) *Suppose we have a short exact sequence $A \overset{i}{\rightarrowtail} B \overset{p}{\twoheadrightarrow} C$ in $(\mathcal{A}, \mathcal{E})$.*

(1) *Given any admissible epic $g \colon B' \twoheadrightarrow B$, its pullback along i yields a commutative diagram with each row and column a short exact sequence:*

$$\begin{array}{ccccc} K & = & K & & \\ \downarrow & & \downarrow & & \\ A' & \overset{i'}{\rightarrowtail} & B' & \overset{pg}{\twoheadrightarrow} & C \\ {\scriptstyle g'}\downarrow & & {\scriptstyle g}\downarrow & & \| \\ A & \overset{i}{\rightarrowtail} & B & \overset{p}{\twoheadrightarrow} & C. \end{array}$$

The central square is bicartesian.

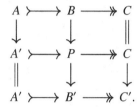

(2) *Given any admissible monic $f\colon B \rightarrowtail B'$, its pushout along p yields a commutative diagram with each row and column a short exact sequence:*

$$
\begin{array}{ccccc}
A & \overset{i}{\rightarrowtail} & B & \overset{p}{\twoheadrightarrow} & C \\
\| & & \downarrow{\scriptstyle f} & & \downarrow{\scriptstyle f'} \\
A & \overset{fi}{\rightarrowtail} & B' & \overset{p'}{\twoheadrightarrow} & C' \\
 & & \downarrow & & \downarrow \\
 & & L & = & L.
\end{array}
$$

Again, the central square is bicartesian.

Proof Again, the statements are dual to one another and we will prove (1). Since pg is the composition of two admissible epics it too must be an admissible epic. So its kernel $i'\colon A' \rightarrowtail B'$ exists and the universal property of the other kernel $i = \ker p$ induces a unique map $g'\colon A' \to A$ making the square commute: $ig' = gi'$. This square is of the type in Lemma 1.10(1), and in particular it must be a pullback. The remainder of the diagram (moving vertically and featuring the kernel K) is now in fact an instance of Lemma 1.10(2). □

Axiom **(Ex4)** states that the composition of two admissible epics is again an admissible epic. The aforementioned shows its kernel to be an extension of the other two kernels. We state this next as another corollary.

Corollary 1.15 (Short Exact Sequence of Admissible Kernels) *Let $f\colon A \twoheadrightarrow B$ and $g\colon B \twoheadrightarrow C$ both be admissible epics in an exact category $(\mathcal{A}, \mathcal{E})$. Denote their respective kernels by K_f and K_g. Then the kernel K_{gf} of the admissible epic $gf\colon A \twoheadrightarrow C$ sits in a short exact sequence*

$$K_f \rightarrowtail K_{gf} \twoheadrightarrow K_g.$$

Proof Take p to be g and g to be f in the statement Corollary 1.14(1). Then the short exact sequence $K_f \rightarrowtail K_{gf} \twoheadrightarrow K_g$ is just the left-most vertical column in the statement of Corollary 1.14(1). □

Of course the dual statement to Corollary 1.15 is equally valid and useful. The reader should formulate it and realize that it is the Noether isomorphism:

$$(C/A)/(B/A) \cong C/B.$$

Next we have the famous (3 × 3)-Lemma in the setting of a general exact category.

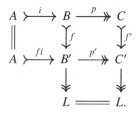

Lemma 1.16 (3 × 3-Lemma) *Consider a commutative diagram*

$$\begin{array}{ccccc}
A' & \xrightarrow{f'} & B' & \xrightarrow{g'} & C' \\
{\scriptstyle i}\downarrow & & {\scriptstyle j}\downarrow & & {\scriptstyle k}\downarrow \\
A & \xrightarrow{f} & B & \xrightarrow{g} & C \\
{\scriptstyle p}\downarrow & & {\scriptstyle q}\downarrow & & {\scriptstyle r}\downarrow \\
A'' & \xrightarrow{f''} & B'' & \xrightarrow{g''} & C''
\end{array}$$

with exact columns in an exact category $(\mathcal{A}, \mathcal{E})$.

(1) *If the bottom two rows, or the top two rows, are short exact sequences, then all three rows are short exact sequences.*

(2) *If* $gf = 0$ *and both the top row and the bottom row are short exact sequences, then the middle row is also a short exact sequence.*

Proof Assume the top two rows are short exact sequences. Then by Corollary 1.13, the morphisms factor through a third short exact sequence:

$$\begin{array}{ccccc}
A' & \overset{f'}{\rightarrowtail} & B' & \overset{g'}{\twoheadrightarrow} & C' \\
{\scriptstyle i}\downarrow & & {\scriptstyle i'}\downarrow & & \| \\
A & \overset{s}{\rightarrowtail} & P & \overset{t}{\twoheadrightarrow} & C' \\
\| & & {\scriptstyle k'}\downarrow & & {\scriptstyle k}\downarrow \\
A & \overset{f}{\rightarrowtail} & B & \overset{g}{\twoheadrightarrow} & C.
\end{array}$$

Here $j = k'i'$. The upper left-hand square is bicartesian and, by Lemma 1.10(1), sits in a commutative diagram with short exact rows and columns as shown next on the left, while the lower right-hand square is bicartesian and, by Corollary 1.14(2), sits in a commutative diagram with short exact rows and columns as shown next on the right:

$$\begin{array}{ccccc}
A' & \overset{f'}{\rightarrowtail} & B' & \overset{g'}{\twoheadrightarrow} & C' \\
{\scriptstyle i}\downarrow & & {\scriptstyle i'}\downarrow & & \| \\
A & \overset{s}{\rightarrowtail} & P & \overset{t}{\twoheadrightarrow} & C' \\
{\scriptstyle p}\downarrow & & {\scriptstyle p'}\downarrow & & \\
A'' & = & A'' & &
\end{array}
\qquad\qquad
\begin{array}{ccccc}
A & \overset{s}{\rightarrowtail} & P & \overset{t}{\twoheadrightarrow} & C' \\
\| & & {\scriptstyle k'}\downarrow & & {\scriptstyle k}\downarrow \\
A & \overset{f}{\rightarrowtail} & B & \overset{g}{\twoheadrightarrow} & C \\
 & & {\scriptstyle rg}\downarrow & & {\scriptstyle r}\downarrow \\
 & & C'' & = & C''.
\end{array}$$

Now we consider the diagram

$$
\begin{array}{ccccc}
B' & \overset{i'}{\rightarrowtail} & P & \overset{p'}{\twoheadrightarrow} & A'' \\
\| & & \downarrow{\scriptstyle k'} & & \downarrow{\scriptstyle f''} \\
B' & \overset{j}{\rightarrowtail} & B & \overset{q}{\twoheadrightarrow} & B'' \\
 & & \downarrow{\scriptstyle rg} & & \downarrow{\scriptstyle g''} \\
 & & C'' & = & C''.
\end{array}
$$

We already know (i) the rows and middle column of this diagram are short exact sequences, and (ii) the two squares containing an equality commute by the previous constructions, and of course (iii) we are trying to *prove* that the right-hand column of the diagram is a short exact sequence. But one can verify that $f''p' = qk'$ by showing that each one becomes equal upon composition with the canonical maps s and i' into the pushout P of $i \colon A' \rightarrowtail A$ and $f' \colon A' \rightarrowtail B'$. For then the universal property of the pushout ensures they are equal. So having verified commutativity of the diagram, it follows from Proposition 1.12(2) that f'' in the right-hand column is an admissible monic and hence the right-hand column is a short exact sequence by Lemma 1.8(2).

We have shown that if the two top rows of the diagram are short exact sequences, then the bottom row is also a short exact sequence. It follows from duality that if the bottom two rows are short exact sequences then so is the top row. We leave it as an exercise to show the remaining statement. □

Exercise 1.4.1 Complete the proof of the (3 × 3)-Lemma 1.16, by proving statement (2). (A full proof appears in Bühler [2010, corollary 3.6].)

We have a few more useful results whose proofs are left as exercises.

Exercise 1.4.2 (Five Lemma) Consider a commutative diagram

$$
\begin{array}{ccccc}
A & \overset{i}{\rightarrowtail} & B & \overset{p}{\twoheadrightarrow} & C \\
\downarrow{\scriptstyle \alpha} & & \downarrow{\scriptstyle \beta} & & \downarrow{\scriptstyle \gamma} \\
A & \underset{i'}{\rightarrowtail} & B' & \underset{p'}{\twoheadrightarrow} & C
\end{array}
$$

with exact rows. Use Corollary 1.13 to prove the following.

(1) If α and γ are admissible monics, then β is also an admissible monic.
(2) If α and γ are admissible epics, then β is also an admissible epic.
(3) If α and γ are isomorphisms, then β is also an isomorphism.

(Again, solutions appear in Bühler [2010, corollary 3.2].)

The following exercise describes the way that a typical exact category arises in practice. A full subcategory $\mathcal{S}$ of $\mathcal{A}$ is called **strictly full** if it is replete, meaning $\mathcal{S}$ is closed under isomorphisms. Explicitly, whenever $S \cong S'$ then we have $S \in \mathcal{S}$ if and only if $S' \in \mathcal{S}$.

Exercise 1.4.3 Assume $(\mathcal{A}, \mathcal{E})$ is an exact category and let $\mathcal{S}$ be a strictly full subcategory of $\mathcal{A}$ that is closed under $\mathcal{E}$-extensions. Show that $(\mathcal{S}, \mathcal{E}_{\mathcal{S}})$ is an exact category, where $\mathcal{E}_{\mathcal{S}}$ consists of all the short exact sequences in $\mathcal{E}$ for which all three terms are objects in $\mathcal{S}$. We call $(\mathcal{S}, \mathcal{E}_{\mathcal{S}})$ the **exact structure on $\mathcal{S}$ inherited from** $(\mathcal{A}, \mathcal{E})$.

1.5 Exact Categories: More Fundamentals and Obscure Axiom

In this section we continue to gather the most basic and useful results concerning exact structures over a general additive category. In particular we prove a result which Quillen had originally listed as an axiom for exact categories. It is now known as the Obscure Axiom.

Again, let $(\mathcal{A}, \mathcal{E})$ be an exact category throughout. We start with a lemma.

Lemma 1.17 $A \xrightarrow{i} B \xrightarrow{p} C$ *is a short exact sequence if and only if*

$$A \bigoplus D \xrightarrow{\left[\begin{smallmatrix} i & 0 \\ 0 & 1_D \end{smallmatrix}\right]} B \bigoplus D \xrightarrow{[p\ 0]} C$$

is a short exact sequence for some (equivalently, for all) object(s) D, and if and only if

$$A \xrightarrow{\left[\begin{smallmatrix} i \\ 0 \end{smallmatrix}\right]} B \bigoplus D \xrightarrow{\left[\begin{smallmatrix} p & 0 \\ 0 & 1_D \end{smallmatrix}\right]} C \bigoplus D$$

is a short exact sequence for some (equivalently, for all) object(s) D.

Proof The following diagram commutes and the columns are short exact sequences for any object D:

$$\begin{array}{ccccc}
A & \xrightarrow{\quad i \quad} & B & \xrightarrow{\quad p \quad} & C \\
{\scriptstyle \left[\begin{smallmatrix} 1_A \\ 0 \end{smallmatrix}\right]} \rightarrowtail & & \rightarrowtail {\scriptstyle \left[\begin{smallmatrix} 1_B \\ 0 \end{smallmatrix}\right]} & & \| \\
A \bigoplus D & \xrightarrow{\left[\begin{smallmatrix} i & 0 \\ 0 & 1_D \end{smallmatrix}\right]} & B \bigoplus D & \xrightarrow{[p\ 0]} & C \\
\twoheadrightarrow {\scriptstyle [0\ 1_D]} & & \twoheadrightarrow {\scriptstyle [0\ 1_D]} & & \\
D & =\!= & D. & &
\end{array}$$

So by Proposition 1.12(1), the upper left square is bicartesian. Being a pushout, the middle row is a short exact sequence whenever the top row is such. On the other hand, suppose there exists an object D for which the middle row is a short exact sequence. Then the following lower left square is again bicartesian, this time by Proposition 1.12(2):

$$
\begin{array}{ccccc}
D & = & D & & \\
{\scriptstyle \left[\begin{smallmatrix} 0 \\ 1_D \end{smallmatrix}\right]} \downarrow & & \downarrow {\scriptstyle \left[\begin{smallmatrix} 0 \\ 1_D \end{smallmatrix}\right]} & & \\
A \bigoplus D & \xrightarrow{\left[\begin{smallmatrix} i & 0 \\ 0 & 1_D \end{smallmatrix}\right]} & B \bigoplus D & \xrightarrow{[p\ 0]} & C \\
\downarrow {\scriptstyle [1_A\ 0]} & & \downarrow {\scriptstyle [1_B\ 0]} & & \| \\
A & \xrightarrow{\quad i \quad} & B & \xrightarrow{\quad p \quad} & C.
\end{array}
$$

Therefore i is a pushout of the admissible monic $\left[\begin{smallmatrix} i & 0 \\ 0 & 1_D \end{smallmatrix}\right]$ and so i too is an admissible monic. This proves the first *if and only if* statement, and the second one is dual. □

We now return to the interplay pointed out in Lemma 1.5, between pushouts and pullbacks and cokernels and kernels. The following result gives a condition for the corresponding null sequence to be a short exact sequence.

Proposition 1.18 *Suppose we have a commutative square*

$$
\begin{array}{ccc}
A & \xrightarrow{g} & C \\
{\scriptstyle f} \downarrow & & \downarrow {\scriptstyle f'} \\
B & \xrightarrow{g'} & D.
\end{array}
$$

If either f or g is an admissible monic and the square is a pushout then we have a short exact sequence

$$
A \overset{\left[\begin{smallmatrix} f \\ -g \end{smallmatrix}\right]}{\rightarrowtail} B \bigoplus C \overset{[g'\ f']}{\twoheadrightarrow} D.
$$

Dually, if either f' or g' is an admissible epic and the square is a pullback then we have the same short exact sequence.

On the other hand, the square is necessarily bicartesian even if $\left(\left[\begin{smallmatrix} f \\ -g \end{smallmatrix}\right], [g'\ f']\right)$ is just a kernel–cokernel pair.

Proof For an arbitrary morphism $g\colon A \to C$, note that the top row of the following commutative diagram must be a short exact sequence since it is isomorphic to the split exact sequence in the bottom row:

$$\begin{array}{ccccc} A & \xrightarrow{\left[\begin{smallmatrix} 1_A \\ -g \end{smallmatrix}\right]} & A \bigoplus C & \xrightarrow{[\,g\ 1_C\,]} & C \\ \| & & \downarrow{\cong} & & \| \\ A & \xrightarrow{\left[\begin{smallmatrix} 1_A \\ 0 \end{smallmatrix}\right]} & A \bigoplus C & \xrightarrow{[\,0\ 1_C\,]} & C. \end{array}$$

The middle isomorphism is $\left[\begin{smallmatrix} 1_A & 0 \\ g & 1_C \end{smallmatrix}\right]$, which has the inverse $\left[\begin{smallmatrix} 1_A & 0 \\ -g & 1_C \end{smallmatrix}\right]$. Now if we have an admissible monic $f\colon A \rightarrowtail B$, then $\left[\begin{smallmatrix} f & 0 \\ 0 & 1_C \end{smallmatrix}\right]\colon A \bigoplus C \rightarrowtail B \bigoplus C$ is also an admissible monic, by Lemma 1.17. Therefore the composite $\left[\begin{smallmatrix} f & 0 \\ 0 & 1_C \end{smallmatrix}\right]\left[\begin{smallmatrix} 1_A \\ -g \end{smallmatrix}\right] = \left[\begin{smallmatrix} f \\ -g \end{smallmatrix}\right]$ is an admissible monic as well. In general, $[\,g'\ f'\,]$ must be a cokernel of $\left[\begin{smallmatrix} f \\ -g \end{smallmatrix}\right]$ whenever the square is a pushout; see Lemma 1.5. This proves that

$$A \xrightarrow{\left[\begin{smallmatrix} f \\ -g \end{smallmatrix}\right]} B \bigoplus C \xrightarrow{[\,g'\ f'\,]} D$$

is a short exact sequence whenever f is an admissible monic. By symmetry we see that this must also be true whenever g is an admissible monic.

Dually, if the square is a pullback and if either f' or g' is an admissible epic, then we have the short exact sequence as well.

By Lemma 1.5, the square is bicartesian if and only if $\left(\left[\begin{smallmatrix} f \\ -g \end{smallmatrix}\right], [\,g'\ f'\,]\right)$ is a kernel–cokernel pair. □

Note that Proposition 1.18 is in agreement with Lemma 1.10 and Proposition 1.12, in the sense that any square obtained by pushout along an admissible monic is always bicartesian.

Corollary 1.19 (Obscure Axiom) *Assume $f\colon A \to B$ is a morphism in $\mathcal{A}$, admitting a cokernel $c\colon B \to C$. If there exists a morphism $g\colon B \to D$ for which gf is an admissible monic, then $A \overset{f}{\rightarrowtail} B \overset{c}{\twoheadrightarrow} C$ is a short exact sequence.*

Proof By Lemma 1.10 we have a pushout diagram

$$\begin{array}{ccccc} A & \overset{gf}{\rightarrowtail} & D & \twoheadrightarrow & E \\ {\scriptstyle f}\downarrow & & \downarrow{\scriptstyle f'} & & \| \\ B & \overset{g'}{\rightarrowtail} & P & \twoheadrightarrow & E \end{array}$$

which, by Proposition 1.18, yields an admissible monic $\left[\begin{smallmatrix} f \\ -gf \end{smallmatrix}\right]\colon A \rightarrowtail B \bigoplus D$. Clearly, $\left[\begin{smallmatrix} 1_B & 0 \\ g & 1_D \end{smallmatrix}\right]$ is an automorphism of $B \bigoplus D$ with inverse $\left[\begin{smallmatrix} 1_B & 0 \\ -g & 1_D \end{smallmatrix}\right]$. So the composite $\left[\begin{smallmatrix} 1_B & 0 \\ g & 1_D \end{smallmatrix}\right]\left[\begin{smallmatrix} f \\ -gf \end{smallmatrix}\right] = \left[\begin{smallmatrix} f \\ 0 \end{smallmatrix}\right]\colon A \rightarrowtail B \bigoplus D$ is an admissible monic. Since

$c = \operatorname{cok} f$, we easily have that $\left[\begin{smallmatrix} c & 0 \\ 0 & 1_D \end{smallmatrix}\right] \colon B \bigoplus D \to C \bigoplus D$ is a cokernel of $\left[\begin{smallmatrix} f \\ 0 \end{smallmatrix}\right]$; see Exercise 1.1.5. Therefore, by Lemma 1.17, we may conclude that $A \stackrel{f}{\rightarrowtail} B \stackrel{c}{\twoheadrightarrow} C$ is a short exact sequence. □

We can use the Obscure Axiom to prove the following result which will be used as a lemma in Chapter 4 to characterize weak equivalences.

Lemma 1.20 *Suppose h is a morphism in an exact category $(\mathcal{A}, \mathcal{E})$ that factorizes as $h = gf$ where f is an admissible monic and g is an admissible epic.*

(1) *If h is an admissible monic, then there is a short exact sequence*

$$\operatorname{Ker} g \rightarrowtail \operatorname{Cok} f \twoheadrightarrow \operatorname{Cok} h.$$

(2) *If h is an admissible epic, then there is a short exact sequence*

$$\operatorname{Ker} h \rightarrowtail \operatorname{Ker} g \twoheadrightarrow \operatorname{Cok} f.$$

Proof The statements are dual and we only prove the first one. From the hypotheses, we easily construct the morphism of short exact sequences shown, where η is obtained from the universal property of the cokernel of f:

$$\begin{array}{ccccc} A & \stackrel{f}{\rightarrowtail} & B & \twoheadrightarrow & \operatorname{Cok} f \\ \Vert & & \downarrow{\scriptstyle g} & & \downarrow{\scriptstyle \eta} \\ A & \stackrel{h}{\rightarrowtail} & C & \twoheadrightarrow & \operatorname{Cok} h. \end{array}$$

By Proposition 1.12(2) we get that the right square is a pullback. It follows from Lemma 1.8(1) that η admits a kernel: $\ker \eta = \operatorname{cok} f \circ \ker g$. Also, since $\eta \circ \operatorname{cok} f$ is equal to the composition of two admissible epics, it too is an admissible epic. So it follows from the dual of Corollary 1.19 that η is an admissible epic. All together we conclude that there is a short exact sequence $\operatorname{Ker} g \rightarrowtail \operatorname{Cok} f \twoheadrightarrow \operatorname{Cok} h$. □

Exercise 1.5.1 (Direct Sums and Summands of Short Exact Sequences) Let $(\mathcal{A}, \mathcal{E})$ be an exact category. Consider the composite

$$A \bigoplus A' \xrightarrow{i \oplus i'} B \bigoplus B' \xrightarrow{p \oplus p'} C \bigoplus C'$$

arising from a pair of composable morphisms $A \xrightarrow{i} B \xrightarrow{p} C$ and $A' \xrightarrow{i'} B' \xrightarrow{p'} C'$. Show that the composite $(i \oplus i', p \oplus p')$ is a short exact sequence if and only if both (i, p) and (i', p') are each short exact sequences. (*Hint*: For the "only if" part, note that $i \oplus i' = \left[\begin{smallmatrix} i & 0 \\ 0 & i' \end{smallmatrix}\right]$, and $\left[\begin{smallmatrix} 1_B \\ 0 \end{smallmatrix}\right] \circ i = \left[\begin{smallmatrix} i & 0 \\ 0 & i' \end{smallmatrix}\right]\left[\begin{smallmatrix} 1_A \\ 0 \end{smallmatrix}\right]$, and use the Obscure Axiom.)

Lemma 1.21 (Homotopy Lemma) *Suppose we have the following commutative diagram in $(\mathcal{A}, \mathcal{E})$ with exact rows:*

$$\begin{array}{ccccc} A_1 & \overset{f_1}{\rightarrowtail} & A_2 & \overset{f_2}{\twoheadrightarrow} & A_3 \\ {\scriptstyle \phi_1}\downarrow & & {\scriptstyle \phi_2}\downarrow & & {\scriptstyle \phi_3}\downarrow \\ B_1 & \underset{g_1}{\rightarrowtail} & B_2 & \underset{g_2}{\twoheadrightarrow} & B_3. \end{array}$$

Then the following statements are equivalent.

(1) *There exists $\alpha\colon A_3 \to B_2$ such that $g_2\alpha = \phi_3$.*
(2) *There exists $\beta\colon A_2 \to B_1$ such that $\beta f_1 = \phi_1$.*
(3) *Regarding the top row as an exact chain complex A, the bottom row as an exact chain complex B, and $\phi = \{\phi_i\}$ as a chain map $\phi\colon A \to B$, then ϕ is null homotopic by a chain homotopy extending α or β.*

Proof (3) $\Rightarrow$ (1) and (3) $\Rightarrow$ (2) are clear. We show (2) $\Rightarrow$ (3). Say $\beta\colon A_2 \to B_1$ has the given property. Then $g_1\beta f_1 = g_1\phi_1 = \phi_2 f_1$. Therefore, $(\phi_2 - g_1\beta)f_1 = 0$. Since $f_2 = \operatorname{cok} f_1$ there exists $\alpha\colon A_3 \to B_2$ such that $\alpha f_2 = \phi_2 - g_1\beta$. Thus we have the desired $\alpha f_2 + g_1\beta = \phi_2$, and it remains to show $g_2\alpha = \phi_3$. For this we note $g_2\alpha f_2 = g_2\phi_2 - g_2 g_1\beta = g_2\phi_2 = \phi_3 f_2$. Since f_2 is an epimorphism we see $g_2\alpha = \phi_3$.

The proof of (1) $\Rightarrow$ (3) is similar. □

Exercise 1.5.2 The existence of an exact structure $\mathcal{E}$ on $\mathcal{A}$ is not needed to obtain a version of the Homotopy Lemma 1.21. State a generalization of the lemma that holds in any additive category.

Exercise 1.5.3 Let $(\mathcal{A}, \mathcal{E})$ and $(\mathcal{B}, \mathcal{E}')$ be exact categories. Assume $F\colon \mathcal{A} \to \mathcal{B}$ is any additive functor admitting either a left or a right adjoint. Let $\mathcal{E}_F \subseteq \mathcal{E}$ be the class of all short exact sequences $A \rightarrowtail B \twoheadrightarrow C$ in $\mathcal{E}$ such that $F(A) \rightarrowtail F(B) \twoheadrightarrow F(C)$ is a short exact sequence in $\mathcal{E}'$. That is, $\mathcal{E}_F$ is the class of all *F-exact sequences* in $\mathcal{E}$. Show that $(\mathcal{A}, \mathcal{E}_F)$ is an exact category. Of course, by duality $(\mathcal{B}, \mathcal{E}'_G)$ is an exact category for the right/left adjoint G.

Exercise 1.5.4 Let $(\mathcal{A}, \mathcal{E})$ and $(\mathcal{B}, \mathcal{E}')$ be exact categories. Suppose that for each object C belonging to some class of objects $\mathcal{C}$ from some category, we have functors $- \otimes C\colon \mathcal{A} \to \mathcal{B}$ with corresponding right adjoints $\mathcal{H}om(C, -)\colon \mathcal{B} \to \mathcal{A}$. Use Exercise 1.5.3 to show that the class $\mathcal{E}_{\otimes\mathcal{C}}$ of all short exact sequences that are $(-\otimes C)$-exact sequences for each $C \in \mathcal{C}$ is an exact structure on $\mathcal{A}$. Similarly, the class $\mathcal{E}'_{\mathcal{H}om\mathcal{C}}$ of all short exact sequences that are $\mathcal{H}om(C, -)$-exact sequences for each $C \in \mathcal{C}$ is an exact structure on $\mathcal{B}$.

1.6 Yoneda Ext Groups

Throughout this section we let $(\mathcal{A}, \mathcal{E})$ be an exact category. The goal is to describe the Yoneda Ext groups, $\mathrm{Ext}^n_{\mathcal{E}}(C, A)$, and the long exact sequences of Ext groups associated to a short exact sequence. The axioms for exact categories are precisely what is needed to define the additive bifunctors $\mathrm{Ext}^n_{\mathcal{E}}(-,-)$ and to prove existence of the long exact sequences as in Theorem 1.22. The good news is that this can all be done in an elementary fashion, using only results of the previous sections. The bad news is that carrying out every detail is a quite long and tedious task. So this is one of only two places in this book where we will be content simply to describe the result (Theorem 1.22) and point the reader to literature that already does a nice job with this. It is okay for the reader encountering this for the first time to take the results on faith and save the details for a future rainy weekend!

To start, one should recall the properties of the hom functor in the following exercise.

Exercise 1.6.1 Let $X \in \mathcal{A}$ be any object and $\mathbb{E}\colon A \stackrel{i}{\rightarrowtail} B \stackrel{p}{\twoheadrightarrow} C$ be any short exact sequence in $\mathcal{E}$. Show the following.

(1) The covariant hom functor $\mathrm{Hom}_{\mathcal{A}}(X, -)\colon \mathcal{A} \to \mathbf{Ab}$ takes $\mathbb{E}$ to a (left) exact sequence

$$0 \to \mathrm{Hom}_{\mathcal{A}}(X, A) \xrightarrow{i_*} \mathrm{Hom}_{\mathcal{A}}(X, B) \xrightarrow{p_*} \mathrm{Hom}_{\mathcal{A}}(X, C)$$

of abelian groups.

(2) The contravariant hom functor $\mathrm{Hom}_{\mathcal{A}}(-, X)\colon \mathcal{A} \to \mathbf{Ab}$ takes $\mathbb{E}$ to a (left) exact sequence

$$0 \to \mathrm{Hom}_{\mathcal{A}}(C, X) \xrightarrow{p^*} \mathrm{Hom}_{\mathcal{A}}(B, X) \xrightarrow{i^*} \mathrm{Hom}_{\mathcal{A}}(A, X)$$

of abelian groups.

The idea of the higher extension functors, $\mathrm{Ext}^n_{\mathcal{E}}(-,-)$, is that they extend these left exact sequences into (possibly infinite) long exact sequences to the right. For example, to extend

$$0 \to \mathrm{Hom}_{\mathcal{A}}(X, A) \xrightarrow{i_*} \mathrm{Hom}_{\mathcal{A}}(X, B) \xrightarrow{p_*} \mathrm{Hom}_{\mathcal{A}}(X, C)$$

to the right, suppose we have an element $f\colon X \to C$ in $\mathrm{Hom}_{\mathcal{A}}(X, C)$. By the pullback axiom for exact categories there is a commutative diagram of short exact sequences:

$$\begin{array}{lccccc}
\delta_1(\mathbb{E}): & A & \overset{i'}{\rightarrowtail} & P & \overset{p'}{\twoheadrightarrow} & X \\
& \| & & \downarrow{\scriptstyle f'} & & \downarrow{\scriptstyle f} \\
\mathbb{E}: & A & \underset{i}{\rightarrowtail} & B & \underset{p}{\twoheadrightarrow} & C.
\end{array}$$

The short exact sequence $\delta_1(\mathbb{E})$ of the top row represents an actual element of $\mathrm{Ext}^1_{\mathcal{E}}(X,A)$. We think of $\mathrm{Ext}^1_{\mathcal{E}}(X,A)$ as the "set of all" such short exact sequences and it turns out that $\mathrm{Ext}^1_{\mathcal{E}}(X,A)$ can be given the structure of an abelian group with $\delta_1\colon \mathrm{Hom}_{\mathcal{A}}(X,C) \to \mathrm{Ext}^1_{\mathcal{E}}(C,A)$ becoming a group homomorphism such that $\mathrm{Im}\, p_* = \mathrm{Ker}\,\delta_1$.

So let us describe the higher Yoneda Ext functors. By an *exact n-sequence* we mean any sequence that can be obtained by splicing together n-many short exact sequences from $\mathcal{E}$. Given any two objects A and C of $\mathcal{A}$, the elements of $\mathrm{Ext}^n_{\mathcal{E}}(C,A)$ are equivalence classes of exact n-sequences of the form

$$\mathbb{E}\colon\ A \rightarrowtail B_{n-1} \to \cdots \to B_1 \to B_0 \twoheadrightarrow C.$$

The equivalence relation is *generated* by a relation $\sim$, where $\mathbb{E}' \sim \mathbb{E}$ means there exists some commutative diagram of the form

$$\begin{array}{lccccccccccc}
\mathbb{E}': & A & \rightarrowtail & B'_{n-1} & \longrightarrow & B'_{n-2} & \longrightarrow & \cdots & \longrightarrow & B'_0 & \twoheadrightarrow & C \\
& \| & & \downarrow & & \downarrow & & & & \downarrow & & \| \\
\mathbb{E}: & A & \rightarrowtail & B_{n-1} & \longrightarrow & B_{n-2} & \longrightarrow & \cdots & \longrightarrow & B_0 & \twoheadrightarrow & C.
\end{array}$$

This is just a morphism of the n-sequences with *fixed ends*. The case $n = 1$ simplifies a great deal, for then by the Short Five Lemma 1.11 any such morphism is necessarily an isomorphism with fixed ends:

$$\begin{array}{lccccc}
\mathbb{E}': & A & \overset{i'}{\rightarrowtail} & B' & \overset{p'}{\twoheadrightarrow} & C \\
& \| & & \downarrow{\scriptstyle v} & & \| \\
\mathbb{E}: & A & \underset{i}{\rightarrowtail} & B & \underset{p}{\twoheadrightarrow} & C.
\end{array}$$

It is clear then that in this (most important) case, the relation $\mathbb{E}' \sim \mathbb{E}$ is already an equivalence relation.

There is a well-defined addition, called the *Baer sum*, on $\mathrm{Ext}^n_{\mathcal{E}}(C,A)$. Briefly, given two n-sequences $\mathbb{E}$ and $\mathbb{F}$ representing elements of $\mathrm{Ext}^n_{\mathcal{E}}(C,A)$, first construct $\mathbb{E} \bigoplus \mathbb{F}$, the n-sequence obtained by taking direct sums termwise. Let $\Delta\colon C \to C \bigoplus C$ denote *diagonal map* given by the matrix $\left[\begin{smallmatrix} 1_C \\ 1_C \end{smallmatrix}\right]$, and let

$\nabla\colon A\bigoplus A \to A$ be the *codiagonal map* (or *fold map*) given by the matrix $[\,1_A\ 1_A\,]$. Then the Baer sum is given by the rule

$$\mathbb{E}+\mathbb{F} := \nabla\left(\mathbb{E}\bigoplus\mathbb{F}\right)\Delta,$$

where this notation means to take the pullback along Δ and the pushout along ∇. (The fact that $\mathcal{E}$ is closed under direct sums, pullbacks, and pushouts is of course crucial here.) It turns out that the pullback and pushout operations can be done in either order here, producing a well-defined element (equivalence class) of $\mathrm{Ext}^n_{\mathcal{E}}(C,A)$. Moreover, this addition gives each $\mathrm{Ext}^n_{\mathcal{E}}(C,A)$ the structure of a (big) abelian group. In particular, the 0 element of $\mathrm{Ext}^1_{\mathcal{E}}(C,A)$ is represented by the canonical split exact sequence

$$A \xrightarrow{\begin{bmatrix}1_A\\0\end{bmatrix}} A\bigoplus C \xrightarrow{[0\ 1_C]} C.$$

Each $\mathrm{Ext}^n_{\mathcal{E}}(C,-)$ becomes an additive covariant functor. Indeed for a given morphism $f\colon A\to A'$,

$$\mathrm{Ext}^n_{\mathcal{E}}(C,f)\colon\ \mathrm{Ext}^n_{\mathcal{E}}(C,A)\to \mathrm{Ext}^n_{\mathcal{E}}(C,A')$$

is a group homomorphism defined by the pushout construction:

$$\begin{array}{ccccccccccc}
A & \rightarrowtail & B_{n-1} & \longrightarrow & B_{n-2} & \longrightarrow & \cdots & \longrightarrow & B_0 & \twoheadrightarrow & C\\
{\scriptstyle f}\downarrow & & \downarrow & & \| & & & & \| & & \|\\
A' & \rightarrowtail & P & \longrightarrow & B_{n-2} & \longrightarrow & \cdots & \longrightarrow & B_0 & \twoheadrightarrow & C.
\end{array}$$

On the other hand, $\mathrm{Ext}^n_{\mathcal{E}}(-,A)$ becomes an additive contravariant functor by way of taking the pullback along a given $g\colon C'\to C$. We have the following fundamental theorem.

Theorem 1.22 (Long Exact Sequence in Ext Groups) *Let $(\mathcal{A},\mathcal{E})$ be an exact category. Then each $\mathrm{Ext}^n_{\mathcal{E}}(-,-)$ is an additive bifunctor, covariant in the second variable and contravariant in the first variable. Given any short exact sequence*

$$A \overset{i}{\rightarrowtail} B \overset{p}{\twoheadrightarrow} C$$

in $\mathcal{E}$, we have the following.

(1) *For each $X\in\mathcal{A}$, there exist group homomorphisms $\{\delta_n\}$ and a long exact sequence of abelian groups*

$$0\to \mathrm{Hom}_{\mathcal{A}}(X,A)\xrightarrow{i_*}\mathrm{Hom}_{\mathcal{A}}(X,B)\xrightarrow{p_*}\mathrm{Hom}_{\mathcal{A}}(X,C)\xrightarrow{\delta_1}\mathrm{Ext}^1_{\mathcal{E}}(X,A)$$

$$\to \mathrm{Ext}^1_{\mathcal{E}}(X,B) \to \mathrm{Ext}^1_{\mathcal{E}}(X,C) \xrightarrow{\delta_2} \mathrm{Ext}^2_{\mathcal{E}}(X,A) \to \mathrm{Ext}^2_{\mathcal{E}}(X,B) \to \cdots$$

$$\to \mathrm{Ext}^{n-1}_{\mathcal{E}}(X,C) \xrightarrow{\delta_n} \mathrm{Ext}^n_{\mathcal{E}}(X,A) \to \mathrm{Ext}^n_{\mathcal{E}}(X,B) \to \mathrm{Ext}^n_{\mathcal{E}}(X,C) \to \cdots.$$

(2) *For each $X \in \mathcal{A}$, there exist group homomorphisms $\{\delta^n\}$ and a long exact sequence of abelian groups*

$$0 \to \mathrm{Hom}_{\mathcal{A}}(C,X) \xrightarrow{p^*} \mathrm{Hom}_{\mathcal{A}}(B,X) \xrightarrow{i^*} \mathrm{Hom}_{\mathcal{A}}(A,X) \xrightarrow{\delta^1} \mathrm{Ext}^1_{\mathcal{E}}(C,X)$$

$$\to \mathrm{Ext}^1_{\mathcal{E}}(B,X) \to \mathrm{Ext}^1_{\mathcal{E}}(A,X) \xrightarrow{\delta^2} \mathrm{Ext}^2_{\mathcal{E}}(C,X) \to \mathrm{Ext}^2_{\mathcal{E}}(B,X) \to \cdots$$

$$\to \mathrm{Ext}^{n-1}_{\mathcal{E}}(A,X) \xrightarrow{\delta^n} \mathrm{Ext}^n_{\mathcal{E}}(C,X) \to \mathrm{Ext}^n_{\mathcal{E}}(B,X) \to \mathrm{Ext}^n_{\mathcal{E}}(A,X) \to \cdots.$$

Proof There are classical sources that carry out the details in the context of abelian categories. For example, there is Mac Lane [1963, chapter III and section XII.4] and Mitchell [1965, chapter VII]. It is a marathon exercise to adapt these proofs to the setting of a general exact category. The author finds the treatment in Mitchell [1965, chapter VII] to be particularly well suited for adapting to exact categories using the results of the previous sections. However, we would be remiss not to refer the reader to Frerick and Sieg [2010, chapter 6]. Here one will find the most complete treatment of the topic, and in the language of exact categories. □

We end this section by noting some properties of $\mathrm{Ext}^1_{\mathcal{E}}$ as it relates to products and coproducts.

Lemma 1.23 *There are canonical isomorphisms*

$$\mathrm{Ext}^1_{\mathcal{E}}\left(C \bigoplus B, A\right) \cong \mathrm{Ext}^1_{\mathcal{E}}(C,A) \bigoplus \mathrm{Ext}^1_{\mathcal{E}}(B,A)$$

and

$$\mathrm{Ext}^1_{\mathcal{E}}\left(C, B \bigoplus A\right) \cong \mathrm{Ext}^1_{\mathcal{E}}(C,B) \bigoplus \mathrm{Ext}^1_{\mathcal{E}}(C,A)$$

of abelian groups.

Proof It is clear from the definition of biproducts (Definition 1.2) that any additive functor F will preserve a biproduct diagram, or in other words, a finite direct sum. In particular, the functors $\mathrm{Ext}^1_{\mathcal{E}}(C,-)$ and $\mathrm{Ext}^1_{\mathcal{E}}(-,A)$ will preserve direct sums, giving us the results. □

Lemma 1.24 $\mathrm{Ext}^1_{\mathcal{E}}$ *satisfies the following properties.*

(1) *Assume* $\left(\bigoplus_{i\in I} C_i, \{\eta_i\}_{i\in I}\right)$ *is an existing coproduct in* $\mathcal{A}$*; so each* $C_i \xrightarrow{\eta_i} \bigoplus_{i\in I} C_i$ *denotes the canonical injection into the ith component. Then for any* $A \in \mathcal{A}$*, there is a canonical monomorphism of abelian groups*

$$\mathrm{Ext}^1_{\mathcal{E}}\left(\bigoplus_{i\in I} C_i, A\right) \to \prod_{i\in I} \mathrm{Ext}^1_{\mathcal{E}}(C_i, A).$$

(2) *Assume* $(\prod_{i\in I} A_i, \{\pi_i\}_{i\in I})$ *is an existing product in* $\mathcal{A}$*; so each* $\prod_{i\in I} A_i \xrightarrow{\pi_i} A_i$ *denotes the canonical projection onto the ith coordinate. Then for any* $C \in \mathcal{A}$*, there is a canonical monomorphism of abelian groups*

$$\mathrm{Ext}^1_{\mathcal{E}}\left(C, \prod_{i\in I} A_i\right) \to \prod_{i\in I} \mathrm{Ext}^1_{\mathcal{E}}(C, A_i).$$

Proof We will prove (1). Applying the additive functor $\mathrm{Ext}^1_{\mathcal{E}}(-, A)$ to the canonical injections $C_i \xrightarrow{\eta_i} \bigoplus_{i\in I} C_i$ we get a collection of morphisms of abelian groups

$$\eta_i^*\colon \mathrm{Ext}^1_{\mathcal{E}}\left(\bigoplus_{i\in I} C_i, A\right) \xrightarrow{\mathrm{Ext}^1_{\mathcal{E}}(\eta_i, A)} \mathrm{Ext}^1_{\mathcal{E}}(C_i, A),$$

which together induce a map into the product

$$(\eta_i^*)_{i\in I}\colon \mathrm{Ext}^1_{\mathcal{E}}\left(\bigoplus_{i\in I} C_i, A\right) \to \prod_{i\in I} \mathrm{Ext}^1_{\mathcal{E}}(C_i, A).$$

Using the Yoneda description of Ext we will prove that this map is always a monomorphism. To do so, suppose we are given a short exact sequence $\mathbb{E} \in \mathrm{Ext}^1_{\mathcal{E}}\left(\bigoplus_{i\in I} C_i, A\right)$ for which $(\eta_i^*)_{i\in I}(\mathbb{E}) = 0$. First, this is equivalent to having $\eta_i^*(\mathbb{E}) = 0$, for all $i \in I$. Second, by the definition of η_i^* we have $\eta_i^*(\mathbb{E}) = \mathbb{E}_i$, where $\mathbb{E}_i$ is the short exact sequence in the top row of the following pushout diagram:

$$\begin{array}{lccccc}
\mathbb{E}_i\colon & A & \overset{j_i}{\rightarrowtail} & P & \overset{q_i}{\twoheadrightarrow} & C_i \\
 & \| & & \downarrow{\scriptstyle m_i} & & \downarrow{\scriptstyle \eta_i} \\
\mathbb{E}\colon & A & \underset{j}{\rightarrowtail} & D & \underset{q}{\twoheadrightarrow} & \bigoplus_{i\in I} C_i.
\end{array}$$

Now having $\eta_i^*(\mathbb{E}) = 0$, for all $i \in I$, means that each q_i has a section s_i, that is $q_i s_i = 1_{C_i}$. The collection $\{m_i s_i\}_{i\in I}$ induces a unique map $s\colon \bigoplus_{i\in I} C_i \to D$ such that $s\eta_i = m_i s_i$ for each $i \in I$. Then s is a section for q. Indeed $(qs)\eta_i = q m_i s_i = \eta_i$ implies that $qs = 1$, by the universal property of the coproduct. □

1.7 Projective and Injective Objects

Let $(\mathcal{A}, \mathcal{E})$ be an exact category throughout this section. Projective and injective objects are fundamental to homological algebra and they make perfect sense in any exact category. An object $P \in \mathcal{A}$ is called **projective** if every short exact sequence $A \rightarrowtail B \twoheadrightarrow P$ splits. If it is necessary to emphasize the exact structure we might call P an $\mathcal{E}$-projective, or an *admissible* projective. An object I is called **injective** if it satisfies the dual.

We have the following characterizations of projective objects. The reader can formulate the dual statements that hold for injective objects.

Proposition 1.25 (Characterizations of Projectives) *The following are equivalent for an object P in an exact category $(\mathcal{A}, \mathcal{E})$.*

(1) *P is projective. That is, every short exact sequence $A \rightarrowtail B \twoheadrightarrow P$ splits and hence admits a direct sum decomposition $B = A \bigoplus P$.*
(2) *The functor* $\operatorname{Hom}_{\mathcal{A}}(P, -)\colon \mathcal{A} \to \mathbf{Ab}$ *is exact. That is, it takes short exact sequences in $\mathcal{E}$ to short exact sequences of abelian groups.*
(3) $\operatorname{Ext}^1_{\mathcal{E}}(P, A) = 0$ *for all $A \in \mathcal{A}$.*

Proof For (1), note that Proposition 1.4 assures us that we have the direct sum decomposition $B = A \bigoplus P$. We prove (1) implies (2). By Exercise 1.6.1, any hom functor $\operatorname{Hom}_{\mathcal{A}}(X, -)\colon \mathcal{A} \to \mathbf{Ab}$ sends short exact sequences to left exact sequences. In the case that $X = P$ is projective, for any short exact sequence $\mathbb{E}\colon A \rightarrowtail B \twoheadrightarrow C$ and morphism $P \to C$, we form the following pullback diagram:

$$\begin{array}{lccccc} \delta_1(\mathbb{E}): & A & \rightarrowtail & Q & \twoheadrightarrow & P \\ & \| & & \downarrow & & \downarrow \\ \mathbb{E}: & A & \rightarrowtail & B & \twoheadrightarrow & C. \end{array}$$

P being projective means the top row splits. This allows for the construction of a morphism $P \to B$, proving $\operatorname{Hom}_{\mathcal{A}}(P, B) \to \operatorname{Hom}_{\mathcal{A}}(P, C)$ is an epimorphism.

To prove (2) implies (1), assume $\operatorname{Hom}_{\mathcal{A}}(P, -)\colon \mathcal{A} \to \mathbf{Ab}$ is exact, and let $A \rightarrowtail B \twoheadrightarrow P$ be a short exact sequence. Using that $\operatorname{Hom}_{\mathcal{A}}(P, B) \to \operatorname{Hom}_{\mathcal{A}}(P, P)$ is an epimorphism, one can see that a preimage of $1_P \in \operatorname{Hom}_{\mathcal{A}}(P, P)$ provides a section (right inverse) for $B \twoheadrightarrow P$. Thus P is projective.

Finally, the zero element of any $\operatorname{Ext}^1_{\mathcal{E}}(C, A)$ group is the (equivalence class of the) canonical split exact sequence. So the definition of a projective object P is *equivalent* to the statement that the Yoneda Ext group $\operatorname{Ext}^1_{\mathcal{E}}(P, A) = 0$ for all A. □

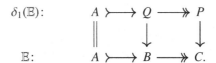

We say that $(\mathcal{A}, \mathcal{E})$ has **enough projectives** if each object $A \in \mathcal{A}$ is an admissible quotient $P \twoheadrightarrow A$ of some projective object P. Dually, we say it has **enough injectives** if each object $A \in \mathcal{A}$ is an admissible subobject $A \rightarrowtail I$ of some injective object I. We leave a few other standard facts concerning projectives and injectives as exercises for the reader.

Lemma 1.26 *Let $\{A_i\}_{i \in I}$ be a collection of objects.*

(1) *If each A_i is projective, then so is their coproduct $\bigoplus_{i \in I} A_i$, assuming it exists.*
(2) *If each A_i is injective, then so is their product $\prod_{i \in I} A_i$, assuming it exists.*

Lemma 1.27 *Projective and injective objects are closed under retracts. (This is the case whether or not the retraction admits a direct sum decomposition, that is, admits a kernel.)*

Exercise 1.7.1 Prove Lemma 1.26.

Exercise 1.7.2 Prove Lemma 1.27.

Exercise 1.7.3 (Schanuel's Lemma) Let $A \in \mathcal{A}$. Given any two short exact sequences $A \rightarrowtail I_1 \twoheadrightarrow Z_1$ and $A \rightarrowtail I_2 \twoheadrightarrow Z_2$ with I_1 and I_2 injective, show that we always have $I_1 \bigoplus Z_2 \cong I_2 \bigoplus Z_1$. Of course, there is a dual statement concerning projectives.

1.8 Weakly Idempotent Complete Additive Categories

There is a minimal amount of kernels and cokernels we need to impose on our additive categories in order to have a nice abstract bedrock to develop the theory of abelian model structures. This is encapsulated within the notion of a weakly idempotent complete additive category. The following corollary, which follows from Proposition 1.4, characterizes these additive categories.

Corollary 1.28 *The following are equivalent for an additive category $\mathcal{A}$.*

(1) *Every split epimorphism $g\colon B \to C$ admits a kernel and hence a direct sum decomposition $B = \operatorname{Ker} g \bigoplus C$.*
(2) *Every split monomorphism $f\colon A \to B$ admits a cokernel and hence a direct sum decomposition $B = A \bigoplus \operatorname{Cok} f$.*

Proof We prove (1) implies (2). So let $f\colon A \to B$ be a split monomorphism, with retraction $r\colon B \to A$. Since r is a split epimorphism, it admits a kernel $s\colon C \to B$ by hypothesis. By Proposition 1.4 we obtain a biproduct diagram $A \underset{f}{\overset{r}{\leftrightarrows}} B \underset{g}{\overset{s}{\leftrightarrows}} C$. In particular, $g = \operatorname{cok} f$; see Exercise 1.1.2. □

The additive categories characterized by Corollary 1.28 go by the following name.

Definition 1.29 We say $\mathcal{A}$ is **weakly idempotent complete**, or **satisfies condition (WIC)**, if every split monic has a cokernel, equivalently, every split epic has a kernel. An exact category $(\mathcal{A}, \mathcal{E})$ is also said to be **weakly idempotent complete** whenever the underlying category $\mathcal{A}$ is such.

The following property is akin to the familiar categorical fact that if a composite gf is monic, then f too is necessarily monic. Note that it is a simplification of the Obscure Axiom (Corollary 1.19) for the case that the underlying category $\mathcal{A}$ is weakly idempotent complete.

Proposition 1.30 (Cancellation Laws) *Let $(\mathcal{A}, \mathcal{E})$ be a weakly idempotent complete exact category. Assume gf is a composite of two morphisms.*

(1) *If gf is an admissible monic, then f must be an admissible monic.*
(2) *If gf is an admissible epic, then g must be an admissible epic.*

Proof We will prove (1). Let $A \xrightarrow{f} B \xrightarrow{g} D$ be the given morphisms. Referring to the Obscure Axiom (Corollary 1.19), we note that all we have to do is show that f admits a cokernel $c\colon B \to C$. We look again at the pushout diagram appearing at the start of the proof of Corollary 1.19. The diagram $B \xrightarrow{g} D \xleftarrow{1} D$ is a "pushout impostor" and hence induces a unique map $h\colon P \to D$ which in particular satisfies $hf' = 1_D$. It means f' is a split monic and so by weak idempotent completeness, f' admits a cokernel $c'\colon P \to C$. But then by Lemma 1.8 the composite $c = c'g'$ is the desired cokernel $c = \operatorname{cok} f$. □

Although weak idempotent completeness is a property of the underlying category $\mathcal{A}$, its importance to exact model structures stems from the next result. Condition (WIC) may be characterized in terms of exact structures as follows.

Proposition 1.31 *Let $(\mathcal{A}, \mathcal{E})$ be an exact category. Then the following are equivalent.*

(1) *$(\mathcal{A}, \mathcal{E})$ is weakly idempotent complete.*
(2) *Every split monomorphism is an admissible monic.*
(3) *The class of admissible monics is closed under retracts.*
(4) *Every split epimorphism is an admissible epic.*
(5) *The class of admissible epics is closed under retracts.*

Proof By Lemma 1.3, any split exact sequence is isomorphic to a canonical split exact sequence, and hence is a short exact sequence in $\mathcal{E}$. So (1), (2), and (4) are equivalent by Proposition 1.4 and Definition 1.29.

Next, assume (2), that every split monomorphism is an admissible monic. To prove (3), suppose we have a commutative diagram

$$\begin{array}{ccccc} A & \overset{f_1}{\rightarrowtail} & X & \xrightarrow{g_1} & A \\ {\scriptstyle j}\downarrow & & \downarrow{\scriptstyle i} & & \downarrow{\scriptstyle j} \\ B & \underset{}{\overset{f_2}{\rightarrowtail}} & Y & \xrightarrow{g_2} & B \end{array}$$

expressing $j\colon A \to B$ as a retract of some admissible monic $i\colon X \rightarrowtail Y$. That is, $g_1 f_1 = 1_A$ and $g_2 f_2 = 1_B$. By hypothesis, f_1 and f_2 are admissible monics. Being a composite of two admissible monics, we have that $if_1 = f_2 j$ is an admissible monic. So by the Cancellation Laws, Proposition 1.30, we get that j is an admissible monic.

We now show (3) implies (2). Suppose that the class of admissible monics is closed under retracts and let $f\colon A \to C$ be a split monomorphism. So there is a map $r\colon C \to A$ with $rf = 1_A$. We wish to show that f is an admissible monic. First, we note that the morphism $\left[\begin{smallmatrix} 1_A \\ -f \end{smallmatrix}\right]\colon A \rightarrowtail A \bigoplus C$ must always be an admissible monic. Indeed this is a special case of Proposition 1.18, by considering the pushout of f along the admissible monic 1_A. So by the hypothesis, we will be done if we can show that $f\colon A \to C$ is a retract of $\left[\begin{smallmatrix} 1_A \\ -f \end{smallmatrix}\right]$. But the following commutative diagram displays that this is indeed the case:

$$\begin{array}{ccccc} A & = & A & = & A \\ {\scriptstyle f}\downarrow & & \downarrow{\scriptstyle \left[\begin{smallmatrix} 1_A \\ -f \end{smallmatrix}\right]} & & \downarrow{\scriptstyle f} \\ C & \xrightarrow[\left[\begin{smallmatrix} r \\ -1_C \end{smallmatrix}\right]]{} & A \bigoplus C & \xrightarrow[\left[\begin{smallmatrix} f & fr-1_C \end{smallmatrix}\right]]{} & C. \end{array}$$

This completes the proof of (3) implies (2), and thus (1)–(4) are all equivalent. By duality, statement (5) is also equivalent. □

In any additive category $\mathcal{A}$, a monomorphism $i\colon S \to A$ represents a subobject $S \subseteq A$. Technically, a subobject is an equivalence class of such monomorphisms. The next couple of exercises specialize this idea to admissible monics, by considering the notion of $\mathcal{E}$-subobjects, denoted by $S \subseteq_{\mathcal{E}} A$.

Exercise 1.8.1 ($\mathcal{E}$-Subobjects) Let $(\mathcal{A}, \mathcal{E})$ be a weakly idempotent complete exact category. Two admissible monics $i\colon S \rightarrowtail A$ and $i'\colon S' \rightarrowtail A$ are *equivalent* if there is an isomorphism $\psi\colon S \rightarrowtail S'$ such that $i'\psi = i$. An equivalence class of such admissible monics is called an *$\mathcal{E}$-subobject*, or an *admissible subobject*, of A, and denoted by $S \subseteq_{\mathcal{E}} A$.

(1) Let $S \subseteq_{\mathcal{E}} A$ and $S' \subseteq_{\mathcal{E}} A$ be $\mathcal{E}$-subobjects. Show that any morphism $S \to S'$ compatible with the inclusions $S \subseteq_{\mathcal{E}} A$ and $S' \subseteq_{\mathcal{E}} A$ is necessarily an

admissible monic and hence determines a compatible subobject inclusion $S \subseteq_{\mathcal{E}} S' \subseteq_{\mathcal{E}} A$.

(2) Let $S \subseteq_{\mathcal{E}} A$ and $S' \subseteq_{\mathcal{E}} A$. Show that if $S \subseteq_{\mathcal{E}} S'$ and $S' \subseteq_{\mathcal{E}} S$, then S and S' represent the same $\mathcal{E}$-subobject of A.

(3) Show that if $S \subseteq_{\mathcal{E}} A$ and $X \subseteq_{\mathcal{E}} A/S$, then X must take the form $X \cong S'/S$ for some $S \subseteq_{\mathcal{E}} S'$.

Exercise 1.8.2 (Properties of $\mathcal{E}$-Subobjects) Let $(\mathcal{A}, \mathcal{E})$ be a weakly idempotent complete exact category and suppose $A \subseteq B \subseteq C$ are subobjects in $\mathcal{A}$.

(1) If $A \subseteq_{\mathcal{E}} B$ and $B \subseteq_{\mathcal{E}} C$, then $A \subseteq_{\mathcal{E}} C$.
(2) If $A \subseteq_{\mathcal{E}} C$ then $A \subseteq_{\mathcal{E}} B$.
(3) If $A \subseteq_{\mathcal{E}} C$ and $B/A \subseteq_{\mathcal{E}} C/A$, then $B \subseteq_{\mathcal{E}} C$. Moreover,

$$(C/A)/(B/A) \cong C/B.$$

1.9 Abelian and Quasi-abelian Categories

An additive category $\mathcal{A}$ is called *pre-abelian* if every morphism has a kernel and a cokernel. This class of categories includes not just all abelian categories but also the more general *quasi-abelian categories* of which relevant examples arise in functional analysis and topology. The main point of this section is to describe the canonical exact structure $(\mathcal{A}, \mathcal{E}_{max})$ that exists on any quasi-abelian category $\mathcal{A}$. If $\mathcal{A}$ is abelian, it is just the class of all the usual short exact sequences.

Throughout this section $\mathcal{A}$ will always denote a pre-abelian category. Note then that $\mathcal{A}$ must also possess all pushouts and pullbacks, by Lemma 1.5.

Utilizing that all morphisms have kernels and cokernels, it is customary to attach the following notation to any given morphism $f\colon A \to B$ of $\mathcal{A}$: The domain of $\ker f$ is denoted by $\operatorname{Ker} f$, and so $\ker f$ might also be denoted by $\operatorname{Ker} f \rightarrowtail A$. On the other hand, $\operatorname{cok} f$, the cokernel of f, may be denoted $B \twoheadrightarrow \operatorname{Cok} f$. We define the *image* of f to be the morphism $\operatorname{im} f := \ker(\operatorname{cok} f)$ and we may also use the notation $\operatorname{Im} f \rightarrowtail B$. The dual notion is the *coimage* of f, defined by $\operatorname{coim} f := \operatorname{cok}(\ker f)$, and we may write $A \twoheadrightarrow \operatorname{Coim} f$.

Lemma 1.32 *Let $f\colon A \to B$ be any morphism in a pre-abelian category $\mathcal{A}$.*

(1) $\ker f = \ker(\operatorname{cok}(\ker f))$.
(2) $\operatorname{cok} f = \operatorname{cok}(\ker(\operatorname{cok} f))$.

Consequently, both $(\ker f, \operatorname{coim} f)$ *and* $(\operatorname{im} f, \operatorname{cok} f)$ *are kernel–cokernel pairs.*

Proof The composite $\operatorname{Ker} f \xrightarrow{\ker f} A \xrightarrow{\operatorname{cok}(\ker f)} \operatorname{Cok}(\ker f)$ is certainly 0. To prove (1), we only need to show that if $t\colon X \to A$ is any other morphism with $0 = \operatorname{cok}(\ker f) \circ t$, then t factors uniquely through $\ker f$. But since $f \circ \ker f = 0$, the universal property of $\operatorname{Cok}(\ker f)$ provides a unique $w\colon \operatorname{Cok}(\ker f) \to B$ such that $f = w \circ \operatorname{cok}(\ker f)$. Therefore, $0 = \operatorname{cok}(\ker f) \circ t = w \circ \operatorname{cok}(\ker f) \circ t = ft$. Thus the universal property of $\operatorname{Ker} f$ does indeed provide a unique factorization of t through $\ker f$. This proves (1) and (2) is dual. □

By a *kernel* (resp. *cokernel*) we mean a morphism that is equal to the kernel (resp. cokernel) of some morphism. Certainly any kernel must be a monomorphism and any cokernel must be an epimorphism. Now by Lemma 1.32, k is a kernel if and only if $(k, \operatorname{cok} k)$ is a kernel–cokernel pair. Dually, c is a cokernel if and only if $(\ker c, c)$ is a kernel–cokernel pair. However, the class of all kernel–cokernel pairs in a pre-abelian category $\mathcal{A}$ need not satisfy the closure properties that define an exact category. It is essentially the definition of a quasi-abelian category that this be the case.

Definition 1.33 A pre-abelian category $\mathcal{A}$ is said to be **quasi-abelian** if the pushout of any kernel along any morphism is again a kernel and, dually, the pullback of any cokernel along any morphism is again a cokernel.

Proposition 1.34 *Let $\mathcal{A}$ be a quasi-abelian category and $\mathcal{E}_{max}$ be the class of all kernel–cokernel pairs in $\mathcal{A}$. Then $(\mathcal{A}, \mathcal{E}_{max})$ is a (weakly idempotent complete) exact category.*

Proof By the previous comments, Axiom **(Ex3)** (*Pushouts and Pullbacks*) follows by the very definition of a quasi-abelian category. So there is essentially just one difficulty, Axiom **(Ex4)** (*Compositions*), in showing that $\mathcal{E}_{max}$ defines an exact structure on $\mathcal{A}$. Let (i, p) and (j, q) be kernel–cokernel pairs. We need to show that $ji = \ker(\operatorname{cok}(ji))$. Set $c = \operatorname{cok}(ji)$ and consider the commutative diagram

$$\begin{array}{ccccc}
A & \xrightarrow{i} & B & \xrightarrow{p} & D \\
\| & & \downarrow{\scriptstyle j} & & \downarrow{\scriptstyle j'} \\
A & \xrightarrow{ji} & C & \xrightarrow{c} & F \\
 & & \downarrow{\scriptstyle q} & & \downarrow{\scriptstyle q'} \\
 & & E & = & E,
\end{array}$$

where j' and q' are obtained by the universal properties of the cokernels involved. By Lemma 1.6(2), the upper right square is a pushout. $\mathcal{A}$ being quasi-abelian guarantees that j' is also a kernel. Since p is right cancellable the right-hand column is certainly a null sequence, and so Lemma 1.6(1) applies and we conclude the same square is also a pullback. But now since $i = \ker p$, Lemma 1.8(1) applies and we conclude $ji = \ker c$, as desired. □

Example 1.35 (Functional Analysis) Jack Kelly has considered abelian model structures on a number of quasi-abelian categories rooted in Functional Analysis. These include certain classes of bornological spaces, non-Archimedian Banach spaces, and locally convex topological vector spaces. See Kelly [2023] and Kelly [2024, chapter 3]. Other good sources of information on exact and quasi-abelian categories arising in functional analysis include the papers of Prosmans and Schneiders. In particular, see Prosmans [2000], Schneiders [1999], and Prosmans and Schneiders [2000]. Several more examples with references are given in Bühler [2010, section 13.2].

Definition 1.36 A pre-abelian category $\mathcal{A}$ is called **abelian** if every monomorphism is a kernel and every epimorphism is a cokernel.

In particular, it follows from Lemma 1.32 that any monomorphism in an abelian category must coincide with the kernel of its cokernel, and dual for any epimorphism. Using this, one can easily prove the next exercise. We recall that a pair of composable morphisms $A \xrightarrow{f} B \xrightarrow{g} C$ in an abelian category is said to be *exact (at B)* if $\operatorname{im} f = \ker g$.

Exercise 1.9.1 Let $f\colon A \to B$ be a morphism in an abelian category $\mathcal{A}$. The following statements are equivalent.

(1) f is monic.
(2) $f = \operatorname{im} f$.
(3) $(f, \operatorname{cok} f)$ is a kernel–cokernel pair.
(4) $\ker f = 0$.
(5) $0 \to A \xrightarrow{f} B$ is exact at A.

Of course there are dual characterizations for epimorphisms.

Exercise 1.9.2 Let $\mathcal{A}$ be an abelian category. By a *left exact sequence* we mean a sequence of morphisms $0 \to A \xrightarrow{f} B \xrightarrow{g} C$ that is exact (at both A and B). Show that this is the case if and only if $f = \ker g$. Of course there is the dual notion of a right exact sequence $A \xrightarrow{f} B \xrightarrow{g} C \to 0$, which is equivalent to the statement that $g = \operatorname{cok} f$.

Definition 1.37 Let $\mathcal{A}$ be an abelian category. By a **(standard) short exact sequence** in $\mathcal{A}$ we mean a sequence of morphisms $0 \to A \xrightarrow{f} B \xrightarrow{g} C \to 0$ that is exact at A, B, and C. Equivalently, (f, g) is a kernel–cokernel pair.

Lemma 1.38 *Let $\mathcal{A}$ be an abelian category. Then the pushout of any monic along any morphism is again a monic and, dually, the pullback of any epic along any morphism is again an epic.*

Proof Suppose that $f\colon A \to B$ is monic and we have a pushout square

$$\begin{array}{ccc} A & \xrightarrow{g} & C \\ {\scriptstyle f}\downarrow & & \downarrow{\scriptstyle f'} \\ B & \xrightarrow{g'} & D. \end{array}$$

By Lemma 1.5(2) and Exercise 1.9.2 we have a right exact sequence

$$A \xrightarrow{\left[\begin{smallmatrix} f \\ -g \end{smallmatrix}\right]} B \bigoplus C \xrightarrow{[g'\ f']} D \to 0.$$

We will show $\left[\begin{smallmatrix} f \\ -g \end{smallmatrix}\right]$ is monic. Indeed let $t\colon X \to A$ be any morphism with $\left[\begin{smallmatrix} f \\ -g \end{smallmatrix}\right] \circ t = 0$. Then in particular we must have $ft = 0$, and since f is given to be monic, $t = 0$. Therefore we have a short exact sequence

$$0 \to A \xrightarrow{\left[\begin{smallmatrix} f \\ -g \end{smallmatrix}\right]} B \bigoplus C \xrightarrow{[g'\ f']} D \to 0.$$

The goal is to show that $f'\colon C \to D$ must also be monic. So let $s\colon Y \to C$ be any morphism for which $f's = 0$. We must show $s = 0$. The morphism $\left[\begin{smallmatrix} 0 \\ s \end{smallmatrix}\right]\colon Y \to B \bigoplus C$ satisfies $[g'\ f']\left[\begin{smallmatrix} 0 \\ s \end{smallmatrix}\right] = 0$, and since $\left[\begin{smallmatrix} f \\ -g \end{smallmatrix}\right]$ is the kernel of $[g'\ f']$ there is a unique morphism $w\colon Y \to A$ satisfying $\left[\begin{smallmatrix} f \\ -g \end{smallmatrix}\right] \circ w = \left[\begin{smallmatrix} 0 \\ s \end{smallmatrix}\right]$. Therefore, $fw = 0$ and $-gw = s$. But f monic implies $w = 0$, and thus $s = 0$ too. □

Proposition 1.39 *Let $\mathcal{A}$ be an abelian category. Then $\mathcal{A}$ is a quasi-abelian category and the class $\mathcal{E}_{max}$ of all kernel–cokernel pairs coincides with the class of all (the standard) short exact sequences as in Definition 1.37.*

Proof It follows immediately from Definition 1.33 and Lemma 1.38 that $\mathcal{A}$ is quasi-abelian. By Definition 1.37, a (standard) short exact sequence is nothing more than a kernel–cokernel pair, that is, an element of the class $\mathcal{E}_{max}$. □

Exercise 1.9.3 (Characterizations of Abelian Categories) Show that the following are equivalent for a pre-abelian category $\mathcal{A}$.

(1) $\mathcal{A}$ is abelian. That is, every monomorphism is a kernel and every epimorphism is a cokernel.

(2) Every morphism f of $\mathcal{A}$ factors as $f = me$ where e is a cokernel and m is a kernel.
(3) For every morphism f of $\mathcal{A}$, the canonical map $\bar{f}\colon \operatorname{Coim} f \to \operatorname{Im} f$ is an isomorphism.

1.10 Proper Classes Are Exact Structures

Let $\mathcal{A}$ be an abelian category. Appearing in Mac Lane's classic text *Homology* [Mac Lane, 1963] is the notion of a proper class of short exact sequences. When Hovey introduced abelian model categories it was indeed in the context of such proper classes. In this brief section we show that the notion of a proper class is equivalent to an exact structure when the underlying category is abelian. In fact, an identical proof works for quasi-abelian categories, and we point this out afterwards.

Let us start with the definition of a proper class as given in Mac Lane's book.

Definition 1.40 Let $\mathcal{A}$ be an abelian category and let $\mathcal{P}$ be some class of (standard) short exact sequences. $\mathcal{P}$ is called a **proper class** of short exact sequences if $\mathcal{P}$ satisfies the following axioms. The axioms utilize the following standard terminology: If (f, g) is some short exact sequence in $\mathcal{P}$, then we say it is a **proper short exact sequence**, f is a **proper monic**, and g is a **proper epic**.
(P1) (*Replete*) $\mathcal{P}$ is closed under isomorphisms.
(P2) (*Split Exact Sequences*) $\mathcal{P}$ contains the canonical split exact sequence

$$A \xrightarrow{\left[\begin{smallmatrix} 1_A \\ 0 \end{smallmatrix}\right]} A \bigoplus C \xrightarrow{[0\ 1_C]} C,$$

for any given pair of objects A and C.
(P3) (*Compositions*) The composition of two proper monics is again a proper monic. Dually, the composition of two proper epics is again a proper epic.
(P4) (*Cancellation Laws*) If f and g are composable monomorphisms and their composition gf is a proper monic, then f too is a proper monic. Dually, if f and g are composable epimorphisms and their composition gf is a proper epic, then g too is a proper epic.

Proposition 1.41 *Let $\mathcal{A}$ be an abelian category. Then $\mathcal{P}$ is a proper class of short exact sequences if and only if $(\mathcal{A}, \mathcal{P})$ is an exact category.*

Proof Note that the statement of the proposition makes sense because (standard) short exact sequences in abelian categories are the same thing as kernel-cokernel pairs. Now comparing Definition 1.9 and Definition 1.40, we note that **(P1)** = **(Ex1)**, **(P2)** = **(Ex2)**, and **(P3)** = **(Ex4)**. So we only need to rectify the equivalence of **(P4)** and **(Ex3)** under the assumption that the underlying category is abelian.

First, suppose $(\mathcal{A}, \mathcal{P})$ is an exact category. Then since every morphism of $\mathcal{A}$ admits a cokernel, the Obscure Axiom (Corollary 1.19), and its dual, immediately imply **(P4)** holds. In fact, since $\mathcal{A}$ is weakly idempotent complete it follows from the simplified version in Proposition 1.30.

On the other hand, suppose we are given a proper class $\mathcal{P}$ on $\mathcal{A}$. Let $\mathcal{E}_{max}$ denote the class of all of the standard short exact sequences, which we will denote here by $0 \to A \xrightarrow{f} B \xrightarrow{g} C \to 0$. Of course we already know from Proposition 1.39 that $(\mathcal{A}, \mathcal{E}_{max})$ is an exact category and we have $\mathcal{P} \subseteq \mathcal{E}_{max}$. We must check that $(\mathcal{A}, \mathcal{P})$ satisfies axiom **(Ex3)** (*Pushouts and Pullbacks*). But any, say pullback, exists, and by applying Lemma 1.10(2) to $(\mathcal{A}, \mathcal{E}_{max})$, we see that the pullback of any epimorphism p yields a commutative diagram with each row a short exact sequence:

$$\begin{array}{ccccccccc} 0 & \longrightarrow & A & \xrightarrow{i'} & P & \xrightarrow{p'} & C' & \longrightarrow & 0 \\ & & \| & & \downarrow{\scriptstyle f'} & & \downarrow{\scriptstyle f} & & \\ 0 & \longrightarrow & A & \xrightarrow{i} & B & \xrightarrow{p} & C & \longrightarrow & 0. \end{array}$$

We must show that p' is a $\mathcal{P}$-proper epic whenever p is a $\mathcal{P}$-proper epic. Since i is a $\mathcal{P}$-monic, so is $i = f'i'$. We wish to "cancel" f' to conclude i' is a $\mathcal{P}$-monic. However, axiom **(P4)** only allows this when f' is monic. Nevertheless, by applying Proposition 1.18 to $(\mathcal{A}, \mathcal{E}_{max})$, we obtain a short exact sequence

$$0 \to P \xrightarrow{\left[\begin{smallmatrix} f' \\ -p' \end{smallmatrix}\right]} B \oplus C' \xrightarrow{[p\ f]} C \to 0.$$

Since $\left[\begin{smallmatrix} 1_B \\ 0 \end{smallmatrix}\right] : B \to B \bigoplus C'$ is a $\mathcal{P}$-monic by **(P2)**, we see that $\left[\begin{smallmatrix} 1_B \\ 0 \end{smallmatrix}\right] \circ i = \left[\begin{smallmatrix} i \\ 0 \end{smallmatrix}\right]$ is a $\mathcal{P}$-monic by **(P3)**. So the composite $\left[\begin{smallmatrix} f' \\ -p' \end{smallmatrix}\right] \circ i' = \left[\begin{smallmatrix} f'i' \\ -p'i' \end{smallmatrix}\right] = \left[\begin{smallmatrix} i \\ 0 \end{smallmatrix}\right]$ is a $\mathcal{P}$-monic. But then by **(P4)** we may now conclude that i' is a $\mathcal{P}$-monic. This proves that p' is a $\mathcal{P}$-proper epic whenever p is a $\mathcal{P}$-proper epic. The dual also holds, so $(\mathcal{A}, \mathcal{P})$ satisfies axiom **(Ex3)** . □

So we need not distinguish between proper classes and exact structures when the underlying category is abelian. In fact, as already mentioned, proper classes and exact structures are one and the same even for quasi-abelian categories. Indeed let $\mathcal{A}$ be a quasi-abelian category and let $\mathcal{E}_{max}$ be the canonical exact structure of Proposition 1.34. $\mathcal{E}_{max}$ consists of all the kernel–cokernel pairs in $\mathcal{A}$. Then the interested reader will easily verify that the previous proof readily adapts to the following statement.

Scholium 1.42 Let $\mathcal{A}$ be a quasi-abelian category and let $\mathcal{P} \subseteq \mathcal{E}_{max}$ be some class of kernel–cokernel pairs. Call $\mathcal{P}$ a **proper class** of short exact sequences if it satisfies the axioms of Definition 1.40. Then $\mathcal{P}$ is a proper class of short exact sequences if and only if $(\mathcal{A}, \mathcal{P})$ is an exact category.

Exercise 1.10.1 Verify that Proposition 1.41 carries over to Scholium 1.42.

1.11 Splitting Idempotents and Idempotent Completeness

The material in this section is more or less optional, and can be referred to later if needed. The only place in the text that calls on this section will be in Chapter 10 where we describe a connection between contractible chain complexes and idempotent completeness. See Theorem 10.15. Assume throughout that $\mathcal{A}$ is an additive category.

An endomorphism $e\colon B \to B$ in $\mathcal{A}$ is called an **idempotent** if $e^2 = e$. Note that if e is an idempotent, then $1_B - e$ is also an idempotent. A special case arises in the situation where we have a retraction $rf = 1_A$ of an object B. In this case the reverse composite $fr\colon B \to B$ is clearly an idempotent. In fact, an idempotent e is said to *split* or to be a **split idempotent** if it factors as $e = fr$ in such a way that there is a retraction $rf = 1_A$. The following result indicates why this terminology is used, for if both e and $1_B - e$ are split idempotents then they are complements within a biproduct diagram.

Corollary 1.43 *Assume $e\colon B \to B$ and its complement $1 - e\colon B \to B$ are each split idempotents. Denote their splittings by $e = fr$ and $1 - e = sg$. Then we have a biproduct diagram*

$$A \underset{f}{\overset{r}{\leftrightarrows}} B \underset{g}{\overset{s}{\leftrightarrows}} C.$$

The kernel, cokernel, and image of e and $1 - e$ all exist; in particular we have

$$\ker e = \ker r = s, \quad \operatorname{cok} e = \operatorname{cok} f = g, \quad \operatorname{im} e = f.$$

Moreover, $\left[\begin{smallmatrix} r \\ g \end{smallmatrix}\right]\colon B \to \operatorname{Im} e \bigoplus \operatorname{Ker} e$ and $\left[\begin{smallmatrix} f & s \end{smallmatrix}\right]\colon \operatorname{Im} e \bigoplus \operatorname{Ker} e \to B$ are inverse isomorphisms between the internal and external direct sum diagrams.

Proof The fact that we have a biproduct diagram follows immidately from Definition 1.2 due to the given information (i) $rf = 1_A$ for some A, (ii) $gs = 1_C$ for some C, and (iii) $fr + sg = e + (1 - e) = 1_B$. Lemma 1.3 provides the inverse isomorphisms between the internal direct sum (biproduct) and the external direct sum. □

Definition 1.44 We say $\mathcal{A}$ is **idempotent complete** if every idempotent in $\mathcal{A}$ splits. We also say that an exact category $(\mathcal{A}, \mathcal{E})$ is **idempotent complete** whenever the underlying category $\mathcal{A}$ is such.

As the terminology suggests we have the following.

Proposition 1.45 *Any idempotent complete additive category $\mathcal{A}$ must in particular be weakly idempotent complete.*

Proof Assume $f: A \to B$ is any split monic with left inverse $rf = 1_A$. We only need to show that $\operatorname{cok} f$ exists. Clearly $e := fr$ is a split idempotent and by assumption its completement $1_B - e$ must also be split. So by Corollary 1.43 we get that $\operatorname{cok} e = \operatorname{cok} f$ exists. □

Proposition 1.46 *The following are equivalent for an additive category $\mathcal{A}$.*

(1) *$\mathcal{A}$ is idempotent complete, that is, every idempotent in $\mathcal{A}$ splits.*
(2) *Every idempotent of $\mathcal{A}$ admits a kernel.*
(3) *Every idempotent of $\mathcal{A}$ admits a cokernel.*

Proof We already have (1) implies (2) and (3) by Corollary 1.43. Conversely, assume every idempotent in $\mathcal{A}$ admits a kernel and let $e: B \to B$ be an idempotent. We must show e splits. We begin by letting $s: C \to B$ be the kernel of e and letting $f: A \to B$ be the kernel of its complement $1 - e$. Since $(1 - e)e = 0$, there exists a unique morphism $r: B \to A$ such that $e = fr$. To finish showing e is a split idempotent it only remains to show that $rf = 1_A$. For this, note

$$frf = ef = ef + 0 = ef + (1 - e)f = f = f1_A.$$

Since f is a kernel it is necessarily monic, and so by left canceling f we get $rf = 1_A$. Therefore, e is a split idempotent, proving (2) implies (1). One can see that (3) implies (1) holds similarly. □

The following result of Freyd will not be used in the text. But it shows that we typically need not distinguish between idempotent complete and weakly idempotent complete additive categories.

Proposition 1.47 (Freyd [1966]) *Let $\mathcal{A}$ be an additive category with countable coproducts. Then $\mathcal{A}$ is idempotent complete if and only if it is weakly idempotent complete.*

2

Cotorsion Pairs

In this chapter we make an initial study of cotorsion pairs with an eye toward developing the theory of abelian model categories. Throughout, we let $(\mathcal{A}, \mathcal{E})$ denote any exact category. It need not be weakly idempotent complete unless explicitly stated otherwise.

2.1 Ext-Pairs and Cotorsion Pairs

A pair of classes $\mathfrak{C} = (\mathcal{X}, \mathcal{Y})$ of objects in $\mathcal{A}$ is called a **cotorsion pair** if $\mathcal{Y} = \mathcal{X}^{\perp}$ and $\mathcal{X} = {}^{\perp}\mathcal{Y}$. Here, given a class of objects $\mathcal{C}$, the right orthogonal $\mathcal{C}^{\perp}$ is defined to be the class of all objects X such that $\mathrm{Ext}^1_{\mathcal{E}}(C, X) = 0$ for all $C \in \mathcal{C}$. Similarly, we define the left orthogonal ${}^{\perp}\mathcal{C}$. We say a cotorsion pair $\mathfrak{C}$ **has enough projectives** if for each $A \in \mathcal{A}$ there exists a short exact sequence (conflation) $Y \rightarrowtail X \twoheadrightarrow A$ with $X \in \mathcal{X}$ and $Y \in \mathcal{Y}$. We call such a short exact sequence an **$\mathcal{X}$-approximation sequence**. The admissible epic $X \twoheadrightarrow A$ is sometimes called a **special $\mathcal{X}$-precover**. On the other hand, we say $\mathfrak{C}$ **has enough injectives** if it satisfies the dual, and we call the dual short exact sequence a **$\mathcal{Y}$-approximation sequence**, and its monic is a **special $\mathcal{Y}$-preenvelope**. A cotorsion pair $\mathfrak{C}$ is called **complete** if it has both enough projectives and enough injectives, or in other words, each object has a special $\mathcal{X}$-precover and a special $\mathcal{Y}$-preenvelope.

There is a weaker notion than that of a cotorsion pair which we will call an *Ext-pair*. Before we define them, we point out a lemma related to the following convenient notation: For a pair $(\mathcal{X}, \mathcal{Y})$ of classes of objects of $\mathcal{A}$, we will write $\mathrm{Ext}^1_{\mathcal{E}}(\mathcal{X}, \mathcal{Y}) = 0$ to mean that $\mathrm{Ext}^1_{\mathcal{E}}(X, Y) = 0$ for all $X \in \mathcal{X}$ and $Y \in \mathcal{Y}$.

Lemma 2.1 *Let $(\mathcal{X}, \mathcal{Y})$ be a pair of classes of objects of $\mathcal{A}$. The following are equivalent.*

(1) $\operatorname{Ext}^1_{\mathcal{E}}(\mathcal{X},\mathcal{Y}) = 0$.
(2) $\mathcal{X} \subseteq {}^{\perp}\mathcal{Y}$.
(3) $\mathcal{Y} \subseteq \mathcal{X}^{\perp}$.

In particular, if $\operatorname{Ext}^1_{\mathcal{E}}(\mathcal{X},\mathcal{Y}) = 0$ *then* $(\mathcal{X},\mathcal{Y})$ *is a cotorsion pair if and only if* $\mathcal{X} \supseteq {}^{\perp}\mathcal{Y}$ *and* $\mathcal{Y} \supseteq \mathcal{X}^{\perp}$.

Proof The statements are immediate from the definitions. □

We now define Ext-pairs, and carry over much of the language associated to cotorsion pairs.

Definition 2.2 Let $\mathfrak{C} = (\mathcal{X},\mathcal{Y})$ be a pair of classes of objects of $\mathcal{A}$. We say $\mathfrak{C}$ is an **Ext-pair** if $\operatorname{Ext}^1_{\mathcal{E}}(\mathcal{X},\mathcal{Y}) = 0$ and if both of the classes $\mathcal{X}$ and $\mathcal{Y}$ contain 0 and are closed under extensions. Just like cotorsion pairs, we say the Ext-pair $\mathfrak{C}$ **has enough projectives** if for each $A \in \mathcal{A}$ there exists a short exact sequence $Y \rightarrowtail X \twoheadrightarrow A$ with $X \in \mathcal{X}$ and $Y \in \mathcal{Y}$. We again call such a short exact sequence an **$\mathcal{X}$-approximation sequence**, and the admissible epic $X \twoheadrightarrow A$ is a **special $\mathcal{X}$-precover**. On the other hand, we say $\mathfrak{C}$ **has enough injectives** if it satisfies the dual, and we call the dual short exact sequence a **$\mathcal{Y}$-approximation sequence**, and its monic is a **special $\mathcal{Y}$-preenvelope**. An Ext-pair $\mathfrak{C}$ is called **complete** if it has both enough projectives and enough injectives.

Of course any cotorsion pair is an Ext-pair and we next explore the converse.

Notation 2.3 Given a class $\mathcal{X}$ of objects containing 0, we let $\overline{\mathcal{X}}$ denote the class of all direct summands of objects in $\mathcal{X}$. Note that $\mathcal{X} \subseteq \overline{\mathcal{X}}$, and that $\overline{\mathcal{X}}$ is closed under direct summands. We call it the **direct summand closure** of $\mathcal{X}$. Of course, $\mathcal{X} = \overline{\mathcal{X}}$ if and only if $\mathcal{X}$ is closed under direct summands.

Proposition 2.4 *Let* $(\mathcal{X},\mathcal{Y})$ *be an Ext-pair. The following statements hold.*

(1) $\operatorname{Ext}^1_{\mathcal{E}}\left(\overline{\mathcal{X}},\overline{\mathcal{Y}}\right) = 0$.
(2) *Assume* $(\mathcal{X},\mathcal{Y})$ *has enough projectives. Then* $\overline{\mathcal{X}} = {}^{\perp}\mathcal{Y}$. *In particular,* $\mathcal{X} = {}^{\perp}\mathcal{Y}$ *if and only if* $\mathcal{X}$ *is closed under direct summands.*
(3) *Assume* $(\mathcal{X},\mathcal{Y})$ *has enough injectives. Then* $\overline{\mathcal{Y}} = \mathcal{X}^{\perp}$. *In particular,* $\mathcal{Y} = \mathcal{X}^{\perp}$ *if and only if* $\mathcal{Y}$ *is closed under direct summands.*

Proof Let $A \in \overline{\mathcal{X}}$ and $B \in \overline{\mathcal{Y}}$. It means there exists $A', B' \in \mathcal{A}$ such that $A \bigoplus A' \in \mathcal{X}$ and $B \bigoplus B' \in \mathcal{Y}$. Hence by Lemma 1.23 we get

$$\begin{aligned} 0 = \operatorname{Ext}^1_{\mathcal{E}}\left(A \bigoplus A', B \bigoplus B'\right) &\cong \operatorname{Ext}^1_{\mathcal{E}}\left(A, B \bigoplus B'\right) \bigoplus \operatorname{Ext}^1_{\mathcal{E}}\left(A', B \bigoplus B'\right) \\ &\cong \operatorname{Ext}^1_{\mathcal{E}}(A, B) \bigoplus \operatorname{Ext}^1_{\mathcal{E}}(A, B') \bigoplus \operatorname{Ext}^1_{\mathcal{E}}(A', B) \bigoplus \operatorname{Ext}^1_{\mathcal{E}}(A', B'). \end{aligned}$$

Thus $\operatorname{Ext}^1_{\mathcal{E}}(A, B) = 0$.

Now assume $(\mathcal{X}, \mathcal{Y})$ has enough projectives. By what we just proved, and Lemma 2.1, we already have $\overline{\mathcal{X}} \subseteq {}^{\perp}\overline{\mathcal{Y}} \subseteq {}^{\perp}\mathcal{Y}$. On the other hand, let $A \in {}^{\perp}\mathcal{Y}$. Since we have enough projectives we may write a short exact sequence $Y \rightarrowtail X \twoheadrightarrow A$ where $X \in \mathcal{X}$ and $Y \in \mathcal{Y}$. But this sequence must be split exact since $A \in {}^{\perp}\mathcal{Y}$. Thus A is a direct summand of X by Proposition 1.4. Hence $A \in \overline{\mathcal{X}}$ as desired. □

Corollary 2.5 *A complete Ext-pair $(\mathcal{X}, \mathcal{Y})$ is a cotorsion pair if and only if $\mathcal{X}$ and $\mathcal{Y}$ are each closed under direct summands.*

Corollary 2.6 *Let $(\mathcal{X}, \mathcal{Y})$ be a complete Ext-pair. Then $\left(\overline{\mathcal{X}}, \overline{\mathcal{Y}}\right)$ is a complete cotorsion pair. We call it the* **closure** *of $(\mathcal{X}, \mathcal{Y})$.*

Proof By Proposition 2.4 we already have $\operatorname{Ext}^1_{\mathcal{E}}\left(\overline{\mathcal{X}}, \overline{\mathcal{Y}}\right) = 0$, so $\overline{\mathcal{X}} \subseteq {}^{\perp}\overline{\mathcal{Y}}$ by Lemma 2.1. On the other hand, using Proposition 2.4 again we get ${}^{\perp}\overline{\mathcal{Y}} \subseteq {}^{\perp}\mathcal{Y} = \overline{\mathcal{X}}$. So $\overline{\mathcal{X}} = {}^{\perp}\overline{\mathcal{Y}}$. Similarly, $\overline{\mathcal{Y}} = \overline{\mathcal{X}}^{\perp}$. □

Lemma 2.7 *Assuming coproducts exist in $\mathcal{A}$, the left side of any cotorsion pair is closed under coproducts. If products exist in $\mathcal{A}$, then the right side of any cotorsion pair is closed under products.*

Proof This follows immediately from Lemma 1.24 □

The following definition will be very important to us later. In Chapter 8 it will be related to the existence of weak generators for homotopy categories. In Chapter 9 we will see that it is crucially linked to the completeness of a cotorsion pair.

Definition 2.8 Let C be any class of objects in an exact category $(\mathcal{A}, \mathcal{E})$, and set $\mathcal{Y} = C^{\perp}$. Then $({}^{\perp}\mathcal{Y}, \mathcal{Y})$ is a cotorsion pair called the cotorsion pair **cogenerated** by C. In the case that C is a set (not a proper class) then we say that $({}^{\perp}\mathcal{Y}, \mathcal{Y})$ is **cogenerated by a set**.

Dually, setting $\mathcal{X} = {}^{\perp}C$ we call $(\mathcal{X}, \mathcal{X}^{\perp})$ the cotorsion pair **generated** by C.

Exercise 2.1.1 Show that $({}^{\perp}\mathcal{Y}, \mathcal{Y})$ is indeed a cotorsion pair, no matter what class C we start with.

Exercise 2.1.2 Assume that $\mathcal{A}$ has coproducts and that $(\mathcal{X}, \mathcal{Y})$ is a cotorsion pair in $(\mathcal{A}, \mathcal{E})$, cogenerated by a set $\mathcal{S}$. Show that the singleton set $\left\{\bigoplus_{X_i \in \mathcal{S}} X_i\right\}$ also cogenerates $(\mathcal{X}, \mathcal{Y})$.

Exercise 2.1.3 Let R be a ring and let R-Mod denote the category of (left) R-modules. Find a set that cogenerates the canonical projective cotorsion pair, $(\mathcal{P}, \mathcal{A})$, and a set that cogenerates the canonical injective cotorsion pair, $(\mathcal{A}, \mathcal{I})$.

Exercise 2.1.4 Let R be a ring.

(1) Show that for any left (resp. right) R-module M, the abelian group $M^+ := \mathrm{Hom}_{\mathbb{Z}}(M, \mathbb{Q}/\mathbb{Z})$ is a right (resp. left) R-module. M^+ is called the *character module* of M. Show that any character module M^+ is pure-injective. That is, the induced map $\mathrm{Hom}_R(N, M^+) \to \mathrm{Hom}_R(P, M^+)$ is an epimorphism for any pure monomorphism $P \hookrightarrow N$.

(2) Let $\mathcal{F}$ denote the class of all flat (left) R-modules and $C := \mathcal{F}^{\perp}$ the class of all cotorsion R-modules. It is known that for any (left) R-module F, and (right) R-module M, there is an isomorphism $\left[\mathrm{Tor}_1^R(M, F)\right]^+ \cong \mathrm{Ext}_R^1(F, M^+)$. Use this to show that $(\mathcal{F}, C)$ is a cotorsion pair.

2.2 Lifting and Factorization Properties of Ext-Pairs

In this section we show that every Ext-pair $\mathfrak{C} = (\mathcal{X}, \mathcal{Y})$ in $(\mathcal{A}, \mathcal{E})$ gives rise to special lifting properties and factorization properties with respect to classes of morphisms we call "$\mathcal{X}$-cofibrations" and "$\mathcal{Y}$-fibrations". These facts are fundamental to the theory of abelian model categories. We start with the following important result characterizing the vanishing of Ext in terms of commutative diagrams.

Lemma 2.9 (Lifting-Extension Lemma) *The following are equivalent for a pair of objects C and K in $\mathcal{A}$.*

(1) $\mathrm{Ext}^1_{\mathcal{E}}(C, K) = 0$.

(2) *For any two given short exact sequences*

$$A \overset{i}{\rightarrowtail} B \overset{c}{\twoheadrightarrow} C \quad \textit{and} \quad K \overset{k}{\rightarrowtail} X \overset{p}{\twoheadrightarrow} Y,$$

and any given commutative square

$$\begin{array}{ccc} A & \xrightarrow{f} & X \\ {\scriptstyle i}\downarrow & \nearrow & \downarrow{\scriptstyle p} \\ B & \xrightarrow[g]{} & Y, \end{array}$$

there exists a lift as suggested by the dashed arrow, making both triangles commute.

Proof (2) implies (1) is the easy direction. For this, let $K \rightarrowtail Z \twoheadrightarrow C$ represent any element of $\mathrm{Ext}^1_{\mathcal{E}}(C, K)$. If (2) were to hold, then this short exact sequence must be split exact as the following diagram proves:

$$\begin{array}{ccc} K & \xrightarrow{1_K} & K \\ \downarrow & \nearrow & \downarrow \\ Z & \longrightarrow & 0. \end{array}$$

(1) implies (2). By Theorem 1.22 we have a commutative diagram of abelian groups with exact rows and columns as shown:

$$\begin{array}{ccccc} \mathrm{Hom}_{\mathcal{A}}(B, K) & \xrightarrow{k_*} & \mathrm{Hom}_{\mathcal{A}}(B, X) & \xrightarrow{p_*} & \mathrm{Hom}_{\mathcal{A}}(B, Y) \\ {\scriptstyle i^*}\downarrow & & {\scriptstyle i^*}\downarrow & & {\scriptstyle i^*}\downarrow \\ \mathrm{Hom}_{\mathcal{A}}(A, K) & \xrightarrow{k_*} & \mathrm{Hom}_{\mathcal{A}}(A, X) & \xrightarrow{p_*} & \mathrm{Hom}_{\mathcal{A}}(A, Y) \\ {\scriptstyle \delta}\downarrow & & {\scriptstyle \delta}\downarrow & & {\scriptstyle \delta}\downarrow \\ \mathrm{Ext}^1_{\mathcal{E}}(C, K) & \xrightarrow{k_*} & \mathrm{Ext}^1_{\mathcal{E}} C, X) & \xrightarrow{p_*} & \mathrm{Ext}^1_{\mathcal{E}}(C, Y). \end{array}$$

(Step 1) As a first step, we will find a map $h\colon B \to X$ making the upper left triangle of the square commute. In terms of the previous commutative diagram we have been given morphisms $f \in \mathrm{Hom}_{\mathcal{A}}(A, X)$ and $g \in \mathrm{Hom}_{\mathcal{A}}(B, Y)$ satisfying $p_*(f) = i^*(g)$, because of the hypothesis $pf = gi$. Since the columns of the aforementioned diagram are exact we get $0 = \delta(i^*(g)) = \delta(p_*(f)) = p_*(\delta(f))$. But we are also given $\mathrm{Ext}^1_{\mathcal{E}}(C, K) = 0$, so this implies $p_*\colon \mathrm{Ext}^1_{\mathcal{E}}(C, X) \to \mathrm{Ext}^1_{\mathcal{E}}(C, Y)$ is one-to-one. Thus $\delta(f) = 0$. This means precisely that the bottom row of the pushout square shown next splits:

$$\begin{array}{ccccc} A & \overset{i}{\rightarrowtail} & B & \overset{c}{\twoheadrightarrow} & C \\ \downarrow{\scriptstyle f} & & \downarrow{\scriptstyle f'} & & \| \\ X & \overset{i'}{\rightarrowtail} & P & \twoheadrightarrow & C. \end{array}$$

We set $h := rf'$, where r is a retraction for i'. Then $hi = rf'i = ri'f = 1_X f = f$. So we found a lift h making the upper left triangle commute:

$$\begin{array}{ccc} A & \xrightarrow{f} & X. \\ {\scriptstyle i}\downarrow & \nearrow{\scriptstyle h} & \\ B & & \end{array}$$

(Step 2) We now adjust the map h, replacing it with a new map $\tilde{h}$, which makes both the upper and lower triangles commute. Noting $(ph-g)i = 0$, the universal property of the cokernel $c\colon B \twoheadrightarrow C$ provides a unique map $\alpha\colon C \to Y$ such that $\alpha c = ph - g$. Now consider the following pullback diagram:

$$\begin{array}{ccccc} K & \rightarrowtail & Q & \overset{p'}{\twoheadrightarrow} & C \\ \| & & \downarrow{\scriptstyle \alpha'} & & \downarrow{\scriptstyle \alpha} \\ K & \overset{k'}{\rightarrowtail} & X & \overset{p}{\twoheadrightarrow} & Y. \end{array}$$

Since $\mathrm{Ext}^1_{\mathcal{E}}(C, K) = 0$, the top sequence is split exact. We let $s\colon C \rightarrowtail Q$ be a section for p' and we set $\beta := \alpha' s$. Then $p\beta = \alpha$.

Finally, we set $\tilde{h} := h - \beta c$ and claim $\tilde{h}\colon B \to X$ is the desired map, making both the upper and lower triangle commute:

$$\begin{array}{ccc} A & \overset{f}{\longrightarrow} & X \\ {\scriptstyle i}\downarrow & \overset{\tilde{h}}{\nearrow} & \downarrow{\scriptstyle p} \\ B & \underset{g}{\longrightarrow} & Y. \end{array}$$

Indeed for the upper left triangle we have $\tilde{h}i = hi - \beta ci = hi = f$. And, for the lower right triangle we have $p\tilde{h} = ph - p\beta c = ph - \alpha c = ph - (ph - g) = g$. □

The next Proposition 2.11 explains how the Lifting-Extension Lemma will be useful in the context of cotorsion pairs, and hence abelian model categories. It relies upon the terminology in the following definition.

Definition 2.10 Let C be a class of objects in $\mathcal{A}$. We call a morphism f a ***C*-cofibration** if f is an admissible monic with some $\mathrm{Cok}\, f \in C$. On the other hand, we call a morphism f a ***C*-fibration** if f is an admissible epic with a $\mathrm{Ker}\, f \in C$.

Proposition 2.11 (Ext-Pairs and Lifting Properties) *Let $(\mathcal{X}, \mathcal{Y})$ be a pair of classes of objects of $\mathcal{A}$. Then $\mathrm{Ext}^1_{\mathcal{E}}(\mathcal{X}, \mathcal{Y}) = 0$ if and only if any commutative diagram as shown here:*

$$\begin{array}{ccc} A & \overset{f}{\longrightarrow} & X \\ {\scriptstyle i}\downarrow & \nearrow & \downarrow{\scriptstyle p} \\ B & \underset{g}{\longrightarrow} & Y, \end{array}$$

where i is an $\mathcal{X}$-cofibration and p is a $\mathcal{Y}$-fibration, possesses a lift as suggested by the dashed arrow, making both triangles commute.

Definition 2.12 In the situation of Proposition 2.11, we say that $\mathcal{X}$-cofibrations have the **left lifting property (LLP)** with respect to $\mathcal{Y}$-fibrations and $\mathcal{Y}$-fibrations have the **right lifting property (RLP)** with respect to $\mathcal{X}$-cofibrations.

We now assume we have a complete Ext-pair $\mathfrak{C} = (\mathcal{X}, \mathcal{Y})$, and show that each morphism in $\mathcal{A}$ factors as an $\mathcal{X}$-cofibration followed by a $\mathcal{Y}$-fibration.

Proposition 2.13 (Factorizations) *Let $\mathfrak{C} = (\mathcal{X}, \mathcal{Y})$ be an Ext-pair. Then $\mathfrak{C}$ is complete if and only if each morphism $f \in \mathcal{A}$ has a factorization $f = qj$ where j is an admissible monic with cokernel in $\mathcal{X}$ and q is an admissible epic with kernel in $\mathcal{Y}$. That is, each f factors as an $\mathcal{X}$-cofibration followed by a $\mathcal{Y}$-fibration.*

Proof The easier direction is the converse, and we will prove that first. So say we are given an object $A \in \mathcal{A}$. Then factoring $0 \to A$ as an admissible monic with cokernel in $\mathcal{X}$ followed by an admissible epic with kernel in $\mathcal{Y}$ provides us with a short exact sequence $Y \rightarrowtail X \twoheadrightarrow A$ with $X \in \mathcal{X}$ and $Y \in \mathcal{Y}$. So $\mathfrak{C}$ has enough projectives. Factoring $A \to 0$ will show that $\mathfrak{C}$ has enough injectives.

Now suppose that the cotorsion pair $\mathfrak{C}$ is complete. We first assume f is an admissible monic, so that we have a short exact sequence

$$A \overset{f}{\rightarrowtail} B \twoheadrightarrow C.$$

Since $\mathfrak{C}$ has enough projectives we can construct a pullback diagram as follows, where $X \in \mathcal{X}$ and $Y \in \mathcal{Y}$:

$$\begin{array}{ccccc} & & Y & = & Y \\ & & \downarrow & & \downarrow i \\ A & \overset{j}{\rightarrowtail} & P & \twoheadrightarrow & X \\ \| & & \downarrow q & & \downarrow p \\ A & \overset{f}{\rightarrowtail} & B & \twoheadrightarrow & C. \end{array}$$

Therefore we get a factorization $f = qj$ in this case, as desired. In the case that f is instead given to be an admissible epic we can argue similarly. We just use that $\mathfrak{C} = (\mathcal{X}, \mathcal{Y})$ has enough injectives applied to $\operatorname{Ker} f$, and then consider the resulting pushout diagram.

Now consider an arbitrary morphism $f : A \to B$. It is a special case of Proposition 1.18 (there, take g to be f and f to be 1_A) that the morphism

$[\, f \; 1_B \,] \colon A \bigoplus B \twoheadrightarrow B$ must always be an admissible epic. We conclude that any morphism $f \colon A \to B$ factors as

$$A \xrightarrow{\left[\begin{smallmatrix} 1_A \\ 0 \end{smallmatrix}\right]} A \bigoplus B \xrightarrow{[\, f \; 1_B \,]} B,$$

where $\left[\begin{smallmatrix} 1_A \\ 0 \end{smallmatrix}\right]$ is an admissible monic and $[\, f \; 1_B \,]$ is an admissible epic.

So then there must be a factorization $[\, f \; 1_B \,] = q'j'$ where j' is an admissible monic with cokernel in $\mathcal{X}$ and q' is an admissible epic with kernel in $\mathcal{Y}$. Since admissible monics are closed under composition, we get that $j' \circ \left[\begin{smallmatrix} 1_A \\ 0 \end{smallmatrix}\right] = q''j$ for some admissible monic j with cokernel in $\mathcal{X}$ and some admissible epic q'' with kernel in $\mathcal{Y}$. Then $f = (q'q'')j$ is the desired factorization. Certainly $q'q''$ is an admissible epic so it is left to show that its kernel is in $\mathcal{Y}$. But it follows from Corollary 1.15 since $\mathcal{Y}$ is closed under extensions. □

Exercise 2.2.1 Similar to Definition 2.12, we say that f has the *left lifting property (LLP)* with respect to g if for any commutative diagram

$$\begin{array}{ccc} A & \longrightarrow & X \\ {\scriptstyle f}\downarrow & \nearrow & \downarrow{\scriptstyle g} \\ B & \longrightarrow & Y, \end{array}$$

a dashed arrow exists as indicated, making both triangles commute. We would also say that g has the *right lifting property (RLP)* with respect to f.

(1) Show that if f has the LLP with respect to g, then the cobase change map for any pushout along f also has the LLP with respect to g. Dually, the base change map for any pullback along g also has the RLP with respect to f.

(2) By a retract of a morphism $A \to B$ we mean a morphism $C \to D$ such that the following diagram commutes and the horizontal compositions are the identity maps:

$$\begin{array}{ccccc} C & \longrightarrow & A & \longrightarrow & C \\ \downarrow & & \downarrow & & \downarrow \\ D & \longrightarrow & B & \longrightarrow & D. \end{array}$$

Show that if f has the LLP with respect to g, then so does any retract of f. Dually, any retract of g has the RLP with respect to f.

2.3 Cotorsion Pairs and Lifting Properties

Let $\mathfrak{C} = (\mathcal{X}, \mathcal{Y})$ be a cotorsion pair in $(\mathcal{A}, \mathcal{E})$. Since $\mathrm{Ext}^1_{\mathcal{E}}(\mathcal{X}, \mathcal{Y}) = 0$, Proposition 2.11 shows that $\mathcal{X}$-cofibrations have the left lifting property (LLP) with respect to $\mathcal{Y}$-fibrations, and $\mathcal{Y}$-fibrations have the right lifting property (RLP) with respect to $\mathcal{X}$-cofibrations. We will now take this one step further for the case that $\mathfrak{C}$ is a cotorsion pair, not just an Ext-pair.

Theorem 2.14 (Cotorsion Pairs and Lifting Properties) *Let $\mathfrak{C} = (\mathcal{X}, \mathcal{Y})$ be a cotorsion pair in $(\mathcal{A}, \mathcal{E})$.*

(1) *An admissible monic f is an $\mathcal{X}$-cofibration if and only if it has the LLP with respect to the class of all $\mathcal{Y}$-fibrations.*
(2) *An admissible epic g is a $\mathcal{Y}$-fibration if and only if it has the RLP with respect to the class of all $\mathcal{X}$-cofibrations.*

Proof We will prove the second statement. We already know from Proposition 2.11 and Definition 2.12 that $\mathcal{Y}$-fibrations have the right lifting property with respect to $\mathcal{X}$-cofibrations. Conversely, suppose g is an admissible epic possessing the RLP with respect to $\mathcal{X}$-cofibrations. We have a short exact sequence

$$K \overset{i}{\rightarrowtail} A \overset{g}{\twoheadrightarrow} B,$$

and we will show $K \in \mathcal{Y}$ by showing $\mathrm{Ext}^1_{\mathcal{E}}(X, K) = 0$ for all $X \in \mathcal{X}$. Let such an $X \in \mathcal{X}$ be given, and consider any short exact sequence

$$K \overset{j}{\rightarrowtail} Z \overset{q}{\twoheadrightarrow} X.$$

Then j is an $\mathcal{X}$-cofibration. So there exists a lift as indicated by the dashed arrow in the following commutative square:

$$\begin{array}{ccc} K & \xrightarrow{i} & A \\ {\scriptstyle j}\downarrow & \nearrow & \downarrow{\scriptstyle g} \\ Z & \xrightarrow[0]{} & B. \end{array}$$

Let us denote the lift by s, so that we have $sj = i$ and $gs = 0$. Since $i = \ker g$, the relation $gs = 0$ yields a unique map $r\colon Z \to K$ such that $s = ir$. Now r is a retraction for j. Indeed $irj = sj = i = i \circ 1_K$ and i being monic lets us left cancel to get $rj = 1_K$. This proves $\mathrm{Ext}^1_{\mathcal{E}}(X, K) = 0$. □

In the special case that $\mathcal{A}$ is weakly idempotent complete, the following corollary provides an even better statement.

Corollary 2.15 *Assume $\mathcal{A}$ is weakly idempotent complete and let $\mathfrak{C} = (\mathcal{X}, \mathcal{Y})$ be a cotorsion pair in $(\mathcal{A}, \mathcal{E})$. Assume also that each object of $\mathcal{A}$ is an admissible subobject of an object in $\mathcal{Y}$ and an admissible quotient of an object in $\mathcal{X}$. (For example, this is the case if $\mathfrak{C}$ is complete.)*

(1) *A morphism f is an $\mathcal{X}$-cofibration if and only if it satisfies the LLP with respect to the class of all $\mathcal{Y}$-fibrations.*
(2) *A morphism g is a $\mathcal{Y}$-fibration if and only if it satisfies the RLP with respect to the class of all $\mathcal{X}$-cofibrations.*

Proof Again, we will just prove the second statement. By Theorem 2.14, it is only left to show that if $g\colon A \to B$ satisfies the RLP with respect to the class of all $\mathcal{X}$-cofibrations, then it must be an admissible epic. What we do, is we write B as an admissible quotient $p\colon X \twoheadrightarrow B$ of some $X \in \mathcal{X}$. Then we have a commutative square as follows, admitting a lift t, since $0 \rightarrowtail X$ is an $\mathcal{X}$-cofibration:

$$\begin{array}{ccc} 0 & \longrightarrow & A \\ \big\downarrow & \overset{t}{\nearrow} & \big\downarrow{\scriptstyle g} \\ X & \underset{p}{\twoheadrightarrow} & B. \end{array}$$

Then since $gt = p$ is an admissible epic, the weakly idempotent complete hypothesis lets us conclude that g too must be an admissible epic; see the Cancellation Laws of Proposition 1.30. □

2.4 Hereditary Cotorsion Pairs

Let $\mathcal{X}$ be a class of objects in the exact category $(\mathcal{A}, \mathcal{E})$. We say that $\mathcal{X}$ is **projectively resolving** if $\mathcal{X}$ contains all the projective objects, $\mathcal{X}$ is closed under extensions, and $\mathcal{X}$ is closed under taking kernels of admissible epics. The latter two conditions mean precisely this: Whenever $A \rightarrowtail B \twoheadrightarrow C$ is a short exact sequence with $C \in \mathcal{X}$, then $A \in \mathcal{X}$ if and only if $B \in \mathcal{X}$. The dual notion is called **injectively coresolving**. We say that an Ext-pair $\mathfrak{C} = (\mathcal{X}, \mathcal{Y})$ is **hereditary** if both $\mathcal{X}$ is projectively resolving and $\mathcal{Y}$ is injectively coresolving. When $\mathfrak{C}$ is a cotorsion pair we have the following convenient result connecting the hereditary property to the vanishing of all $\mathrm{Ext}^n_{\mathcal{E}}$.

Theorem 2.16 (Hereditary Test) *Assume $(\mathcal{X}, \mathcal{Y})$ is a cotorsion pair in $(\mathcal{A}, \mathcal{E})$ and consider the following statements.*

(1) $\mathcal{X}$ *is projectively resolving.*
(2) $\mathcal{Y}$ *is injectively coresolving.*
(3) $\mathrm{Ext}^2_{\mathcal{E}}(X, Y) = 0$ *for all* $X \in \mathcal{X}$ *and* $Y \in \mathcal{Y}$.
(4) $\mathrm{Ext}^n_{\mathcal{E}}(X, Y) = 0$ *for all* $X \in \mathcal{X}$ *and* $Y \in \mathcal{Y}$, *and all* $n \geq 2$.

Then we have the following implications.

- *(3) implies both (1) and (2).*
- *If each object of* $\mathcal{A}$ *is an admissible subobject of an object in* $\mathcal{Y}$*, then (2) implies (3) and (4).*
- *If each object of* $\mathcal{A}$ *is an admissible quotient of an object in* $\mathcal{X}$*, then (1) implies (3) and (4).*

In particular, whenever each object of $\mathcal{A}$ *is both an admissible subobject of one in* $\mathcal{Y}$ *and an admissible quotient of one in* $\mathcal{X}$*, then conditions (1)–(4) are all equivalent.*

Proof (3) implies (2). Let $Y \rightarrowtail Y' \twoheadrightarrow Z$ be a short exact sequence with $Y, Y' \in \mathcal{Y}$. For any $X \in \mathcal{X}$, applying $\mathrm{Hom}_{\mathcal{A}}(X, -)$, the corresponding long exact sequence of Theorem 1.22 shows $\mathrm{Ext}^1_{\mathcal{E}}(X, Z) = 0$, proving $Z \in \mathcal{X}^{\perp} = \mathcal{Y}$. A similar argument shows (3) implies (1).

We now assume that each object of $\mathcal{A}$ is an admissible quotient of an object of $\mathcal{X}$ and prove (1) implies (3). We use the Yoneda description of the group $\mathrm{Ext}^2_{\mathcal{E}}(X, Y)$. Recall that its elements are equivalence classes of exact 2-sequences of the form

$$\mathbb{E}: \qquad Y \overset{k}{\rightarrowtail} E_2 \xrightarrow{f} E_1 \overset{c}{\twoheadrightarrow} X,$$

where the equivalence relation is *generated* by the relation $\sim$, where $\mathbb{E}' \sim \mathbb{E}$ means there exists some commutative diagram of the following form:

$$\begin{array}{lccccccc} \mathbb{E}': & Y & \overset{k'}{\rightarrowtail} & E'_2 & \xrightarrow{f'} & E'_1 & \overset{c'}{\twoheadrightarrow} & X \\ & \| & & \downarrow & & \downarrow & & \| \\ \mathbb{E}: & Y & \overset{k}{\rightarrowtail} & E_2 & \xrightarrow{f} & E_1 & \overset{c}{\twoheadrightarrow} & X. \end{array}$$

Now let $\mathbb{E} \in \mathrm{Ext}^2_{\mathcal{E}}(X, Y)$ be arbitrary and assume (1). Our goal is to show that $\mathbb{E}$ is equivalent to the split 2-sequence

$$\mathbb{S}: \qquad Y \rightarrowtail Y \bigoplus Z \longrightarrow Z \bigoplus X \twoheadrightarrow X.$$

Given any $\mathbb{E}$ as previously, use the hypothesis to write an admissible epic $p\colon X' \twoheadrightarrow E_1$ where $X' \in \mathcal{X}$. Letting E_2' denote the pullback of $E_2 \xrightarrow{f} E_1 \xleftarrow{p} X'$, one constructs a morphism of exact 2-sequences:

$$\begin{array}{lccccccc}
\mathbb{E}': & Y & \overset{k'}{\rightarrowtail} & E_2' & \xrightarrow{f'} & X' & \overset{cp}{\twoheadrightarrow} & X \\
 & \| & & \downarrow{\scriptstyle p'} & & \downarrow{\scriptstyle p} & & \| \\
\mathbb{E}: & Y & \overset{k}{\rightarrowtail} & E_2 & \xrightarrow{f} & E_1 & \overset{c}{\twoheadrightarrow} & X.
\end{array}$$

Looking at the short exact sequence $Y \overset{k'}{\rightarrowtail} E_2' \overset{f'}{\twoheadrightarrow} \operatorname{Im} f'$, we note that $\operatorname{Im} f' = \operatorname{Ker}(cp) \in \mathcal{X}$, by hypothesis (1). Since $Y \in \mathcal{Y}$, this means the short exact sequence splits. Using this fact, one can now easily construct a morphism of exact 2-sequences, showing $\mathbb{E}' \sim \mathbb{S}$. This means that $\mathbb{S}$, $\mathbb{E}'$, and $\mathbb{E}$ all represent the same element, namely zero, in the Yoneda description of $\operatorname{Ext}^2_{\mathcal{E}}(X, Y)$. This proves (3). A generalization of this argument, with exact n-sequences, will show (1) implies (4).

In the case that each object of $\mathcal{A}$ is an admissible subobject of one in $\mathcal{Y}$, a similar argument shows (2) implies (3), and more generally (2) implies (4). □

The next corollary provides two very special instances in which the four conditions of Theorem 2.16 are guaranteed to be equivalent.

Corollary 2.17 *Assume $\mathfrak{C} = (\mathcal{X}, \mathcal{Y})$ is a cotorsion pair in $(\mathcal{A}, \mathcal{E})$. The following conditions are equivalent if $\mathfrak{C}$ is complete, or, if $(\mathcal{A}, \mathcal{E})$ has both enough projectives and enough injectives (as defined in Section 1.7).*

(1) *$(\mathcal{X}, \mathcal{Y})$ is hereditary. That is, $\mathcal{X}$ is projectively resolving and $\mathcal{Y}$ is injectively coresolving.*
(2) *$\mathcal{X}$ is projectively resolving.*
(3) *$\mathcal{Y}$ is injectively coresolving.*
(4) *$\operatorname{Ext}^2_{\mathcal{E}}(X, Y) = 0$ for all $X \in \mathcal{X}$ and $Y \in \mathcal{Y}$.*
(5) *$\operatorname{Ext}^n_{\mathcal{E}}(X, Y) = 0$ for all $X \in \mathcal{X}$ and $Y \in \mathcal{Y}$, and all $n \geq 2$.*

Next we have the following useful generalization of the standard horseshoe lemma from homological algebra.

Lemma 2.18 (Generalized Horseshoe Lemma) *Let $(\mathcal{X}, \mathcal{Y})$ be an hereditary cotorsion pair and assume $A \overset{f}{\rightarrowtail} B \overset{g}{\twoheadrightarrow} C$ is a short exact sequence in $(\mathcal{A}, \mathcal{E})$. Given two other short exact sequences*

$$Y_A \overset{i_A}{\rightarrowtail} X_A \overset{p_A}{\twoheadrightarrow} A$$

$$Y_C \overset{i_C}{\rightarrowtail} X_C \overset{p_C}{\twoheadrightarrow} C$$

with $Y_A, Y_C \in \mathcal{Y}$ and $X_A, X_C \in \mathcal{X}$, we may construct all the dashed arrows as follows, so that the entire diagram commutes, each row and column is a short exact sequence, and such that $Y_B \in \mathcal{Y}$ and $X_B \in \mathcal{X}$:

$$\begin{array}{ccccc}
Y_A & \dashrightarrow & Y_B & \dashrightarrow & Y_C \\
{\scriptstyle i_A}\downarrow & & \downarrow & & \downarrow{\scriptstyle i_C} \\
X_A & \dashrightarrow & X_B & \dashrightarrow & X_C \\
{\scriptstyle p_A}\downarrow & & \downarrow & & \downarrow{\scriptstyle p_C} \\
A & \overset{f}{\rightarrowtail} & B & \overset{g}{\twoheadrightarrow} & C.
\end{array}$$

Proof We start by constructing a pullback diagram as shown:

$$\begin{array}{ccccc}
 & & Y_C & = & Y_C \\
 & & \downarrow & & \downarrow{\scriptstyle i_C} \\
A & \rightarrowtail & P & \twoheadrightarrow & X_C \\
\| & & {\scriptstyle p'}\downarrow & & \downarrow{\scriptstyle p_C} \\
A & \rightarrowtail & B & \twoheadrightarrow & C.
\end{array}$$

Applying $\operatorname{Hom}(X_C, -)$ to $Y_A \overset{i_A}{\rightarrowtail} X_A \overset{p_A}{\twoheadrightarrow} A$ yields an exact sequence

$$\operatorname{Ext}^1_{\mathcal{E}}(X_C, X_A) \to \operatorname{Ext}^1_{\mathcal{E}}(X_C, A) \to \operatorname{Ext}^2_{\mathcal{E}}(X_C, Y_A).$$

But $\operatorname{Ext}^2_{\mathcal{E}}(X_C, Y_A) = 0$ since the cotorsion pair is hereditary. This means that

$$\operatorname{Ext}^1_{\mathcal{E}}(X_C, X_A) \xrightarrow{\operatorname{Ext}^1_{\mathcal{E}}(X_C, p_A)} \operatorname{Ext}^1_{\mathcal{E}}(X_C, A)$$

is an epimorphism (it is even an isomorphism), which in using the Yoneda characterization of $\operatorname{Ext}^1_{\mathcal{E}}$ implies there exists a pushout diagram:

$$\begin{array}{ccccc}
X_A & \rightarrowtail & Z & \twoheadrightarrow & X_C \\
{\scriptstyle p_A}\downarrow & & {\scriptstyle p''}\downarrow & & \| \\
A & \rightarrowtail & P & \twoheadrightarrow & X_C.
\end{array}$$

By Corollary 1.14(2), the morphism $p'' : Z \to P$ must be an admissible epic, and we obtain a short exact sequence $Y_A \rightarrowtail Z \twoheadrightarrow P$. By composing the two commutative diagrams, as shown next, the desired result follows from Corollary 1.15 and the fact that both $\mathcal{X}$ and $\mathcal{Y}$ are each closed under extensions:

$$\begin{array}{ccccc}
X_A & \rightarrowtail & Z & \twoheadrightarrow & X_C \\
{\scriptstyle p_A}\big\downarrow & & {\scriptstyle p''}\big\downarrow & & \big\| \\
A & \rightarrowtail & P & \twoheadrightarrow & X_C \\
\big\| & & {\scriptstyle p'}\big\downarrow & & \big\downarrow{\scriptstyle p_C} \\
A & \rightarrowtail & B & \twoheadrightarrow & C.
\end{array}$$

□

Example 2.19 Let R be a ring and let $\mathcal{F}$ be the class of all flat (left) R-modules. By well-known properties of flat modules, $\mathcal{F}$ is projectively resolving. Since the category R-Mod has enough projectives and injectives, Enochs' flat cotorsion pair $(\mathcal{F}, \mathcal{C})$ is hereditary. In fact, all of the cotorsion pairs discussed in the introductory chapter are hereditary.

Note that Lemma 2.18 and Corollary 1.15 have duals which we leave for the reader to formulate. We also would like to give the following alternate proof of the Generalized Horseshoe Lemma.

Alternate Proof of Generalized Horseshoe Lemma An alternate proof, assuming that $(\mathcal{X}, \mathcal{Y})$ is complete, can be summarized by the following commutative diagram:

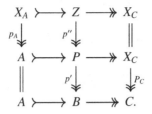

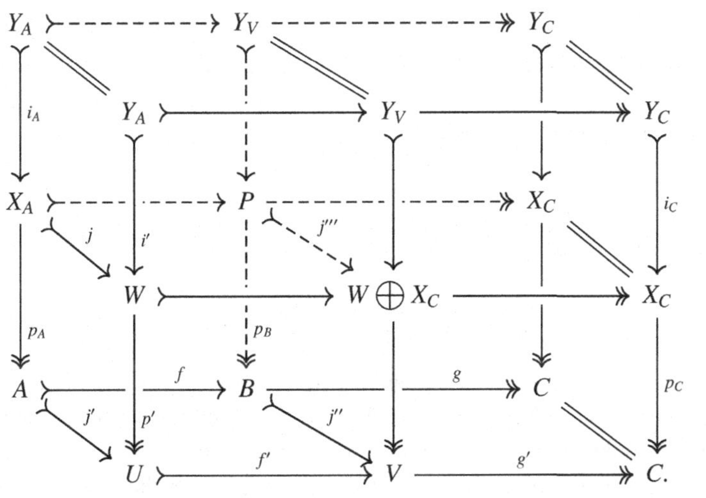

To construct the diagram, we first let $X_A \overset{j}{\rightarrowtail} W \twoheadrightarrow X'$ be a $\mathcal{Y}$-approximation of X_A. Note then that $W \in \omega := \mathcal{X} \cap \mathcal{Y}$. We then take the pushout, denoted by U, of the pair (j, p_A). It follows from Corollary 1.14(2) that p' is an admissible epic with kernel i' and that $\operatorname{Cok} j' = X'$. We then take the pushout of the pair (f, j') to get the square with V and we note that $\operatorname{Cok} j'' = X'$ too. Since the class $\mathcal{Y}$ is coresolving we get that $U \in \mathcal{Y}$. Thus the morphism p_C lifts over g'. This lift allows us to construct, in a manner analogous to the usual horseshoe lemma for projective resolutions, the entire front face of the diagram. (The lower two horizontal rows are already short exact sequences, so it follows from the (3×3)-Lemma 1.16 that the top row is one too.) We finally obtain all of the back dashed arrows by pullback of the newly constructed front face along the original short exact sequence (f, g). We have $Y_V \in \mathcal{Y}$, and $W \bigoplus X_C \in \mathcal{X}$, and it follows that $P \in \mathcal{X}$ too because $\mathcal{X}$ is resolving and since the construction produces $\operatorname{Cok} j''' = \operatorname{Cok} j'' = \operatorname{Cok} j' = \operatorname{Cok} j = X' \in \mathcal{X}$. The pullback of the pair (j', p') coincides with X_A, by Proposition 1.12. Also the (3×3)-Lemma 1.16 applies, and shows $X_A \rightarrowtail P \twoheadrightarrow X_C$ to be a short exact sequence. □

Exercise 2.4.1 Let $(\mathcal{A}, \mathcal{E})$ be an exact category with enough projectives and injectives; see Section 1.7. A class of objects $\mathcal{X}$ is said to be *syzygy closed* if for any short exact sequence $X' \rightarrowtail P \twoheadrightarrow X$ with $X \in \mathcal{X}$ and P a projective object, then $X' \in \mathcal{X}$ too. The dual notion is called *cosyzygy closed.* Show that a cotorsion pair $(\mathcal{X}, \mathcal{Y})$ in $(\mathcal{A}, \mathcal{E})$ is hereditary if and only if $\mathcal{X}$ is syzygy closed if and only if $\mathcal{Y}$ is cosyzygy closed.

Exercise 2.4.2 (Functoriality of Generalized Horseshoe Lemma) Suppose $(\mathcal{X}, \mathcal{Y})$ is a complete hereditary cotorsion pair. Show that the Generalized Horseshoe Lemma is functorial. That is, suppose

$$\begin{array}{llcccccc} \mathbb{E}: & A & \overset{f}{\rightarrowtail} & B & \overset{g}{\twoheadrightarrow} & C \\ & \downarrow{\scriptstyle\alpha} & & \downarrow{\scriptstyle\beta} & & \downarrow{\scriptstyle\gamma} \\ \mathbb{E}': & A' & \overset{f'}{\rightarrowtail} & B' & \overset{g'}{\twoheadrightarrow} & C' \end{array}$$

is a morphism of short exact sequences. Show that for any choice of morphisms of $\mathcal{X}$-approximation sequences extending α and γ there is another such morphism extending β in such a way that every square, on every cube of the resulting diagram, commutes. (It is a marathon exercise in pushouts and pullbacks to check that each step of the alternate proof given previously is functorial.)

2.5 Projective and Injective Cotorsion Pairs

In this section we define what is meant by projective and injective cotorsion pairs. Such cotorsion pairs are ubiquitous in applications of homological algebra. We will also see later that such cotorsion pairs arise intrinsically in the theory of *hereditary* abelian model structures. So these are fundamental structures that we will encounter throughout the book. Recall that we are still allowing $(\mathcal{A}, \mathcal{E})$ to be any exact category, not necessarily weakly idempotent complete. However, in applications $\mathcal{A}$ will typically be weakly idempotent complete and we will see later that in this setting any injective (or projective) cotorsion pair is equivalent, through Hovey's correspondence, to an abelian model structure on $(\mathcal{A}, \mathcal{E})$.

First, recall from Section 1.7 that projective and injective objects, as well as the concept of enough projectives or injectives, are defined relative to the exact structure $\mathcal{E}$. A class of objects $\mathcal{W}$ is called **thick** if it is closed under direct summands and satisfies the 2 out of 3 property on short exact sequences.

Definition 2.20 Assume $(\mathcal{A}, \mathcal{E})$ has enough projectives and let $\mathcal{P}$ denote the class of all projectives. A complete cotorsion pair $(\mathcal{Q}, \mathcal{W})$ is said to be a **projective cotorsion pair** if $\mathcal{W}$ is thick and $\mathcal{P} = \mathcal{Q} \cap \mathcal{W}$.

On the other hand, assume $(\mathcal{A}, \mathcal{E})$ has enough injectives and let $\mathcal{I}$ denote the class of all injectives. By an **injective cotorsion pair** we mean a complete cotorsion pair $(\mathcal{W}, \mathcal{R})$ for which $\mathcal{W}$ is thick and $\mathcal{I} = \mathcal{W} \cap \mathcal{R}$.

We now give a convenient characterization of the complete cotorsion pairs that are injective (or projective).

Proposition 2.21 (Characterizations of Injective Cotorsion Pairs) *Assume $(\mathcal{A}, \mathcal{E})$ has enough injectives and $(\mathcal{W}, \mathcal{R})$ is a complete cotorsion pair. Then the following statements are equivalent.*

(1) *$(\mathcal{W}, \mathcal{R})$ is an injective cotorsion pair.*
(2) *$(\mathcal{W}, \mathcal{R})$ is hereditary and $\mathcal{W} \cap \mathcal{R}$ equals the class $\mathcal{I}$ of injective objects.*
(3) *$(\mathcal{W}, \mathcal{R})$ is hereditary and for any $R \in \mathcal{R}$ we can find a short exact sequence $R' \rightarrowtail I \twoheadrightarrow R$ where I is injective and $R' \in \mathcal{R}$.*
(4) *$\mathcal{W}$ is thick and contains the injective objects.*

Proof First, (1) implies (2) by Definition 2.20 and Corollary 2.17. For (2) implies (3) note that for any $R \in \mathcal{R}$ using that the cotorsion pair $(\mathcal{W}, \mathcal{R})$ has enough projectives we can write $R' \rightarrowtail W \twoheadrightarrow R$ where $W \in \mathcal{W}$ and $R' \in \mathcal{R}$. Then since $\mathcal{R}$ is closed under extensions we have $W \in \mathcal{R}$. So by hypothesis we have W is injective.

Next, we show (3) implies (4). The assumptions imply that $\mathcal{W}$ is already closed under direct summands and is projectively resolving. So it remains to show that if $V \rightarrowtail W \twoheadrightarrow X$ is an exact sequence with $V, W \in \mathcal{W}$ then $X \in \mathcal{W}$ too. Note that by applying $\mathrm{Hom}_{\mathcal{A}}(-, R)$ to this short exact sequence, for any $R \in \mathcal{R}$, it follows from Corollary 2.17 that $\mathrm{Ext}^2_{\mathcal{E}}(X, R) = 0$. To see that $\mathrm{Ext}^1_{\mathcal{E}}(X, R) = 0$ for every $R \in \mathcal{R}$, pick a short exact sequence $R' \rightarrowtail I \twoheadrightarrow R$, where I is injective and $R' \in \mathcal{R}$. Applying $\mathrm{Hom}_{\mathcal{A}}(X, -)$ to this sequence gives $\mathrm{Ext}^1_{\mathcal{E}}(X, R) \cong \mathrm{Ext}^2_{\mathcal{E}}(X, R')$, which is zero by what we just proved.

For (4) implies (1), first note that the hypothesis makes clear that the injective objects are in $\mathcal{W} \cap \mathcal{R}$. So we just wish to show that everything in $\mathcal{W} \cap \mathcal{R}$ is injective. So suppose $X \in \mathcal{W} \cap \mathcal{R}$. Then using that $(\mathcal{A}, \mathcal{E})$ has enough injectives find a short exact sequence $X \rightarrowtail I \twoheadrightarrow I/X$ where I is injective. By hypothesis $I \in \mathcal{W}$ and also $\mathcal{W}$ is assumed to be thick, which means $I/X \in \mathcal{W}$. But now since $(\mathcal{W}, \mathcal{R})$ is a cotorsion pair the exact sequence splits. Therefore X is a direct summand of I, proving $X \in \mathcal{I}$. □

Everything has a projective dual. In particular we have the following.

Proposition 2.22 (Characterizations of Projective Cotorsion Pairs) *Assume $(\mathcal{A}, \mathcal{E})$ has enough projectives and $(\mathcal{Q}, \mathcal{W})$ is a complete cotorsion pair. Then the following statements are equivalent.*

(1) *$(\mathcal{Q}, \mathcal{W})$ is a projective cotorsion pair.*
(2) *$(\mathcal{Q}, \mathcal{W})$ is hereditary and $\mathcal{Q} \cap \mathcal{W}$ equals the class $\mathcal{P}$ of projective objects.*
(3) *$(\mathcal{Q}, \mathcal{W})$ is hereditary and for any $Q \in \mathcal{Q}$ we can find a short exact sequence $Q \rightarrowtail P \twoheadrightarrow Q'$ where P is projective and $Q' \in \mathcal{Q}$.*
(4) *$\mathcal{W}$ is thick and contains the projective objects.*

Exercise 2.5.1 Suppose $\{(\mathcal{X}_i, \mathcal{Y}_i)\}_{i \in I}$ is a collection of cotorsion pairs in $(\mathcal{A}, \mathcal{E})$, each cogenerated by some class $\mathcal{S}_i$ and generated by some class $\mathcal{T}_i$.

(1) Show that $({}^{\perp}(\cap_{i \in I} \mathcal{Y}_i), \cap_{i \in I} \mathcal{Y}_i)$ is a cotorsion pair cogenerated by the class $\cup_{i \in I} \mathcal{S}_i$. In particular, if each $\mathcal{S}_i$ is a *set*, then this cotorsion pair is also cogenerated by a set. Note that if each $\mathcal{Y}_i$ is thick and contains all projective objects, then $\cap_{i \in I} \mathcal{Y}_i$ has these same properties.
(2) Dually, $(\cap_{i \in I} \mathcal{X}_i, (\cap_{i \in I} \mathcal{X}_i)^{\perp})$ is a cotorsion pair generated by the class $\cup_{i \in I} \mathcal{T}_i$. If each $\mathcal{X}_i$ is thick and contains all injectives, then the same goes for $\cap_{i \in I} \mathcal{X}_i$.

Exercise 2.5.2 Let $\mathcal{A}$ be an abelian category with enough injectives and let $M \in \mathcal{A}$. We call M *Gorenstein injective* if $M = Z_0 J$ for some exact complex J of injective objects which remains exact upon applying $\mathrm{Hom}_{\mathcal{A}}(I, -)$ for any injective object I. Let $\mathcal{GI}$ be the class of all Gorenstein injectives.

(1) Show that $\mathcal{R} \subseteq \mathcal{GI}$ whenever $(\mathcal{W}, \mathcal{R})$ is an injective cotorsion pair.
(2) Show that if $({}^{\perp}\mathcal{GI}, \mathcal{GI})$ is a complete cotorsion pair then it must be an injective cotorsion pair.

Note that we have the dual results concerning *Gorenstein projective* objects whenever $\mathcal{A}$ has enough projectives.

3

Stable Categories from Cotorsion Pairs

In this chapter we define stable categories and construct, from any complete cotorsion pair (or even Ext-pair), four additive functors between stable categories. When we specialize to the cotorsion pairs making up an abelian model structure these will provide the cofibrant and fibrant replacement functors, denoted Q and R, as well as the loop and suspension functors, Ω and Σ.

Throughout this chapter, $(\mathcal{A}, \mathcal{E})$ may be any exact category; it need not be weakly idempotent complete. We let $\mathfrak{C} = (\mathcal{X}, \mathcal{Y})$ denote an Ext-pair in the sense of Definition 2.2. This of course includes the more-likely scenario that $\mathfrak{C}$ is a cotorsion pair.

3.1 The Stable Category of an Additive Category

Before attaching stable categories to $\mathfrak{C} = (\mathcal{X}, \mathcal{Y})$, we first will describe the general construction of stable catgories. Given a class of objects υ in any additive category $\mathcal{A}$ the idea is to "kill" the objects of υ, forcing them to be the zero objects in a new additive category $\mathrm{St}_\upsilon(\mathcal{A})$.

Recall that A is a *retract* of an object B if there exist morphisms $A \xrightarrow{f} B$ and $B \xrightarrow{g} A$ such that $gf = 1_A$. Note that any nonempty class υ closed under retracts must always contain 0 and be replete (i.e. isomorphism closed).

Proposition 3.1 *Let $\mathcal{A}$ be any additive category and assume υ is a nonempty class of objects closed under finite direct sums (biproducts). For each pair of objects $A, B \in \mathcal{A}$, define a relation $\sim$ on $\mathrm{Hom}_{\mathcal{A}}(A, B)$ by $f \sim g$ if and only if $g - f$ factors through an object of υ. We denote this relation by $\sim^\upsilon$ whenever it is necessary to emphasize υ. The following hold.*

(1) *The relation* $\sim$ *is an equivalence relation. We let* $[f]$, *or* $[f]_{\upsilon}$ *whenever it is necessary to emphasize* υ, *denote the equivalence class containing* $f \in \mathrm{Hom}_{\mathcal{A}}(A, B)$.

(2) *There is a category* $\mathrm{St}_{\upsilon}(\mathcal{A})$ *whose objects are the same as* $\mathcal{A}$ *and whose morphism sets are defined and denoted by* $\underline{\mathrm{Hom}}_{\upsilon}(A, B) := \mathrm{Hom}_{\mathcal{A}}(A, B)/\sim$. *Composition of maps is well defined by* $[g] \circ [f] = [gf]$ *and the identity map for any given object* A *is given by the equivalence class* $[1_A]$.

(3) *Defining* $[f] + [g] = [f + g]$ *makes* $\mathrm{St}_{\upsilon}(\mathcal{A})$ *into an additive category. There is a canonical (full) additive functor* $\mathcal{A} \xrightarrow{\gamma} \mathrm{St}_{\upsilon}(\mathcal{A})$, $f \mapsto [f]$, *whose kernel is the class of all retracts of objects in* υ. *That is, the zero objects of* $\mathrm{St}_{\upsilon}(\mathcal{A})$ *are precisely the retracts, in* $\mathcal{A}$, *of an object in* υ. *Of course in the case* υ *is closed under retracts then* υ *is precisely the class of zero objects.*

Proof For (1), check the conditions to be an equivalence relation. (Reflexive) Since $f - f = 0$ factors through $0 \in \upsilon$, we get $f \sim f$. (In fact, the zero morphism factors through *any* object of υ, we just need that υ is nonempty.) (Symmetric) Say $f \sim g$. Then $g - f$ factors through some $W \in \upsilon$ as, say, $g - f = \beta\alpha$. Then $-(g - f) = -(\beta\alpha)$. So $f - g = (-\beta)\alpha$. This proves $f - g$ factors through $W \in \upsilon$, so $g \sim f$. (Transitive) Say $f \sim g$ and $g \sim h$. So $g - f = \beta_1\alpha_1$ factors through some $W_1 \in \upsilon$ and $h - g = \beta_2\alpha_2$ factors through some $W_2 \in \upsilon$. Adding these equations gives us $h - f = \beta_2\alpha_2 + \beta_1\alpha_1$. But this last map equals the composition

$$A \xrightarrow{\left[\begin{smallmatrix}\alpha_1 \\ \alpha_2\end{smallmatrix}\right]} W_1 \bigoplus W_2 \xrightarrow{[\beta_1\ \beta_2]} B.$$

Since $W_1 \bigoplus W_2 \in \upsilon$ we get $f \sim h$.

For (2), we first need to show that composition of maps is well defined. That is, we must show $f_1 \sim f_2$ and $g_1 \sim g_2$ implies $g_1f_1 \sim g_2f_2$. But first, it is easy to see that for a fixed g, we have $f_1 \sim f_2$ implies $gf_1 \sim gf_2$. Indeed if $f_2 - f_1 = \beta\alpha$ factors through some $W \in \upsilon$, then $gf_2 - gf_1 = g(f_2 - f_1) = g(\beta\alpha) = (g\beta)\alpha$ factors through the same W. This proves $gf_2 \sim gf_1$. Second, one similarly can see that for a fixed f, we have $g_1 \sim g_2$ implies $g_1f \sim g_2f$. Now going back to the general case, suppose $f_1 \sim f_2$ and $g_1 \sim g_2$. Then $g_1f_1 \sim g_2f_1$ and $g_2f_1 \sim g_2f_2$. By transitivity of $\sim$ we get $g_1f_1 \sim g_2f_2$. This completes the proof that composition is well defined and clearly it is associative, making $\mathrm{St}_{\upsilon}(\mathcal{A})$ a category with identity maps of the form $[1_A]$.

To show (3), that the category is additive, we must show $f_1 \sim f_2$ and $g_1 \sim g_2$ implies $f_1 + g_1 \sim f_2 + g_2$. So say $f_2 - f_1 = \beta_1\alpha_1$ is a factorization through W_1 and $g_2 - g_1 = \beta_2\alpha_2$ is a factorization through W_2. Then $(f_2 + g_2) - (f_1 + g_1) = (f_2 - f_1) + (g_2 - g_1) = \beta_1\alpha_1 + \beta_2\alpha_2$ and this last map equals the composition

$$A \xrightarrow{\left[\begin{smallmatrix}\alpha_1 \\ \alpha_2\end{smallmatrix}\right]} W_1 \bigoplus W_2 \xrightarrow{[\beta_1\ \beta_2]} B.$$

It now follows immediately that each $\underline{\mathrm{Hom}}_{\upsilon}(A, B)$ inherits the structure of an abelian group from $\mathrm{Hom}_{\mathcal{A}}(A, B)$, and composition of maps is bilinear. Thus the category $\mathrm{St}_{\upsilon}(\mathcal{A})$ is a preadditive category. It is clear that there is a canonical additive functor $\mathcal{A} \xrightarrow{\gamma} \mathrm{St}_{\upsilon}(\mathcal{A})$, identity on objects, and mapping $f \mapsto [f]$. It is also clear that biproducts exist in $\mathrm{St}_{\upsilon}(\mathcal{A})$, and are taken as in $\mathcal{A}$, since the additive functor γ will preserve biproduct diagrams. Lastly, a 0 object in $\mathcal{A}$ is also as 0 object in $\mathrm{St}_{\upsilon}(\mathcal{A})$, making it an additive category.

It is only left to prove that the kernel of γ is precisely the class of all $\mathcal{A}$-retracts of objects in υ. That is, that $A = 0$ in $\mathrm{St}_{\upsilon}(\mathcal{A})$ if and only if A is a retract of an object in υ. For this we will use the fact that, in any additive category, an object is 0 if and only if its identity map coincides with the zero morphism. So here it means it is enough to show $[1_A] = [0_A]$ if and only if A is a retract of an object in υ. But the condition $[1_A] = [0_A]$ translates to mean the identity factors as $1_A = gf$, through some $W \in \upsilon$. This happens if and only if A is a retract of some $W \in \upsilon$. □

As mentioned in the proof of Proposition 3.1, the functor γ takes biproducts in $\mathcal{A}$ to biproducts in $\mathrm{St}_{\upsilon}(\mathcal{A})$. But we also have the following.

Proposition 3.2 *Let $\mathcal{A}$ and υ be as in Proposition 3.1.*

(1) *If coproducts exist in $\mathcal{A}$ and υ is closed under coproducts, then coproducts also exist in* $\mathrm{St}_{\upsilon}(\mathcal{A})$ *and are taken as in $\mathcal{A}$. More precisely, let $\{A_i\}_{i\in I}$ be a collection of objects and let $\left(\bigoplus_{i\in I} A_i, \{\eta_i\}_{i\in I}\right)$ denote their coproduct in $\mathcal{A}$; so each $A_i \xrightarrow{\eta_i} \bigoplus_{i\in I} A_i$ is the canonical injection into the ith component. Then $\left(\bigoplus_{i\in I} A_i, \{[\eta_i]\}_{i\in I}\right)$ is a coproduct of $\{A_i\}_{i\in I}$ in* $\mathrm{St}_{\upsilon}(\mathcal{A})$.
(2) *If products exist in $\mathcal{A}$ and υ is closed under products, then products also exist in* $\mathrm{St}_{\upsilon}(\mathcal{A})$ *and are taken as in $\mathcal{A}$.*

Proof We leave the proof as an exercise for the reader. □

Exercise 3.1.1 Prove Proposition 3.2.

3.2 The Stable Category of a Complete Cotorsion Pair

We now return to let $\mathfrak{C} = (\mathcal{X}, \mathcal{Y})$ denote a cotorsion pair in an exact category $(\mathcal{A}, \mathcal{E})$. As noted in the chapter introduction, $\mathfrak{C}$ may even just be an Ext-pair in the sense of Definition 2.2. We set $\omega = \mathcal{X} \cap \mathcal{Y}$ and call this the **core** of $\mathfrak{C}$. Since ω is closed under biproducts, we may apply Proposition 3.1 (taking $\upsilon = \omega$) to obtain the additive category $\mathrm{St}_{\omega}(\mathcal{A})$. We call this the **stable category** of $\mathfrak{C}$. But the more general setup that follows will be useful as we develop the theory of abelian model categories, starting in Chapter 4. Indeed we then will specialize ρ in Setup 3.2.1 to be the *very good* ***right*** homotopy relation, denoted $\rho = r$. On the other hand, λ in the dual Setup 3.2.2 will be taken to be $\lambda = \ell$, the *very good* ***left*** homotopy relation.

Setup 3.2.1 Assume $\mathfrak{C} = (\mathcal{X}, \mathcal{Y})$ has enough projectives and that we are given a class of objects ρ, closed under biproducts, satisfying $\omega \subseteq \rho \subseteq \mathcal{X}$. Note that we have stable categories and canonical functors sitting in the commutative diagram:

$$\begin{array}{ccc} \mathcal{A} & & \\ \downarrow{\scriptstyle \gamma_\omega} & \searrow^{\gamma_\rho} & \\ \mathrm{St}_\omega(\mathcal{A}) & \xrightarrow[\bar{\gamma}_\rho]{} & \mathrm{St}_\rho(\mathcal{A}). \end{array}$$

The following lemma concerning Setup 3.2.1 is crucial and will be used often.

Lemma 3.3 (Fundamental Lemma) *Suppose $(\mathcal{X}, \mathcal{Y})$ and ρ are as in Setup 3.2.1. Consider the following solid arrow diagram whose rows are short exact sequences such that $X_A \in \mathcal{X}$ and $Y_B \in \mathcal{Y}$:*

$$\begin{array}{lccccc} \mathbb{X}_A: & Y_A & \overset{i_A}{\rightarrowtail} & X_A & \overset{p_A}{\twoheadrightarrow} & A \\ & {\scriptstyle k}\downarrow & & {\scriptstyle h}\downarrow & & \downarrow{\scriptstyle f} \\ \mathbb{X}_B: & Y_B & \overset{i_B}{\rightarrowtail} & X_B & \overset{p_B}{\twoheadrightarrow} & B. \end{array}$$

(In particular, the two rows may be $\mathcal{X}$-approximation sequences.) Then there exist dashed arrows k and h making the diagram commute and the pair (k, h) satisfies the following uniqueness conditions.

(1) *The lifting pair (k, h) is unique in the stable category $\mathrm{St}_\rho(\mathcal{A})$ and it depends only on the ρ-equivalence class of f. Precisely, for all $f' : A \to B$ and any corresponding lifting pair (k', h'), if $[f]_\rho = [f']_\rho$ then both $[h]_\rho = [h']_\rho$ and $[k]_\rho = [k']_\rho$ (in fact, even $[k]_\omega = [k']_\omega$).*
(2) *For all $f' : A \to B$ and any corresponding lifting pair (k', h'), if $[f]_\mathcal{Y} = [f']_\mathcal{Y}$ then $[h]_\omega = [h']_\omega$.*

Proof It is easy to find a pair (k, h) completing the diagram since we have $\mathrm{Ext}^1_\mathcal{E}(X_A, Y_B) = 0$ and Y_B is a kernel. For (1), assume (h, k) is a lifting pair for f and that $f' : A \to B$ is another map in $\mathcal{A}$ with its own lifting pair (k', h'). Then we get a commutative diagram as follows:

$$\begin{array}{ccccc} Y_A & \overset{i_A}{\rightarrowtail} & X_A & \overset{p_A}{\twoheadrightarrow} & A \\ {\scriptstyle k'-k}\downarrow & & {\scriptstyle h'-h}\downarrow & & {\scriptstyle f'-f}\downarrow \\ Y_B & \underset{}{\overset{i_B}{\rightarrowtail}} & X_B & \overset{p_B}{\twoheadrightarrow} & B. \end{array}$$

From this we see it is enough to prove $f \sim^{\rho} 0$ implies both $h \sim^{\rho} 0$ and $k \sim^{\rho} 0$ in the original statement. So suppose $f = \beta\alpha$ where $\alpha\colon A \to W_1$ and $\beta\colon W_1 \to B$ and $W_1 \in \rho$. Since $W_1 \in \mathcal{X}$, we see that β lifts over p_B, so we get a map $A \to X_B$ (factoring through $W_1 \in \rho$) making the lower right triangle commute. By the Homotopy Lemma 1.21, this map extends to a chain homotopy from the the top row to the bottom row. However, the map $X_A \to Y_B$ in this homotopy factors through another object $W_2 \in \omega \subseteq \rho$. Indeed this W_2 is found by taking an $\mathcal{X}$-approximation of $Y_B \in \mathcal{Y}$, as $\mathcal{Y}$ is closed under extensions. All told, both maps in the chain homotopy factor through an object of ρ, and by composing with p_A and i_B we may construct maps $X_A \xrightarrow{s_1} W_1 \xrightarrow{t_1} X_B$ and $X_A \xrightarrow{s_2} W_2 \xrightarrow{t_2} X_B$ such that $h = t_1 s_1 + t_2 s_2$. Thus h equals the composition

$$X_A \overset{\left[\begin{smallmatrix} s_1 \\ s_2 \end{smallmatrix}\right]}{\rightarrowtail} W_1 \bigoplus W_2 \overset{[t_1\ t_2]}{\twoheadrightarrow} X_B.$$

We are done showing $f \sim^{\rho} 0 \implies h \sim^{\rho} 0$, as we have shown h factors through $W_1 \bigoplus W_2 \in \rho$.

As for showing, $k \sim^{\rho} 0$, this is the easy part. We have already seen that the map $X_A \to Y_B$ in the chain homotopy from the top to the bottom row factors through $W_2 \in \omega \subseteq \rho$.

To prove (2), for the same reason as previously, it is enough to show that if $f \sim^{\mathcal{Y}} 0$ then $h \sim^{\omega} 0$. So suppose $f\colon A \to B$ factors as $f = \beta\alpha$ through an object $Y \in \mathcal{Y}$. We may construct a commutative diagram as shown where the middle row is an $\mathcal{X}$-approximation sequence, and hence $X_Y \in \omega$:

$$\begin{array}{ccccc} Y_A & \overset{i_A}{\rightarrowtail} & X_A & \overset{p_A}{\twoheadrightarrow} & A \\ {\scriptstyle k_\alpha}\downarrow & & {\scriptstyle h_\alpha}\downarrow & & \downarrow{\scriptstyle \alpha} \\ \Omega Y & \overset{i_Y}{\rightarrowtail} & X_Y & \overset{p_Y}{\twoheadrightarrow} & Y \\ {\scriptstyle k_\beta}\downarrow & & {\scriptstyle h_\beta}\downarrow & & \downarrow{\scriptstyle \beta} \\ Y_B & \overset{i_B}{\rightarrowtail} & X_B & \overset{p_B}{\twoheadrightarrow} & B. \end{array}$$

By what we just proved in (1), and taking $\rho = \omega$, we have

$$f = \beta\alpha \implies f \sim^{\omega} \beta\alpha \implies h \sim^{\omega} h_\beta h_\alpha \implies h \sim^{\omega} 0,$$

where the last implication holds since $X_Y \in \omega$ means $h_\beta h_\alpha \sim^{\omega} 0$. □

3.2.1 The Dual Setup

We now state the dual of Setup 3.2.1 and the lemma that follows it. Here we use the notation λ as it will be applicable to the (very good) *left* homotopy relation associated to an abelian model structure.

Setup 3.2.2 Here we assume the $\mathfrak{C} = (\mathcal{X}, \mathcal{Y})$ has enough injectives and that we are given a class of objects λ, closed under biproducts, satisfying $\mathcal{Y} \supseteq \lambda \supseteq \omega$. Note that we have stable categories and canonical functors sitting in the commutative diagram:

$$\begin{array}{ccc} & & \mathcal{A} \\ & \swarrow^{\gamma_\lambda} & \downarrow{\scriptstyle \gamma_\omega} \\ \mathrm{St}_\lambda(\mathcal{A}) & \xleftarrow{\bar{\gamma}_\lambda} & \mathrm{St}_\omega(\mathcal{A}). \end{array}$$

Lemma 3.4 (Dual Fundamental Lemma) *Suppose $(\mathcal{X}, \mathcal{Y})$ and λ are as in Setup 3.2.2. Consider the following solid arrow diagram whose rows are short exact sequences such that $Y_B \in \mathcal{Y}$ and $X_A \in \mathcal{X}$:*

$$\begin{array}{lccccc} \mathbb{Y}_A: & A & \xrightarrow{j_A} & Y_A & \xrightarrow{q_A} & X_A \\ & {\scriptstyle f}\downarrow & & {\scriptstyle h}\downarrow & & {\scriptstyle c}\downarrow \\ \mathbb{Y}_B: & B & \xrightarrow{j_B} & Y_B & \xrightarrow{q_B} & X_B. \end{array}$$

(In particular, the two rows may be $\mathcal{Y}$-approximation sequences.) Then there exist dashed arrows h and c making the diagram commute and the pair (h, c) satisfies the following uniqueness conditions.

(1) *The lifting pair (h, c) is unique in the stable category $\mathrm{St}_\lambda(\mathcal{A})$ and it depends only on the λ-equivalence class of f. Precisely, for all $f' \colon A \to B$ and any corresponding lifting pair (h', c'), if $[f]_\lambda = [f']_\lambda$ then both $[h]_\lambda = [h']_\lambda$ and $[c]_\lambda = [c']_\lambda$ (in fact, even $[c]_\omega = [c']_\omega$).*
(2) *For all $f' \colon A \to B$ and any corresponding lifting pair (h', c'), if $[f]_{\mathcal{X}} = [f']_{\mathcal{X}}$ then $[h]_\omega = [h']_\omega$.*

Exercise 3.2.1 Let $\mathfrak{C} = (\mathcal{X}, \mathcal{Y})$ be a cotorsion pair, or even just an Ext-pair, and assume it has enough projectives or enough injectives. Show that whenever $X \in \mathcal{X}$ and $Y \in \mathcal{Y}$, we have $\underline{\mathrm{Hom}}_\omega(X, Y) = 0$.

Exercise 3.2.2 Let $\mathfrak{C} = (\mathcal{X}, \mathcal{Y})$ be an hereditary cotorsion pair in an exact category $(\mathcal{A}, \mathcal{E})$ with enough projectives and enough injectives. Show that $X \in \mathcal{X}$ (resp. $Y \in \mathcal{Y}$) if and only if $\underline{\mathrm{Hom}}_\omega(X, Y) = 0$ for all $Y \in \mathcal{Y}$ (resp.

$\underline{\mathrm{Hom}}_{\omega}(X,Y) = 0$ for all $X \in \mathcal{X}$). *Hint: Given any* $Y \in \mathcal{Y}$, *write a short exact sequence* $Y \rightarrowtail I \twoheadrightarrow \Sigma Y$ *with* I *injective. For any other object* X, *there is an epimorphism of abelian groups* $\delta\colon \mathrm{Hom}_{\mathcal{A}}(X,\Sigma Y) \twoheadrightarrow \mathrm{Ext}^1_{\mathcal{E}}(X,Y)$ *which in fact descends to an epimorphism* $\bar{\delta}\colon \underline{\mathrm{Hom}}_{\omega}(X,\Sigma Y) \twoheadrightarrow \mathrm{Ext}^1_{\mathcal{E}}(X,Y)$.

3.3 Approximation Functors

Here we will develop the basic properties of the approximation functors associated to $\mathfrak{C} = (\mathcal{X},\mathcal{Y})$. A special instance of these will be the cofibrant and fibrant replacement functors when we turn our focus to abelian model categories. Throughout this section, up until Section 3.3.1 where we state the dual results, we continue the underlying assumptions of Setup 3.2.1 from the previous section. That is, we assume again that $(\mathcal{X},\mathcal{Y})$ has enough projectives and that we are given a class of objects ρ, closed under biproducts, satisfying $\omega \subseteq \rho \subseteq \mathcal{X}$.

Let us now write an $\mathcal{X}$-approximation sequence as $\Omega A \rightarrowtail XA \twoheadrightarrow A$, where $XA \in \mathcal{X}$ and $\Omega A \in \mathcal{Y}$. Then XA is unique in $\mathrm{St}_\rho(\mathcal{A})$ in the following sense: For any two choices of $\mathcal{X}$-approximations of $A \in \mathcal{A}$, say $p_A\colon XA \twoheadrightarrow A$ and $p'_A\colon X'A \twoheadrightarrow A$, there is a unique isomorphism $[h_A]_\rho\colon XA \to X'A$, in $\mathrm{St}_\rho(\mathcal{A})$, satisfying $[p_A]_\rho = [p'_A]_\rho \circ [h_A]_\rho$. The reader can readily verify this using the Fundamental Lemma 3.3, by taking $f = 1_A$ in the statement of that lemma. So in the same way that we may discuss *the* kernel of a morphism in an abelian category (it satisfies a property and is unique up to a canonical isomorphism) we may discuss *the* $\mathcal{X}$**-approximation** of A, denoted $\left(XA \xrightarrow{[p_A]_\rho} A\right) \in \mathrm{St}_\rho(\mathcal{A})$. By abusing the language we might sometimes also refer to XA as the $\mathcal{X}$-approximation of A. Note that $\mathcal{X}$ determines a strictly full subcategory (i.e. an isomorphism-closed, or *replete*, full subcategory) of $\mathcal{A}$. Since it is closed under finite biproducts, it is an additive subcategory of $\mathcal{A}$. It also follows from Proposition 3.1 that $\mathrm{St}_\rho(\mathcal{X})$ is an additive category, and we let $I\colon \mathrm{St}_\rho(\mathcal{X}) \to \mathrm{St}_\rho(\mathcal{A})$ denote the inclusion functor.

Exercise 3.3.1 Use the Fundamental Lemma 3.3 to show that we get additive functors and natural transformations as described in the following definition. However, see Remark 3.6 as well.

Definition 3.5 The process of taking $\mathcal{X}$-approximations defines an additive functor $X\colon \mathcal{A} \to \mathrm{St}_\rho(\mathcal{X})$. Its value on a morphism $f\colon A \to B$ is computed by constructing any morphism of $\mathcal{X}$-approximation sequences:

$$\begin{array}{llcccc} \mathbb{X}_A: & \Omega A & \overset{i_A}{\rightarrowtail} & XA & \overset{p_A}{\twoheadrightarrow} & A \\ & \Omega(f)\downarrow & & X(f)\downarrow & & f\downarrow \\ \mathbb{X}_B: & \Omega B & \overset{i_B}{\rightarrowtail} & XB & \overset{p_B}{\twoheadrightarrow} & B. \end{array}$$

So X is defined by $f \mapsto [X(f)]_\rho$, but it is also compatible with $\mathcal{A} \xrightarrow{\gamma_\rho} \mathrm{St}_\rho(\mathcal{A})$, and so descends via the rule $X([f]_\rho) = [X(f)]_\rho$ to an additive functor $\mathrm{St}_\rho(\mathcal{A}) \xrightarrow{X} \mathrm{St}_\rho(\mathcal{X})$. We will call this the **$\mathcal{X}$-approximation functor**. Moreover, considering X as an endofunctor on $\mathrm{St}_\rho(\mathcal{A})$, the family of morphisms $\{[p_A]_\rho\}$ determines a natural transformation $\{[p_A]_\rho\}\colon X \to 1_{\mathrm{St}_\rho(\mathcal{A})}$ which we will call the **$\mathcal{X}$-approximation transformation**. As a matter of convenience, we set the convention that $XA = A$ whenever $A \in \mathcal{X}$. That is, we assign $[1_A]_\rho\colon A \to A$ to be the $\mathcal{X}$-approximation, where we have set $XA = A$ and $p_A = 1_A$, for all $A \in \mathcal{X}$.

Remark 3.6 To be more precise, in order for $\mathcal{A} \xrightarrow{X} \mathrm{St}_\rho(\mathcal{X})$ to be an honest functor there should be assigned, to each $A \in \mathcal{A}$, a particular $\mathcal{X}$-approximation sequence $\mathbb{X}_A\colon \Omega A \rightarrowtail XA \twoheadrightarrow A$, (so $XA \in \mathcal{X}, \Omega A \in \mathcal{Y}$). Then we get a functor X and a natural transformation $\{[p_A]_\rho\}\colon X \xrightarrow{p} 1_{\mathrm{St}_\rho(\mathcal{A})}$. Given any other assignment $A \mapsto \mathbb{X}'_A$ of such short exact sequences, we get by the Fundamental Lemma 3.3, a canonical isomorphism of functors $\{[h_A]_\rho\}\colon X \xrightarrow{h} X'$ satisfying $p = p' \circ h$. These compatibilities justify our nonchalance in defining the $\mathcal{X}$-approximation functor as we have done in Definition 3.5. Similar definitions will appear on occasion throughout the book. In a few instances (mainly in Chapter 6) we will make further comment to cast away potential confusion that might arise.

Note that we may take $\rho = \omega$, giving us an additive endofunctor on $\mathrm{St}_\omega(\mathcal{A})$, the stable category of $\mathfrak{C}$. For a possibly different ρ, we get a commutative diagram of functors:

$$\begin{array}{ccc} \mathrm{St}_\omega(\mathcal{A}) & \xrightarrow{\bar{\gamma}_\rho} & \mathrm{St}_\rho(\mathcal{A}) \\ X\downarrow & & \downarrow X \\ \mathrm{St}_\omega(\mathcal{X}) & \xrightarrow{\bar{\gamma}_\rho} & \mathrm{St}_\rho(\mathcal{X}). \end{array} \tag{3.1}$$

We now show that $\mathrm{St}_\rho(\mathcal{X})$ is what is often called a *coreflective* subcategory of $\mathrm{St}_\rho(\mathcal{A})$.

Proposition 3.7 *The $\mathcal{X}$-approximation functor* $\mathrm{St}_\rho(\mathcal{A}) \xrightarrow{X} \mathrm{St}_\rho(\mathcal{X})$ *is right adjoint to the inclusion* $\mathrm{St}_\rho(\mathcal{X}) \xrightarrow{I} \mathrm{St}_\rho(\mathcal{A})$*. That is,* $\mathrm{St}_\rho(\mathcal{X})$ *is a coreflective subcategory of* $\mathrm{St}_\rho(\mathcal{A})$ *with X being the coreflector functor.*

Proof We need to find a natural bijection

$$\mathrm{St}_\rho(\mathcal{A})(IA, B) \cong \mathrm{St}_\rho(\mathcal{X})(A, XB)$$

for all $A \in \mathrm{St}_\rho(\mathcal{X})$ and $B \in \mathrm{St}_\rho(\mathcal{A})$. This boils down to a natural bijection

$$\underline{\mathrm{Hom}}_\rho(A, B) \cong \underline{\mathrm{Hom}}_\rho(A, XB).$$

But composition with the $\mathcal{X}$-approximation transformation, $\{[p_B]_\rho\}$, induces a natural bijection

$$\underline{\mathrm{Hom}}_\rho(A, XB) \xrightarrow{[p_B]_\rho \circ -} \underline{\mathrm{Hom}}_\rho(A, B),$$

whose inverse is given by

$$\underline{\mathrm{Hom}}_\rho(A, B) \xrightarrow{[X(-)]_\rho} \underline{\mathrm{Hom}}_\rho(A, XB),$$

where $X(-)$ represents a lift as in the diagram appearing in Definition 3.5, using our convention that $p_A = 1_A$ since $A \in \mathcal{X}$. Indeed for a map $f\colon A \to B$ we have $[f]_\rho \mapsto [X(f)]_\rho \mapsto [p_B]_\rho \circ [X(f)]_\rho = [f]_\rho$. In the other direction, given a map $h\colon A \to XB$, we have $[h]_\rho \mapsto [p_B]_\rho \circ [h]_\rho = [p_B \circ h]_\rho \mapsto [X(p_B \circ h)]_\rho$. It is immediate from the Fundamental Lemma 3.3 that $[X(p_B \circ h)]_\rho = [h]_\rho$. □

Now again take $\rho = \omega$, so that we have the $\mathcal{X}$-approximation functor

$$X\colon \mathrm{St}_\omega(\mathcal{A}) \to \mathrm{St}_\omega(\mathcal{X}).$$

Note that for the other class of objects, $\mathcal{Y}$, the stable category $\mathrm{St}_\mathcal{Y}(\mathcal{A})$ sits in the following commutative diagram with the canonical functors:

$$\begin{array}{ccc} & & \mathrm{St}_\omega(\mathcal{A}) \\ & \overset{\gamma_\omega}{\nearrow} & \downarrow{\scriptstyle \bar{\gamma}_\mathcal{Y}} \\ \mathcal{A} & \xrightarrow[\gamma_\mathcal{Y}]{} & \mathrm{St}_\mathcal{Y}(\mathcal{A}). \end{array}$$

We will now see that X descends via $\mathrm{St}_\omega(\mathcal{A}) \xrightarrow{\bar{\gamma}_\mathcal{Y}} \mathrm{St}_\mathcal{Y}(\mathcal{A})$ to an additive functor $\bar{X}\colon \mathrm{St}_\mathcal{Y}(\mathcal{A}) \to \mathrm{St}_\omega(\mathcal{X})$.

Proposition 3.8 *The $\mathcal{X}$-approximation functor $X\colon \mathrm{St}_\omega(\mathcal{A}) \to \mathrm{St}_\omega(\mathcal{X})$ descends via $\mathrm{St}_\omega(\mathcal{A}) \xrightarrow{\bar{\gamma}_\mathcal{Y}} \mathrm{St}_\mathcal{Y}(\mathcal{A})$ to an additive functor $\bar{X}\colon \mathrm{St}_\mathcal{Y}(\mathcal{A}) \to \mathrm{St}_\omega(\mathcal{X})$. That is, $\bar{X}$ is defined the same way as X is defined on objects and the rule $\bar{X}([f]_\mathcal{Y}) := [X(f)]_\omega$ is well defined on morphisms. This provides a commutative diagram of functors:*

$$\begin{array}{ccccc} & & \mathrm{St}_\omega(\mathcal{A}) & & \\ & \overset{\gamma_\omega}{\nearrow} & \downarrow{\scriptstyle \bar{\gamma}_{\mathcal{Y}}} & \overset{X}{\searrow} & \\ \mathcal{A} & \xrightarrow[\gamma_{\mathcal{Y}}]{} & \mathrm{St}_{\mathcal{Y}}(\mathcal{A}) & \xrightarrow[\bar{X}]{} & \mathrm{St}_\omega(\mathcal{X}). \end{array}$$

Proof The rule $\bar{X}([f]_{\mathcal{Y}}) := [X(f)]_\omega$ is well defined due to the Fundamental Lemma 3.3(2). The reader can easily check that $\bar{X}$ defines an additive functor $\mathrm{St}_{\mathcal{Y}}(\mathcal{A}) \xrightarrow{\bar{X}} \mathrm{St}_\omega(\mathcal{X})$ and that $X = \bar{X} \circ \bar{\gamma}_{\mathcal{Y}}$. □

The adjunction of Proposition 3.7 also descends through $\bar{\gamma}_{\mathcal{Y}}$ as described in the following exercise.

Exercise 3.3.2 Show that by restricting the domain of $\bar{\gamma}_{\mathcal{Y}}$ to $\mathrm{St}_\omega(\mathcal{X})$, we get a functor $\mathrm{St}_\omega(\mathcal{X}) \xrightarrow{\bar{\gamma}_{\mathcal{Y}}} \mathrm{St}_{\mathcal{Y}}(\mathcal{A})$ which is left adjoint to $\bar{X}\colon \mathrm{St}_{\mathcal{Y}}(\mathcal{A}) \to \mathrm{St}_\omega(\mathcal{X})$.

Exercise 3.3.3 Note that if $A \in \mathcal{X}$ and $\Omega A \overset{i_A}{\rightarrowtail} XA \overset{p_A}{\twoheadrightarrow} A$ is an $\mathcal{X}$-approximation sequence, then $[p_A]_\rho$ itself provides the canonical isomorphism in $\mathrm{St}_\rho(\mathcal{A})$ between the two $\mathcal{X}$-approximations $[p_A]_\rho$ and $[1_A]_\rho$. Generalize the proof of Proposition 3.7 to explicitly show that our convention of setting $XA = A$ whenever $A \in \mathcal{X}$, is unnecessary. Verify that the proof Exercise 3.3.2 also generalizes so that the convention can be avoided.

3.3.1 Dual Statements

On the other hand, let us assume the hypotheses of the dual Setup 3.2.2. That is, assume for the remainder of this section that $\mathfrak{C} = (\mathcal{X}, \mathcal{Y})$ has enough injectives and λ is a class of objects, closed under biproducts, satisfying $\mathcal{Y} \supseteq \lambda \supseteq \omega$. Then for any $A \in \mathcal{A}$, we may discuss the **$\mathcal{Y}$-approximation,** $\left(A \xrightarrow{[j_A]_\lambda} YA\right) \in \mathrm{St}_\lambda(\mathcal{A})$. For future reference, we state its precise definition here along with the dual of Proposition 3.7.

Definition 3.9 The process of taking $\mathcal{Y}$-approximations defines an additive functor $\mathcal{A} \xrightarrow{Y} \mathrm{St}_\lambda(\mathcal{Y})$. Its value on a morphism $f\colon A \to B$ is computed by constructing a morphism of $\mathcal{Y}$-approximation sequences:

$$\begin{array}{lccccc} \Upsilon_A: & A & \overset{j_A}{\rightarrowtail} & YA & \overset{q_A}{\twoheadrightarrow} & \Sigma A \\ & {\scriptstyle f}\downarrow & & {\scriptstyle Y(f)}\downarrow & & {\scriptstyle \Sigma(f)}\downarrow \\ \Upsilon_B: & B & \overset{j_B}{\rightarrowtail} & YB & \overset{q_B}{\twoheadrightarrow} & \Sigma B. \end{array}$$

So Y is defined by $f \mapsto [Y(f)]_\lambda$, but it is also compatible with $\mathcal{A} \xrightarrow{\gamma_\lambda} \mathrm{St}_\lambda(\mathcal{A})$, and so descends via the rule $Y([f]_\lambda) = [Y(f)]_\lambda$ to an additive functor $\mathrm{St}_\lambda(\mathcal{A}) \xrightarrow{Y}$

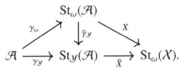

$\mathrm{St}_\lambda(\mathcal{Y})$. We will call this the **$\mathcal{Y}$-approximation functor**. Moreover, considering Y as an endofunctor on $\mathrm{St}_\lambda(\mathcal{A})$, the family of morphisms $\{[j_A]_\lambda\}$ determines a natural transformation $\{[j_A]_\lambda\}\colon 1_{\mathrm{St}_\lambda(\mathcal{A})} \to Y$ which we will call the **$\mathcal{Y}$-approximation transformation**. For convenience, we assign $YA = A$ and $[1_A]_\lambda\colon A \to A$ to be the $\mathcal{Y}$-approximation whenever $A \in \mathcal{Y}$.

In the case that λ is distinct from ω, we get a commutative diagram of functors:

$$\begin{array}{ccc} \mathrm{St}_\lambda(\mathcal{A}) & \xleftarrow{\bar{\gamma}_\lambda} & \mathrm{St}_\omega(\mathcal{A}) \\ {\scriptstyle Y}\downarrow & & \downarrow{\scriptstyle Y} \\ \mathrm{St}_\lambda(\mathcal{Y}) & \xleftarrow{\bar{\gamma}_\lambda} & \mathrm{St}_\omega(\mathcal{Y}). \end{array} \tag{3.2}$$

Proposition 3.10 *The $\mathcal{Y}$-approximation functor $\mathrm{St}_\lambda(\mathcal{A}) \xrightarrow{Y} \mathrm{St}_\lambda(\mathcal{Y})$ is left adjoint to the inclusion $\mathrm{St}_\lambda(\mathcal{Y}) \xrightarrow{I} \mathrm{St}_\lambda(\mathcal{A})$. That is, $\mathrm{St}_\lambda(\mathcal{Y})$ is a reflective subcategory of $\mathrm{St}_\lambda(\mathcal{A})$ with Y being the reflector functor.*

Taking $\lambda = \omega$, the $\mathcal{Y}$-approximation functor $Y\colon \mathrm{St}_\omega(\mathcal{A}) \to \mathrm{St}_\omega(\mathcal{Y})$ factors through $\mathrm{St}_\mathcal{X}(\mathcal{A})$.

Proposition 3.11 *The $\mathcal{Y}$-approximation functor $Y\colon \mathrm{St}_\omega(\mathcal{A}) \to \mathrm{St}_\omega(\mathcal{Y})$ descends via $\mathrm{St}_\omega(\mathcal{A}) \xrightarrow{\bar{\gamma}_\mathcal{X}} \mathrm{St}_\mathcal{X}(\mathcal{A})$ to an additive functor $\bar{Y}\colon \mathrm{St}_\mathcal{X}(\mathcal{A}) \to \mathrm{St}_\omega(\mathcal{Y})$. That is, $\bar{Y}$ is defined the same way as Y on objects and the rule $\bar{Y}([f]_\mathcal{X}) := [Y(f)]_\omega$ is well defined on morphisms, providing a commutative diagram of functors:*

$$\begin{array}{ccccc} & & \mathrm{St}_\omega(\mathcal{A}) & & \\ & {\scriptstyle \gamma_\omega}\nearrow & \downarrow{\scriptstyle \bar{\gamma}_\mathcal{X}} & \searrow{\scriptstyle Y} & \\ \mathcal{A} & \xrightarrow[\gamma_\mathcal{X}]{} & \mathrm{St}_\mathcal{X}(\mathcal{A}) & \xrightarrow[\bar{Y}]{} & \mathrm{St}_\omega(\mathcal{Y}). \end{array}$$

3.4 Loop and Suspension Functors

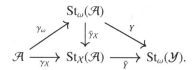

All together, we can associate four additive functors to $\mathfrak{C} = (\mathcal{X}, \mathcal{Y})$, as long as it is complete. In the previous section we looked at the $\mathcal{X}$-approximation functor and its dual. We now look at the other two, which are also dual to one another: the *loop* functor and the *suspension* functor. They will be fundamental later when we construct the triangulated structure on the homotopy category of an abelian model structure. For now, we simply point out their existence for future reference.

Consider again the diagram, from Definition 3.5, where each row here is an $\mathcal{X}$-approximation sequence:

$$\begin{array}{ccccc} \Omega A & \xrightarrow{i_A} & XA & \xrightarrow{p_A} & A \\ {\scriptstyle \Omega(f)}\downarrow & & {\scriptstyle X(f)}\downarrow & & {\scriptstyle f}\downarrow \\ \Omega B & \xrightarrow{i_B} & XB & \xrightarrow{p_B} & B. \end{array}$$

Any object ΩA obtained this way from an $\mathcal{X}$-approximation is unique as an object of $\mathrm{St}_\rho(\mathcal{A})$, up to a canonical isomorphism in $\mathrm{St}_\rho(\mathcal{A})$. In fact, better yet, it is unique as an object in $\mathrm{St}_\omega(\mathcal{A})$ up to a canonical isomorphism. The reader can again verify this with the Fundamental Lemma 3.3 and go on to see that we obtain an additive functor as in the next definition and remark that follows.

Definition 3.12 The following functors are induced by $\mathfrak{C} = (\mathcal{X}, \mathcal{Y})$.

(1) Suppose $\mathfrak{C}$ and ρ are as in Setup 3.2.1. Then the process of taking kernels of $\mathcal{X}$-approximations defines an additive functor $\Omega\colon \mathcal{A} \to \mathrm{St}_\omega(\mathcal{Y})$. Its value on a morphism $f\colon A \to B$ is computed by constructing a morphism of $\mathcal{X}$-approximation sequences:

$$\begin{array}{lccccc} \mathbb{X}_A\colon & \Omega A & \xrightarrow{i_A} & XA & \xrightarrow{p_A} & A \\ & {\scriptstyle \Omega(f)}\downarrow & & {\scriptstyle X(f)}\downarrow & & {\scriptstyle f}\downarrow \\ \mathbb{X}_B\colon & \Omega B & \xrightarrow{i_B} & XB & \xrightarrow{p_B} & B. \end{array}$$

So Ω is defined by $f \mapsto [\Omega(f)]_\omega$. It is also compatible with $\mathcal{A} \xrightarrow{\gamma_\rho} \mathrm{St}_\rho(\mathcal{A})$, meaning it descends via the rule $\Omega([f]_\rho) := [\Omega(f)]_\omega$ to an additive functor $\mathrm{St}_\rho(\mathcal{A}) \xrightarrow{\Omega} \mathrm{St}_\omega(\mathcal{Y})$. We will call this the **loop functor for** $\mathfrak{C}$.

(2) Suppose $\mathfrak{C}$ and λ are as in Setup 3.2.2. Then the process of taking cokernels of $\mathcal{Y}$-approximations defines an additive functor $\Sigma\colon \mathcal{A} \to \mathrm{St}_\omega(\mathcal{X})$. Its value on a morphism $f\colon A \to B$ is computed by constructing a morphism of $\mathcal{Y}$-approximation sequences:

$$\begin{array}{lccccc} \mathbb{Y}_A\colon & A & \xrightarrow{j_A} & YA & \xrightarrow{q_A} & \Sigma A \\ & \downarrow{\scriptstyle f} & & \downarrow{\scriptstyle Y(f)} & & \downarrow{\scriptstyle \Sigma(f)} \\ \mathbb{Y}_B\colon & B & \xrightarrow{j_B} & YB & \xrightarrow{q_B} & \Sigma B. \end{array}$$

So Σ is defined by $f \mapsto [\Sigma(f)]_\omega$. It is also compatible with $\mathcal{A} \xrightarrow{\gamma_\lambda} \mathrm{St}_\lambda(\mathcal{A})$, meaning it descends via the rule $\Sigma([f]_\lambda) := [\Sigma(f)]_\omega$ to an additive functor $\mathrm{St}_\lambda(\mathcal{A}) \xrightarrow{\Sigma} \mathrm{St}_\omega(\mathcal{Y})$. We will call this the **suspension functor for** $\mathfrak{C}$.

Remark 3.13 Considering Ω as an endofunctor on $\mathrm{St}_\rho(\mathcal{A})$, the family of morphisms $\{[i_A]_\rho\}$ determines a natural transformation $\Omega \xrightarrow{\{[i_A]_\rho\}} X$ which we will call the **loop transformation**. In the dual situation, we have the **suspension transformation** $Y \xrightarrow{\{[q_A]_\lambda\}} \Sigma$. Of course the points made in Remark 3.6 apply here as well: While the universal properties allow us to abuse the concept of functor, an honest functor arises from a particular assignment of approximation sequences. When considering triangulated structures arising from cotorsion pairs (in Chapter 6) it will become more important for the compatibilities between different suspension or loop functors to receive explicit mention.

4

Hovey Triples and Abelian Model Structures

In this chapter we begin the study of abelian model structures, which are essentially equivalent to what we call *Hovey triples*. Indeed under the (very mild) assumption that we are working in a weakly idempotent complete exact category $(\mathcal{A}, \mathcal{E})$, there is a one-to-one correspondence between Hovey triples and abelian model structures on $(\mathcal{A}, \mathcal{E})$. This is described in Hovey's Correspondence, Theorem 4.25. However, the simpler of the two ideas is that of a Hovey triple; it is just a triple of classes of objects which link together to form two complete cotorsion pairs. We will define the most fundamental ideas surrounding abelian model categories, for example (co)fibrant replacements and homotopy relations, directly in terms of these cotorsion pairs. Such ideas also make good sense regardless of whether or not $\mathcal{A}$ is weakly idempotent complete. Therefore, in this chapter and throughout the book, we approach abelian model structures by way of Hovey triples. For completeness, we will of course define (closed) abelian model structures and prove Hovey's Correspondence. But this doesn't appear until the final section (Section 4.6), and it will mainly be definitions and formalities by that point. With some care, we can even remove the weakly idempotent complete hypothesis and still have a version of Hovey's Correspondence. Appendix A shows how this can be done, and also extends Hovey's Correspondence to Quillen's original definition of a (not necessarily closed) model structure.

Throughout this chapter we let $(\mathcal{A}, \mathcal{E})$ denote any exact category. It need not be weakly idempotent complete unless explicitly stated otherwise.

4.1 Hovey Triples and Hovey Ext-Triples

In addition to Hovey triples it is just as easy to introduce the weaker notion of a Hovey Ext-triple. They will be shown to correspond to Quillen's original definition of a model structure, while Hovey triples correspond to what Quillen called a *closed* model structure.

Definition 4.1 Let $\mathfrak{M} = (\mathcal{Q}, \mathcal{W}, \mathcal{R})$ be a triple of objects of $\mathcal{A}$. We set the notation $\mathcal{Q}_\mathcal{W} := \mathcal{Q} \cap \mathcal{W}$ and $\mathcal{R}_\mathcal{W} := \mathcal{W} \cap \mathcal{R}$. We call $\mathfrak{M}$ a **Hovey triple** if it satisfies each of the following.

(1) $(\mathcal{Q}_\mathcal{W}, \mathcal{R})$ and $(\mathcal{Q}, \mathcal{R}_\mathcal{W})$ are each complete cotorsion pairs relative to $(\mathcal{A}, \mathcal{E})$. Objects in $\mathcal{Q}$ are called **cofibrant** and objects in $\mathcal{R}$ are called **fibrant**.
(2) $\mathcal{W}$ satisfies the 2 out of 3 property on short exact sequences. That is, whenever two out of three terms in a short exact sequence are in $\mathcal{W}$, then so is the third. Objects in $\mathcal{W}$ are called **trivial**. Thus objects in $\mathcal{Q}_\mathcal{W}$ are called **trivially cofibrant** and objects in $\mathcal{R}_\mathcal{W}$ are **trivially fibrant**.

More generally, we call $\mathfrak{M}$ a **Hovey Ext-triple** if the pairs $(\mathcal{Q}_\mathcal{W}, \mathcal{R})$ and $(\mathcal{Q}, \mathcal{R}_\mathcal{W})$ are complete Ext-pairs but perhaps not actually cotorsion pairs. In this case we still use the terms *fibrant*, *cofibrant*, etc.

If $\mathfrak{M}$ is a Hovey triple, then by Corollary 2.5, each of $\mathcal{Q}$, $\mathcal{Q}_\mathcal{W}$, $\mathcal{R}$, and $\mathcal{R}_\mathcal{W}$ are closed under direct summands. It is less obvious that the class $\mathcal{W}$ of trivial objects is also automatically closed under direct summands. This is shown in Proposition 4.4. Along with the 2 out of 3 property, this makes $\mathcal{W}$ a **thick class** in $(\mathcal{A}, \mathcal{E})$. The key to proving this is the following easy characterization of $\mathcal{W}$.

Lemma 4.2 (Characterization of Trivial Objects) *Suppose $\mathfrak{M} = (\mathcal{Q}, \mathcal{W}, \mathcal{R})$ is any Hovey Ext-triple in $(\mathcal{A}, \mathcal{E})$. Then the class $\mathcal{W}$ of trivial objects is characterized in two ways as follows:*

$$\begin{aligned}\mathcal{W} &= \{ A \in \mathcal{A} \mid \exists \textit{ short exact sequence } A \rightarrowtail \tilde{R} \twoheadrightarrow \tilde{Q} \textit{ with } \tilde{R} \in \mathcal{R}_\mathcal{W}, \tilde{Q} \in \mathcal{Q}_\mathcal{W} \} \\ &= \{ A \in \mathcal{A} \mid \exists \textit{ short exact sequence } \tilde{R} \rightarrowtail \tilde{Q} \twoheadrightarrow A \textit{ with } \tilde{R} \in \mathcal{R}_\mathcal{W}, \tilde{Q} \in \mathcal{Q}_\mathcal{W} \}.\end{aligned}$$

Consequently, if $\mathfrak{M}$ is a Hovey triple, and $\mathfrak{M} = (\mathcal{Q}, \mathcal{V}, \mathcal{R})$ is another Hovey triple, then $\mathcal{V} = \mathcal{W}$.

The lemma is easy to prove. Before doing so, we note that it is basically saying that the trivial objects are precisely those whose fibrant approximations (or cofibrant approximations) are trivially fibrant (resp. trivially cofibrant). Cofibrant and fibrant approximation sequences are discussed in Section 4.4.

Proof ($\supseteq$) It is clear that each class described is contained in $\mathcal{W}$ since $\mathcal{W}$ satisfies the 2 out of 3 property on short exact sequences. ($\subseteq$) Let $W \in \mathcal{W}$. Then apply enough injectives of $(\mathcal{Q}, \mathcal{R}_\mathcal{W})$ to get a short exact sequence $W \rightarrowtail \tilde{R} \twoheadrightarrow \tilde{Q}$ where $\tilde{R} \in \mathcal{R}_\mathcal{W}$ and $\tilde{Q} \in \mathcal{Q}$. But indeed $\tilde{Q} \in \mathcal{Q} \cap \mathcal{W}$ since $\mathcal{W}$ satisfies the 2 out of 3 property. So W is in the top class described. On the other hand, W is also in the bottom class described by a similar argument using enough projectives of the cotorsion pair $(\mathcal{Q}_{\mathcal{W}}, \mathcal{R})$.

Now assume $\mathfrak{M}$ is a Hovey triple. Then if $(\mathcal{Q}, \mathcal{V}, \mathcal{R})$ is any other Hovey triple we must have $\mathcal{V} \cap \mathcal{R} = \mathcal{R}_{\mathcal{W}}$ (since each equal $\mathcal{Q}^{\perp}$) and $\mathcal{Q} \cap \mathcal{V} = \mathcal{Q}_{\mathcal{W}}$ (since each equal ${}^{\perp}\mathcal{R}$). So by what we just proved in the last paragraph it immediately follows that $\mathcal{V} = \mathcal{W}$. □

Remark 4.3 Since $\mathcal{W}$ is closed under extensions it naturally inherits an exact structure from $(\mathcal{A}, \mathcal{E})$; see Example 1.4.3. Lemma 4.2 implies that $(\mathcal{Q}_{\mathcal{W}}, \mathcal{R}_{\mathcal{W}})$ is a complete Ext-pair (resp. cotorsion pair) in $\mathcal{W}$ whenever $\mathfrak{M}$ is a Hovey Ext-triple (resp. Hovey triple).

The difference between Hovey triples and Hovey Ext-triples is clarified in the following statement.

Proposition 4.4 *Let $\mathfrak{M} = (\mathcal{Q}, \mathcal{W}, \mathcal{R})$ be any Hovey Ext-triple. Then the following are equivalent.*

(1) *$\mathfrak{M}$ is a Hovey triple.*
(2) *$\mathcal{Q}$, $\mathcal{R}$, $\mathcal{Q}_{\mathcal{W}}$, and $\mathcal{R}_{\mathcal{W}}$ are each closed under direct summands.*
(3) *$\mathcal{Q}$, $\mathcal{R}$, and $\mathcal{W}$ are each closed under direct summands.*

Proof The equivalence of (1) $\iff$ (2) follows from Proposition 2.4. The implication (3) $\implies$ (2) is trivial. We show (2) $\implies$ (3). So suppose $W \in \mathcal{W}$ has a direct sum decomposition $W = A \bigoplus B$. Using enough projectives we find short exact sequences $\tilde{R}_A \rightarrowtail Q_A \twoheadrightarrow A$ and $\tilde{R}_B \rightarrowtail Q_B \twoheadrightarrow B$ where $\tilde{R}_A, \tilde{R}_B \in \mathcal{R}_{\mathcal{W}}$ and $Q_A, Q_B \in \mathcal{Q}$. By Exercise 1.5.1 we get another short exact sequence

$$\tilde{R}_A \bigoplus \tilde{R}_B \rightarrowtail Q_A \bigoplus Q_B \twoheadrightarrow W$$

and still $\tilde{R}_A \bigoplus \tilde{R}_B \in \mathcal{R}_{\mathcal{W}}$ and $Q_A \bigoplus Q_B \in \mathcal{Q}$, since each class is closed under extensions. Thus $Q_A \bigoplus Q_B \in \mathcal{Q} \cap \mathcal{W} = \mathcal{Q}_{\mathcal{W}}$. So by hypothesis, Q_A and Q_B are each in $\mathcal{Q}_{\mathcal{W}}$ too. The short exact sequences $\tilde{R}_A \rightarrowtail Q_A \twoheadrightarrow A$ and $\tilde{R}_B \rightarrowtail Q_B \twoheadrightarrow B$ now prove that both A and B are each in $\mathcal{W}$, by Lemma 4.2. □

Corollary 4.5 *For any Hovey triple $\mathfrak{M} = (\mathcal{Q}, \mathcal{W}, \mathcal{R})$, it is automatic that $\mathcal{W}$ is closed under direct summands, and hence is a thick class.*

4.2 Fibrations, Cofibrations, and Weak Equivalences

Throughout this section, we let $\mathfrak{M} = (\mathcal{Q}, \mathcal{W}, \mathcal{R})$ denote a Hovey triple on $(\mathcal{A}, \mathcal{E})$. In fact, $\mathfrak{M}$ may even just be a Hovey Ext-triple in the sense of Definition 4.1.

Definition 4.6 Attached to $\mathfrak{M} = (\mathcal{Q}, \mathcal{W}, \mathcal{R})$, we define the following classes of morphisms.

(1) A morphism f is called a **cofibration**, or more precisely an $\mathfrak{M}$-**cofibration**, if it is an admissible monic with $\operatorname{Cok} f \in \mathcal{Q}$.
(2) A morphism f is called a **fibration**, or an $\mathfrak{M}$-**fibration**, if it is an admissible epic with $\operatorname{Ker} f \in \mathcal{R}$.
(3) A morphism f is a **weak equivalence**, or an $\mathfrak{M}$-**weak equivalence**, if it has a factorization $f = pi$ where i is a cofibration with $\operatorname{Cok} i \in \mathcal{Q}_{\mathcal{W}}$ and p is a fibration with $\operatorname{Ker} p \in \mathcal{R}_{\mathcal{W}}$.

There are some trivial observations to make. First, the admissible monic $0 \rightarrowtail Q$ is an $\mathfrak{M}$-cofibration if and only if $Q \in \mathcal{Q}$. This is consistent with our terminology that objects of $\mathcal{Q}$ are called cofibrant. Similarly, the admissible epic $R \twoheadrightarrow 0$ is an $\mathfrak{M}$-fibration if and only if $R \in \mathcal{R}$. Finally, one verifies with Lemma 4.2 that $0 \rightarrowtail W$ is an $\mathfrak{M}$-weak equivalence if and only if $W \in \mathcal{W}$ if and only if $W \twoheadrightarrow 0$ is an $\mathfrak{M}$-weak equivalence.

Proposition 4.7 *The following hold for* $\mathfrak{M} = (\mathcal{Q}, \mathcal{W}, \mathcal{R})$.

(1) *An admissible monic f is a $\mathfrak{M}$-weak equivalence if and only if* $\operatorname{Cok} f \in \mathcal{W}$. *We will call such morphisms* ***trivial monics***.
(2) *An admissible epic f is a $\mathfrak{M}$-weak equivalence if and only if* $\operatorname{Ker} f \in \mathcal{W}$. *We will call such morphisms* ***trivial epics***.

Proof The statements are dual, and we will prove the statement about admissible monics. First, suppose f is an admissible monic and it factors as $f = pi$ where i is some admissible monic with $\operatorname{Cok} i \in \mathcal{Q}_{\mathcal{W}}$ and p is some admissible epic with $\operatorname{Ker} p \in \mathcal{R}_{\mathcal{W}}$. By Lemma 1.20 we obtain a short exact sequence

$$\operatorname{Ker} p \rightarrowtail \operatorname{Cok} i \twoheadrightarrow \operatorname{Cok} f.$$

Since $\mathcal{W}$ satisfies the 2 out of 3 property on short exact sequences we get that $\operatorname{Cok} f \in \mathcal{W}$ too.

On the other hand, suppose $\operatorname{Cok} f \in \mathcal{W}$. Using Proposition 2.13, we obtain a factorization $f = pi$, where i is some admissible monic with $\operatorname{Cok} i \in \mathcal{Q}$ and p is some admissible epic with $\operatorname{Ker} p \in \mathcal{R}_{\mathcal{W}}$. Lemma 1.20 again provides a short exact sequence

$$\operatorname{Ker} p \rightarrowtail \operatorname{Cok} i \twoheadrightarrow \operatorname{Cok} f.$$

Again, since $\mathcal{W}$ satisfies the 2 out of 3 property on short exact sequences, we get that $\operatorname{Cok} i \in \mathcal{W}$. Thus $\operatorname{Cok} i \in \mathcal{Q}_{\mathcal{W}}$. □

Corollary 4.8 *The following hold for* $\mathfrak{M} = (\mathcal{Q}, \mathcal{W}, \mathcal{R})$.

(1) *A cofibration f is a $\mathfrak{M}$-weak equivalence if and only if* $\operatorname{Cok} f \in \mathcal{Q}_{\mathcal{W}}$. *Such morphisms are called* ***trivial cofibrations***.
(2) *A fibration f is a $\mathfrak{M}$-weak equivalence if and only if* $\operatorname{Ker} f \in \mathcal{R}_{\mathcal{W}}$. *Such morphisms are called* ***trivial fibrations***.

Note that the terminology is in agreement with the previous terminology we set: That the objects of $\mathcal{Q}_{\mathcal{W}}$ are called *trivially cofibrant* while the objects of $\mathcal{R}_{\mathcal{W}}$ are called *trivially fibrant*.

We now also obtain the following corollary by applying Corollary 1.15 and its dual.

Corollary 4.9 *The following hold for* $\mathfrak{M} = (\mathcal{Q}, \mathcal{W}, \mathcal{R})$.

(1) *The class of all cofibrations (resp. trivial cofibrations, resp. trivial monics) determines a subcategory of $\mathcal{A}$ containing all the objects of $\mathcal{A}$.*
(2) *The class of all fibrations (resp. trivial fibrations, resp. trivial epics) determines a subcategory of $\mathcal{A}$ containing all the objects of $\mathcal{A}$.*

Exercise 4.2.1 Let $(\mathcal{A}, \mathcal{E})$ be an exact category and let $\mathfrak{M} = (\mathcal{Q}, \mathcal{W}, \mathcal{R})$ be a triple of classes of objects in $\mathcal{A}$. Define the *opposite* of $\mathfrak{M}$ to be the triple defined by $\mathfrak{M}^{\text{op}} := (\mathcal{R}, \mathcal{W}, \mathcal{Q})$. Show that $\mathfrak{M}$ is a Hovey triple on $(\mathcal{A}, \mathcal{E})$ if and only if $\mathfrak{M}^{\text{op}}$ is a Hovey triple on $(\mathcal{A}^{\text{op}}, \mathcal{E}^{\text{op}})$, that is, on the opposite category with the opposite exact structure. Check that a (trivial) $\mathfrak{M}^{\text{op}}$-cofibration is precisely a (trivial) $\mathfrak{M}$-fibration, and vice versa. The weak equivalences of $\mathfrak{M}^{\text{op}}$ are the morphisms that equal an opposite of a weak equivalence of $\mathfrak{M}$.

4.3 Left and Right Homotopy and Stable Categories

Throughout this section, we again let $\mathfrak{M} = (\mathcal{Q}, \mathcal{W}, \mathcal{R})$ denote a given Hovey triple on $(\mathcal{A}, \mathcal{E})$, and again it may even just be a Hovey Ext-triple in the sense of Definition 4.1. Our goal here is to study the (very good) left and right homotopy relations and define the left and right stable category of $\mathfrak{M}$. First we introduce some further terminology.

Definition 4.10 We attach the following language to $\mathfrak{M}$.

- Objects in $\mathcal{Q} \cap \mathcal{R}$ are called **bifibrant**.
- We define the **core** of $\mathfrak{M}$ to be the class $\omega := \mathcal{Q} \cap \mathcal{W} \cap \mathcal{R}$. Objects in ω are called **trivially bifibrant**.
- We set $\mathfrak{L} = (\mathcal{Q}, \mathcal{R}_{\mathcal{W}})$ and call it the **left cotorsion pair**. Similarly, we set $\mathfrak{R} = (\mathcal{Q}_{\mathcal{W}}, \mathcal{R})$ and call it the **right cotorsion pair**.

In the general case that $\mathfrak{M}$ is just a Hovey Ext-triple, then we would call $\mathfrak{L}$ the **left Ext-pair** and $\mathfrak{R}$ the **right Ext-pair**.

We also define the following relations on the morphism sets $\mathrm{Hom}_{\mathcal{A}}(A, B)$. Let $f, g\colon A \to B$ be two morphisms in $\mathcal{A}$.

- We will say f and g are **homotopic**, written $f \sim g$, if $g - f$ factors through some trivial object $W \in \mathcal{W}$.
- We will say f and g are **ℓ-homotopic** (or **very good left homotopic**), written $f \sim^{\ell} g$, if $g - f$ factors through a trivially fibrant object $W \in \mathcal{R}_{\mathcal{W}}$.
- We will say f and g are **r-homotopic** (or **very good right homotopic**), written $f \sim^{r} g$, if $g - f$ factors through a trivially cofibrant object $W \in \mathcal{Q}_{\mathcal{W}}$.
- We will say f and g are **ω-homotopic**, written $f \sim^{\omega} g$, if $g - f$ factors through a trivially bifibrant object $W \in \omega$.

It may help one to remember the difference between the definitions of $\sim^{\ell}$ and $\sim^{r}$ as follows: Note that $\sim^{\ell}$ is defined in terms of morphisms factoring through the class of trivial objects appearing in the *left* pair $\mathfrak{L} = (\mathcal{Q}, \mathcal{R}_{\mathcal{W}})$. On the other hand, $\sim^{r}$ is defined in terms of morphisms factoring through the class of trivial objects appearing in the *right* pair $\mathfrak{R} = (\mathcal{Q}_{\mathcal{W}}, \mathcal{R})$.

It is easy to see, but also follows from Proposition 3.1, that $\sim$, $\sim^{\ell}$, $\sim^{r}$, and $\sim^{\omega}$ are each equivalence relations on $\mathrm{Hom}_{\mathcal{A}}(A, B)$. By the same proposition we may associate a stable category to each of these relations. We will denote them by $\mathrm{St}(\mathcal{A})$, $\mathrm{St}_{\ell}(\mathcal{A})$, $\mathrm{St}_{r}(\mathcal{A})$, and $\mathrm{St}_{\omega}(\mathcal{A})$ and respectively call them the **stable category of** $\mathfrak{M}$, the **left stable category of** $\mathfrak{M}$, the **right stable category of** $\mathfrak{M}$, and the **ω-stable category of** $\mathfrak{M}$. In each case, the objects are the same as $\mathcal{A}$, but the morphism sets are homotopy equivalence classes of maps, denoted by $[f]$, $[f]_{\ell}$, $[f]_{r}$, and $[f]_{\omega}$. Then $\underline{\mathrm{Hom}}(A, B)$, $\underline{\mathrm{Hom}}_{\ell}(A, B)$, $\underline{\mathrm{Hom}}_{r}(A, B)$, and $\underline{\mathrm{Hom}}_{\omega}(A, B)$ respectively denote the additive groups of morphisms from A to B. Finally, the canonical functors will be denoted by $\gamma\colon \mathcal{A} \to \mathrm{St}(\mathcal{A})$, $\gamma_{\ell}\colon \mathcal{A} \to \mathrm{St}_{\ell}(\mathcal{A})$, $\gamma_{r}\colon \mathcal{A} \to \mathrm{St}_{r}(\mathcal{A})$, and $\gamma_{\omega}\colon \mathcal{A} \to \mathrm{St}_{\omega}(\mathcal{A})$.

We clearly have the implications $f \sim^{\omega} g \implies f \sim^{\ell} g \implies f \sim g$ and $f \sim^{\omega} g \implies f \sim^{r} g \implies f \sim g$. The following provides instances for the converses to hold.

Lemma 4.11 *Let $f, g\colon A \to B$ be two morphisms in $\mathcal{A}$.*

(1) *Assume the domain A is cofibrant. Then $f \sim g$ if and only if $f \sim^{r} g$, and $f \sim^{\ell} g$ if and only if $f \sim^{\omega} g$. In particular, we have equality of the hom-sets $\underline{\mathrm{Hom}}(A, B) = \underline{\mathrm{Hom}}_{r}(A, B)$ and $\underline{\mathrm{Hom}}_{\ell}(A, B) = \underline{\mathrm{Hom}}_{\omega}(A, B)$ whenever A is cofibrant.*

(2) *Assume the codomain B is fibrant. Then $f \sim g$ if and only if $f \sim^{\ell} g$, and $f \sim^{r} g$ if and only if $f \sim^{\omega} g$. In particular, we have equality of the hom-sets*

$\underline{\mathrm{Hom}}(A,B) = \underline{\mathrm{Hom}}_{\ell}(A,B)$ and $\underline{\mathrm{Hom}}_{r}(A,B) = \underline{\mathrm{Hom}}_{\omega}(A,B)$ whenever B is fibrant.

(3) *Assume the domain A is cofibrant and the codomain B is fibrant. Then $\sim = \sim^r = \sim^\ell = \sim^\omega$ and we may denote the common hom-sets simply by $\underline{\mathrm{Hom}}(A,B)$.*

Proof The first two statements are dual and we will prove the first one. So suppose that A is cofibrant and $f, g\colon A \to B$ are homotopic (resp. ℓ-homotopic), so that $g - f$ factors through some object $W \in \mathcal{W}$ (resp. $W \in \mathcal{R}_{\mathcal{W}}$). Using enough projectives, write a short exact sequence $K \rightarrowtail QW \twoheadrightarrow W$ where $QW \in \mathcal{Q}$ and $K \in \mathcal{R}_{\mathcal{W}}$. Since $\mathcal{W}$ (resp. $\mathcal{R}_{\mathcal{W}}$) is closed under extensions we see $QW \in \mathcal{Q}_{\mathcal{W}}$ (resp. $QW \in \omega := \mathcal{Q} \cap \mathcal{R} \cap \mathcal{W}$). Then since A is cofibrant, the map $A \to W$ in the factorization of $g - f$, lifts over $QW \twoheadrightarrow W$. Thus the factorization of $g - f$ extends to a factorization of $g - f$ through QW, showing $f \sim^r g$ (resp. $f \sim^\omega g$).

The third statement clearly follows from the first two statements. □

Definition 4.12 Let $f\colon A \to B$ be a morphism.

(1) We say f is a **homotopy equivalence**, if it represents an isomorphism in $\mathrm{St}(\mathcal{A})$. That is, there exists a morphism $g\colon B \to A$ such that $gf \sim 1_A$ and $fg \sim 1_B$.
(2) We say f is a **right stable equivalence**, or an r**-homotopy equivalence**, if it represents an isomorphism in $\mathrm{St}_r(\mathcal{A})$. That is, there exists a morphism $g\colon B \to A$ such that $gf \sim^r 1_A$ and $fg \sim^r 1_B$.
(3) We say f is a **left stable equivalence**, or an ℓ**-homotopy equivalence**, if it represents an isomorphism in $\mathrm{St}_\ell(\mathcal{A})$. That is, there exists a morphism $g\colon B \to A$ such that $gf \sim^\ell 1_A$ and $fg \sim^\ell 1_B$.
(4) We say f is an ω**-stable equivalence**, or ω**-homotopy equivalence**, if it represents an isomorphism in $\mathrm{St}_\omega(\mathcal{A})$. That is, there exists a morphism $g\colon B \to A$ such that $gf \sim^\omega 1_A$ and $fg \sim^\omega 1_B$.

Lemma 4.13 *We have the following conditions guaranteeing that a morphism represents an isomorphism in one of the stable categories attached to $\mathfrak{M}$.*

(1) *If A is fibrant and $j\colon A \rightarrowtail B$ is a trivial cofibration, then j is a right stable equivalence. If in addition, B is also fibrant, then j is an ω-stable equivalence.*
(2) *If B is cofibrant and $q\colon A \twoheadrightarrow B$ is a trivial fibration, then q is a left stable equivalence. If in addition, A is also cofibrant, then q is an ω-stable equivalence.*

(3) *If A and B are each bifbrant and $f\colon A \to B$ is any weak equivalence, then f is a homotopy equivalence, equivalently, an ω-stable equivalence.*

Proof (1) Suppose A is fibrant and $j\colon A \to B$ is a trivial cofibration. Then we have a short exact sequence $A \overset{j}{\rightarrowtail} B \overset{q}{\twoheadrightarrow} Q$, where $Q \in \mathcal{Q}_{\mathcal{W}}$. Since $\mathrm{Ext}^1_{\mathcal{E}}(Q, A) = 0$, it follows that the sequence splits. That is, we have maps $B \xrightarrow{t} A$ and $Q \xrightarrow{s} B$ satisfying (i) $tj = 1_A$, (ii) $qs = 1_Q$, and (iii) $sq + jt = 1_B$. By (iii), we have $1_B - jt = sq$ factors through Q, a trivially cofibrant object. So $jt \sim^r 1_B$. Along with (i), this proves $[j]_r$ is an isomorphism in $\mathrm{St}_r(\mathcal{A})$ with inverse $[t]_r$, proving the first part of statement (1). In the case that B is fibrant, then $jt \sim^r 1_B$ is equivalent to $jt \sim^\omega 1_B$, by Lemma 4.11. So in this case, $[j]_\omega$ is an isomorphism in $\mathrm{St}_\omega(\mathcal{A})$ with inverse $[t]_\omega$.

The proof of (2) is similar to (1). To prove (3), suppose A and B are each bifibrant and $A \xrightarrow{f} B$ is a weak equivalence. We factor $f = qj$ into a trivial cofibration j with cokernel $Q \in \mathcal{Q}_{\mathcal{W}}$, followed by a trivial fibration q with kernel $R \in \mathcal{R}_{\mathcal{W}}$ as shown:

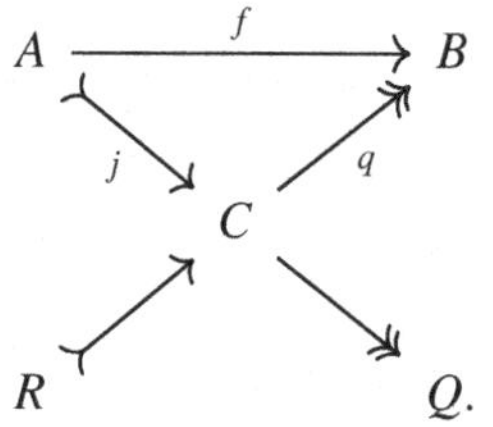

Note that C must also be bifibrant. It follows from (1) that j is an ω-stable equivalence and from (2) that q is, too. Hence the composition f must also be an ω-stable equivalence. This proves (3), for by Lemma 4.11, an ω-stable equivalence is nothing more than a homotopy equivalence whenever A and B are each bifibrant. □

The following theorem will not be used in the sequel, but is included for completeness. It clarifies the connection between our relations, $\sim$, $\sim^r$, $\sim^\ell$, and $\sim^\omega$, and the standard notion of right (and left) homotopic maps from the general theory of model categories. Its proof, as well as the standard model category language that is referenced, is deferred to Appendix B.

Theorem 4.14 *Assume $\mathcal{A}$ is weakly idempotent complete and let $f, g\colon A \to B$ be any morphisms.*

(1) *f is formally right homotopic to g if and only if $f \sim g$ in the sense of Definition 4.10. Dually, f is formally left homotopic to g if and only if*

$f \sim g$. In particular, regardless of the domain A and codomain B, the formal right and left homotopy relations coincide and this common relation is always an equivalence relation. This justifies why we simply call such maps ***homotopic*** *and call $\sim$ the* ***homotopy relation****.*

(2) *There is a very good right homotopy from f to g if and only if $f \sim^r g$ in the sense of Definition 4.10. Thus we again obtain an equivalence relation on all hom-sets called the* ***very good right homotopy relation****. Dually, there is a very good left homotopy from f to g if and only if $f \sim^\ell g$ in the sense of Definition 4.10. Thus we call this equivalence relation the* ***very good left homotopy relation****.*

(3) *Whenever the domain A is cofibrant we have $\sim\, = \,\sim^r$ and $\sim^\ell\, = \,\sim^\omega$. Dually, $\sim\, = \,\sim^\ell$ and $\sim^r\, = \,\sim^\omega$ whenever the codomain B is fibrant.*

(4) *Suppose the domain A is cofibrant and the codomain B is fibrant. Then the equivalence relations $\sim\, = \,\sim^r\, = \,\sim^\ell\, = \,\sim^\omega$ all coincide.*

4.4 Cofibrant and Fibrant Approximations

Once again, $\mathfrak{M} = (Q, \mathcal{W}, \mathcal{R})$ will denote a Hovey triple throughout this entire section. And again, the reader may relax $\mathfrak{M}$ to just be a Hovey Ext-triple in the sense of Definition 4.1.

Using completeness of the left pair $\mathfrak{L} = (Q, \mathcal{R}_\mathcal{W})$ we can find, for any object A, a Q-approximation sequence

$$R_A \overset{i_A}{\rightarrowtail} QA \overset{p_A}{\twoheadrightarrow} A$$

with $QA \in Q$ and $R_A \in \mathcal{R}_\mathcal{W}$. We will call any such sequence a **cofibrant approximation sequence (of A)**. On the other hand, using completeness of the right pair $\mathfrak{R} = (Q_\mathcal{W}, \mathcal{R})$ we can find $RA \in \mathcal{R}$ and $Q_A \in Q_\mathcal{W}$ fitting into a **fibrant approximation sequence**

$$A \overset{j_A}{\rightarrowtail} RA \overset{q_A}{\twoheadrightarrow} Q_A.$$

Note that in the language of Definition 2.2, any such p_A is a special Q-precover while j_A is a special $\mathcal{R}$-preenvelope.

Although cofibrant and fibrant approximation sequences are not unique, they are unique in $\mathrm{St}_\omega(\mathcal{A})$ up to a canonical isomorphism. This is all a special case of results from Sections 3.2 and 3.3; in particular, see Definition 3.5 and Remark 3.6.

Definition 4.15 $\mathfrak{M} = (\mathcal{Q}, \mathcal{W}, \mathcal{R})$ induces the following endofunctors on the ω-stable category $\mathrm{St}_\omega(\mathcal{A})$. (In fact, the domains and codomains of the functors may be modified as described in Exercise 4.4.1.)

(1) The process of taking cofibrant approximation sequences with the left pair $\mathfrak{L} = (\mathcal{Q}, \mathcal{R}_{\mathcal{W}})$ defines a functor $\mathcal{A} \xrightarrow{Q} \mathrm{St}_\omega(\mathcal{Q})$ descending via $\mathcal{A} \xrightarrow{\gamma_\omega} \mathrm{St}_\omega(\mathcal{A})$ to a functor $\mathrm{St}_\omega(\mathcal{A}) \xrightarrow{Q} \mathrm{St}_\omega(\mathcal{Q})$. This is called the **cofibrant approximation functor**. The value of Q on a representative morphism $f\colon A \to B$ is computed by constructing any morphism of cofibrant approximation sequences:

$$\begin{array}{lccccc}
\mathbb{Q}_A: & R_A & \overset{i_A}{\rightarrowtail} & QA & \overset{p_A}{\twoheadrightarrow} & A \\
 & \downarrow & & \downarrow{\scriptstyle Q(f)} & & \downarrow{\scriptstyle f} \\
\mathbb{Q}_B: & R_B & \overset{i_A}{\rightarrowtail} & QB & \overset{p_B}{\twoheadrightarrow} & B.
\end{array} \qquad (\dagger)$$

That is, $Q([f]_\omega) = [Q(f)]_\omega$ is a well-defined endofunctor on $\mathrm{St}_\omega(\mathcal{A})$ and there is an associated natural transformation $Q \xrightarrow{\{[p_A]\}} 1_{\mathrm{St}_\omega(\mathcal{A})}$, called the **cofibrant approximation transformation.**

(2) The process of taking fibrant approximation sequences with the right pair $\mathfrak{R} = (\mathcal{Q}_{\mathcal{W}}, \mathcal{R})$ defines a functor $\mathcal{A} \xrightarrow{R} \mathrm{St}_\omega(\mathcal{R})$ descending via $\mathcal{A} \xrightarrow{\gamma_\omega} \mathrm{St}_\omega(\mathcal{A})$ to a functor $\mathrm{St}_\omega(\mathcal{A}) \xrightarrow{R} \mathrm{St}_\omega(\mathcal{R})$. This is called the **fibrant approximation functor**. The value of R on a representative morphism $f\colon A \to B$ is computed by constructing any morphism of fibrant approximation sequences:

$$\begin{array}{lccccc}
\mathbb{R}_A: & A & \overset{j_A}{\rightarrowtail} & RA & \overset{q_A}{\twoheadrightarrow} & Q_A \\
 & \downarrow{\scriptstyle f} & & \downarrow{\scriptstyle R(f)} & & \downarrow \\
\mathbb{R}_B: & B & \overset{j_B}{\rightarrowtail} & RB & \overset{q_B}{\twoheadrightarrow} & Q_B.
\end{array} \qquad (\dagger\dagger)$$

That is, $R([f]_\omega) = [R(f)]_\omega$ is a well-defined endofunctor on $\mathrm{St}_\omega(\mathcal{A})$ and there is an associated natural transformation $1_{\mathrm{St}_\omega(\mathcal{A})} \xrightarrow{\{[j_A]\}} R$, called the **fibrant approximation transformation.**

Remark 4.16 We use the term **(co)fibrant replacement** synonymously with *(co)fibrant approximation.* We also may sometimes refer to $Q(f)$ in Diagram $(\dagger)$ as a *cofibrant approximation of* f and similarly $R(f)$ in Diagram $(\dagger\dagger)$ as a *fibrant approximation of* f. Although it is convenient to denote these maps by $Q(f)$ and $R(f)$, we are not asserting for example that $f \mapsto Q(f)$ is a functor. $Q(f)$ only represents some (or any) morphism making the diagram commute. The point is that, independent of this choice, the association $f \mapsto [Q(f)]_\omega$ is a well-defined functor. Indeed even $[f]_\omega \mapsto Q[f]_\omega := [Q(f)]_\omega$ is well defined.

Exercise 4.4.1 Referring to Section 3.3, verify that we may modify the domains and codomains for each of the functors Q and R as follows. (We are using superscripts to indicate the domains and subscripts for codomains.)

(1) Q modifies to an endofunctor $Q^r_r\colon \mathrm{St}_r(\mathcal{A}) \to \mathrm{St}_r(\mathcal{A})$.

(2) Q modifies to a functor $Q^\ell_\omega\colon \mathrm{St}_\ell(\mathcal{A}) \to \mathrm{St}_\omega(\mathcal{A})$ and consequently to an endofunctor $Q^\ell_\ell\colon \mathrm{St}_\ell(\mathcal{A}) \to \mathrm{St}_\ell(\mathcal{A})$.

(3) R modifies to an endofunctor $R^\ell_\ell\colon \mathrm{St}_\ell(\mathcal{A}) \to \mathrm{St}_\ell(\mathcal{A})$.

(4) R modifies to a functor $R^r_\omega\colon \mathrm{St}_r(\mathcal{A}) \to \mathrm{St}_\omega(\mathcal{A})$ and consequently to an endofunctor $R^r_r\colon \mathrm{St}_r(\mathcal{A}) \to \mathrm{St}_r(\mathcal{A})$.

With this notation, the compatibilities of the functor Q in Definition 4.15 may be expressed as $Q_\omega = Q^\omega_\omega \circ \gamma_\omega$, and similarly, we have $Q_r = Q^r_r \circ \gamma_r$, $Q_\omega = Q^\ell_\omega \circ \gamma_\ell$, etc. We also have all of the dual statements for R.

Exercise 4.4.1 shows in particular that Q and R are both endofunctors on each of $\mathrm{St}_\omega(\mathcal{A})$, $\mathrm{St}_r(\mathcal{A})$, and $\mathrm{St}_\ell(\mathcal{A})$. In fact, the next exercise asks us to show that they are even endofunctors on the stable category $\mathrm{St}(\mathcal{A}) := \mathrm{St}_{\mathcal{W}}(\mathcal{A})$.

Exercise 4.4.2 Show that Q modifies to a functor $Q^{\mathcal{W}}_r\colon \mathrm{St}(\mathcal{A}) \to \mathrm{St}_r(\mathcal{A})$ and consequently to an endofunctor on $\mathrm{St}(\mathcal{A})$. Dually, we have $R^{\mathcal{W}}_\ell\colon \mathrm{St}(\mathcal{A}) \to \mathrm{St}_\ell(\mathcal{A})$ and consequently an endofunctor on $\mathrm{St}(\mathcal{A})$.

4.5 Trivial Morphisms and the 2 out of 3 Axiom

We continue to let $\mathfrak{M} = (\mathcal{Q}, \mathcal{W}, \mathcal{R})$ denote a Hovey triple on $(\mathcal{A}, \mathcal{E})$, or even a Hovey Ext-triple in the sense of Definition 4.1. In this section we examine the 2 out of 3 Axiom on $\mathfrak{M}$-weak equivalences, reducing what is necessary to verify the axiom. Using this we will then see that the 2 out and 3 Axiom is *automatic* whenever $\mathcal{A}$ is weakly idempotent complete. Throughout this section, $\mathcal{A}$ need not be weakly idempotent complete unless explicitly stated otherwise.

Recall that, by definition, an $\mathfrak{M}$-weak equivalence is a morphism which factors as a trivial cofibration followed by a trivial fibration. Consider the following condition on the class of $\mathfrak{M}$-weak equivalences.

(2 out of 3 Axiom) Let f and g be morphisms in $\mathcal{A}$ such that the composition gf is defined. Whenever two out of three of f, g, and gf are $\mathfrak{M}$-weak equivalences, then the third morphism is also an $\mathfrak{M}$-weak equivalence.

We will define the notion of a *trivial morphism* and show that the 2 out of 3 Axiom is equivalent to the statement that all trivial morphisms are $\mathfrak{M}$-weak equivalences. This is the condition that is automatic whenever the underlying category $\mathcal{A}$ is weakly idempotent complete. But this approach allows one to broaden the theory so that an arbitrary Frobenius category may be considered to have a "model structure" $\mathfrak{M} = (Q, \mathcal{W}, \mathcal{R}) = (\mathcal{A}, \mathcal{W}, \mathcal{A})$ where $\mathcal{W}$ is the class of all projective–injective objects. See Appendix A.

A morphism $f\colon A \to B$ in $(\mathcal{A}, \mathcal{E})$ is called an **admissible morphism** (or a **strict morphism**) if it factors as a composition $f = me$ where m is an admissible monic and e is an admissible epic. If such a factorization exists, it is unique in the sense that if there is another such factorization $f = m'e'$, then there is a unique isomorphism α making both triangles of the diagram commute:

$$\begin{array}{ccc} A & \xrightarrow{e'} & I' \\ {\scriptstyle e}\downarrow & \overset{\alpha}{\nearrow} & \downarrow{\scriptstyle m'} \\ I & \xrightarrow[m]{} & B. \end{array}$$

Any admissible morphism f has a kernel and a cokernel. In fact, $\ker f = \ker e$ and $\operatorname{cok} f = \operatorname{cok} m$. So the morphism $m\colon I \to B$ is the *image* of f and the morphism $e\colon A \to I$ is the *coimage* of f.

Exercise 4.5.1 Verify each of these claims, in particular the following.

(1) The factorization of an admissible morphism $f = me$ is unique up to a canonical isomorphism α making the triangles commute.
(2) $\ker f = \ker e$ and $\operatorname{cok} f = \operatorname{cok} m$.

Recall that a trivial monic is an admissble monic with cokernel in $\mathcal{W}$ and a trivial epic is the dual notion; see Proposition 4.7. These are special types of trivial morphisms in the sense defined in the following. Note that the definition depends only on the exact structure $\mathcal{E}$ and the class $\mathcal{W}$ of trivial objects.

Definition 4.17 We call a morphism $f\colon A \to B$ in $\mathcal{A}$ a **trivial morphism** if it factors as a composition $f = me$ where m is a trivial monic and e is a trivial epic.

Thus a trivial morphism is precisely an admissible morphism with $\operatorname{Ker} f \in \mathcal{W}$ and $\operatorname{Cok} f \in \mathcal{W}$. We now show that $\mathfrak{M} = (Q, \mathcal{W}, \mathcal{R})$ satisfies the 2 out of 3 Axiom if and only if every trivial morphism is an $\mathfrak{M}$-weak equivalence. The proof will rely on the following series of lemmas.

Lemma 4.18 *Assume that every trivial morphism is an $\mathfrak{M}$-weak equivalence. Then the class of $\mathfrak{M}$-weak equivalences is closed under composition, and so determines a subcategory of $\mathcal{A}$ containing all the objects of $\mathcal{A}$.*

Proof Let gf be a composition of two $\mathfrak{M}$-weak equivalences. This means we have factorizations $f = pi$ and $g = qj$ where i, j are trivial cofibrations and p, q are trivial fibrations. Then the composition jp is defined and by definition is a trivial morphism. So by hypothesis we have $jp = \beta\alpha$ where α is a trivial cofibration and β is a trivial fibration. So now we have $gf = q(jp)i = (q\beta)(\alpha i)$. But trivial fibrations are closed under composition, by Corollary 4.9. So $q\beta$ is a trivial fibration, and similarly αi is a trivial cofibration. Thus gf is a $\mathfrak{M}$-weak equivalence in this case. □

The following lemma concerns further special cases of the 2 out of 3 property.

Lemma 4.19 *Assume that every trivial morphism is an $\mathfrak{M}$-weak equivalence and let*

$$h = gf : A \xrightarrow{f} B \xrightarrow{g} C$$

be a composition of morphisms. Then we have the following special cases of the 2 out of 3 Axiom.

(1) *If h and f are trivial cofibrations then g is a weak equivalence.*
(2) *If h and g are trivial fibrations then f is a weak equivalence.*
(3) *If h is a trivial cofibration and g is a trivial fibration then f is a weak equivalence.*
(4) *If h is a trivial fibration and f is a trivial cofibration then g is a weak equivalence.*
(5) *If h and g are trivial cofibrations then f is a weak equivalence.*
(6) *If h and f are trivial fibrations then g is a weak equivalence.*

Proof First we prove (1) and (2). Using Proposition 2.13 (Factorizations), we factor g as qj where j is a trivial cofibration and q is a fibration. We must show that $\operatorname{Ker} q \in \mathcal{W}$. We now have $h = q(jf)$, and jf is a composition of two trivial cofibrations and so itself is a trivial cofibration, by Corollary 4.9. By Lemma 1.20 we get a short exact sequence

$$\operatorname{Ker} q \rightarrowtail \operatorname{Cok}(jf) \twoheadrightarrow \operatorname{Cok} h,$$

and since $\operatorname{Cok}(jf) \in \mathcal{W}$ and $\operatorname{Cok} h \in \mathcal{W}$ we get $\operatorname{Ker} q \in \mathcal{W}$ by the 2 out of 3 property on $\mathcal{W}$. This proves (1) and statement (2) is dual.

Next we prove (3) and (4). To prove (3), using Proposition 2.13 (Factorizations) we factor f as pi where i is a cofibration and p is a trivial fibration.

The composition gp is then a trivial fibration. We then apply Lemma 1.20 to $h = (gp)i$ to get a short exact sequence

$$\operatorname{Ker}(gp) \rightarrowtail \operatorname{Cok} i \twoheadrightarrow \operatorname{Cok} h.$$

Since $\mathcal{W}$ satisfies the 2 out of 3 property on short exact sequence, and since $\operatorname{Ker}(gp), \operatorname{Cok} h \in \mathcal{W}$, we conclude $\operatorname{Cok} i \in \mathcal{W}$. This proves i is a trivial cofibration and hence f is a weak equivalence. The proof of (4) is dual.

Finally we prove (5) and (6). To prove (5), we again start by using Proposition 2.13 (Factorizations), to factor f as $f = pi$ where i is a cofibration and p is a trivial fibration. We need to show that i is in fact a trivial cofibration. We have $h = (gp)i$, and we note gp is a trivial morphism So by hypothesis we can replace $gp = qj$, where j is a trivial cofibration and q is a trivial fibration. Thus $h = (gp)i = q(ji)$, and ji is certainly a cofibration by Corollary 4.9. By what we just proved in (3) we now may conclude that ji is a weak equivalence, so necessarily a trivial cofibration by Corollary 4.8. But now (the dual of) Corollary 1.15 applies to the composition ji, and the corresponding short exact sequence implies that $\operatorname{Cok} i$ is trivial. Thus i is a trivial cofibration. The proof of (6) is dual. □

Theorem 4.20 *The following statements are equivalent for* $\mathfrak{M} = (\mathcal{Q}, \mathcal{W}, \mathcal{R})$.

(1) *The class of* $\mathfrak{M}$*-weak equivalences satisfies the 2 out of 3 Axiom.*
(2) *The class of* $\mathfrak{M}$*-weak equivalences is closed under compositions.*
(3) *Each trivial morphism is an* $\mathfrak{M}$*-weak equivalence.*

Proof (1) $\implies$ (2) $\implies$ (3) is easy, for if $f = me$ is a trivial morphism, then it is already the composition of two weak equivalences by Proposition 4.7. Now (3) $\implies$ (2) was proven in Lemma 4.18, so it is left to show (2) $\implies$ (1). So let $A \xrightarrow{f} B \xrightarrow{g} C$ be a composition of morphisms. We will prove that if gf and g are weak equivalences, then f also is a weak equivalence. (The other statement is dual.) Note that by factorizing f as a cofibration followed by a trivial fibration we may assume that f is a cofibration. So assume that f is a cofibration, and we will argue that it must be a trivial cofibration.

We begin by factoring $g = pi$ where i is a trivial cofibration and p is a trivial fibration. Then we may factor $if = qj$ where j is a cofibration and q is a trivial fibration. We have a commutative diagram:

$$\begin{array}{ccccc} A & \xrightarrow{f} & B & \xrightarrow{g} & C \\ {\scriptstyle j}\downarrow & & \downarrow{\scriptstyle i} & & \| \\ A' & \xrightarrow{q} & B' & \xrightarrow{p} & C. \end{array}$$

Now gf may be factored as $gf = rk$ where k is a trivial cofibration and r is a trivial fibration. Thus the outer square of the following diagram commutes, and since $\mathrm{Ext}^1_{\mathcal{E}}(\mathrm{Cok}\, j, \mathrm{Ker}\, r) = 0$, the Lifting-Extension Lemma 2.9 provides a map α making both triangles of the diagram commute:

$$\begin{array}{ccc} A & \overset{k}{\rightarrowtail} & A'' \\ {\scriptstyle j}\downarrow & \overset{\alpha}{\nearrow} & \downarrow{\scriptstyle r} \\ A' & \underset{pq}{\twoheadrightarrow} & C. \end{array}$$

By applying Lemma 4.19 we make the following series of observations which result in proof that if is a trivial cofibration.

- With $pq = r\alpha$ and pq, r being trivial fibrations we conclude α is a weak equivalence (Lemma 4.19(2)). So $\alpha = sl$, where l is a trivial cofibration and s is a trivial fibration.
- $\alpha j = k \implies (sl)j = k \implies s(lj) = k$ and s being a trivial fibration and k being a trivial cofibration implies lj is a weak equivalence (Lemma 4.19(3)). This implies lj is a trivial cofibration since it is already the composition of two cofibrations (Corollaries 4.8 and 4.9). But then by Lemma 4.19(5) we conclude that j is a weak equivalence. Since j too is a cofibration this means j is a trivial cofibration.
- So this proves that $if = qj$ is a weak equivalence since j is a trivial cofibration and q is a trivial fibration. But if is already a cofibration, so if is a trivial cofibration. (Again Corollaries 4.8 and 4.9.)

Finally, since if and i are trivial cofibrations, we conclude f is a weak equivalence by Lemma 4.19(5). But again f is already a cofibration, so f is a trivial cofibration. □

As promised, we now show that the 2 out of 3 Axiom on $\mathfrak{M}$-weak equivalences is automatic whenever $\mathcal{A}$ is weakly idempotent complete. Recall that weak idempotent completeness just means that every monomorphism possessing a left inverse in $\mathcal{A}$ must have a cokernel. Or equivalently, that every epimorphism possessing a right inverse in $\mathcal{A}$ must have a kernel; see Section 1.8.

Theorem 4.21 *Assume the underlying category $\mathcal{A}$ is weakly idempotent complete. Then the class of $\mathfrak{M}$-weak equivalences associated to any Hovey triple $\mathfrak{M} = (\mathcal{Q}, \mathcal{W}, \mathcal{R})$ automatically satisfies the 2 out of 3 Axiom. The same is true if $\mathfrak{M}$ is just a Hovey Ext-triple.*

Proof We show that any trivial morphism is a weak equivalence. So let f be a trivial morphism, meaning that it factors as $f = me$, a trivial monic m composed with a trivial epic e. Using Proposition 2.13 (Factorizations), we may factor this as $me = \beta\alpha$ where α is a cofibration and β is a trivial fibration. We will now prove that $\operatorname{Cok}\alpha \in \mathcal{W}$. Indeed we construct the following commutative diagram, where r must be an admissible epic by the weak idempotent completeness assumption; see Proposition 1.30:

$$\begin{array}{ccccc} B & \overset{\alpha}{\rightarrowtail} & X & \overset{c}{\twoheadrightarrow} & \operatorname{Cok}\alpha \\ {\scriptstyle e}\downarrow\!\!\!\downarrow & & \downarrow\!\!\!\downarrow{\scriptstyle \beta} & & \downarrow\!\!\!\downarrow{\scriptstyle r} \\ C & \underset{m}{\rightarrowtail} & D & \twoheadrightarrow & \operatorname{Cok} m. \end{array}$$

By the (3×3)-Lemma 1.16 we get a short exact sequence

$$\operatorname{Ker} e \rightarrowtail \operatorname{Ker}\beta \twoheadrightarrow \operatorname{Ker} r.$$

Since $\mathcal{W}$ satisfies the 2 out of 3 property on short exact sequences, we conclude $\operatorname{Ker} r \in \mathcal{W}$, and hence $\operatorname{Cok}\alpha \in \mathcal{W}$. This proves α is a trivial cofibration. □

A fundamental consequence of the 2 out of 3 Axiom is that the class of $\mathfrak{M}$-weak equivalences can be characterized in terms of only $\mathcal{W}$, without mention of Q and $\mathcal{R}$. In particular we have the following useful statement.

Proposition 4.22 *Assume $\mathfrak{M} = (Q, \mathcal{W}, \mathcal{R})$ satisfies the 2 out of 3 Axiom on $\mathfrak{M}$-weak equivalences; for instance this is always the case whenever $\mathcal{A}$ is weakly idempotent complete. Then a morphism f is an $\mathfrak{M}$-weak equivalence if and only if f factors as a trivial monic followed by a trivial epic. That is, if f admits a factorization $f = st$, where t is a trivial monic and s is a trivial epic.*

Note that the order of the composition in Proposition 4.22 is reversed from the case that f is a trivial morphism.

Proof Indeed, by the very definition we certainly have that any $\mathfrak{M}$-weak equivalence f has a factorization $f = st$, where t is a trivial cofibration (so trivial monic) and s is a trivial fibration (so trivial epic). On the other hand, by Proposition 4.7 we know that all trivial monics and trivial epics are $\mathfrak{M}$-weak equivalences. So any morphism with such a factorization $f = st$ is an $\mathfrak{M}$-weak equivalence, by the 2 out of 3 Axiom (see Lemma 4.18 in particular). □

4.6 Hovey's Correspondence

The point of this section is to show that, when $(\mathcal{A}, \mathcal{E})$ is a weakly idempotent complete exact catgory, a Hovey triple $\mathfrak{M} = (Q, \mathcal{W}, \mathcal{R})$ is equivalent to an abelian model structure. It means $\mathfrak{M}$ is equivalent to a Quillen model structure that is compatible with the exact structure $\mathcal{E}$. This equivalence is known as *Hovey's correspondence*. In fact, Hovey's correspondence generalizes to a bijective correspondence between Hovey Ext-triples and (abelian) model structures in Quillen's original sense, that they are not necessarily *closed*. Moreover, Appendix A gives a generalization of Hovey's correspondence that works for an arbitrary exact category $(\mathcal{A}, \mathcal{E})$.

Let us start by giving the definition of a model category, which corresponds to what Quillen called a *closed* model category.

Definition 4.23 (Quillen [1969]) A **(closed) model category** is a category C along with a triple $\mathcal{M} = (Cof, We, Fib)$ of classes of morphisms in C, respectively called *cofibrations*, *weak equivalences*, and *fibrations*, all satisfying the axioms that follow. By definition, we call a map a *trivial cofibration* if it is both a cofibration and a weak equivalence. Similarly a *trivial fibration* is both a fibration and a weak equivalence.

(M0) (Finite Limits and Colimits Axiom) All finite limits and colimits exist in C.

(M1) (Subcategories Axiom) Each of *Cof*, *We*, and *Fib* are closed under compositions and contain all isomorphisms in C. In particular, each is a subcategory containing all objects of C.

(M2) (Lifting Axiom) Assume we have a commutative diagram as follows where i is a cofibration and p is a fibration:

$$\begin{array}{ccc} A & \longrightarrow & X \\ {\scriptstyle i}\downarrow & \nearrow & \downarrow{\scriptstyle p} \\ B & \longrightarrow & Y. \end{array}$$

Then the dashed arrow exists, making both triangles commute, whenever i or p is also a weak equivalence. (That is, whenever i is a trivial cofibration or p is a trivial fibration.)

(M3) (Factorization Axiom) Any map f may be factored as $f = pi$ in two ways. First, where p is a fibration and i is a trivial cofibration. Second, where p is a trivial fibration and i is a cofibration.

(M4) (2 out of 3 Axiom) Let gf be a composition of two morphisms in C. If two out of three of the morphisms f, g, and gf are a weak equivalence then so is the third.

(M5) (Retracts Axiom) The class of fibrations, cofibrations, and weak equivalences are each closed under retracts. By a retract of a morphism f we mean a morphism g for which there is a commutative diagram

$$\begin{array}{ccccc} C & \longrightarrow & A & \longrightarrow & C \\ \downarrow g & & \downarrow f & & \downarrow g \\ D & \longrightarrow & B & \longrightarrow & D \end{array}$$

in which the horizontal compositions are both the identity.

Note that axiom (M0) is an assumption on the category C and does not have anything to do with the class of cofibrations, fibrations, or weak equivalences. For this reason we say that the triple $\mathcal{M} = (Cof, We, Fib)$ is a **(closed) model structure** on C if it satisfies (M1)–(M5). So technically we think of a model category as a category C possessing all finite limits and colimits, along with a model structure.

Now suppose $\mathcal{M} = (Cof, We, Fib)$ is a model structure on an exact category $(\mathcal{A}, \mathcal{E})$. We set the following terminology for objects $A \in \mathcal{A}$.

- We say A is *trivial* if the admissible monic $0 \rightarrowtail A$ is a weak equivalence. By the 2 out of 3 Axiom, and considering the composition $0 \rightarrowtail A \twoheadrightarrow 0$, we see that A is trivial if and only if the admissible epic $A \twoheadrightarrow 0$ is a weak equivalence.
- We say A is *cofibrant* if $0 \rightarrowtail A$ is a cofibration.
- We say A is *fibrant* if $A \twoheadrightarrow 0$ is a fibration.
- We say A is *trivially cofibrant* if it is both trivial and cofibrant. In other words, $0 \rightarrowtail A$ is a trivial cofibration.
- We say A is *trivially fibrant* if it is both trivial and fibrant. In other words, $A \twoheadrightarrow 0$ is a trivial fibration.

The basic idea of an abelian model category is that the model structure axioms should be compatible with the exact category axioms on $(\mathcal{A}, \mathcal{E})$. Now cofibrations (and trivial cofibrations) in model categories are typically "nice monomorphisms" and it turns out that they are always closed under pushouts. (See Exercise A.0.1.) In exact categories, the "nice monomorphisms" are the admissible monics and their cokernels *are* pushouts, along the 0 morphism out of the domain. So we are led to define (trivial) cofibrations to be admissible monics with (trivially) cofibrant cokernels. Similar comments apply to fibrations and admissible epics, so we make the following definition.

Definition 4.24 Let $(\mathcal{A}, \mathcal{E})$ be any exact category possessing a model structure $\mathcal{M} = (Cof, We, Fib)$. We say $\mathcal{M}$ is **abelian** (or **exact**, or **admissible**) if the following hold.

(1) A morphism f is a (trivial) cofibration if and only if it is an admissible monic with (trivially) cofibrant cokernel, $\operatorname{Cok} f$.

(2) A morphism f is a (trivial) fibration if and only if it is an admissible epic with (trivially) fibrant kernel, $\operatorname{Ker} f$.

Note that the definition of an abelian model structure shifts all of the attention from morphisms to objects. Hovey's correspondence makes this precise.

Theorem 4.25 (Hovey's Correspondence for Weakly Idempotent Complete Exact Categories) *Assume $\mathcal{A}$ is weakly idempotent complete. There is a bijective correspondence between Hovey triples $\mathfrak{M} = (\mathcal{Q}, \mathcal{W}, \mathcal{R})$ and (closed) abelian model structures on $(\mathcal{A}, \mathcal{E})$. The correspondence acts by*

$$\mathfrak{M} = (\mathcal{Q}, \mathcal{W}, \mathcal{R}) \hookrightarrow (\mathfrak{M}\text{-cofibrations}, \mathfrak{M}\text{-weak equivalences}, \mathfrak{M}\text{-fibrations})$$

and the inverse is given by

$$\mathcal{M} = (Cof, We, Fib) \hookrightarrow (\text{Cofibrant objects}, \text{Trivial objects}, \text{Fibrant objects}).$$

In fact, this correspondence extends to a bijective correspondence between Hovey Ext-triples and (not necessarily closed) abelian model structures on $(\mathcal{A}, \mathcal{E})$.

Proof First, let $\mathfrak{M} = (\mathcal{Q}, \mathcal{W}, \mathcal{R})$ be a given Hovey triple. We will argue that ($\mathfrak{M}$-cofibrations, $\mathfrak{M}$-weak equivalences, $\mathfrak{M}$-fibrations), as defined in Definition 4.6, determines a (closed) abelian model structure. First, Corollary 4.8 tells us that a morphism is both an $\mathfrak{M}$-cofibration and an $\mathfrak{M}$-weak equivalence if and only if it is an admissible monic with cokernel in $\mathcal{Q}_{\mathcal{W}}$. (And similar for the trivial fibrations.) Thus the alleged model structure will be abelian in the sense of Definition 4.24 once we verify Axioms (M1)–(M5). Now the Lifting Axiom (M2) is immediate from Lemma 2.9, since $\operatorname{Ext}^1_{\mathcal{E}}(Q, R) = 0$ whenever $Q \in \mathcal{Q}_{\mathcal{W}}$ and $R \in \mathcal{R}$, or, $Q \in \mathcal{Q}$ and $R \in \mathcal{R}_{\mathcal{W}}$. It follows from Proposition 2.13 that the Factorization Axiom (M3) holds since $(\mathcal{Q}_{\mathcal{W}}, \mathcal{R})$ and $(\mathcal{Q}, \mathcal{R}_{\mathcal{W}})$ are both complete.

Turning to the Subcategories Axiom (M1), recall that all isomorphisms are both admissible monics and admssible epics. So isomorphisms are easily seen to be $\mathfrak{M}$-fibrations, $\mathfrak{M}$-cofibrations, and $\mathfrak{M}$-weak equivalences. The fact that these three classes are all closed under compositions follows from Corollary 4.9 and Theorem 4.21. Moreover, the 2 out of 3 Axiom (M4) follows immediately from what we already proved in Theorem 4.21. This completes the proofs of axioms (M1)–(M4).

Let us prove the Retracts Axiom (M5). Suppose g is a retract of a map f. In the case that f is a (trivial) cofibration we have from Proposition 1.31 that g is also an admissible monic, and consequently we see that it too is a (trivial) cofibration since $\mathcal{Q}$ and $\mathcal{Q}_{\mathcal{W}}$ are closed under retracts. Similarly, retracts of (trivial) fibrations are again (trivial) fibrations. Now assume that f is a weak equivalence. The hypothesis that $\mathcal{A}$ is weakly idempotent complete implies that s and s' are admissible monics and r and r' are admissible epics in the retraction diagram:

$$\begin{array}{ccccc} C & \overset{s}{\rightarrowtail} & A & \overset{r}{\twoheadrightarrow} & C \\ \downarrow{\scriptstyle g} & & \downarrow{\scriptstyle f} & & \downarrow{\scriptstyle g} \\ D & \overset{s'}{\rightarrowtail} & B & \overset{r'}{\twoheadrightarrow} & D. \end{array}$$

So by Corollary 1.28 we may rewrite this f as a direct sum:

$$f : C \bigoplus C' \xrightarrow{g \oplus g'} D \bigoplus D'.$$

Using the Factorization Axiom (M3), we may factor $g = pi$ and $g' = p'i'$ where i, i' are trivial cofibrations and p, p' are fibrations. Because $g \oplus g' = (p \oplus p') \circ (i \oplus i')$, by the 2 out of 3 Axiom we know that $p \oplus p'$ must be a weak equivalence, hence a trivial fibration. Therefore, $\operatorname{Ker}(p \oplus p') = \operatorname{Ker} p \bigoplus \operatorname{Ker} p' \in \mathcal{R}_{\mathcal{W}}$. Being closed under direct summands we conclude $\operatorname{Ker} p \in \mathcal{R}_{\mathcal{W}}$. Therefore p is a trivial fibration, proving that g is a weak equivalence. This completes the proof that ($\mathfrak{M}$-cofibrations, $\mathfrak{M}$-weak equivalences, $\mathfrak{M}$-fibrations) is an abelian model structure in the sense of Definition 4.24.

We now turn to prove the converse. So let us now assume we are given a (closed) model structure $\mathcal{M} = (Cof, We, Fib)$ that is abelian in the sense of Definition 4.24, and let $(\mathcal{Q}, \mathcal{W}, \mathcal{R})$ be the associated triple of cofibrant objects, trivial objects, and fibrant objects. We must show that $(\mathcal{Q}, \mathcal{W}, \mathcal{R})$ is a Hovey triple. So set $\mathcal{Q}_{\mathcal{W}} := \mathcal{Q} \cap \mathcal{W}$ and $\mathcal{R}_{\mathcal{W}} := \mathcal{W} \cap \mathcal{R}$; these are precisely the classes of trivially cofibrant and trivially fibrant objects of the model structure. Let us first see why $(\mathcal{Q}_{\mathcal{W}}, \mathcal{R})$ and $(\mathcal{Q}, \mathcal{R}_{\mathcal{W}})$ are each complete cotorsion pairs. The proofs are similar so we pick to prove this for $(\mathcal{Q}_{\mathcal{W}}, \mathcal{R})$. Here, the fact that $\operatorname{Ext}^1_{\mathcal{E}}(\mathcal{Q}_{\mathcal{W}}, \mathcal{R}) = 0$ follows immediately from the Lifting Axiom (M2) and Proposition 2.11. By the Factorization Axiom (M3) along with Proposition 2.13, the alleged cotorsion pair will be complete once we know it is a cotorsion pair. Note that we already know $\mathcal{Q}_{\mathcal{W}} \subseteq {}^{\perp}\mathcal{R}$. On the other hand, given $A \in {}^{\perp}\mathcal{R}$, we may write a short exact sequence $R \rightarrowtail Q \twoheadrightarrow A$ where $R \in \mathcal{R}$ and $Q \in \mathcal{Q}_{\mathcal{W}}$. But this sequence must be split exact since $A \in {}^{\perp}\mathcal{R}$. Thus A is a direct summand of Q by Proposition 1.4. It is clear from the Retracts Axiom (M5) that $\mathcal{Q}_{\mathcal{W}}$ is closed under direct summands. So $A \in \mathcal{Q}_{\mathcal{W}}$. This proves $\mathcal{Q}_{\mathcal{W}} = {}^{\perp}\mathcal{R}$. A similar argument shows $\mathcal{Q}_{\mathcal{W}}^{\perp} = \mathcal{R}$ and so $(\mathcal{Q}_{\mathcal{W}}, \mathcal{R})$ is a complete cotorsion pair.

Our final goal is to show that $\mathcal{W}$ satisfies the 2 out of 3 property on short exact sequences. That is, we wish to prove that if two out of three of the terms in an exact sequence are in $\mathcal{W}$, then so is the third. We will make repeated use of the following observations which follow easily from the 2 out of 3 Axiom (M4): If $X \to Y$ is a weak equivalence and either X or Y is in $\mathcal{W}$, then so is the other. On the other hand, if $X, Y \in \mathcal{W}$, then *any* map $X \to Y$ is a weak equivalence.

So now suppose $A \overset{f}{\rightarrowtail} B \overset{g}{\twoheadrightarrow} C$ is a short exact sequence. Using the Factorization Axioms, write $g = pi$ where p is a fibration (so an admissible epic with kernel in $\mathcal{R}$) and i is a trivial cofibration (so an admissible monic with cokernel in $\mathcal{Q}_{\mathcal{W}}$). Let $A' = \operatorname{Ker} p$ and use the universal property of $\ker p$ to get the following commutative diagram:

$$\begin{array}{ccccc} A & \overset{f}{\rightarrowtail} & B & \overset{g}{\twoheadrightarrow} & C \\ {\scriptstyle j}\downarrow & & \downarrow{\scriptstyle i} & & \| \\ A' & \overset{k}{\rightarrowtail} & B' & \overset{p}{\twoheadrightarrow} & C. \end{array} \qquad (\star)$$

Since i and f are admissible monics, it follows that the composition $kj = if$ is an admissible monic. It then follows from Proposition 1.30 that j is also an admissible monic. So by Proposition 1.12 and Lemma 1.10 we get that $\operatorname{Cok} j \in \mathcal{Q}_{\mathcal{W}}$. Therefore j is a trivial cofibration. In particular j is a weak equivalence.

Now if $A \in \mathcal{W}$, then $A' \in \mathcal{W}$ since j is a weak equivalence. In this case p is a trivial fibration. So then $g = pi$ is a weak equivalence. So in this case $B \in \mathcal{W}$ if and only if $C \in \mathcal{W}$.

On the other hand, if $B, C \in \mathcal{W}$, then any map $B \to C$ is a weak equivalence and in particular $g = pi$ is a weak equivalence. In this case, by the 2 out of 3 Axiom, p must be a trivial fibration. Therefore $A' \in \mathcal{W}$. Since we already proved that j is a weak equivalence it follows that $A \in \mathcal{W}$.

This completes the proof that $\mathfrak{M} = (\mathcal{Q}, \mathcal{W}, \mathcal{R})$ forms a Hovey triple. However, we should verify that the weak equivalences of $\mathcal{M}$ coincide with the $\mathfrak{M}$-weak equivalences. To see this, note that for any model structure on any category, by the Factorization Axiom (M3) and the 2 out of 3 Axiom (M4), a map f is a weak equivalence if and only if it has a factorization $f = pi$ where i is a trivial cofibration and p is a trivial fibration. So in the case of an abelian model structure, a map f is a weak equivalence if and only if it has a factorization $f = pi$ where i is an admissible monic with trivially cofibrant cokernel and p is an admissible epic with trivially fibrant kernel. This is precisely the definition of an $\mathfrak{M}$-weak equivalence.

We refer the reader to Appendix A for the proof that Hovey's correspondence extends to a bijective correspondence between Hovey Ext-triples and (not necessarily closed) abelian model structures. □

Exercise 4.6.1 Let $\mathcal{M} = (\mathit{Cof}, \mathit{We}, \mathit{Fib})$ be a (closed) model structure on a category $\mathcal{C}$. Show each of the following. (See Definition 2.12 to understand the meaning of the left lifting property [LLP] and the right lifting property [RLP].)

(1) f is a cofibration if and only if it satisfies the LLP with respect to the class of trivial fibrations.
(2) f is a trivial cofibration if and only if it satisfies the LLP with respect to the class of fibrations.
(3) f is a fibration if and only if it satisfies the RLP with respect to the class of trivial cofibrations.
(4) f is a trivial fibration if and only if it satisfies the RLP with respect to the class of cofibrations.

Hint: In each case, use the Factorization Axiom followed by the Retracts Axiom.

5

The Homotopy Category of an Abelian Model Structure

We see in Hovey's Correspondence that an abelian model structure, on a weakly idempotent complete exact category $(\mathcal{A}, \mathcal{E})$, is equivalent to a Hovey triple $\mathfrak{M} = (\mathcal{Q}, \mathcal{W}, \mathcal{R})$. In this chapter we define and deduce the first principles of $\mathrm{Ho}(\mathfrak{M})$, the homotopy category of $\mathfrak{M}$. The most fundamental result is that $\mathrm{Ho}(\mathfrak{M})$ is the localization of $\mathcal{A}$ with respect to the class of all $\mathfrak{M}$-weak equivalences. $\mathrm{Ho}(\mathfrak{M})$ also carries a significant amount of structure. Most notably, we will see in Chaper 6 that $\mathrm{Ho}(\mathfrak{M})$ is a triangulated category.

For convenience we assume throughout this chapter that the underlying category $\mathcal{A}$ is weakly idempotent complete. The end of Appendix A explains how the results of this chapter may be extended to a Hovey Ext-triple on a general exact category $(\mathcal{A}, \mathcal{E})$.

5.1 Homotopy Equivalences

Throughout this section we let $\mathfrak{M} = (\mathcal{Q}, \mathcal{W}, \mathcal{R})$ denote a Hovey triple on $(\mathcal{A}, \mathcal{E})$. The goal here is to describe when $\mathfrak{M}$-weak equivalences are (left or right) stable equivalences, and vice versa. Recall from Definition 4.12 that a morphism is called a *right stable equivalence* if it represents an isomorphism in the right stable category, $\mathrm{St}_r(\mathcal{A})$. A *left stable equivalence* is defined similarly. Recall too that $\omega := \mathcal{Q} \cap \mathcal{W} \cap \mathcal{R}$.

Proposition 5.1 *Let $f \colon A \to B$ be a morphism.*

(1) *If f is both a cofibration and a left stable equivalence, then f must be a trivial cofibration with cokernel in ω.*
(2) *If f is a fibration and also a right stable equivalence, then f must be a trivial fibration with kernel in ω.*

Proof The statements are dual and we will prove (1). So let us assume that $f: A \rightarrowtail B$ is a cofibration and that it has a left stable inverse $g: B \to A$. This means we have $gf \sim^{\ell} 1_A$ and $fg \sim^{\ell} 1_B$. Since $C := \operatorname{Cok} f$ is already cofibrant our goal is to show $C \in \mathcal{R}_{\mathcal{W}}$. From $gf \sim^{\ell} 1_A$ we have an object $W \in \mathcal{R}_{\mathcal{W}}$ and morphisms $t: A \to W$ and $s: W \to A$ such that $1_A - gf = st$. By considering the pushout of f along $-t$, it follows from Proposition 1.18 that $\begin{bmatrix} f \\ t \end{bmatrix}: A \rightarrowtail B \bigoplus W$ is an admissible monic and we will denote its cokernel by $h: B \bigoplus W \twoheadrightarrow J$. We construct a morphism of short exact sequences as follows, where η is obtained from the universal property of the cokernel h:

$$\begin{array}{ccccc} A & \xrightarrow{\begin{bmatrix} f \\ t \end{bmatrix}} & B \bigoplus W & \xrightarrow{h} & J \\ \| & & \downarrow{\scriptstyle [\, 1_B \;\, 0 \,]} & & \downarrow{\scriptstyle \eta} \\ A & \xrightarrow{f} & B & \longrightarrow & C. \end{array}$$

Since the kernel of $[\, 1_B \;\, 0 \,]$ is W, Lemma 1.20(1) implies (by taking the composition to be $f = [\, 1_B \;\, 0 \,]\begin{bmatrix} f \\ t \end{bmatrix}$ in that lemma) there is a short exact sequence

$$W \rightarrowtail J \xrightarrow{\eta} C.$$

This sequence must be split exact since C is cofibrant and $W \in \mathcal{R}_{\mathcal{W}}$. Thus C is a direct summand of J and so we will be done once we show that $J \in \mathcal{R}_{\mathcal{W}}$. To see this, observe that $\begin{bmatrix} f \\ t \end{bmatrix}$ is a left stable equivalence since both f and $[\, 1_B \;\, 0 \,]$ are such and since stable equivalences always satisfy the 2 out of 3 property. (They are just isomorphisms in the stable category.) So since $h \circ \begin{bmatrix} f \\ t \end{bmatrix} = 0$ it follows that $[h]_{\ell} = 0$. Thus h factors as $h = \beta\alpha$ through some object $W' \in \mathcal{R}_{\mathcal{W}}$. On the other hand, the admissible monic $\begin{bmatrix} f \\ t \end{bmatrix}$ splits because $1_A - gf = st$ means $[\, g \;\, s \,]\begin{bmatrix} f \\ t \end{bmatrix} = 1_A$. This implies there exists a morphism m such that $hm = 1_J$. Thus $(\beta)(\alpha m) = 1_J$ which shows that J is a retract of W' and thus is itself in the class $\mathcal{R}_{\mathcal{W}}$. □

Theorem 5.2 *Assume $\mathfrak{M}$ is a Hovey triple and let $f: A \to B$ be a morphism.*

(1) *If B is cofibrant and f is a left stable equivalence, then f must be a weak equivalence.*
(2) *If A is fibrant and f is a right stable equivalence, then f must be a weak equivalence.*

Proof The statements are dual and we will prove (1). So assume B is cofibrant and f is a left stable equivalence. By the Factorization Axiom (which comes from Proposition 2.13) we may factor $f = qi$ into a cofibration i, so with a cokernel $Q \in \mathcal{Q}$, followed by a trivial fibration q, so with a kernel $R \in \mathcal{R}_{\mathcal{W}}$ as shown:

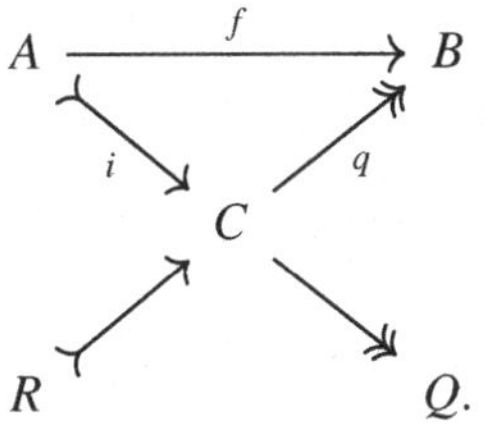

Since B is cofibrant, q must be a left stable equivalence by Lemma 4.13(2). Being isomorphisms in $\mathrm{St}_\ell(\mathcal{A})$, the class of left stable equivalences must satisfy the 2 out of 3 property. In particular, i must be a left stable equivalence. But then by Proposition 5.1(1) we get that i is a trivial cofibration. So f is a weak equivalence. □

Since any ω-homotopy equivalence is clearly both a left and right stable equivalence we get the following.

Corollary 5.3 *Assume $\mathfrak{M}$ is a Hovey triple. If $f: A \to B$ is an ω-homotopy equivalence and either A is fibrant or B is cofibrant, then f must be a weak equivalence.*

We also recover the following result which is basic in all model category theory.

Corollary 5.4 *Assume $\mathfrak{M}$ is a Hovey triple and let $f: A \to B$ be a morphism between bifibrant objects. Then f is a weak equivalence if and only if f is a homotopy equivalence (equivalently, a left or right or ω-stable equivalence).*

Proof First note that $\sim \, = \, \sim^r \, = \, \sim^\ell \, = \, \sim^\omega$, by Lemma 4.11(3). The "if" part follows from Corollary 5.3, and our arguments require that $\mathfrak{M}$ be a Hovey triple. The "only if" part is from Lemma 4.13(3) and only requires $\mathfrak{M}$ to be a Hovey Ext-triple. □

5.2 Bifibrant Approximations

Throughout this section we again let $\mathfrak{M} = (\mathcal{Q}, \mathcal{W}, \mathcal{R})$ be a Hovey triple on $(\mathcal{A}, \mathcal{E})$. In fact, here $\mathfrak{M}$ may even just be a Hovey Ext-triple on $(\mathcal{A}, \mathcal{E})$.

Recall that we have the cofibrant and fibrant approximation functors of Section 4.4. Of course we can restrict the functor Q to $\mathrm{St}_\omega(\mathcal{R})$, and its image is

$\mathrm{St}_\omega(\mathcal{Q}\cap\mathcal{R})$. We will again just denote this functor by $Q\colon \mathrm{St}_\omega(\mathcal{R}) \to \mathrm{St}_\omega(\mathcal{Q}\cap\mathcal{R})$. Similarly, the functor R restricts to $R\colon \mathrm{St}_\omega(\mathcal{Q}) \to \mathrm{St}_\omega(\mathcal{Q}\cap\mathcal{R})$. Since cofibrant and fibrant approximations are unique, up to a canonical isomorphism in $\mathrm{St}_\omega(\mathcal{A})$, one wants to know if the same is true for bifibrant approximations. The only issue is whether or not $RQ \cong QR$, and we show in Theorem 5.6 that there is indeed a natural isomorphism from RQ to QR.

Lemma 5.5 *The following hold by the 2 out of 3 Axiom.*

(1) *Any choice of morphism $Q(f)$ completing Diagram (†) of Section 4.4 is a weak equivalence if and only if f is a weak eqivalence.*
(2) *Any choice of morphism $R(f)$ completing Diagram (††) of Section 4.4 is a weak equivalence if and only if f is a weak eqivalence.*

Proof This follows immediately from the 2 out of 3 Axiom applied to the cofibrant and fibrant replacement diagrams (†) and (††) of Section 4.4. □

Theorem 5.6 (Commutativity $RQ \cong QR$) *There is a diagram of functors as shown satisfying the properties that follow:*

$$\begin{array}{ccc} \mathrm{St}_\omega(\mathcal{A}) & \underset{I}{\overset{R}{\rightleftarrows}} & \mathrm{St}_\omega(\mathcal{R}) \\ {\scriptstyle I}\uparrow\downarrow{\scriptstyle Q} & & {\scriptstyle I}\uparrow\downarrow{\scriptstyle Q} \\ \mathrm{St}_\omega(\mathcal{Q}) & \underset{I}{\overset{R}{\rightleftarrows}} & \mathrm{St}_\omega(\mathcal{Q}\cap\mathcal{R}). \end{array}$$

(1) *Each I denotes an inclusion functor and each of the four pairs of functors is an adjunction. Each functor appearing on the top or left side is a left adjoint while the right adjoints are written on the bottom or right side.*
(2) *The diagram commutes up to an isomorphism. That is, there is a natural isomorphism of functors $\tau = \{[\tau_A]\}\colon RQ \cong QR$. Moreover, we have $[\tau_A] = [1_{QA}]$ whenever A is fibrant and $[\tau_A] = [1_{RA}]$ whenever A is cofibrant.*

Proof The adjoint pair (I, Q) on the left side of the diagram is a special case of Proposition 3.7. It is easy to see that it restricts as in the right side of the diagram. Similar statements hold for the pair (R, I).

Now for any A, we construct an isomorphism $RQA \xrightarrow{\tau_A} QRA$, in $\mathrm{St}_\omega(\mathcal{Q}\cap\mathcal{R})$, as follows. First, take the $\mathcal{R}$-approximation $j_A\colon A \rightarrowtail RA$. We may then take a $\mathcal{Q}$-approximation of j_A, followed by another $\mathcal{R}$-approximation as suggested in the commutative diagram:

$$\begin{array}{ccc} RQA & \overset{R(Q(j_A))}{\dashrightarrow} & QRA \\ {\scriptstyle j_{QA}}\big\uparrow & & \big\| {\scriptstyle j_{RQA}} \\ QA & \overset{Q(j_A)}{\dashrightarrow} & QRA \\ {\scriptstyle p_A}\big\downarrow & & \big\downarrow {\scriptstyle p_{RA}} \\ A & \overset{j_A}{\rightarrowtail} & RA. \end{array} \tag{5.1}$$

It follows from Lemmas 4.13 and 5.5 that $RQA \xrightarrow{R(Q(j_A))} QRA$ represents an isomorphism in $\mathrm{St}_\omega(\mathcal{Q}\cap\mathcal{R})$. One can also see here that we can take $R(Q(j_A)) = 1_{QA}$ whenever A is fibrant and $R(Q(j_A)) = 1_{RA}$ whenever A is cofibrant. So we set $\tau_A := R(Q(j_A))$, and it is left as Exercise 5.2.1 to show that the isomorphism $[\tau_A]$ is natural in A. □

Exercise 5.2.1 Complete the proof of Theorem 5.6 by showing that the isomorphism τ_A is natural in A. (It all goes back to the uniqueness of the lifts in the Fundamental Lemmas 3.3 and 3.4 and hence in Definition 4.15.)

Exercise 5.2.2 The natural isomorphism $\tau = \{[\tau_A]\}\colon RQ \cong QR$ of Theorem 5.6 was defined by $\tau_A := R(Q(j_A))$ where $\left\{A \xrightarrow{[j_A]} RA\right\}$ is the fibrant approximation transformation. Verify that we get another natural isomorphism $\delta = \{[\delta_A]\}\colon RQ \cong QR$ by starting with the cofibrant approximation transformation $\left\{QA \xrightarrow{[p_A]} A\right\}$ and setting $\delta_A := Q(R(p_A))$. It can be shown, see Exercise 5.5.2, that $[\tau_A] = [\delta_A]$.

The isomorphism $\tau = \{[\tau_A]\}\colon RQ \cong QR$ of Theorem 5.6, together with the following easy lemma, immediately give us the corollary that follows.

Lemma 5.7 *Suppose we have two functors $F, G\colon \mathcal{A} \to \mathcal{B}$ and a natural isomorphism $\tau = \{\tau_A\}\colon F \to G$. Then the action*

$$h \mapsto \tau_{A'} \circ h \circ \tau_A^{-1}$$

induces a bijection, natural in both A and A':

$$\mathrm{Hom}_{\mathcal{B}}(FA, FA') \cong \mathrm{Hom}_{\mathcal{B}}(GA, GA').$$

Exercise 5.2.3 Prove Lemma 5.7.

Recall that we may denote the hom-sets $\underline{\mathrm{Hom}}_{\omega}(Q, R)$ simply by $\underline{\mathrm{Hom}}(Q, R)$ whenever Q is cofibrant and R is fibrant; see Lemma 4.11(3).

Corollary 5.8 *For morphisms* $[h] \in \underline{\mathrm{Hom}}(RQA, RQB)$*, the action*

$$[h] \mapsto [\tau_B] \circ [h] \circ [\tau_A]^{-1}$$

defines an isomorphism, natural in both A and B:

$$\underline{\mathrm{Hom}}(RQA, RQB) \cong \underline{\mathrm{Hom}}(QRA, QRB).$$

We call this isomorphism **conjugation with** τ. *In the case that A and B are each either fibrant or cofibrant, then conjugation with* τ *is the identity.*

We also get the following result.

Corollary 5.9 *We have isomorphisms as follows, natural in both A and B:*

$$\underline{\mathrm{Hom}}(QRA, QRB) \cong \underline{\mathrm{Hom}}(QA, RB) \cong \underline{\mathrm{Hom}}(RQA, RQB).$$

Proof Theorem 5.6 also makes it clear that the functors $\underline{\mathrm{Hom}}_{\omega}(QRA, -)$ and $\underline{\mathrm{Hom}}_{\omega}(RQA, -)$ are naturally isomorphic via "composition with τ". Similarly the contravariant functors are isomorphic $\underline{\mathrm{Hom}}_{\omega}(-, QRA) \cong \underline{\mathrm{Hom}}_{\omega}(-, RQA)$. Now since Q is right adjoint to the inclusion we get a natural isomorphism

$$\underline{\mathrm{Hom}}_{\omega}(QRA, QRB) \cong \underline{\mathrm{Hom}}_{\omega}(QRA, RB).$$

The latter is naturally isomorphic to $\underline{\mathrm{Hom}}_{\omega}(RQA, RB)$. But now this is naturally isomorphic to $\underline{\mathrm{Hom}}_{\omega}(QA, RB)$, since R is left adjoint to the inclusion. □

5.3 The Homotopy Category of an Abelian Model Structure

We continue to let $\mathfrak{M} = (Q, \mathcal{W}, \mathcal{R})$ be a Hovey triple on $(\mathcal{A}, \mathcal{E})$, and again, $\mathfrak{M}$ may even just be a Hovey Ext-triple in this section. We now define the *homotopy category*, $\mathrm{Ho}(\mathfrak{M})$, associated to $\mathfrak{M} = (Q, \mathcal{W}, \mathcal{R})$. The construction of $\mathrm{Ho}(\mathfrak{M})$ that we give in Definition 5.11 will employ the following lemma.

Lemma 5.10 *Given any functor* $F : C \to \mathcal{D}$*, there is a category, which we will denote by* $\mathrm{Ho}(C_F)$*, defined as follows: The objects of* $\mathrm{Ho}(C_F)$ *are the same as the objects of C, but for two objects X and Y we set* $\mathrm{Ho}(C_F)(X, Y) := \mathcal{D}(FX, FY)$*. Composition is defined in the obvious way – it is inherited from* $\mathcal{D}$*. This makes* $\mathrm{Ho}(C_F)$ *a category, for associativity and identities are also inherited from* $\mathcal{D}$*. We have the following.*

(1) *There is a canonical functor* $\gamma_F : C \to \mathrm{Ho}(C_F)$ *defined to be the identity on objects but F on arrows.*

(2) *If $F: C \to \mathcal{D}$ is an additive functor between additive categories then* $\mathrm{Ho}(C_F)$ *inherits the structure of an additive category as well. In particular, since F preserves biproduct diagrams, biproducts in* $\mathrm{Ho}(C_F)$ *coincide with those in C but with F applied to the natural inclusion and projection morphisms. In other words, γ_F preserves biproduct diagrams and is an additive functor in this case.*

The proofs are straightforward and left to the reader as an exercise. Recall the definition of *biproduct* is given in Definition 1.2.

Exercise 5.3.1 Prove Lemma 5.10.

We now define the homotopy category of $\mathfrak{M}$ by applying Lemma 5.10 to the composition functor $\mathrm{St}_\omega(\mathcal{A}) \xrightarrow{Q} \mathrm{St}_\omega(Q) \xrightarrow{R} \mathrm{St}_\omega(Q \cap \mathcal{R})$. Because of Theorem 5.6, we could instead use the composite $\mathrm{St}_\omega(\mathcal{A}) \xrightarrow{R} \mathrm{St}_\omega(\mathcal{R}) \xrightarrow{Q} \mathrm{St}_\omega(Q \cap \mathcal{R})$. The two equivalent approaches are compared more closely in Section 5.5.

Definition 5.11 We define the **homotopy category**, $\mathrm{Ho}(\mathfrak{M})$, by applying Lemma 5.10 to the bifibrant replacement functor $\mathrm{St}_\omega(\mathcal{A}) \xrightarrow{RQ} \mathrm{St}_\omega(Q \cap \mathcal{R})$ of Theorem 5.6. Thus $\mathrm{Ho}(\mathfrak{M})$ satisfies each of the following properties.

- The objects of $\mathrm{Ho}(\mathfrak{M})$ are the same as the objects in $\mathrm{St}_\omega(\mathcal{A})$, and hence the same as those in $\mathcal{A}$.
- For two objects A and B, the morphism sets are

$$\mathrm{Ho}(\mathfrak{M})(A, B) := \underline{\mathrm{Hom}}(RQA, RQB).$$

  That is, a morphism $A \to B$ is, by definition, a homotopy equivalence class of morphisms $RQA \xrightarrow{[f]} RQB$. The identity morphisms and (bilinear) composition are inherited from $\mathrm{St}_\omega(Q \cap \mathcal{R})$ and make $\mathrm{Ho}(\mathfrak{M})$ an additive category.
- There is an additive functor $\gamma_{RQ}: \mathrm{St}_\omega(\mathcal{A}) \to \mathrm{Ho}(\mathfrak{M})$ defined to be the identity on objects but RQ on arrows. By composing with the canonical projection, $\mathcal{A} \xrightarrow{\gamma_\omega} \mathrm{St}_\omega(\mathcal{A})$, we get an additive functor

$$\gamma = \gamma_{\mathfrak{M}} := \mathcal{A} \xrightarrow{\gamma_\omega} \mathrm{St}_\omega(\mathcal{A}) \xrightarrow{\gamma_{RQ}} \mathrm{Ho}(\mathfrak{M}),$$

  called the **localization functor**. Thus γ is well defined by the mapping $f \mapsto [f]_\omega \mapsto [R(Q(f))]_\omega$, where $R(Q(f))$ is any morphism constructed as in the following commutative diagram:

$$\begin{array}{ccc} RQA & \xrightarrow{R(Q(f))} & RQB \\ {\scriptstyle j_{QA}}\big\uparrow & & \big\uparrow{\scriptstyle j_{QB}} \\ QA & \xrightarrow{Q(f)} & QB \\ {\scriptstyle p_A}\big\downarrow & & \big\downarrow{\scriptstyle p_B} \\ A & \xrightarrow{f} & B. \end{array}$$

- Since $\gamma_\omega\colon \mathcal{A} \to \mathrm{St}_\omega(\mathcal{A})$ preserves biproduct diagrams, the localization functor $\gamma\colon \mathcal{A} \to \mathrm{Ho}(\mathfrak{M})$ does too, by Lemma 5.10(2).

This definition of $\mathrm{Ho}(\mathfrak{M})$ is dependent on the choice of cofibrant and fibrant approximation sequences assigned to the objects of $\mathcal{A}$. However, different assignments result in canonically isomorphic homotopy categories. (See Exercise 5.3.3.) Exercise 5.4.1 outlines another standard way to define $\mathrm{Ho}(\mathfrak{M})$.

We will see in Section 5.4 that $\mathcal{A} \xrightarrow{\gamma} \mathrm{Ho}(\mathfrak{M})$ is a "localization" with respect to the weak equivalences. In particular, γ sends each $j_A\colon A \rightarrowtail RA$, and each $p_A\colon QA \twoheadrightarrow A$, to an isomorphism. However, it is easy to see this directly from our definition of $\mathrm{Ho}(\mathfrak{M})$. We do this in the following lemma which also gives useful characterizations of these isomorphisms.

Lemma 5.12 *The localization functor $\mathcal{A} \xrightarrow{\gamma} \mathrm{Ho}(\mathfrak{M})$ sends any fibrant approximation, $j_A\colon A \rightarrowtail RA$, or any cofibrant approximation, $p_A\colon QA \twoheadrightarrow A$, to an isomorphism. The following identities are satisfied, where $\tau_A\colon RQA \to QRA$ is the homotopy equivalence constructed in Theorem 5.6.*

(1) $\gamma(p_A) = [1_{RQA}] \in \mathrm{Ho}(\mathfrak{M})(QA, A) := \underline{\mathrm{Hom}}(RQA, RQA)$.
(2) $[\gamma(p_A)]^{-1} = [1_{RQA}] \in \mathrm{Ho}(\mathfrak{M})(A, QA) := \underline{\mathrm{Hom}}(RQA, RQA)$.
(3) $\gamma(j_{QA}) = [1_{RQA}] \in \mathrm{Ho}(\mathfrak{M})(QA, RQA) := \underline{\mathrm{Hom}}(RQA, RQA)$.
(4) $[\gamma(j_{QA})]^{-1} = [1_{RQA}] \in \mathrm{Ho}(\mathfrak{M})(RQA, QA) := \underline{\mathrm{Hom}}(RQA, RQA)$.
(5) $\gamma(j_A) = [\tau_A] \in \mathrm{Ho}(\mathfrak{M})(A, RA) := \underline{\mathrm{Hom}}(RQA, QRA)$.
(6) $[\gamma(j_A)]^{-1} = [\tau_A]^{-1} \in \mathrm{Ho}(\mathfrak{M})(RA, A) := \underline{\mathrm{Hom}}(QRA, RQA)$.
(7) $\gamma(p_{RA}) = [1_{QRA}] \in \mathrm{Ho}(\mathfrak{M})(QRA, RA) := \underline{\mathrm{Hom}}(QRA, QRA)$.
(8) $[\gamma(p_{RA})]^{-1} = [1_{QRA}] \in \mathrm{Ho}(\mathfrak{M})(RA, QRA) := \underline{\mathrm{Hom}}(QRA, QRA)$.

Remark 5.13 We note that (7) and (8) are just special cases of (1) and (2). Moreover, (3) and (4) are special cases of (5) and (6), for the proof shows that $\gamma(j_A) = [\tau_A] = [1_{RA}]$ whenever A is cofibrant.

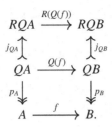

Proof The first four statements can be seen immediately from the following commutative diagram:

$$\begin{array}{ccccc}
RQA & \xleftarrow{1_{RQA}} & RQA & \xrightarrow{1_{RQA}} & RQA \\
{\scriptstyle 1_{RQA}}\big\uparrow & & {\scriptstyle j_{QA}}\big\uparrow & & \big\uparrow{\scriptstyle j_{QA}} \\
RQA & \xleftarrow{j_{QA}} & QA & \xrightarrow{1_{QA}} & QA \\
{\scriptstyle 1_{RQA}}\big\downarrow & & \big\downarrow{\scriptstyle 1_{QA}} & & \big\downarrow{\scriptstyle p_A} \\
RQA & \xleftarrow{j_{QA}} & QA & \xrightarrow{p_A} & A.
\end{array}$$

The last four statements come from the fact that we have the following commutative diagram, where the left side is just as in the proof of Theorem 5.6:

$$\begin{array}{ccccc}
RQA & \xrightarrow{\tau_A} & QRA & \xleftarrow{1_{QRA}} & QRA \\
{\scriptstyle j_{QA}}\big\uparrow & & {\scriptstyle 1_{QRA}}\big\uparrow & & \big\uparrow{\scriptstyle 1_{QRA}} \\
QA & \xrightarrow{Q(j_A)} & QRA & \xleftarrow{1_{QRA}} & QRA \\
{\scriptstyle p_A}\big\downarrow & & {\scriptstyle p_{RA}}\big\downarrow & & \big\downarrow{\scriptstyle 1_{QRA}} \\
A & \xrightarrow{j_A} & RA & \xleftarrow{p_{RA}} & QRA.
\end{array}$$

Again, as noted in Theorem 5.6, we have $[\tau_A] = [1_{RA}]$ whenever A is cofibrant. So in fact $\gamma(j_A) = [\tau_A] = [1_{RA}]$ whenever A is cofibrant. □

It can be a source of confusion that the set $\underline{\mathrm{Hom}}(RQA, RQB)$ represents a variety of morphism sets in $\mathrm{Ho}(\mathfrak{M})$. For example, a single morphism

$$[h] \in \underline{\mathrm{Hom}}(RQA, RQB)$$

might represent an element of $\mathrm{Ho}(\mathfrak{M})(A, B)$, or $\mathrm{Ho}(\mathfrak{M})(A, QB)$, or even $\mathrm{Ho}(\mathfrak{M})(RQA, QB)$, etc. But these sets are *distinct*. In other words, they are disjoint copies of the set $\underline{\mathrm{Hom}}(RQA, RQB)$. For this reason it will sometimes be useful to "tag" a morphism $[h] \in \underline{\mathrm{Hom}}(RQA, RQB)$ with its domain and codomain as defined next.

Definition 5.14 We sometimes will write ${}_A[h]_B$ to mean that $[h]$ represents a morphism in $\mathrm{Ho}(\mathfrak{M})$ from A to B. That is, the notation ${}_A[h]_B$ is shorthand for

$$[h] \in \mathrm{Ho}(\mathfrak{M})(A, B) := \underline{\mathrm{Hom}}(RQA, RQB).$$

So then, for example, ${}_{QA}[h]_B$ means $[h] \in \mathrm{Ho}(\mathfrak{M})(QA, B) := \underline{\mathrm{Hom}}(RQA, RQB)$, etc.

Let $f \in \mathrm{Ho}(\mathfrak{M})(A,B) := \underline{\mathrm{Hom}}(RQA, RQB)$, so that $f = {}_A[h]_B$ is represented by some morphism $h\colon RQA \to RQB$ in $\mathcal{A}$. Evidently we now have two morphisms from A to B in the category $\mathrm{Ho}(\mathfrak{M})$. Namely, f, and the composition

$$\gamma(p_B) \circ [\gamma(j_{QB})]^{-1} \circ \gamma(h) \circ \gamma(j_{QA}) \circ [\gamma(p_A)]^{-1}.$$

The following proposition says that these two morphisms are exactly the same!

Proposition 5.15 *Let $f \in \mathrm{Ho}(\mathfrak{M})(A,B) := \underline{\mathrm{Hom}}(RQA, RQB)$, so that $f = {}_A[h]_B$ is represented by some morphism $h\colon RQA \to RQB$ in $\mathcal{A}$. Then we have*

$$f = \gamma(p_B) \circ [\gamma(j_{QB})]^{-1} \circ \gamma(h) \circ \gamma(j_{QA}) \circ [\gamma(p_A)]^{-1}.$$

Proof The definition of γ makes it clear that $\gamma(h) = {}_{RQA}[h]_{RQB}$. The result now follows immediately from the identities in Lemma 5.12. □

Corollary 5.16 *Suppose $F\colon \mathrm{Ho}(\mathfrak{M}) \to \mathcal{B}$ and $G\colon \mathrm{Ho}(\mathfrak{M}) \to \mathcal{B}$ are functors and $\eta = \left\{F(A) \xrightarrow{\eta_A} G(A)\right\}_{A\in\mathcal{A}}$ is a family of morphisms in $\mathcal{B}$. Then $\eta\colon F \to G$ is a natural transformation if and only if $\eta\colon F \circ \gamma \to G \circ \gamma$ is a natural transformation.*

Proof Since γ is the identity on objects the statement makes sense. Let's prove the statement. First, if $\eta = \left\{F(A) \xrightarrow{\eta_A} G(A)\right\}_{A\in\mathcal{A}}$ determines a natural transformation $F \xrightarrow{\eta} G$ then it also determines a natural transformation $F \circ \gamma \xrightarrow{\eta} G \circ \gamma$, for we have commutative squares:

$$\begin{array}{ccc} F\gamma(A) & \xrightarrow{\eta_A} & G\gamma(A) \\ {\scriptstyle F\gamma(f)}\downarrow & & \downarrow{\scriptstyle G\gamma(f)} \\ F\gamma(B) & \xrightarrow{\eta_B} & G\gamma(B). \end{array} \qquad (\star\star)$$

On the other hand, suppose $\eta = \left\{F(A) \xrightarrow{\eta_A} G(A)\right\}_{A\in\mathcal{A}}$ determines a natural transformation $F \circ \gamma \xrightarrow{\eta} G \circ \gamma$. As in Proposition 5.15, any morphism $f \in \mathrm{Ho}(\mathfrak{M})(A,B)$ takes the form

$$f = \gamma(p_B) \circ [\gamma(j_{QB})]^{-1} \circ \gamma(h) \circ \gamma(j_{QA}) \circ [\gamma(p_A)]^{-1}.$$

So $F(f)$ and $G(f)$ are the horizontal compositions shown next

$$\begin{array}{ccccccc} F\gamma(A) & \xrightarrow{F\gamma(p)^{-1}} & F\gamma(QA) & \xrightarrow{F\gamma(j)} & F\gamma(RQA) & \xrightarrow{F\gamma(h)} & F\gamma(RQB)\ \cdots \\ {\scriptstyle \eta_A}\downarrow & & {\scriptstyle \eta_{QA}}\downarrow & & {\scriptstyle \eta_{RQA}}\downarrow & & {\scriptstyle \eta_{RQB}}\downarrow \\ G\gamma(A) & \xrightarrow[G\gamma(p)^{-1}]{} & G\gamma(QA) & \xrightarrow[G\gamma(j)]{} & G\gamma(RQA) & \xrightarrow[G\gamma(h)]{} & G\gamma(RQB)\ \cdots \end{array}$$

$$\begin{array}{ccccc}
\cdots F\gamma(RQB) & \xrightarrow{F\gamma(j)^{-1}} & F\gamma(QB) & \xrightarrow{F\gamma(p)} & F\gamma(B) \\
\eta_{RQB}\big\downarrow & & \eta_{QB}\big\downarrow & & \eta_{B}\big\downarrow \\
\cdots G\gamma(RQB) & \xrightarrow[G\gamma(j)^{-1}]{} & G\gamma(QB) & \xrightarrow[G\gamma(p)]{} & G\gamma(B).
\end{array}$$

The vertical arrows are given and provide commutative diagrams in $\mathcal{B}$. Thus the entire outer rectangle is a commutative diagram in $\mathcal{B}$, as desired. □

Another immediate consequence of Definition 5.11 is that it combines with Corollary 5.9 to give the following fundamental result about the homotopy category.

Corollary 5.17 (Morphism Sets in $\mathrm{Ho}(\mathfrak{M})$) *There is a natural isomorphism*

$$\mathrm{Ho}(\mathfrak{M})(A, B) \cong \underline{\mathrm{Hom}}(QA, RB).$$

We give an explicit formula for the isomorphism as follows.

Let us first describe the isomorphism of Corollary 5.17 for the particular case of when A is cofibrant and B is fibrant. Here the identification

$$\mathrm{Ho}(\mathfrak{M})(A, B) \cong \underline{\mathrm{Hom}}(A, B)$$

is given (left to right) by the action ${}_A[h]_B \mapsto [p_B \circ h \circ j_A]$. Its inverse is given by $[f] \mapsto \gamma(f) := {}_A[R(Q(f))]_B$. This is summarized by the following commutative diagram which is a special case of the one from Definition 5.11:

$$\begin{array}{ccc}
RA & \xrightarrow[R(Q(f))]{h} & QB \\
j_A\big\uparrow & & \big\| \\
A & \xrightarrow[Q(f)]{h \circ j_A} & QB \\
\big\| & & \big\downarrow p_B \\
A & \xrightarrow[f]{p_B \circ h \circ j_A} & B.
\end{array}$$

Now for the general case, let A and B be arbitrary. Then we pass to the previous case by using the isomorphisms $\gamma(p_A)$ and $\gamma(j_B)$ of Lemma 5.12. For example, $\gamma(j_B)$ induces a natural isomorphism

$$\mathrm{Ho}(\mathfrak{M})(A, B) \xrightarrow{[\gamma(j_B)]_*} \mathrm{Ho}(\mathfrak{M})(A, RB)$$

explicitly given by $[\gamma(j_B)]_*({}_A[h]_B) = {}_B[\tau_B]_{RB} \circ {}_A[h]_B = {}_A[\tau_B \circ h]_{RB}$. On the other hand, $\gamma(p_A)$ induces a natural isomorphism

$$\mathrm{Ho}(\mathfrak{M})(A, B) \xrightarrow{[\gamma(p_A)]^*} \mathrm{Ho}(\mathfrak{M})(QA, B)$$

explicitly given by $[\gamma(p_A)]^*({}_A[h]_B) = {}_A[h]_B \circ {}_{QA}[1_{RQA}]_A = {}_{QA}[h]_B$.

Putting all of this together, we get an explicit formula for a natural isomorphism

$$\mathrm{Ho}(\mathfrak{M})(A, B) \cong \underline{\mathrm{Hom}}(QA, RB).$$

It is given by the rule ${}_A[h]_B \mapsto [p_{RB} \circ \tau_B \circ h \circ j_{QA}]$.

Exercise 5.3.2 Assume $\mathfrak{M} = (\mathcal{Q}, \mathcal{W}, \mathcal{R})$ is an abelian model structure on $(\mathcal{A}, \mathcal{E})$ and $\mathfrak{M}' = (\mathcal{Q}', \mathcal{W}', \mathcal{R}')$ is an abelian model structure on $(\mathcal{A}', \mathcal{E}')$. Show that

$$\mathfrak{M} \times \mathfrak{M}' := (\mathcal{Q} \times \mathcal{Q}', \mathcal{W} \times \mathcal{W}', \mathcal{R} \times \mathcal{R}')$$

is an abelian model structure, called the *product model structure*, on the product of the exact categories $(\mathcal{A} \times \mathcal{A}', \mathcal{E} \times \mathcal{E}')$. Verify that cofibrations, fibrations, and weak equivalences are inherited componentwise, and that there is a canonical isomorphism $\mathrm{Ho}(\mathfrak{M} \times \mathfrak{M}') \cong \mathrm{Ho}(\mathfrak{M}) \times \mathrm{Ho}(\mathfrak{M}')$.

Exercise 5.3.3 Recall that a choice $\{p_A : QA \twoheadrightarrow A\}$ of cofibrant approximations is implicit in the definition of the functor $Q : \mathrm{St}_\omega(\mathcal{A}) \to \mathrm{St}_\omega(\mathcal{A})$, and similar for the functor $R : \mathrm{St}_\omega(\mathcal{A}) \to \mathrm{St}_\omega(\mathcal{A})$. Show that if Q' and R' are cofibrant and fibrant approximation functors arising from different choices, then the homotopy category $\mathrm{Ho}(\mathfrak{M})$ defined by using the composite $R'Q'$ is canonically isomorphic to the one defined using the composite RQ. (See Remark 3.6 and Lemma 5.7.)

5.4 Localization of Weak Equivalences

Unless explicitly stated otherwise, throughout this section $\mathfrak{M} = (\mathcal{Q}, \mathcal{W}, \mathcal{R})$ denotes a Hovey Ext-triple on $(\mathcal{A}, \mathcal{E})$. We will show that the localization functor $\gamma : \mathcal{A} \to \mathrm{Ho}(\mathfrak{M})$ of Definition 5.11 is a localization of $\mathcal{A}$ with respect to the $\mathfrak{M}$-weak equivalences. We start with the following.

Proposition 5.18 *If $f \in \mathcal{A}$ is a weak equivalence, then $\gamma(f)$ is an isomorphism in* $\mathrm{Ho}(\mathfrak{M})$.

See Proposition 5.23 for a converse of the present proposition.

Proof It follows right from Definition 5.11 that a morphism ${}_A[h]_B$ in Ho($\mathfrak{M}$) is an isomorphism if and only if $h\colon RQA \to RQB$ is a homotopy equivalence. In particular, given any morphism $f\colon A \to B$ in $\mathcal{A}$, we have $\gamma(f)$ is an isomorphism if and only if any bifibrant replacement representative $R(Q(f))\colon RQA \to RQB$ is a homotopy equivalence.

Now since $\mathfrak{M}$ satisfies the 2 out of 3 Axiom, Lemma 5.5 implies that $R(Q(f))$ will indeed be a weak equivalence whenever $f\colon A \to B$ is a weak equivalence. Then Lemma 4.13(3) (which only requires $\mathfrak{M}$ to be a Hovey Ext-triple) applies, and guarantees that $R(Q(f))$ is a homotopy equivalence in this case. □

Lemma 5.19 *Suppose $F\colon \mathcal{A} \to \mathcal{B}$ is an additive functor mapping all weak equivalences to isomorphisms. Then F identifies all homotopic (i.e. $\mathcal{W}$-homotopic) maps. In particular, it identifies all ℓ-homotopic, r-homotopic, and ω-homotopic maps.*

Proof Suppose f and g are homotopic, that is, $f - g$ factors through some trivial object $W \in \mathcal{W}$. By definition, trivial objects are weakly equivalent to 0. So the hypothesis implies that $F(f) - F(g)$ factors through a zero object of $\mathcal{B}$. Hence $F(f) = F(g)$. □

For each $* \in \{\mathcal{W}, \ell, r, \omega\}$, let $\gamma_*\colon \mathcal{A} \to \mathrm{St}_*(\mathcal{A})$ denote the canonical functor to the stable category $\mathrm{St}(\mathcal{A})$, $\mathrm{St}_\ell(\mathcal{A})$, $\mathrm{St}_r(\mathcal{A})$, or $\mathrm{St}_\omega(\mathcal{A})$. Then any functor F as in Lemma 5.19 descends to a functor $\bar{F}^*\colon \mathrm{St}_*(\mathcal{A}) \to \mathcal{B}$ satisfying $F = \bar{F}^* \circ \gamma_*$. In particular, Proposition 5.18 allows us to make the following definition.

Definition 5.20 The localization functor $\gamma\colon \mathcal{A} \to \mathrm{Ho}(\mathfrak{M})$ factors through each of the stable categories $\mathrm{St}(\mathcal{A})$, $\mathrm{St}_\ell(\mathcal{A})$, $\mathrm{St}_r(\mathcal{A})$, and $\mathrm{St}_\omega(\mathcal{A})$ as follows: For each $* \in \{\mathcal{W}, \ell, r, \omega\}$, γ descends to a functor $\bar{\gamma}^*\colon \mathrm{St}_*(\mathcal{A}) \to \mathrm{Ho}(\mathfrak{M})$ satisfying $\gamma = \bar{\gamma}^* \circ \gamma_*$.

Theorem 5.21 (Localization Theorem) *The canonical functor $\gamma\colon \mathcal{A} \to$ Ho($\mathfrak{M}$) is universally initial among all (additive) functors out of $\mathcal{A}$ mapping weak equivalences to isomorphisms. That is, given another (additive) functor $F\colon \mathcal{A} \to \mathcal{B}$ mapping all weak equivlalences to isomorphisms there exists a unique (additive) functor* Ho(F) *such that $F =$ Ho(F) $\circ\, \gamma$. In fact, for each $* \in \{\mathcal{W}, \ell, r, \omega\}$, the following diagram commutes:*

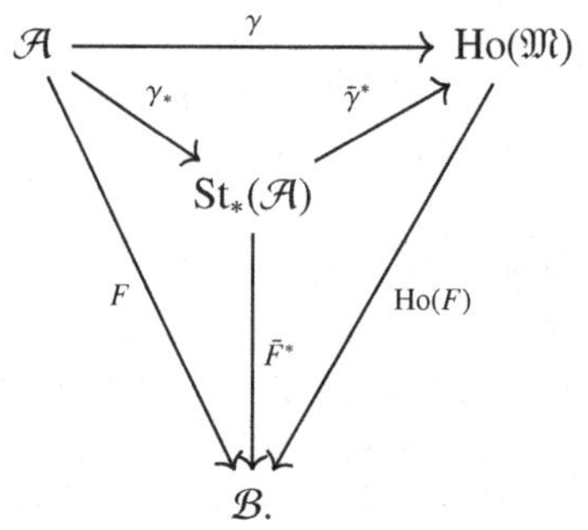

Given any morphism $f = {}_A[h]_B \in \mathrm{Ho}(\mathfrak{M})(A,B)$*, so with representative* $h\colon RQA \to RQB$*, the functor* $\mathrm{Ho}(F)$ *is defined by the rule:*

$$\mathrm{Ho}(F)(f) := F(p_B) \circ [F(j_{QB})]^{-1} \circ F(h) \circ F(j_{QA}) \circ [F(p_A)]^{-1}.$$

Remark 5.22 The word *additive* is in parentheses because it is not required that the given functor F be additive. Indeed $\mathcal{B}$ may be any category in order to get a functor $\mathrm{Ho}(F)$ such that $F = \mathrm{Ho}(F) \circ \gamma$. In our case γ is always additive and the proof shows that $\mathrm{Ho}(F)$ is additive whenever F is additive. So γ satisfies the universal property with respect to all categories but the universal property restricts to the world of additive categories and functors.

Proof The fact that γ is an additive functor mapping all weak equivalences to isomorphisms follows from Proposition 5.18. The top triangle commutes from the comments made before Definition 5.20.

Given F, in order for the left triangle to commute we must define $\bar{F}^*$ by the rule $\bar{F}^*([f]_*) := F(f)$. The rule is well defined by Lemma 5.19.

Now we define $\mathrm{Ho}(F)$. But for the purpose of simplifying the notation we will write $\widehat{F}$ instead of $\mathrm{Ho}(F)$ for the rest of the proof. Recall that

$$\mathrm{Ho}(\mathfrak{M})(A,B) := \underline{\mathrm{Hom}}(RQA, RQB)$$

and we showed in Proposition 5.15 that any morphism $f = {}_A[h]_B \in \mathrm{Ho}(\mathfrak{M})(A,B)$ with representative $h\colon RQA \to RQB$ in $\mathcal{A}$ must be *equal* to the composition

$${}_A[h]_B = \gamma(p_B) \circ [\gamma(j_{QB})]^{-1} \circ \gamma(h) \circ \gamma(j_{QA}) \circ [\gamma(p_A)]^{-1}.$$

Thus in order for $\widehat{F}$ to be a functor we must have

$$\widehat{F}({}_A[h]_B) = \widehat{F}(\gamma(p_B)) \circ \left[\widehat{F}(\gamma(j_{QB}))\right]^{-1} \circ \widehat{F}(\gamma(h)) \circ \widehat{F}(\gamma(j_{QA})) \circ \left[\widehat{F}(\gamma(p_A))\right]^{-1}.$$

And, in order to have $F = \widehat{F}\gamma$, we must have

$$\widehat{F}({}_A[h]_B) := F(p_B) \circ [F(j_{QB})]^{-1} \circ F(h) \circ F(j_{QA}) \circ [F(p_A)]^{-1}.$$

Therefore, this equation *must* be the definition of $\widehat{F}$ and we have shown uniqueness once we confirm that this rule actually defines a functor. Note that the rule is well defined since F is assumed to map weak equivalences to isomorphisms and since, by Lemma 5.19, the choice of representative that is used, $h\colon RQA \to RQB$, is irrelevant.

Next, one easily checks that the rule for $\widehat{F}$ respects identities and compositions. Indeed $1_A \in \mathrm{Ho}(\mathfrak{M})(A, A) := \underline{\mathrm{Hom}}(RQA, RQA)$ is the equivalence class $[1_{RQA}]$, represented by the identity $1_{RQA} \in \mathcal{A}$. Thus

$$\widehat{F}([1_{RQA}]) := F(p_A) \circ [F(j_{QA})]^{-1} \circ F(1_{RQA}) \circ F(j_{QA}) \circ [F(p_A)]^{-1} = 1_{FA}.$$

Similarly, one checks compositions and we conclude $\widehat{F}$ defines a functor.

We now check that $\widehat{F}$ is an additive functor whenever F is additive. For this, we recall that the addition on $\mathrm{Ho}(\mathfrak{M})(A, B) := \underline{\mathrm{Hom}}(RQA, RQB)$ is well defined by $[h_1] + [h_2] := [h_1 + h_2]$. So if F is an additive functor we get

$$\widehat{F}([h_1 + h_2]) = F(p_B) \circ [F(j_{QB})]^{-1} \circ (F(h_1) + F(h_2)) \circ F(j_{QA}) \circ [F(p_A)]^{-1}.$$

Now one just uses that composition of morphisms is bilinear to conclude that this coincides with $\widehat{F}([h_1]) + \widehat{F}([h_2])$, showing $\widehat{F}$ to be an additive functor.

Let us verify that $F = \widehat{F}\gamma$. For this we first recall that $\gamma\colon \mathcal{A} \to \mathrm{Ho}(\mathfrak{M})$ is well defined by the mapping $f \mapsto {}_A[R(Q(f))]_B$, where $R(Q(f))$ is constructed as in the following commutative diagram:

$$\begin{array}{ccc}
RQA & \xrightarrow{R(Q(f))} & RQB \\
{\scriptstyle j_{QA}}\big\uparrow & & \big\uparrow{\scriptstyle j_{QB}} \\
QA & \xrightarrow{Q(f)} & QB \\
{\scriptstyle p_A}\big\downarrow & & \big\downarrow{\scriptstyle p_B} \\
A & \xrightarrow{f} & B.
\end{array}$$

Therefore the composition $\widehat{F}\gamma$ is well defined by

$$\widehat{F}(\gamma(f)) := F(p_B) \circ [F(j_{QB})]^{-1} \circ F(R(Q(f))) \circ F(j_{QA}) \circ [F(p_A)]^{-1}.$$

The reader easily verifies that this composition equals $F(f)$ by applying F to the commutative diagram and using that the vertical morphisms become isomorphisms.

Finally, we check immediately that the remaining triangle on the right side of the diagram also commutes since those functors are well defined. Indeed $\widehat{F}\bar{\gamma}^*([f]_*) = \widehat{F}(\bar{\gamma}^*(\gamma_*(f))) = \widehat{F}(\gamma(f)) = F(f) = \bar{F}^*([f]_*)$. □

The following exercise outlines a direct way to define the homotopy category $\mathrm{Ho}(\mathfrak{M})$, avoiding the use of either class $\mathcal{Q}$ or $\mathcal{R}$; recall Proposition 4.22. It relies on using "zig-zags" of morphisms to formally invert the weak equivalences.

Exercise 5.4.1 (Formal Localization of Weak Equivalences) Let We denote the class of all $\mathfrak{M}$-weak equivalences. Follow the steps here to construct $\mathrm{Ho}(\mathfrak{M})$ as the formal localization of $\mathcal{A}$ by the class We.

(1) Construct the free category, denoted $\left(\mathcal{A}, We^{-1}\right)$, on the arrows of $\mathcal{A}$ and formal inverses of arrows of We as follows: The objects of $\left(\mathcal{A}, We^{-1}\right)$ are the same as the objects of $\mathcal{A}$. An arrow $A \to B$ of $\left(\mathcal{A}, We^{-1}\right)$ is a formal finite string of composable arrows ${}_A(f_1, f_2, \ldots, f_n)_B$, starting at A and ending at B, where each f_i may be either an arrow of $\mathcal{A}$, or the reversal w_i^{-1} of some weak equivalence $w_i \in We$. Composition is defined by

$${}_B(g_1, g_2, \ldots, g_m)_C \circ {}_A(f_1, f_2, \ldots, f_n)_B := {}_A(f_1, f_2, \ldots, f_n, g_1, g_2, \ldots, g_m)_C.$$

For each object A, there is an empty string $()_A$ which serves as the identity on A.

(2) Define $\mathcal{A}\left[We^{-1}\right]$ to be the quotient category of $\left(\mathcal{A}, We^{-1}\right)$ obtained by imposing the following relations.

(a) $(1_A) = ()_A$ for all objects A.

(b) ${}_A(f, g)_C = {}_A(gf)_C$ whenever $A \xrightarrow{f} B \xrightarrow{g} C$ are composable arrows of $\mathcal{A}$, including the possibility that these are the reversals of two composable morphisms of We. That is, ${}_A\left(v^{-1}, w^{-1}\right)_C = {}_A\left((vw)^{-1}\right)_C$ for $A \xleftarrow{v} B \xleftarrow{w} C$.

(c) ${}_A\left(w, w^{-1}\right)_A = ()_A$ and ${}_B\left(w^{-1}, w\right)_B = ()_B$ whenever $A \xrightarrow{w} B$ is in We.

We think of an arrow $A \to B$ of $\mathcal{A}\left[We^{-1}\right]$ as a formal "zig-zag"

$$A \xrightarrow{f_1} X_1 \xleftarrow{w_2} X_2 \xrightarrow{f_3} X_3 \xleftarrow{w_4} \cdots \xrightarrow{f_{n-1}} X_{n-1} \xleftarrow{w_n} B$$

(though the first and last arrow of the zig-zag may point in either direction).

(3) Define a functor $\gamma' \colon \mathcal{A} \to \mathcal{A}\left[We^{-1}\right]$ to be the identity on objects and by the rule $\gamma'(f) = {}_A(f)_B$ whenever $f \colon A \to B$ is an arrow of $\mathcal{A}$. Prove that γ' is universally initial among all functors out of $\mathcal{A}$ taking weak equivalences to isomorphisms.

(4) Conclude that there is a *unique* isomorphism $\mathrm{Ho}(\mathfrak{M}) \cong \mathcal{A}\left[We^{-1}\right]$ that is compatible with $\gamma \colon \mathcal{A} \to \mathrm{Ho}(\mathfrak{M})$ and $\gamma' \colon \mathcal{A} \to \mathcal{A}\left[We^{-1}\right]$. *Note:* Because of Remark 5.22 we may ignore additivity of the functors.

5.4.1 Hovey Triples and Localization of Weak Equivalences

Again, $\mathfrak{M}$ only needs to be a Hovey Ext-triple to obtain the previous results. However, we now wish to point out that Proposition 5.18 is an "if and only if" in most cases.

Proposition 5.23 *Assume that $\mathfrak{M}$ is a Hovey triple, not just a Hovey Ext-triple. Then, $f \in \mathcal{A}$ is a weak equivalence if and only if $\gamma(f)$ is an isomorphism in* $\mathrm{Ho}(\mathfrak{M})$.

Proof The "only if" part was proved in Proposition 5.18. As noted there, for any given morphism $f \colon A \to B$ in $\mathcal{A}$, we have $\gamma(f)$ is an isomorphism if and only if any bifibrant replacement representative $R(Q(f)) \colon RQA \to RQB$ is a homotopy equivalence. So now assume that $R(Q(f))$ is a homotopy equivalence. Then Corollary 5.4 (which required the retract closure properties of a Hovey triple) applies, so that we may conclude $R(Q(f))$ must be a weak equivalence. Then in turn Lemma 5.5 implies that f is a weak equivalence. □

Exercise 5.4.2 Assume $\mathfrak{M}$ is a Hovey triple.

(1) Show that if f and g are homotopic maps, then f is a weak equivalence if and only if g is a weak equivalence. In particular this is the case when $f \sim^{\omega} g$, $f \sim^{\ell} g$, or $f \sim^{r} g$.

(2) Show that any homotopy equivalence must be a weak equivalence. In particular, any ω-stable (or just right stable or left stable) equivalence must be a weak equivalence.

5.5 The Equivalent Subcategories of Cofibrant and Fibrant Objects

As in the previous sections, $\mathfrak{M} = (Q, \mathcal{W}, \mathcal{R})$ only needs to be a Hovey Ext-triple on $(\mathcal{A}, \mathcal{E})$. The definition of $\mathrm{Ho}(\mathfrak{M})$, given in Definition 5.11, is based on the functor RQ. Of course one could use the reverse composition QR. To make this precise, we should really denote the construction given in Definition 5.11 by $\mathrm{Ho}_{RQ}(\mathfrak{M})$ and call it the ***RQ*-representation of the homotopy category**. In the same way, we may use QR to construct another category, $\mathrm{Ho}_{QR}(\mathfrak{M})$, which we call the ***QR*-representation of the homotopy category**. Theorem 5.6 assures us that we get canonically isomorphic categories either way. We make this formal in Corollary 5.24.

But first, let $\mathrm{Ho}_{RQ}(Q)$ denote the full subcategory of $\mathrm{Ho}_{RQ}(\mathfrak{M})$ consisting of all the cofibrant objects. Similarly, let $\mathrm{Ho}_{QR}(Q) \subseteq \mathrm{Ho}_{QR}(\mathfrak{M})$ denote this category in the QR-representation. Because we have set $QA = A$ whenever A is cofibrant, we see at once that $\mathrm{Ho}_{RQ}(Q) = \mathrm{Ho}_{QR}(Q)$; they are the exact same category. Hence we denote this common subcategory simply by $\mathrm{Ho}(Q)$. Similarly, we have $\mathrm{Ho}_{RQ}(\mathfrak{M}) \supseteq \mathrm{Ho}(\mathcal{R}) \subseteq \mathrm{Ho}_{QR}(\mathfrak{M})$ for the full subcategory of fibrant objects.

Corollary 5.24 *The "conjugation with τ" isomorphism of Corollary 5.8 induces an isomorphism (not just an equivalence) of categories*

$$\mathrm{Ho}(\tau)\colon \mathrm{Ho}_{RQ}(\mathfrak{M}) \to \mathrm{Ho}_{QR}(\mathfrak{M}).$$

$\mathrm{Ho}(\tau)$ *is defined to be the identity on objects and "conjugation by τ"*

$$[h] \mapsto [\tau_B] \circ [h] \circ [\tau_A]^{-1}$$

on morphisms. Moreover, $\mathrm{Ho}(\tau)$ *acts as the identity functor when restricted to either* $\mathrm{Ho}(Q)$ *or* $\mathrm{Ho}(\mathcal{R})$.

Proof It is trivial to verify that $\mathrm{Ho}(\tau)$ preserves identities and compositions and so defines a functor, and so it is an isomorphism of categories with the stated property, by Corollary 5.8. (In fact, $\mathrm{Ho}(\tau)$ is the canonical functor guaranteed by the Localization Theorem 5.21. See Exercise 5.5.2.) □

Now let $I\colon \mathrm{Ho}(Q) \to \mathrm{Ho}_{RQ}(\mathfrak{M})$ denote the inclusion functor. There is a functor $\mathrm{Ho}(Q)\colon \mathrm{Ho}_{RQ}(\mathfrak{M}) \to \mathrm{Ho}(Q)$, in the reverse direction, induced by the functor $Q\colon \mathrm{St}_\omega(\mathcal{A}) \to \mathrm{St}_\omega(Q)$ as follows: On objects, it sends A to its assigned cofibrant replacement via the functor Q. As for morphisms, using the notation of Definition 5.14, it acts by ${}_A[h]_B \mapsto {}_{QA}[h]_{QB}$. We next show that the pair $(I, \mathrm{Ho}(Q))$ is an equivalence of categories.

But note that $\mathrm{Ho}(Q)$ also contains the full subcategory of all bifibrant objects, which is nothing more than the category $\mathrm{St}_\omega(Q \cap \mathcal{R})$. We again let $I\colon \mathrm{St}_\omega(Q \cap \mathcal{R}) \to \mathrm{Ho}(Q)$ denote the inclusion functor, and we again have a functor going backwards $\mathrm{Ho}(R)\colon \mathrm{Ho}(Q) \to \mathrm{St}_\omega(Q \cap \mathcal{R})$. On objects, it sends a cofibrant object A to its bifibrant replacement via the functor R. On morphisms, it acts by ${}_A[h]_B \mapsto {}_{RA}[h]_{RB}$.

Proposition 5.25 *Each functor that follows is an equivalence of categories, and the diagram displays* $\mathrm{Ho}(Q)$ *as a full equivalent subcategory of* $\mathrm{Ho}_{RQ}(\mathfrak{M})$, *and* $\mathrm{St}_\omega(Q \cap \mathcal{R})$ *as a full equivalent subcategory of* $\mathrm{Ho}(Q)$*:*

$$\mathrm{Ho}_{RQ}(\mathfrak{M}) \underset{I}{\overset{\mathrm{Ho}(Q)}{\rightleftarrows}} \mathrm{Ho}(Q) \underset{I}{\overset{\mathrm{Ho}(R)}{\rightleftarrows}} \mathrm{St}_\omega(Q \cap \mathcal{R}).$$

Proof It is clear that $\mathrm{Ho}(Q) \circ I = 1_{\mathrm{Ho}(Q)}$ since $QA = A$ for all cofibrant objects A. Similarly, $\mathrm{Ho}(R) \circ I = 1_{\mathrm{St}_\omega(Q \cap \mathcal{R})}$ since $RA = A$ for all bifibrant objects A.

There is an isomorphism $1_{\mathrm{Ho}_{RQ}(\mathfrak{M})} \cong I \circ \mathrm{Ho}(Q)$, defined by

$$\{\gamma(p_A)\}\colon I \circ \mathrm{Ho}(Q) \to 1_{\mathrm{Ho}_{RQ}(\mathfrak{M})},$$

where $p_A\colon QA \to A$ is the assigned cofibrant approximation. Indeed we saw in Lemma 5.12 that each $\gamma(p_A)$ is an isomorphism in $\mathrm{Ho}_{RQ}(\mathfrak{M})$, and that $\gamma(p_A) = [1_{RQA}]$. So given a morphism ${}_A[h]_B \in \mathrm{Ho}_{RQ}(\mathfrak{M})$, one easily checks that the following square commutes, in $\mathrm{Ho}_{RQ}(\mathfrak{M})$:

$$\begin{array}{ccc} QA & \xrightarrow{\gamma(p_A)} & A \\ {\scriptstyle {}_{QA}[h]_{QB}}\big\downarrow & & \big\downarrow{\scriptstyle {}_A[h]_B} \\ QB & \xrightarrow{\gamma(p_B)} & B. \end{array}$$

So $\{\gamma(p_A)\}$ is a natural isomorphism.

One can see a similar situation for the inclusion $I\colon \mathrm{St}_\omega(Q \cap \mathcal{R}) \to \mathrm{Ho}(Q)$ and its inverse $\mathrm{Ho}(R)\colon \mathrm{Ho}(Q) \to \mathrm{St}_\omega(Q \cap \mathcal{R})$. That is, one can verify that there is an isomorphism $1_{\mathrm{Ho}(Q)} \cong I \circ \mathrm{Ho}(R)$, defined by

$$\{\gamma(j_A)\}\colon 1_{\mathrm{Ho}(Q)} \to I \circ \mathrm{Ho}(R),$$

where $j_A\colon A \to RA$ is a fibrant approximation of a cofibrant A. This time, the corresponding square commutes because, as noted in Theorem 5.6 and Lemma 5.12, we have $\gamma(j_A) = [\tau_A] = [1_{RA}]$ whenever A is cofibrant. □

Using instead the QR-representation of the homotopy category, we get functors

$$\mathrm{Ho}_{QR}(\mathfrak{M}) \xrightarrow{\mathrm{Ho}(R)} \mathrm{Ho}(\mathcal{R}) \xrightarrow{\mathrm{Ho}(Q)} \mathrm{St}_\omega(Q \cap \mathcal{R}).$$

We obtain the following result, which is the dual of Proposition 5.25.

Proposition 5.26 *Each functor that follows is an equivalences of categories, and the diagram displays* $\mathrm{Ho}(\mathcal{R})$ *as a full equivalent subcategory of* $\mathrm{Ho}_{QR}(\mathfrak{M})$*, and* $\mathrm{St}_\omega(Q \cap \mathcal{R})$ *as a full equivalent subcategory of* $\mathrm{Ho}(\mathcal{R})$*:*

$$\mathrm{Ho}_{QR}(\mathfrak{M}) \underset{I}{\overset{\mathrm{Ho}(R)}{\rightleftarrows}} \mathrm{Ho}(\mathcal{R}) \underset{I}{\overset{\mathrm{Ho}(Q)}{\rightleftarrows}} \mathrm{St}_\omega(Q \cap \mathcal{R}).$$

The next exercise shows that the diagram in Theorem 5.6 induces a diagram of adjoint equivalences of homotopy categories. We just sew together the diagrams from Propositions 5.25 and 5.26 via the canonical isomorphism $\mathrm{Ho}(\tau)$ of Corollary 5.24.

Exercise 5.5.1 Show that the following diagram commutes up to isomorphism, and the natural isomorphism is again $\tau = \{[\tau_A]\}$ of Theorem 5.6:

$$
\begin{array}{ccccc}
 & \mathrm{Ho}_{RQ}(\mathfrak{M}) & \xrightarrow{\mathrm{Ho}(\tau)} & \mathrm{Ho}_{QR}(\mathfrak{M}) & \\
 {\scriptstyle \mathrm{Ho}(Q)}\swarrow & & & & \searrow{\scriptstyle \mathrm{Ho}(R)} \\
\mathrm{Ho}(\mathcal{Q}) & & & & \mathrm{Ho}(\mathcal{R}). \\
 & {\scriptstyle \mathrm{Ho}(R)}\searrow & & \swarrow{\scriptstyle \mathrm{Ho}(Q)} & \\
 & & \mathrm{St}_\omega(\mathcal{Q} \cap \mathcal{R}) & &
\end{array}
$$

This indicates that the fibrant replacement equivalence $\mathrm{Ho}_{RQ}(\mathfrak{M}) \to \mathrm{Ho}(\mathcal{R})$ should be defined to be the composite $\mathrm{Ho}_{RQ}(R) := \mathrm{Ho}_{QR}(R) \circ \mathrm{Ho}(\tau)$. (Here the subscripts on the functors are indicating the domain, and similarly we could denote the functor $\mathrm{Ho}(Q) \colon \mathrm{Ho}_{RQ}(\mathfrak{M}) \to \mathrm{Ho}(\mathcal{Q})$ with the more precise notation $\mathrm{Ho}_{RQ}(Q)$.) Note that this makes the following diagram commute on the nose, not just up to a natural isomorphism:

$$
\begin{array}{ccc}
\mathrm{Ho}_{RQ}(\mathfrak{M}) & \xrightarrow[\cong]{\mathrm{Ho}(\tau)} & \mathrm{Ho}_{QR}(\mathfrak{M}) \\
{\scriptstyle \mathrm{Ho}_{RQ}(R)}\downarrow & & \downarrow{\scriptstyle \mathrm{Ho}_{QR}(R)} \\
\mathrm{Ho}(\mathcal{R}) & \xrightarrow[\text{Identity}]{\mathrm{Ho}(\tau)} & \mathrm{Ho}(\mathcal{R}).
\end{array}
$$

Exercise 5.5.2 To emphasize that we are using the RQ-representation, denote the canonical functor γ by $\gamma_{RQ} \colon \mathcal{A} \to \mathrm{Ho}_{RQ}(\mathfrak{M})$. In the same way, we can let $\gamma_{QR} \colon \mathcal{A} \to \mathrm{Ho}_{QR}(\mathfrak{M})$ denote the canonical functor in the QR-representation.

(1) Following up on Exercise 5.2.2, let $\delta = \{[\delta_A]\} \colon RQ \cong QR$ be the natural isomorphism described there, defined by $\delta_A \colon RQA \xrightarrow{Q(R(p_A))} QRA$. State the QR-version of Lemma 5.12, giving the analog of the eight identities for the functor γ_{QR}.

(2) Show that $[\tau_A] = [\delta_A]$ for any object A. Therefore, $\tau = \delta$ is the canonical isomorphism $RQ \cong QR$. (*Hint*: Apply γ_{QR} to the commutative diagram (5.1) appearing in the proof of Theorem 5.6. Then use the identities from part (1).)

(3) Show that the canonical functor $\mathrm{Ho}(\gamma_{QR}) \colon \mathrm{Ho}(\mathfrak{M}) \to \mathrm{Ho}_{QR}(\mathfrak{M})$ satisfying $\gamma_{QR} = \mathrm{Ho}(\gamma_{QR}) \circ \gamma$ (guaranteed by the Localization Theorem 5.21) is exactly the isomorphism $\mathrm{Ho}(\tau) \colon \mathrm{Ho}_{RQ}(\mathfrak{M}) \to \mathrm{Ho}_{QR}(\mathfrak{M})$ of Corollary 5.24.

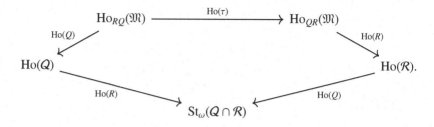

(4) Verify that in the QR-analog of Theorem 5.21, the functor $\mathrm{Ho}_{QR}(F)$ is defined by the rule:

$$\mathrm{Ho}_{QR}(F)(f) := [F(j_B)]^{-1} \circ F(p_{RB}) \circ F(h) \circ [F(p_{RA})]^{-1} \circ F(j_A).$$

Use this and the identities in Lemma 5.12 to show that

$$\mathrm{Ho}_{QR}(\gamma_{RQ}) = [\mathrm{Ho}(\delta)] = [\mathrm{Ho}(\tau)]^{-1}.$$

Exercise 5.5.3 Continuing with the idea of *opposite* abelian model structures from Exercise 4.2.1, show that we have a canonical isomorphism of categories, $\mathrm{Ho}(\mathfrak{M}^{\mathrm{op}}) \cong [\mathrm{Ho}(\mathfrak{M})]^{\mathrm{op}}$. In more detail, show $\mathrm{Ho}_{\hat{R}\hat{Q}}(\mathfrak{M}^{\mathrm{op}}) = [\mathrm{Ho}_{QR}(\mathfrak{M})]^{\mathrm{op}}$, where $\hat{R}$ is fibrant replacement in $\mathfrak{M}^{\mathrm{op}}$ and $\hat{Q}$ is cofibrant replacement in $\mathfrak{M}^{\mathrm{op}}$.

5.6 Homotopy Coproducts and Products

In this section we look at coproducts and products in the homotopy category. We assume throughout that $\mathfrak{M} = (\mathcal{Q}, \mathcal{W}, \mathcal{R})$ is a Hovey triple.

First we have the following lemma regarding (co)products in the left and right stable categories defined in Section 4.3.

Lemma 5.27 *Assuming coproducts exist in $\mathcal{A}$, then coproducts also exist in the right stable category $\mathrm{St}_r(\mathcal{A})$, and are taken as in $\mathcal{A}$. More precisely, let $\{A_i\}_{i\in I}$ be a collection of objects and let $\left(\bigoplus_{i\in I} A_i, \{\eta_i\}_{i\in I}\right)$ denote their coproduct in $\mathcal{A}$; so each $A_i \xrightarrow{\eta_i} \bigoplus_{i\in I} A_i$ is the canonical injection into the ith component. Then $\left(\bigoplus_{i\in I} A_i, \{[\eta_i]_r\}_{i\in I}\right)$ is a coproduct of $\{A_i\}_{i\in I}$ in $\mathrm{St}_r(\mathcal{A})$.*

On the other hand, assuming products exist in $\mathcal{A}$, then products also exist in the left stable category $\mathrm{St}_l(\mathcal{A})$, and are taken as in $\mathcal{A}$.

Proof The left side of any cotorsion pair is closed under coproducts, by Lemma 2.7. So by taking $\upsilon = \mathcal{Q}_{\mathcal{W}}$ in Proposition 3.2, we immediately get that coproducts in $\mathrm{St}_r(\mathcal{A}) := \mathrm{St}_{\mathcal{Q}_{\mathcal{W}}}(\mathcal{A})$ are taken as described. We take $\upsilon = \mathcal{R}_{\mathcal{W}}$ to get the second statement concerning products. □

We have the following proposition which basically states that "the homotopy coproduct is the coproduct of cofibrant replacements" while "the homotopy product is the product of fibrant replacements".

Proposition 5.28 *Coproducts and products exist in* $\mathrm{Ho}(\mathfrak{M})$ *as follows.*

(1) *If coproducts exist in $\mathcal{A}$ then coproducts also exist in $\mathrm{Ho}(\mathfrak{M})$. For a collection of objects $\{A_i\}_{i\in I}$ we may compute their coproduct as follows: Given*

cofibrant approximations QA_i, *let* $QA_i \xrightarrow{\eta_{QA_i}} \bigoplus_{i\in I} QA_i$ *denote the canonical injections into a coproduct of* $\{QA_i\}_{i\in I}$ *in* $\mathcal{A}$. *Then a coproduct of* $\{A_i\}_{i\in I}$, *in* $\mathrm{Ho}(\mathfrak{M})$, *is given by the pair* $\left(\bigoplus_{i\in I} QA_i, \{R\,[\eta_{QA_i}]\}_{i\in I}\right)$.

(2) *If products exist in* $\mathcal{A}$ *then products also exist in* $\mathrm{Ho}(\mathfrak{M})$. *For a collection of objects* $\{A_i\}_{i\in I}$ *we may compute their product as follows: Given fibrant approximations* RA_i, *let* $\prod_{i\in I} RA_i \xrightarrow{\pi_{RA_i}} RA_i$ *denote the canonical projections out of a product of* $\{RA_i\}_{i\in I}$ *in* $\mathcal{A}$. *Then* $\left(\prod_{i\in I} RA_i, \{Q\,[\pi_{RA_i}]\}_{i\in I}\right)$ *is a product of* $\{A_i\}_{i\in I}$, *in* $\mathrm{Ho}_{QR}(\mathfrak{M})$, *the QR-representation of the homotopy category. It follows that* $\left(\prod_{i\in I} RA_i, \left\{[\tau_{A_i}]^{-1} \circ Q\,[\pi_{RA_i}]\right\}_{i\in I}\right)$ *is a product of* $\{A_i\}_{i\in I}$, *in* $\mathrm{Ho}(\mathfrak{M})$, *where* $[\tau_{A_i}] : RQA_i \to QRA_i$ *is the canonical isomorphism from Theorem 5.6(2).*

Remark 5.29 Previously, and in the following, $R\,[\eta_{QA_i}]$ denotes the homotopy class $[R(\eta_{QA_i})]$. It is unambiguous, by Lemma 4.11; see also Exercises 4.4.1 and 4.4.2.

Proof Note that $\bigoplus_{i\in I} QA_i$ is already cofibrant by Lemma 2.7. So $R\,[\eta_{QA_i}]$ is indeed a morphism in $\mathrm{Ho}(\mathfrak{M})\left(A_i, \bigoplus QA_i\right) := \underline{\mathrm{Hom}}\left(RQA_i, R\left(\bigoplus QA_i\right)\right)$. Thus $\{R\,[\eta_{QA_i}]\}_{i\in I}$ is a "cocone above $\{A_i\}_{i\in I}$", in $\mathrm{Ho}(\mathfrak{M})$, which we now show is a coproduct. We start by noting that for any object B, applying $\mathrm{Ho}(\mathfrak{M})(-, B)$ yields a "cone above $\{\mathrm{Ho}(\mathfrak{M})\,(A_i, B)\}_{i\in I}$", in the category of abelian groups. Hence the universal property of products of abelian groups provides a canonical map

$$\mathrm{Ho}(\mathfrak{M})\left(\bigoplus_{i\in I} QA_i, B\right) \to \prod_{i\in I} \mathrm{Ho}(\mathfrak{M})(A_i, B).$$

One verifies that to show this map is an isomorphism for each B is equivalent to our goal of proving $\{R\,[\eta_{QA_i}]\}_{i\in I}$ is a coproduct.

So now using some of the natural isomorphisms we have accumulated so far we will show that the canonical map is an isomorhpism. By definition of hom-sets in $\mathrm{Ho}(\mathfrak{M})$ the canonical map becomes the first row of the following commutative diagram:

$$\begin{array}{ccc}
\underline{\mathrm{Hom}}_{\omega}\left(R\left(\bigoplus_{i\in I} QA_i\right), RQB\right) & \longrightarrow & \prod_{i\in I} \underline{\mathrm{Hom}}_{\omega}(RQA_i, RQB) \\
\cong\downarrow & & \downarrow\cong \\
\underline{\mathrm{Hom}}_{\omega}\left(\bigoplus_{i\in I} QA_i, RB\right) & \longrightarrow & \prod_{i\in I} \underline{\mathrm{Hom}}_{\omega}(QA_i, RB) \\
\| & & \| \\
\underline{\mathrm{Hom}}_{r}\left(\bigoplus_{i\in I} QA_i, RB\right) & \longrightarrow & \prod_{i\in I} \underline{\mathrm{Hom}}_{r}(QA_i, RB).
\end{array}$$

The natural isomorphism of the top and middle rows is from Corollary 5.9, while the equality of the middle and bottom rows is from Lemma 4.11. But

Lemma 5.27 says that coproducts in the right stable category, $\mathrm{St}_r(\mathcal{A})$, are taken as in $\mathcal{A}$. This means the canonical morphism making up the bottom row is an isomorphism and hence the top row is too.

We now turn to statement (2). Note that we have proved statement (1) using the RQ-representation of $\mathrm{Ho}(\mathfrak{M})$. So the dual argument gives us the analogous result for products, but using the QR-representation of $\mathrm{Ho}(\mathfrak{M})$. Of course the two representations are canonically isomorphic via the functor

$$\mathrm{Ho}(\tau) : \mathrm{Ho}_{RQ}(\mathfrak{M}) \to \mathrm{Ho}_{QR}(\mathfrak{M})$$

from Corollary 5.24, which acts on morphisms by $[h] \mapsto [\tau_B] \circ [h] \circ [\tau_A]^{-1}$. Thus the inverse $\mathrm{Ho}(\tau)^{-1}$, which acts by $[h] \mapsto [\tau_B]^{-1} \circ [h] \circ [\tau_A]$, will preserve the product. This process maps the morphisms $Q[\pi_{RA_i}] \mapsto [\tau_{A_i}]^{-1} \circ Q[\pi_{RA_i}] \circ \left[\tau_{\prod_{i\in I} RA_i}\right]$. However, since $\prod_{i\in I} RA_i$ is fibrant, we have $\left[\tau_{\prod_{i\in I} RA_i}\right] = \left[1_{Q(\prod_{i\in I} RA_i)}\right]$, by Theorem 5.6(2). This proves the final claim. □

We just saw that the coproduct of $\{A_i\}_{i\in I}$ in $\mathrm{Ho}(\mathfrak{M})$ coincides, as objects, with the coproduct $\bigoplus_{i\in I} QA_i$ in $\mathcal{A}$. The next proposition gives conditions guaranteeing that, moreover, $\bigoplus_{i\in I} QA_i$ is the cofibrant approximation of a coproduct $\bigoplus_{i\in I} A_i$ in $\mathcal{A}$. We note that one of the needed conditions is an analog of Grothendieck's **(AB4)** Axiom. Indeed one could say that an exact category $\mathcal{A}$ satisfies **(AB4)** if set-indexed coproducts exist in $\mathcal{A}$ and if coproducts of short exact sequences (conflations) are again short exact sequences. The dual **(AB4*)** would be the statement that set-indexed products exist in $\mathcal{A}$ and products of short exact sequences are again short exact sequences.

Proposition 5.30 *Assume that coproducts exist in $\mathcal{A}$ and that coproducts of short exact sequences are again short exact sequences. If the class $\mathcal{R}_W$ of trivially fibrant objects is closed under coproducts then the following hold.*

(1) *For a coproduct object, $\bigoplus_{i\in I} A_i \in \mathcal{A}$, a cofibrant approximation can be taken as $Q\left(\bigoplus_{i\in I} A_i\right) = \bigoplus_{i\in I} QA_i$.*

(2) *Given a collection of objects $\{A_i\}_{i\in I}$ in $\mathcal{A}$, let $A_i \xrightarrow{\eta_i} \bigoplus_{i\in I} A_i$ denote the canonical injections into a coproduct of $\{A_i\}_{i\in I}$ in $\mathcal{A}$. Then*

$$\left(\bigoplus_{i\in I} A_i, \{\gamma(\eta_i) = RQ[\eta_i]\}_{i\in I}\right)$$

is a coproduct of $\{A_i\}_{i\in I}$, in $\mathrm{Ho}(\mathfrak{M})$.

(3) *The class $\mathcal{W}$ of trivial objects is closed under coproducts.*

Of course Proposition 5.30 has a dual. We leave the formulation of the statement to the reader.

Proof For each A_i we find a cofibrant approximation sequence

$$R'_i \rightarrowtail QA_i \twoheadrightarrow A_i,$$

so each $R'_i \in \mathcal{R}_{\mathcal{W}}$ and $QA_i \in \mathcal{Q}$. The hypothesis tells us that

$$\bigoplus_{i\in I} R'_i \rightarrowtail \bigoplus_{i\in I} QA_i \twoheadrightarrow \bigoplus_{i\in I} A_i$$

is again a short exact sequence. We have $\bigoplus_{i\in I} QA_i \in \mathcal{Q}$ by Lemma 2.7, and we have $\bigoplus_{i\in I} R'_i \in \mathcal{R}_{\mathcal{W}}$, by hypothesis. This means $\bigoplus_{i\in I} QA_i$ is a cofibrant approximation of $\bigoplus_{i\in I} A_i$ proving (1).

Now (2) is a corollary of Proposition 5.28 along with the following observations. We have a diagram of short exact sequences:

$$\begin{array}{ccccc} R'_i & \rightarrowtail & QA_i & \xrightarrow{p_{A_i}} & A_i \\ \downarrow & & \downarrow{\scriptstyle \eta_{QA_i}} & & \downarrow{\scriptstyle \eta_i} \\ \bigoplus R'_i & \rightarrowtail & \bigoplus QA_i & \xrightarrow{\oplus p_{A_i}} & \bigoplus A_i. \end{array}$$

The diagram shows that we can take $Q(\eta_i) = \eta_{QA_i}$. Hence $\gamma(\eta_i) = RQ[\eta_i] = R[\eta_{QA_i}]$, and from this one can see that the cofibrant approximation transformation $\bigoplus QA_i \xrightarrow{\oplus p_{A_i}} \bigoplus A_i$ is an isomorphism in $\mathrm{Ho}(\mathfrak{M})$ from the "cocone" $\{R[\eta_{QA_i}]\}_{i\in I}$, of Proposition 5.28, to the "cocone" $\{\gamma(\eta_i)\}$.

We prove (3). By Lemma 4.2 we know that the class of trivial objects coincides with the class of all $W \in \mathcal{A}$ such that there exists a short exact sequence $U \rightarrowtail V \twoheadrightarrow W$ with $V \in \mathcal{Q}_{\mathcal{W}}$ and $U \in \mathcal{R}_{\mathcal{W}}$. Since $\mathcal{Q}_{\mathcal{W}}$ is always closed under coproducts by Lemma 2.7, and since $\mathcal{R}_{\mathcal{W}}$ is too by hypothesis, we use the assumption that short exact sequences are closed under coproducts to conclude that $\mathcal{W}$ is also closed under coproducts. □

Exercise 5.6.1 Assume $\mathcal{A}$ has coproducts and let $\{f_i \colon A_i \to B_i\}_{i\in I}$ be a collection of morphisms in $\mathcal{A}$. Show that their coproduct morphism $\coprod_{i\in I} \gamma(f_i)\colon \coprod_{i\in I} A_i \to \coprod_{i\in I} B_i$ in $\mathrm{Ho}(\mathfrak{M})$ is the morphism $\bigoplus_{i\in I} QA_i \to \bigoplus_{i\in I} QB_i$ represented by $[R(\oplus Q(f_i))]$.

Remark 5.31 The results of this section generalize to Hovey Ext-triples as follows. First, as noted in Section 4.3 the constructions of the left and right stable categories hold for Hovey Ext-triples. Then observe that Lemma 5.27, Proposition 5.28, and Proposition 5.30 will hold as long as one makes the additional assumption that $\mathcal{Q}_{\mathcal{W}}$ is closed under coproducts and $\mathcal{R}_{\mathcal{W}}$ is closed under products. For example, any Ext-pair $(\mathrm{Filt}(\mathcal{S}), \mathcal{S}^{\perp})$ cogenerated by a set $\mathcal{S}$ as in Corollary 9.8.2 always satisfies these properties.

6

The Triangulated Homotopy Category

This chapter is devoted to studying the triangulated structures that can be attached to an abelian model structure. The homotopy category does not just have compatible left and right triangulations, Nakaoka and Palu showed it to always be an actual triangulated category [Nakaoka and Palu, 2019]. The approach we take here to show this may be summarized as follows: We see in Section 6.3 that we can attach both a left and right triangulated structure to any complete cotorsion pair. It will follow from this that $\mathrm{Ho}(\mathfrak{M})$ is both a left and right triangulated category. Moreover, the associated cotorsion pairs also induce a suspension functor, Σ, and loop functor, Ω. It turns out that these are always inverse auto equivalences of $\mathrm{Ho}(\mathfrak{M})$, at least when the underlying category is weakly idempotent complete. Thus $\mathrm{Ho}(\mathfrak{M})$ is a triangulated category, and we will give natural characterizations of the distinguished triangles. Ultimately, we show in Section 6.7 that the canonical functor $\gamma\colon \mathcal{A} \to \mathrm{Ho}(\mathfrak{M})$ may be thought of as the triangulated localization of $(\mathcal{A}, \mathcal{E})$ with respect to the class $\mathcal{W}$ of trivial objects.

6.1 Right and Left Triangulated Categories

Let $\Sigma\colon \mathcal{T} \to \mathcal{T}$ be an additive endofunctor on an additive category $\mathcal{T}$. Let Δ be a class of **right triangles**, that is, diagrams of the form

$$A \xrightarrow{f} B \xrightarrow{g} C \xrightarrow{h} \Sigma A.$$

Morphisms between right triangles are triples (α,β,γ) such that the diagram commutes:

$$\begin{array}{ccccccc} A & \xrightarrow{f} & B & \xrightarrow{g} & C & \xrightarrow{h} & \Sigma A \\ \alpha\downarrow & & \downarrow\beta & & \downarrow\gamma & & \downarrow\Sigma(\alpha) \\ A' & \xrightarrow{f'} & B' & \xrightarrow{g'} & C' & \xrightarrow{h'} & \Sigma A'. \end{array}$$

The definition of a one-sided triangulated category we give here is extracted from Beligiannis and Marmaridis [1994] and Assem et al. [1998], although our statement of Verdier's Octahedral Axiom (R-Tr4) is slightly different. In fact, there are many ways to state (Tr4) and every author seems to have their favorite, leading to several equivalent versions of (Tr4) throughout standard literature on triangulated categories.

Definition 6.1 The pair (Σ, Δ) is called a **right triangulated structure** on $\mathcal{T}$, and we say that $\mathcal{T}$ is a **right triangulated category**, if Δ is closed under isomorphisms and satisfies the following four axioms.

(R-Tr1) (Existence) For any morphism $f\colon A \to B$ in $\mathcal{T}$, there is some right triangle $A \xrightarrow{f} B \to C \to \Sigma A$ in Δ. Also, for any object A in $\mathcal{T}$, the right triangle $0 \to A \xrightarrow{1_A} A \to 0$ is in Δ.

(R-Tr2) (Right Rotations) For any right triangle $A \xrightarrow{f} B \xrightarrow{g} C \xrightarrow{h} \Sigma A$ in Δ the right triangle $B \xrightarrow{g} C \xrightarrow{h} \Sigma A \xrightarrow{-\Sigma f} \Sigma B$ is also in Δ.

(R-Tr3) (Right Fill-Ins) Given any two right triangles in Δ,

$$A \xrightarrow{f} B \xrightarrow{g} C \xrightarrow{h} \Sigma A \quad \text{and} \quad A' \xrightarrow{f'} B' \xrightarrow{g'} C' \xrightarrow{h'} \Sigma A',$$

along with a commutative diagram $f'\alpha = \beta f$ in $\mathcal{T}$, there exists a morphism $\gamma \in \mathcal{T}$ such that the entire diagram commutes:

$$\begin{array}{ccccccc} A & \xrightarrow{f} & B & \xrightarrow{g} & C & \xrightarrow{h} & \Sigma A \\ \alpha\downarrow & & \downarrow\beta & & \downarrow\gamma & & \downarrow\Sigma(\alpha) \\ A' & \xrightarrow{f'} & B' & \xrightarrow{g'} & C' & \xrightarrow{h'} & \Sigma A'. \end{array}$$

(R-Tr4) (Right Octahedrals) Consider the composition $h = g \circ f$ of any two morphisms $A \xrightarrow{f} B$ and $B \xrightarrow{g} C$ in $\mathcal{T}$. Given any three right triangles in Δ on these morphisms as shown:

$$A \xrightarrow{f} B \xrightarrow{f'} D \xrightarrow{f''} \Sigma A, \quad B \xrightarrow{g} C \xrightarrow{g'} E \xrightarrow{g''} \Sigma B$$

and

$$A \xrightarrow{h} C \xrightarrow{h'} F \xrightarrow{h''} \Sigma A,$$

there exist morphisms $s\colon D \to F$ and $t\colon F \to E$ such that the following diagram commutes and the third column is also in Δ:

$$\begin{array}{ccccccc}
A & \xrightarrow{f} & B & \xrightarrow{f'} & D & \xrightarrow{f''} & \Sigma A \\
\| & & {\scriptstyle g}\downarrow & & \downarrow{\scriptstyle \exists!\, s} & & \| \\
A & \xrightarrow{h} & C & \xrightarrow{h'} & F & \xrightarrow{h''} & \Sigma A \\
 & & {\scriptstyle g'}\downarrow & & \downarrow{\scriptstyle \exists!\, t} & & \downarrow{\scriptstyle \Sigma(f)} \\
 & & E & = & E & \xrightarrow{g''} & \Sigma B \\
 & & {\scriptstyle g''}\downarrow & & \downarrow{\scriptstyle \Sigma(f')\circ g''} & & \\
 & & \Sigma B & \xrightarrow{\Sigma(f')} & \Sigma D. & &
\end{array}$$

If only axioms (R-Tr1)–(R-Tr3) are satisfied, we say the pair (Σ, Δ) is a **right pretriangulated structure** on $\mathcal{T}$, and we say that $\mathcal{T}$ is a **right pre-triangulated category**. Finally, given a right (pre)triangulated category $\mathcal{T} = (\mathcal{T}, \Sigma, \Delta)$, we call Δ the class of **distinguished right triangles** or **exact right triangles** for $\mathcal{T}$, and we call Σ the **suspension functor**.

Let us make some basic observations regarding the axioms. First, note that any right triangle $A \xrightarrow{f} B \xrightarrow{g} C \xrightarrow{h} \Sigma A$ gives an infinite string of rotated right triangles:

$$A \xrightarrow{f} B \xrightarrow{g} C \xrightarrow{h} \Sigma A \xrightarrow{-\Sigma f} \Sigma B \xrightarrow{-\Sigma g} \Sigma C \xrightarrow{-\Sigma h} \Sigma^2 A \xrightarrow{\Sigma^2 f} \Sigma^2 B \xrightarrow{\Sigma^2 g} \Sigma^2 C \to \cdots.$$

The right rotation axiom (R-Tr2) states that every three consecutive arrows is in Δ whenever the original is in Δ. We can think of this as a *right long exact sequence*.

The Octahedral Axiom (R-Tr4) is admittedly unattractive. One way to feel better about it is to compare it to the statement and proof of (the dual of) Corollary 1.15. That corollary says that the cokernel of a composite of admissible monics is an extension of the two cokernels. The proof proceeds by constructing a diagram exactly foreshadowing the one in (R-Tr4), but with $\Sigma = 0$. The point here is that distinguished right triangles are to be thought of as substitutes for short exact sequences, and (R-Tr4) is putting the third terms in each triangle into a fourth distinguished right triangle. We will see that (R-Tr1)–(R-Tr3) imply that the third term of any distinguished right triangle depends, up to non-unique isomorphism, only on the first morphism f. In some situations we will write the third term as $C(f)$ and call it the "cone on f" and think of it as a substitute for the notion of the cokernel of f. Thus (R-Tr4) is saying that the three cones sit in a distinguished right triangle $C(f) \xrightarrow{s} C(h) \xrightarrow{t} C(g) \to \Sigma C(f)$, and in such a way that "everything commutes". Note that the two top rows and the two middle columns in (R-Tr4) are clearly morphisms of distinguished right triangles. The square in the lower right-hand corner is expressing the hidden fact that there is a third morphism of distinguished right triangles. It is made explicit in the bottom two rows of the following diagram which is an equivalent way to express the axiom, but which now "hides" one of the morphisms made explicit in the original diagram:

$$\begin{array}{ccccccc}
A & \xrightarrow{f} & B & \xrightarrow{f'} & D & \xrightarrow{f''} & \Sigma A \\
\| & & {\scriptstyle g}\downarrow & & \downarrow{\scriptstyle \exists! s} & & \| \\
A & \xrightarrow{h} & C & \xrightarrow{h'} & F & \xrightarrow{h''} & \Sigma A \\
{\scriptstyle f}\downarrow & & \| & & \downarrow{\scriptstyle \exists! t} & & \downarrow{\scriptstyle \Sigma(f)} \\
B & \xrightarrow{g} & C & \xrightarrow{g'} & E & \xrightarrow{g''} & \Sigma B \\
 & & & & \downarrow{\scriptstyle \Sigma(f')\circ g''} & & \\
 & & & & \Sigma D. & &
\end{array}$$

From this perspective the analogy to Corollary 1.15 is revealing a connection to the (3×3)-Lemma (aka Nine Lemma). Another way to appreciate axiom (R-Tr4) is to attempt the "open problem" in Exercise 6.1.1, and to compare with the exercise that follows it on right exact sequences in abelian categories (Exercise 6.1.2). It seems that nobody yet has been able to find an example of a (right) pretriangulated category that is not (right) triangulated.

Exercise 6.1.1 (Open Problem) See how close you can get to proving (R-Tr4) from axioms (R-Tr1)–(R-Tr3).

Exercise 6.1.2 (Right Exact Sequences) Let $\mathcal{A}$ be an abelian category. Take the endofunctor $\Sigma\colon \mathcal{A} \to \mathcal{A}$ to be the zero functor, and let Δ be the class of all right exact sequences $A \to B \to C \to 0$. Show that $(\mathcal{A}, \Sigma, \Delta)$ is a right triangulated category.

Exercise 6.1.3 Show that if you change the sign of any *two* morphisms in a distinguished right triangle you still have another distinguished right triangle. This is not necessarily true for changing just one sign – a counterexample can be found in Iverson [1986, example 4.21, p. 32].

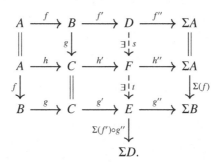

Proposition 6.2 *Assume $\mathcal{T} = (\mathcal{T}, \Sigma, \Delta)$ is a right pretriangulated category. Let $A \xrightarrow{f} B \xrightarrow{g} C \xrightarrow{h} \Sigma A$ be in Δ. Then $gf = 0$ and $hg = 0$. In fact, the following hold.*

(1) *The morphism g is a weak cokernel of f and h is a weak cokernel of g. That is, in each case the existence, but not necessarily the uniqueness, of the universal property of cokernel holds.*
(2) *If the suspension functor Σ is fully faithful, then f is a weak kernel of g and g is a weak kernel of h. That is, in each case the existence, but not necessarily the uniqueness, of the universal property of kernel holds. Thus (f, g) and (g, h) each form a "weak kernel–cokernel pair" in this case.*

Proof By (R-Tr1) and (R-Tr2), $A \xrightarrow{1_A} A \to 0 \to \Sigma A$ is a distinguished right triangle. By (R-Tr3) we get an arrow as indicated in the following diagram:

$$\begin{array}{ccccccc}
A & \xrightarrow{1_A} & A & \longrightarrow & 0 & \longrightarrow & \Sigma A \\
{\scriptstyle 1_A}\downarrow & & \downarrow{\scriptstyle f} & & \downarrow & & \downarrow{\scriptstyle 1_{\Sigma A}} \\
A & \xrightarrow{f} & B & \xrightarrow{g} & C & \xrightarrow{h} & \Sigma A.
\end{array}$$

So $gf = 0$. It follows from (R-Tr2) that $hg = 0$ too.

To prove (1), suppose $s\colon B \to X$ satisfies $sf = 0$. Then by (R-Tr3) we get an arrow t as indicated in the following diagram:

$$\begin{array}{ccccccc}
A & \xrightarrow{f} & B & \xrightarrow{g} & C & \xrightarrow{h} & \Sigma A \\
\downarrow & & \downarrow{\scriptstyle s} & & \downarrow{\scriptstyle t} & & \downarrow \\
0 & \longrightarrow & X & \xrightarrow{1_X} & X & \longrightarrow & 0.
\end{array}$$

This proves g is a weak cokernel of f. By (R-Tr2) it follows that h is also a weak cokernel of g.

Now to prove (2), suppose $s\colon X \to B$ satisfies $gs = 0$. By (R-Tr1) and (R-Tr2), $X \to 0 \to \Sigma X \xrightarrow{-1_{\Sigma X}} \Sigma X$ is a distinguished right triangle. By (R-Tr3) we get a "fill-in" morphism as indicated in the following diagram:

$$\begin{array}{ccccccc}
X & \longrightarrow & 0 & \longrightarrow & \Sigma X & \xrightarrow{-1_{\Sigma X}} & \Sigma X \\
{\scriptstyle s}\downarrow & & \downarrow & & \downarrow & & \downarrow{\scriptstyle \Sigma s} \\
B & \xrightarrow{g} & C & \xrightarrow{h} & \Sigma A & \xrightarrow{-\Sigma f} & \Sigma B.
\end{array}$$

Since Σ is a full (and faithful) functor there is a (unique) morphism $t\colon X \to A$ such that this "fill-in" morphism takes the form Σt. Therefore

$$(-\Sigma f) \circ \Sigma t = -\Sigma s \implies -\Sigma(ft) = -\Sigma s \implies \Sigma(ft) = \Sigma s.$$

Since Σ is faithful we get $ft = s$, proving f is a weak kernel of g. It follows from (R-Tr2) that g is a weak kernel of h too. □

Corollary 6.3 *Assume $\mathcal{T} = (\mathcal{T}, \Sigma, \Delta)$ is a right pretriangulated category. Let $A \xrightarrow{f} B \xrightarrow{g} C \xrightarrow{h} \Sigma A$ be in Δ.*

(1) *For each object $X \in \mathcal{T}$, applying the functor* $\mathrm{Hom}_{\mathcal{T}}(-, X)$ *yields a long exact sequence of abelian groups*

$$\cdots \xrightarrow{(\Sigma^2 f)^*} \mathrm{Hom}_{\mathcal{T}}\left(\Sigma^2 A, X\right) \xrightarrow{(-\Sigma h)^*} \mathrm{Hom}_{\mathcal{T}}(\Sigma C, X) \xrightarrow{(-\Sigma g)^*} \mathrm{Hom}_{\mathcal{T}}(\Sigma B, X)$$

$$\xrightarrow{(-\Sigma f)^*} \mathrm{Hom}_{\mathcal{T}}(\Sigma A, X) \xrightarrow{h^*} \mathrm{Hom}_{\mathcal{T}}(C, X) \xrightarrow{g^*} \mathrm{Hom}_{\mathcal{T}}(B, X) \xrightarrow{f^*} \mathrm{Hom}_{\mathcal{T}}(A, X).$$

(2) *Assume the suspension functor Σ is fully faithful. Then for each object $X \in \mathcal{T}$, applying the functor* $\mathrm{Hom}_{\mathcal{T}}(X, -)$ *yields a long exact sequence of abelian groups*

$$\mathrm{Hom}_{\mathcal{T}}(X, A) \xrightarrow{f_*} \mathrm{Hom}_{\mathcal{T}}(X, B) \xrightarrow{g_*} \mathrm{Hom}_{\mathcal{T}}(X, C) \xrightarrow{h_*} \mathrm{Hom}_{\mathcal{T}}(X, \Sigma A) \xrightarrow{(-\Sigma f)_*}$$

$$\mathrm{Hom}_{\mathcal{T}}(X, \Sigma B) \xrightarrow{(-\Sigma g)_*} \mathrm{Hom}_{\mathcal{T}}(X, \Sigma C) \xrightarrow{(-\Sigma h)_*} \mathrm{Hom}_{\mathcal{T}}(X, \Sigma^2 A) \xrightarrow{(\Sigma^2 f)_*} \cdots.$$

Proof We just prove (1) as (2) is similar. Recall that by (R-Tr2), the Right Rotation Axiom, every three consecutive arrows of the following sequence is in Δ:

$$A \xrightarrow{f} B \xrightarrow{g} C \xrightarrow{h} \Sigma A \xrightarrow{-\Sigma f} \Sigma B \xrightarrow{-\Sigma g} \Sigma C \xrightarrow{-\Sigma h} \Sigma^2 A \xrightarrow{\Sigma^2 f} \Sigma^2 B \xrightarrow{\Sigma^2 g} \Sigma^2 C \to \cdots.$$

So we only need to show that $\mathrm{Hom}_{\mathcal{T}}(C, X) \xrightarrow{g^*} \mathrm{Hom}_{\mathcal{T}}(B, X) \xrightarrow{f^*} \mathrm{Hom}_{\mathcal{T}}(A, X)$ is exact. But this is just a retranslation of the weak cokernel propery of Proposition 6.2. □

In light of Exercise 6.1.2, we note that Corollary 6.3(1) generalizes the standard fact that $\mathrm{Hom}_{\mathcal{A}}(-, X)$ sends right exact sequences to left exact sequences whenever $\mathcal{A}$ is abelian.

Corollary 6.4 (Five Lemma for Right Pretriangulated Categories) *Assume $\mathcal{T} = (\mathcal{T}, \Sigma, \Delta)$ is a right pretriangulated category. Suppose both rows in the following diagram are in Δ and α and β are both isomorphisms satisfying*

$f'\alpha = \beta f$ in $\mathcal{T}$. Then any fill-in map γ as guaranteed by (R-Tr3) must be an isomorphism:

$$\begin{array}{ccccccc} A & \xrightarrow{f} & B & \xrightarrow{g} & C & \xrightarrow{h} & \Sigma A \\ {\scriptstyle\alpha}\downarrow & & \downarrow{\scriptstyle\beta} & & \downarrow{\scriptstyle\gamma} & & \downarrow{\scriptstyle\Sigma(\alpha)} \\ A' & \xrightarrow{f'} & B' & \xrightarrow{g'} & C' & \xrightarrow{h'} & \Sigma A'. \end{array}$$

Remark 6.5 We emphasize that (R-Tr4), the Octahedral Axiom, is not needed to obtain the Five Lemma. We only need Δ to satisfy axioms (R-Tr1)–(R-Tr3).

Proof Let X be arbitrary and denote the contravariant functor $\mathrm{Hom}_{\mathcal{T}}(-, X)$ by h^X. Applying this functor yields the following commutative diagram, whose rows are exact sequences of abelian groups by Corollary 6.3:

$$\begin{array}{ccccccccc} h^X(\Sigma B') & \longrightarrow & h^X(\Sigma A') & \longrightarrow & h^X(C') & \longrightarrow & h^X(B') & \longrightarrow & h^X(A') \\ {\scriptstyle h^X(\Sigma\beta)}\downarrow & & {\scriptstyle h^X(\Sigma\alpha)}\downarrow & & \downarrow{\scriptstyle h^X(\gamma)} & & \downarrow{\scriptstyle h^X(\beta)} & & \downarrow{\scriptstyle h^X(\alpha)} \\ h^X(\Sigma B) & \longrightarrow & h^X(\Sigma A) & \longrightarrow & h^X(C) & \longrightarrow & h^X(B) & \longrightarrow & h^X(A). \end{array}$$

Since α and β are isomorphisms, certainly $h^X(\alpha)$, $h^X(\beta)$, $h^X(\Sigma\alpha)$, and $h^X(\Sigma\beta)$ are each isomorphisms. So the usual Five Lemma (for abelian groups) applies and gives us that $h^X(\gamma)$ is an isomorphism for all X. But ranging over all objects X, the $\{h^X(\gamma)\}$ assemble to a natural transformation

$$\left\{h^X(\gamma)\right\} : \mathrm{Hom}_{\mathcal{T}}(C', -) \xrightarrow{\gamma^*} \mathrm{Hom}_{\mathcal{T}}(C, -),$$

which we just showed to be a natural isomorphism. Yoneda's Lemma implies that γ itself must be an isomorphism. □

The statements in the following exercise will be useful to us in the next section.

Exercise 6.1.4 (Rotations of (R-Tr3) Axiom and Five Lemma) Assume $\mathcal{T} = (\mathcal{T}, \Sigma, \Delta)$ is a right pretriangulated category with fully faithful suspension functor $\Sigma \colon \mathcal{T} \to \mathcal{T}$. Show that the following variations of Axiom (R-Tr3) (Right Fill-Ins) hold for any two given right triangles in Δ,

$$A \xrightarrow{f} B \xrightarrow{g} C \xrightarrow{h} \Sigma A \quad \text{and} \quad A' \xrightarrow{f'} B' \xrightarrow{g'} C' \xrightarrow{h'} \Sigma A'.$$

(1) Given any commutative square $g'\beta = \gamma g$ in $\mathcal{T}$, there exists a morphism

$\alpha \in \mathcal{T}$ such that the entire diagram commutes:

$$\begin{array}{ccccccc} A & \xrightarrow{f} & B & \xrightarrow{g} & C & \xrightarrow{h} & \Sigma A \\ {\scriptstyle\alpha}\downarrow & & \downarrow{\scriptstyle\beta} & & \downarrow{\scriptstyle\gamma} & & \downarrow{\scriptstyle\Sigma(\alpha)} \\ A' & \xrightarrow{f'} & B' & \xrightarrow{g'} & C' & \xrightarrow{h'} & \Sigma A'. \end{array}$$

Moreover, any such fill-in morphism α must be an isomorphism whenever β and γ are isomorphisms.

(2) Given any commutative square $h'\gamma = \Sigma(\alpha)h$ in $\mathcal{T}$, there exists a morphism $\beta \in \mathcal{T}$ such that the entire diagram commutes:

$$\begin{array}{ccccccc} A & \xrightarrow{f} & B & \xrightarrow{g} & C & \xrightarrow{h} & \Sigma A \\ {\scriptstyle\alpha}\downarrow & & \downarrow{\scriptstyle\beta} & & \downarrow{\scriptstyle\gamma} & & \downarrow{\scriptstyle\Sigma(\alpha)} \\ A' & \xrightarrow{f'} & B' & \xrightarrow{g'} & C' & \xrightarrow{h'} & \Sigma A'. \end{array}$$

Moreover, any such fill-in morphism β must be an isomorphism whenever α and γ are isomorphisms.

Exercise 6.1.5 Assume $\mathcal{T} = (\mathcal{T}, \Sigma, \Delta)$ is a right pretriangulated category.

(1) Show that if $A \xrightarrow{g} B \to 0 \to \Sigma A$ is a distinguished right triangle, then g is an epimorphism (that is, right cancellable).

(2) Assume the suspension functor Σ is fully faithful. Show that the previous right triangle is distinguished if and only if g is an isomorphism.

Left Triangulated Categories We leave it to the reader to dualize each concept and result in this section. To do so, we let $\Omega\colon \mathcal{T} \to \mathcal{T}$ denote the additive endofunctor and we impose axioms on a class of **left triangles**, which are diagrams of the form

$$\Omega A \xrightarrow{h} C \xrightarrow{g} B \xrightarrow{f} A.$$

We dualize to obtain the axioms: **(L-Tr1)** (Existence), **(L-Tr2)** (Left Rotations), **(L-Tr3)** (Left Fill-Ins), and **(L-Tr4)** (Left Octahedrals). The pair (Ω, ∇) denotes a **left (pre)triangulated structure** on $\mathcal{T}$ while we call the triple $\mathcal{T} = (\mathcal{T}, \Omega, \nabla)$ a **left (pre)triangulated category**. Elements of ∇ are called **distinguished left triangles** or **exact left triangles**, and Ω is called the **loop functor**.

Right and Left Triangulated Functors A morphism of triangulated structures ought to be an additive functor that preserves the distinguished triangles. But note that if we are given a functor $F\colon \mathcal{T} \to \mathcal{T}'$ between (say right) triangulated categories, then a distinguished triangle $A \xrightarrow{f} B \xrightarrow{g} C \xrightarrow{h} \Sigma A$ in $\mathcal{T}$ is mapped by F to a diagram $FA \xrightarrow{F(f)} FB \xrightarrow{F(g)} FC \xrightarrow{F(h)} F\Sigma A$. If we do not have $F\Sigma A = \Sigma' FA$, then this is not an honest right triangle in $\mathcal{T}'$. So we require that F commutes with suspension, at least up to a natural isomorphism as in the following definition.

Definition 6.6 Let $F\colon \mathcal{T} \to \mathcal{T}'$ be an additive functor from a right pretriangulated category $\mathcal{T} = (\mathcal{T}, \Sigma, \Delta)$ to another right pretriangulated category $\mathcal{T}' = (\mathcal{T}', \Sigma', \Delta')$. We say that F is a **right triangulated functor**, or **right exact functor**, if there exists a natural isomorphism, $\xi\colon F\Sigma \cong \Sigma' F$, and for all right triangles $A \xrightarrow{f} B \xrightarrow{g} C \xrightarrow{h} \Sigma A$ in Δ, the following right triangle is in Δ':

$$FA \xrightarrow{F(f)} FB \xrightarrow{F(g)} FC \xrightarrow{\xi_A \circ F(h)} \Sigma' FA.$$

We define a **left triangulated functor**, or **left exact functor**, similarly.

Exercise 6.1.6 Let $F\colon \mathcal{A} \to \mathcal{B}$ be a functor between abelian categories. Verify that F is a right exact functor in the usual sense if and only if F is a right triangulated functor where $\mathcal{A}$ and $\mathcal{B}$ are given the right triangulated structures from Exercise 6.1.2.

6.2 Triangulated Categories

A triangulated category is simply a right triangulated category $\mathcal{T}$ in which the suspension functor $\Sigma\colon \mathcal{T} \to \mathcal{T}$ is an autoequivalence (equivalence between itself). It is equivalent to say that $\mathcal{T}$ along with the inverse equivalence $\Omega\colon \mathcal{T} \to \mathcal{T}$ is a left triangulated category. The same ideas apply to just pretriangulated categories. The point of this section is to make this self-duality clear to the reader. We will still favor working with *right* (pre)triangulated structures throughout.

Let us point out right away the useful fact that the converse of (R-Tr2) holds when the suspension functor $\Sigma\colon \mathcal{T} \to \mathcal{T}$ is fully faithful. This includes of course the possibility that Σ is an autoequivalence.

Lemma 6.7 (Converse of (R-Tr2) Axiom) *Assume $\mathcal{T} = (\mathcal{T}, \Sigma, \Delta)$ is a right pretriangulated category with fully faithful suspension functor $\Sigma\colon \mathcal{T} \to \mathcal{T}$. Let $A \xrightarrow{f} B \xrightarrow{g} C \xrightarrow{h} \Sigma A$ be any (not assumed distinguished) right triangle. Then* *this is a distinguished right triangle whenever its right rotation* $B \xrightarrow{g} C \xrightarrow{h} \Sigma A \xrightarrow{-\Sigma(f)} \Sigma B$ *is distinguished.*

Proof Two applications of (R-Tr2) gives us a distinguished right triangle

$$\Sigma A \xrightarrow{-\Sigma(f)} \Sigma B \xrightarrow{-\Sigma(g)} \Sigma C \xrightarrow{-\Sigma(h)} \Sigma^2 A.$$

On the other hand, by (R-Tr1), there is a distinguished right triangle

$$A \xrightarrow{f} B \xrightarrow{g'} C' \xrightarrow{h'} \Sigma A.$$

Rotating it three times with (R-Tr2) and then applying (R-Tr3) yields a commutative diagram as follows, with rows in Δ:

$$\begin{array}{ccccccc}
\Sigma A & \xrightarrow{-\Sigma(f)} & \Sigma B & \xrightarrow{-\Sigma(g')} & \Sigma C' & \xrightarrow{-\Sigma(h')} & \Sigma^2 A \\
\| & & \| & & \downarrow{\scriptstyle \gamma} & & \| \\
\Sigma A & \xrightarrow{-\Sigma(f)} & \Sigma B & \xrightarrow{-\Sigma(g)} & \Sigma C & \xrightarrow{-\Sigma(h)} & \Sigma^2 A.
\end{array}$$

By the Five Lemma, Corollary 6.4, we conclude that γ is an isomorphism. However, the fully faithful hypothesis implies $\gamma = \Sigma(\xi)$ for some unique $\xi\colon C' \to C$ which must also be an isomorphism. We have $\Sigma(g) = \Sigma(\xi) \circ \Sigma(g') = \Sigma(\xi \circ g')$, and because Σ is faithful we infer $g = \xi \circ g'$. Similarly, $h' = h \circ \xi$. This allows us to construct an isomorphism from the distinguished right triangle $A \xrightarrow{f} B \xrightarrow{g'} C' \xrightarrow{h'} \Sigma A$, to the originally given right triangle. Since Δ is closed under isomorphisms, we conclude the originally given right triangle is distinguished. □

Our goal is to prove Proposition 6.11 and this will require us to fix some notation. To do so, we first recall some basic facts concerning category equivalences. For a further discussion of the following one can consult a standard reference on category theory; we have followed Mac Lane [1998, sections IV.1 and IV.4].

Throughout the rest of this section we let (Σ, Ω) be an autoequivalence of a right pretriangulated category $\mathcal{T} = (\mathcal{T}, \Sigma, \Delta)$. The *unit* natural isomorphism will be denoted $\eta\colon 1_{\mathcal{T}} \to \Omega\Sigma$ and the *counit* natural isomorphism denoted $\epsilon\colon \Sigma\Omega \to 1_{\mathcal{T}}$. So our default is that we are viewing Σ as left adjoint to Ω. We have adjunction isomorphisms

$$\mathrm{Hom}_{\mathcal{T}}(\Sigma A, B) \xrightarrow{\cong} \mathrm{Hom}_{\mathcal{T}}(A, \Omega B)$$

which work from left to right via composition as follows:

$$\Sigma A \xrightarrow{s} B \quad \rightsquigarrow \quad \begin{array}{ccc} \Omega\Sigma A & \xrightarrow{\Omega(s)} & \Omega B. \\ {\scriptstyle \eta_A}\uparrow & \nearrow & \\ A & & \end{array}$$

Inversely, from right to left the function works by forming the following composition:

$$A \xrightarrow{t} \Omega B \quad \rightsquigarrow \quad \begin{array}{ccc} \Sigma A & \xrightarrow{\Sigma(t)} & \Sigma\Omega B \\ & \searrow & \downarrow{\scriptstyle \epsilon_B} \\ & & B. \end{array}$$

Passing from left to right and then back left again (and vice versa) leads us to the following commutative square identities which will be important for us:

$$\text{(i)} \quad \begin{array}{ccc} \Sigma\Omega\Sigma A & \xrightarrow{\Sigma\Omega(s)} & \Sigma\Omega B \\ {\scriptstyle \Sigma(\eta_A)}\uparrow & & \downarrow{\scriptstyle \epsilon_B} \\ \Sigma A & \xrightarrow{\quad s \quad} & B, \end{array} \qquad \text{(ii)} \quad \begin{array}{ccc} \Omega\Sigma A & \xrightarrow{\Omega\Sigma(t)} & \Omega\Sigma\Omega B \\ {\scriptstyle \eta_A}\uparrow & & \downarrow{\scriptstyle \Omega(\epsilon_B)} \\ A & \xrightarrow{\quad t \quad} & \Omega B. \end{array}$$

But at the same time it is a standard fact that the reverse pairing (Ω, Σ) is also an adjoint equivalence, with unit $\epsilon^{-1}\colon 1_{\mathcal{T}} \to \Sigma\Omega$ and counit $\eta^{-1}\colon \Omega\Sigma \to 1_{\mathcal{T}}$. That is, we also have adjunction isomorphisms

$$\mathrm{Hom}_{\mathcal{T}}(\Omega A, B) \xrightarrow{\cong} \mathrm{Hom}_{\mathcal{T}}(A, \Sigma B),$$

which work from left to right via composition as follows:

$$\Omega A \xrightarrow{u} B \quad \rightsquigarrow \quad \begin{array}{ccc} \Sigma\Omega A & \xrightarrow{\Sigma(u)} & \Sigma B. \\ {\scriptstyle \epsilon_A^{-1}}\uparrow & \nearrow & \\ A & & \end{array}$$

Inversely, from right to left the function works by forming the following composition:

$$A \xrightarrow{v} \Sigma B \quad \rightsquigarrow \quad \begin{array}{ccc} \Omega A & \xrightarrow{\Omega(v)} & \Omega\Sigma B \\ & \searrow & \downarrow{\scriptstyle \eta_B^{-1}} \\ & & B. \end{array}$$

Thus we also have the following commutative square identities:

$$\text{(iii)} \quad \begin{array}{ccc} \Omega\Sigma\Omega A & \xrightarrow{\Omega\Sigma(u)} & \Omega\Sigma B \\ {\scriptstyle \Omega(\epsilon_A^{-1})}\uparrow & & \downarrow{\scriptstyle \eta_B^{-1}} \\ \Omega A & \xrightarrow{\quad u \quad} & B, \end{array} \qquad \text{(iv)} \quad \begin{array}{ccc} \Sigma\Omega A & \xrightarrow{\Sigma\Omega(v)} & \Sigma\Omega\Sigma B \\ {\scriptstyle \epsilon_A^{-1}}\uparrow & & \downarrow{\scriptstyle \Sigma(\eta_B^{-1})} \\ A & \xrightarrow{\quad v \quad} & \Sigma B. \end{array}$$

Definition 6.8 Given any right triangle $A \xrightarrow{f} B \xrightarrow{g} C \xrightarrow{h} \Sigma A$, define its **left adjunct rotation** to be the left triangle $\Omega C \xrightarrow{-\check{h}} A \xrightarrow{f} B \xrightarrow{g} C$ where $\check{h} := \eta_A^{-1} \circ \Omega(h)$ is the "left adjunct" of h. Similarly, given any left triangle $\Omega C \xrightarrow{h} A \xrightarrow{f} B \xrightarrow{g} C$, define its **right adjunct rotation** to be the right triangle $A \xrightarrow{f} B \xrightarrow{g} C \xrightarrow{-\hat{h}} \Sigma A$ where $\hat{h} := \Sigma(h) \circ \epsilon_C^{-1}$ is the "right adjunct" of h.

Identities (iii) and (iv) say that $\hat{\check{h}} = h$ and $\check{\hat{h}} = h$. Therefore it is immediate that the left and right adjunct rotation operations also reverse each other.

Definition 6.9 We define ∇ to be the class of all left triangles $\Omega C \xrightarrow{h} A \xrightarrow{f} B \xrightarrow{g} C$ which equal the left adjunct rotation of a right triangle in Δ. Equivalently, ∇ is the class of all left triangles whose right adjunct rotation is in Δ. In other words, we define a **distinguished left triangle** to be a left triangle that corresponds to a distinguished right triangle through an adjunct rotation.

The next lemma gives us a third way to describe the left triangles in ∇. As motivation, note that in the case of when Σ is an automorphism of $\mathcal{T}$, so that $\Omega = \Sigma^{-1}$, each left triangle is also a right triangle and vice versa. For example, a left triangle $\Omega C \xrightarrow{h} A \xrightarrow{f} B \xrightarrow{g} C$ may be viewed as the right triangle $\Omega C \xrightarrow{h} A \xrightarrow{f} B \xrightarrow{g} \Sigma\Omega C$. In our case, (Σ, Ω) is only an autoequivalence, and so we must identify the given left triangle with the right triangle $\Omega C \xrightarrow{h} A \xrightarrow{f} B \xrightarrow{\epsilon_C^{-1} \circ g} \Sigma\Omega C$ obtained by composition with the unit isomorphism $\epsilon_C^{-1} : C \to \Sigma\Omega C$. On the other hand, right triangles should be identified with left triangles by composition with the counit isomorphism η^{-1}.

Lemma 6.10 *The following are equivalent for a given left triangle*

$$\Omega C \xrightarrow{h} A \xrightarrow{f} B \xrightarrow{g} C.$$

(1) *It is distinguished, that is, an element of* ∇.

(2) *The right triangle* $\Omega C \xrightarrow{h} A \xrightarrow{f} B \xrightarrow{\epsilon_C^{-1} \circ g} \Sigma\Omega C$ *obtained by composition with the unit isomorphism* $\epsilon_C^{-1} : C \to \Sigma\Omega C$ *is in* Δ.

Proof Note that there is an isomorphism between the two triangles as shown:

$$\begin{array}{ccccccc}
\Omega C & \xrightarrow{h} & A & \xrightarrow{f} & B & \xrightarrow{\epsilon_C^{-1}\circ g} & \Sigma\Omega C \\
\| & & \| & & \| & & \downarrow{\scriptstyle \epsilon_C} \\
\Omega C & \xrightarrow{h} & A & \xrightarrow{f} & B & \xrightarrow{g} & C.
\end{array}$$

There is also an isomorphism from the right rotation of the top row to the right adjunct rotation of the bottom row as indicated:

$$\begin{array}{ccccccc}
A & \xrightarrow{f} & B & \xrightarrow{\epsilon_C^{-1}\circ g} & \Sigma\Omega C & \xrightarrow{-\Sigma(h)} & \Sigma A \\
\| & & \| & & \downarrow{\scriptstyle \epsilon_C} & & \| \\
A & \xrightarrow{f} & B & \xrightarrow{g} & C & \xrightarrow{-\hat{h}} & \Sigma A.
\end{array}$$

From these diagrams, one can see that the lemma is true by (R-Tr2) and its converse in Lemma 6.7. □

Proposition 6.11 *Let $\mathcal{T} = (\mathcal{T}, \Sigma, \Delta)$ be a right (pre)triangulated category with suspension functor Σ an autoequivalence. Let (Σ, Ω) denote the adjoint equivalence with unit $\eta\colon 1_{\mathcal{T}} \to \Omega\Sigma$ and counit $\epsilon\colon \Sigma\Omega \to 1_{\mathcal{T}}$ and let ∇ be the class of distinguished left triangles characterized by Lemma 6.10. Then $\mathcal{T} = (\mathcal{T}, \Omega, \nabla)$ is also a left (pre)triangulated category.*

Proof The class ∇ of distinguished left triangles is clearly closed under isomorphisms. We prove (L-Tr2) first.

(L-Tr2) (Left Rotations) Let $\Omega C \xrightarrow{h} A \xrightarrow{f} B \xrightarrow{g} C$ be in ∇. We must show that its left rotation $\Omega B \xrightarrow{-\Omega(g)} \Omega C \xrightarrow{h} A \xrightarrow{f} B$ is also in ∇. This is equivalent to showing that its right adjunct rotation,

$$\Omega C \xrightarrow{h} A \xrightarrow{f} B \xrightarrow{+\widehat{\Omega(g)}} \Sigma\Omega C,$$

is in Δ. Following Definition 6.8 we see that the right adjunct of $\Omega(g)$ is

$$\widehat{\Omega(g)} := \Sigma\Omega(g) \circ \epsilon_B^{-1} = \epsilon_C^{-1} \circ g.$$

Now it is immediate from Lemma 6.10(2).

(L-Tr1) (Existence) Given $A \in \mathcal{T}$, we want to show that $\Omega 0 \to A \xrightarrow{1_A} A \to 0$ is in ∇. But this is immediate from Lemma 6.10(2) and (R-Tr1). Next, given any morphism $f\colon A \to B$, we may use (R-Tr1) to write a distinguished right triangle $A \xrightarrow{f} B \xrightarrow{g} C \xrightarrow{h} \Sigma A$. Therefore its left adjunct rotation $\Omega C \xrightarrow{-\check{h}} A \xrightarrow{f}$

$B \xrightarrow{g} C$ is in ∇. But now we can use the already proved (L-Tr2) to complete the proof of (L-Tr1).

(L-Tr3) (Left Fill-ins) Consider a commutative diagram in $\mathcal{T}$, with rows in ∇:

$$\begin{array}{ccccccc} \Omega C & \xrightarrow{h} & A & \xrightarrow{f} & B & \xrightarrow{g} & C \\ & & & & \downarrow & & \downarrow \\ \Omega C' & \xrightarrow{h'} & A' & \xrightarrow{f'} & B' & \xrightarrow{g'} & C'. \end{array}$$

Right adjunct rotation induces a commutative diagram with rows in Δ:

$$\begin{array}{ccccccc} A & \xrightarrow{f} & B & \xrightarrow{g} & C & \xrightarrow{-\hat{h}} & \Sigma A \\ \downarrow & & \downarrow & & \downarrow & & \downarrow \\ A' & \xrightarrow{f'} & B' & \xrightarrow{g'} & C' & \xrightarrow{-\widehat{h'}} & \Sigma A'. \end{array}$$

The dashed arrows exist by Exercise 6.1.4(1) and then (L-Tr3) is proved by taking left adjuct rotations to return to the original diagram.

(L-Tr4) (Left Octahedrals) Let $A \xrightarrow{f} B \xrightarrow{g} C$ be given, and set $h = gf$, and assume further that we are given three left triangles in ∇ as shown here:

$$\Omega B \xrightarrow{f''} D \xrightarrow{f'} A \xrightarrow{f} B, \qquad \Omega C \xrightarrow{g''} E \xrightarrow{g'} B \xrightarrow{g} C$$

and

$$\Omega C \xrightarrow{h''} F \xrightarrow{h'} A \xrightarrow{h} C.$$

The task is to find morphisms $j' \colon D \to F$ and $j \colon F \to E$ such that the following diagram commutes and so that the second column is also in ∇:

$$\begin{array}{ccccccc} & & \Omega E & \xrightarrow{\Omega g'} & \Omega B & & \\ & & \downarrow{\scriptstyle f''\circ\Omega g'} & & \downarrow{\scriptstyle f''} & & \\ \Omega B & \xrightarrow[f'']{} & D & = & D & & \\ \downarrow{\scriptstyle \Omega g} & & \downarrow{\scriptstyle j'} & & \downarrow{\scriptstyle f'} & & \\ \Omega C & \xrightarrow{h''} & F & \xrightarrow{h'} & A & \xrightarrow{h} & C \\ \| & & \downarrow{\scriptstyle j} & & \downarrow{\scriptstyle f} & & \| \\ \Omega C & \xrightarrow{g''} & E & \xrightarrow{g'} & B & \xrightarrow{g} & C. \end{array} \tag{6.1}$$

We take the right adjunct rotations of the given left triangles to obtain the three right triangles in Δ shown:

$$D \xrightarrow{f'} A \xrightarrow{f} B \xrightarrow{\widehat{f''}} \Sigma D, \quad E \xrightarrow{g'} B \xrightarrow{g} C \xrightarrow{\widehat{g''}} \Sigma E$$

and

$$F \xrightarrow{h'} A \xrightarrow{h} C \xrightarrow{\widehat{h''}} \Sigma F.$$

We then take their right rotations, which will also be in Δ, by (R-Tr2):

$$A \xrightarrow{f} B \xrightarrow{\widehat{f''}} \Sigma D \xrightarrow{-\Sigma f'} \Sigma A, \quad B \xrightarrow{g} C \xrightarrow{\widehat{g''}} \Sigma E \xrightarrow{-\Sigma g'} \Sigma B$$

and

$$A \xrightarrow{h} C \xrightarrow{\widehat{h''}} \Sigma F \xrightarrow{-\Sigma h'} \Sigma A.$$

We now have a diagram as in the Right Octahedral Axiom, (R-Tr4), and since Σ is fully faithful we get maps j' and j such that the following diagram commutes and the third column is also in Δ:

$$\begin{array}{ccccccc}
A & \xrightarrow{f} & B & \xrightarrow{\widehat{f''}} & \Sigma D & \xrightarrow{-\Sigma f'} & \Sigma A \\
\| & & {\scriptstyle g}\downarrow & & \downarrow{\scriptstyle \exists!\,\Sigma j'} & & \| \\
A & \xrightarrow{h} & C & \xrightarrow{\widehat{h''}} & \Sigma F & \xrightarrow{-\Sigma h'} & \Sigma A \\
 & & {\scriptstyle \widehat{g''}}\downarrow & & \downarrow{\scriptstyle \exists!\,\Sigma j} & & \downarrow{\scriptstyle \Sigma f} \\
 & & \Sigma E & = & \Sigma E & \xrightarrow{-\Sigma g'} & \Sigma B \\
 & & {\scriptstyle -\Sigma g'}\downarrow & & \downarrow{\scriptstyle -\Sigma(\widehat{f''} g')} & & \\
 & & \Sigma B & \xrightarrow{\Sigma \widehat{f''}} & \Sigma^2 D. & &
\end{array}$$

Now using the converse of Axiom (R-Tr2), Lemma 6.7, along with the fact that Σ is fully faithful, we get the following commutative diagram with its middle two rows and two rightmost columns in Δ:

$$
\begin{array}{ccccccc}
 & & F & \xrightarrow{j} & E & = & E \\
 & & {\scriptstyle h'}\big\downarrow & & \big\downarrow{\scriptstyle g'} & & \big\downarrow{\scriptstyle \widehat{f''}\circ g'} \\
D & \xrightarrow{f'} & A & \xrightarrow{f} & B & \xrightarrow{\widehat{f''}} & \Sigma D \\
{\scriptstyle j'}\big\downarrow & & \Vert & & \big\downarrow{\scriptstyle g} & & \big\downarrow{\scriptstyle \Sigma j'} \\
F & \xrightarrow{h'} & A & \xrightarrow{h} & C & \xrightarrow{\widehat{h''}} & \Sigma F \\
 & & & & \big\downarrow{\scriptstyle \widehat{g''}} & & \big\downarrow{\scriptstyle \Sigma j} \\
 & & & & \Sigma E & = & \Sigma E.
\end{array}
$$

The far right column, with repeated left rotations, an application of Exercise 6.1.3, and finally a left adjunct rotation, are shown next:

$$\Omega E \xrightarrow{f''\circ\Omega g'} D \xrightarrow{+j'} F \xrightarrow{+j} E \xrightarrow{\widehat{f''}\circ g'} \Sigma D \xrightarrow{\Sigma j'} \Sigma F \xrightarrow{\Sigma j} \Sigma E.$$

Now taking left adjunct rotations of the previous diagram leads us to a left octahedral diagram, but the version that is dual to the one in the discussion of Axiom (R-Tr4), after Definition 6.1. (This version of (L-Tr4) appears for instance in Beligiannis and Marmaridis [1994, definition 2.2].) Alternatively, we reinterpret the previous diagram to be the following commutative diagram whose middle two rows and first two columns are in Δ:

$$
\begin{array}{ccccccc}
D & = & D & & B & \xrightarrow{\widehat{f''}} & \Sigma D \\
{\scriptstyle j'}\big\downarrow & & \big\downarrow{\scriptstyle f'} & & {\scriptstyle g}\big\downarrow & & \big\downarrow{\scriptstyle \Sigma j'} \\
F & \xrightarrow{h'} & A & \xrightarrow{h} & C & \xrightarrow{\widehat{h''}} & \Sigma F \\
{\scriptstyle j}\big\downarrow & & \big\downarrow{\scriptstyle f} & & \Vert & & \big\downarrow{\scriptstyle \Sigma j} \\
E & \xrightarrow{g'} & B & \xrightarrow{g} & C & \xrightarrow{\widehat{g''}} & \Sigma E \\
{\scriptstyle \widehat{f''}\circ g'}\big\downarrow & & \big\downarrow{\scriptstyle \widehat{f''}} & & & & \\
\Sigma D & = & \Sigma D. & & & &
\end{array}
$$

Now taking the left adjunct rotation of this diagram leads us back to the initially desired diagram in (6.1); the upper right square corresponds to the top left square in Diagram (6.1). □

We have worked throughout with right (pre)triangulated categories with Σ an autoequivalence. But everything, in particular Proposition 6.11, has a dual.

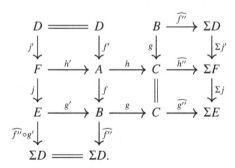

This shows that left and right triangulated categories are the same things when (Σ, Ω) is an autoequivalence. So we make the following definitions.

Definition 6.12 By a **pretriangulated category** $\mathcal{T}$ we mean a right pretriangulated category $\mathcal{T} = (\mathcal{T}, \Sigma, \Delta)$ in which the endofunctor $\Sigma \colon \mathcal{T} \to \mathcal{T}$ is part of an autoequivalence (Σ, Ω). Equivalently, $\mathcal{T} = (\mathcal{T}, \Omega, \nabla)$ is a left pretriangulated category where the elements of ∇ and Δ are in one-to-one correspondence by adjunct rotations. If $(\mathcal{T}, \Sigma, \Delta)$ also satisfies (R-Tr4), or equivalently if $(\mathcal{T}, \Omega, \nabla)$ satisfies (L-Tr4), then we simply say $\mathcal{T}$ is a **triangulated category**. Moreover, any left or right triangle (resp. left or right distinguished triangle) may simply be called a **triangle** (resp. **distinguished triangle**). Functors as in Definition 6.6 are simply called **triangulated functors**, or **exact functors**. Finally, whenever it is known that (Σ, Ω) is an autoequivalence, we may simply denote the axioms of a triangulated category by **(Tr1)**, **(Tr2)**, **(Tr3)**, and **(Tr4)**.

A standard fact is that $\mathrm{Hom}_{\mathcal{T}}(X, -)$ and $\mathrm{Hom}_{\mathcal{T}}(-, X)$ send distinguished triangles to long exact sequences. For example, given any distinguished triangle $A \xrightarrow{f} B \xrightarrow{g} C \xrightarrow{h} \Sigma A$, we get the right long exact sequence starting with

$$\mathrm{Hom}_{\mathcal{T}}(X, A) \xrightarrow{f_*} \mathrm{Hom}_{\mathcal{T}}(X, B) \xrightarrow{g_*} \mathrm{Hom}_{\mathcal{T}}(X, C) \xrightarrow{h_*} \cdots,$$

as in Corollary 6.3(2). But the dual of Corollary 6.3(1) applied to the adjunct rotation $\Omega C \xrightarrow{-\check{h}} A \xrightarrow{f} B \xrightarrow{g} C$ yields a left long exact sequence ending with

$$\cdots \xrightarrow{(-\check{h})_*} \mathrm{Hom}_{\mathcal{T}}(X, A) \xrightarrow{f_*} \mathrm{Hom}_{\mathcal{T}}(X, B) \xrightarrow{g_*} \mathrm{Hom}_{\mathcal{T}}(X, C).$$

Splicing the two sequences together gives us the long exact sequence in the following Corollary.

Corollary 6.13 *Assume $\mathcal{T}$ is a pretriangulated category and let*

$$A \xrightarrow{f} B \xrightarrow{g} C \xrightarrow{h} \Sigma A$$

be a distinguished right triangle. For each object $X \in \mathcal{T}$, applying the functor $\mathrm{Hom}_{\mathcal{T}}(X, -)$ *yields a long exact sequence of abelian groups*

$$\cdots \xrightarrow{(\Omega\check{h})_*} \mathrm{Hom}_{\mathcal{T}}(X, \Omega A) \xrightarrow{(-\Omega f)_*} \mathrm{Hom}_{\mathcal{T}}(X, \Omega B) \xrightarrow{(-\Omega g)_*} \mathrm{Hom}_{\mathcal{T}}(X, \Omega C) \xrightarrow{(-\check{h})_*}$$

$$\mathrm{Hom}_{\mathcal{T}}(X, A) \xrightarrow{f_*} \mathrm{Hom}_{\mathcal{T}}(X, B) \xrightarrow{g_*} \mathrm{Hom}_{\mathcal{T}}(X, C) \xrightarrow{h_*} \mathrm{Hom}_{\mathcal{T}}(X, \Sigma A) \xrightarrow{(-\Sigma f)_*}$$

$$\mathrm{Hom}_{\mathcal{T}}(X, \Sigma B) \xrightarrow{(-\Sigma g)_*} \mathrm{Hom}_{\mathcal{T}}(X, \Sigma C) \xrightarrow{(-\Sigma h)_*} \mathrm{Hom}_{\mathcal{T}}(X, \Sigma^2 A) \xrightarrow{(\Sigma^2 f)_*} \cdots.$$

Applying the contravariant functor $\mathrm{Hom}_{\mathcal{T}}(-, X)$ *yields a similar long exact sequence. Moreover, similar statements hold given any distinguished triangle of the form* $\Omega C \xrightarrow{h} A \xrightarrow{f} B \xrightarrow{g} C$.

6.3 One-Sided Triangulations from Complete Cotorsion Pairs

Let $(\mathcal{X}, \mathcal{Y})$ be a complete cotorsion pair in an exact category $(\mathcal{A}, \mathcal{E})$. The point of this section is to show that the stable category $\mathrm{St}_{\mathcal{X}}(\mathcal{A})$ is a left triangulated category while the stable category $\mathrm{St}_{\mathcal{Y}}(\mathcal{A})$ is a right triangulated category. The immediate consequence we are interested in is made explicit in Section 6.4: If $\mathfrak{M}$ is a Hovey triple on $(\mathcal{A}, \mathcal{E})$, then its left and right stable categories, $\mathrm{St}_{\ell}(\mathcal{A})$ and $\mathrm{St}_{r}(\mathcal{A})$ from Section 4.3, each inherit a one-sided triangulated structure.

The constructions of $\mathrm{St}_{\mathcal{X}}(\mathcal{A})$ and $\mathrm{St}_{\mathcal{Y}}(\mathcal{A})$ are dual to one another and we choose to construct the right triangulation on $\mathrm{St}_{\mathcal{Y}}(\mathcal{A})$. To this end we only need to assume that $(\mathcal{X}, \mathcal{Y})$ is an Ext-pair with enough injectives. *So throughout this section we let* $\mathfrak{C} = (\mathcal{X}, \mathcal{Y})$ *denote an Ext-pair with enough injectives, as defined in Section 2.1. We note that* $\mathcal{A}$ *does not even need to be weakly idempotent complete in this section.*

By Definition 3.12, and by taking $\lambda = \mathcal{Y}$ in the reference there to Setup 3.2.2, the process of taking cokernels of $\mathcal{Y}$-approximations defines an additive functor $\Sigma\colon \mathcal{A} \xrightarrow{\Sigma} \mathrm{St}_{\omega}(\mathcal{A}) \xrightarrow{\bar{\gamma}_{\mathcal{Y}}} \mathrm{St}_{\mathcal{Y}}(\mathcal{A})$. However, as noted in Remarks 3.6 and 3.13, Σ is not an honest functor unless we have assigned, to each $A \in \mathcal{A}$, a particular short exact sequence $\mathbb{S}_A\colon A \rightarrowtail YA \twoheadrightarrow \Sigma A$. So now we make explicit that Σ arises from some particular assignment $\mathbb{S}\colon A \mapsto \mathbb{S}_A$ of such suspension sequences (i.e. $\mathcal{Y}$-approximation sequences). The value of Σ on a morphism $f\colon A \to B$ is computed by constructing any morphism of $\mathcal{Y}$-approximation sequences

$$\begin{array}{ccccc} A & \overset{j_A}{\rightarrowtail} & YA & \overset{q_A}{\twoheadrightarrow} & \Sigma A \\ \downarrow{\scriptstyle f} & & \downarrow{\scriptstyle Y(f)} & & \downarrow{\scriptstyle \Sigma(f)} \\ B & \overset{j_B}{\rightarrowtail} & YB & \overset{q_B}{\twoheadrightarrow} & \Sigma B. \end{array}$$

So Σ is defined by $f \mapsto [\Sigma(f)]_{\mathcal{Y}}$. It is also compatible with $\mathcal{A} \xrightarrow{\gamma_{\mathcal{Y}}} \mathrm{St}_{\mathcal{Y}}(\mathcal{A})$, meaning it descends via the rule $\Sigma([f]_{\mathcal{Y}}) := [\Sigma(f)]_{\mathcal{Y}}$ to an additive endofunctor

$$\Sigma\colon \mathrm{St}_{\mathcal{Y}}(\mathcal{A}) \to \mathrm{St}_{\mathcal{Y}}(\mathcal{A}).$$

This is the **suspension functor** on $\mathrm{St}_{\mathcal{Y}}(\mathcal{A})$; the endofunctor we use to construct the right triangulated structure. We now define the class $\Delta_{\mathcal{Y}}$ of distinguished right triangles in $\mathrm{St}_{\mathcal{Y}}(\mathcal{A})$.

Definition 6.14 Let $f\colon A \to B$ be any morphism in the category $\mathcal{A}$. Taking the suspension sequence, $\mathbb{S}_A\colon A \rightarrowtail YA \twoheadrightarrow \Sigma A$, we define the **cone on f**, denoted $C(f)$, by the pushout construction:

$$\begin{array}{ccccc} A & \overset{j_A}{\rightarrowtail} & YA & \overset{q_A}{\twoheadrightarrow} & \Sigma A \\ \downarrow f & & \downarrow w_f & & \| \\ B & \overset{j_f}{\rightarrowtail} & C(f) & \overset{q_f}{\twoheadrightarrow} & \Sigma A. \end{array}$$

Note that the bottom row is then also a short exact sequence. We obtain a right triangle in $\mathrm{St}_{\mathcal{Y}}(\mathcal{A})$,

$$A \xrightarrow{[f]_{\mathcal{Y}}} B \xrightarrow{[j_f]_{\mathcal{Y}}} C(f) \xrightarrow{[q_f]_{\mathcal{Y}}} \Sigma A,$$

called the **strict right triangle on f**. Any such right triangle is called a **strict right triangle**. Any right triangle in $\mathrm{St}_{\mathcal{Y}}(\mathcal{A})$ that is isomorphic to some strict right triangle is called a **distinguished right triangle** (or a **right exact triangle**). We let $\Delta_{\mathcal{Y}}$ denote the class of all distinguished right triangles in $\mathrm{St}_{\mathcal{Y}}(\mathcal{A})$.

Remark 6.15 The concepts defined in Definition 6.14 are relative to a choice of suspension sequence. However, as long as we are careful, the choice of suspension sequence used is not important. This is clarified in more detail in Section 6.3.1. The main point is that different assignments of suspension sequences, $\mathbb{S}\colon A \mapsto \mathbb{S}_A$ and $\mathbb{S}'\colon A \mapsto \mathbb{S}'_A$, yield naturally isomorphic functors $\Sigma' \cong_{\xi^{\mathbb{S}'}_{\mathbb{S}}} \Sigma$. Indeed for each A, the other sequence, $\mathbb{S}'_A\colon A \rightarrowtail Y'A \twoheadrightarrow \Sigma' A$, induces a canonical isomorphism $[\Sigma(1_A)]_{\mathcal{Y}}\colon \Sigma' A \to \Sigma A$, as in the Dual Fundamental Lemma 3.4. We set $\xi^{\mathbb{S}'_A}_{\mathbb{S}_A} := [\Sigma(1_A)]_{\mathcal{Y}}$, and these morphisms assemble to a natural isomorphism $\xi^{\mathbb{S}'}_{\mathbb{S}} := \{\xi^{\mathbb{S}'_A}_{\mathbb{S}_A}\}$ that we call the **suspension connecting isomorphism** from $\Sigma'\colon \mathcal{A} \to \mathrm{St}_{\mathcal{Y}}(\mathcal{A})$ to $\Sigma\colon \mathcal{A} \to \mathrm{St}_{\mathcal{Y}}(\mathcal{A})$. The following points are clarified in Section 6.3.1: First, the strict right triangle on $f\colon A \to B$ obtained by pushout along $\mathbb{S}'_A$ yields, upon composing with $\xi^{\mathbb{S}'_A}_{\mathbb{S}_A}$, a distinguished triangle relative to $\mathbb{S}_A$ (Corollary 6.18). Second, the suspension connecting isomorphism, $\xi^{\mathbb{S}'}_{\mathbb{S}}$, makes the identity functor $1_{\mathrm{St}_{\mathcal{Y}}(\mathcal{A})}\colon (\mathrm{St}_{\mathcal{Y}}(\mathcal{A}), \Sigma', \Delta'_{\mathcal{Y}}) \to (\mathrm{St}_{\mathcal{Y}}(\mathcal{A}), \Sigma, \Delta_{\mathcal{Y}})$ an isomorphism of right triangulated categories.

We now prove the main result of this section, that $\mathrm{St}_{\mathcal{Y}}(\mathcal{A})$ is a right triangulated category. The ideas here are adapted from the work of Beligiannis and Marmaridis [1994], with some modifications made by Zhi-Wei Li [2015].

Theorem 6.16 (Beligiannis–Marmaridis [1994] and Zhi-Wei Li [2015]) *Let* $(\mathcal{X}, \mathcal{Y})$ *be an Ext-pair with enough injectives in any exact category* $(\mathcal{A}, \mathcal{E})$. *Let* $\Sigma \colon \mathrm{St}_{\mathcal{Y}}(\mathcal{A}) \to \mathrm{St}_{\mathcal{Y}}(\mathcal{A})$ *be the suspension functor arising from some assignment* $\mathbb{S} \colon A \mapsto \mathbb{S}_A$ *of suspension sequences. Let* $\Delta_{\mathcal{Y}}$ *denote the associated class of all distinguished right triangles. Then* $(\Sigma, \Delta_{\mathcal{Y}})$ *is a right triangulated structure on* $\mathrm{St}_{\mathcal{Y}}(\mathcal{A})$.

Proof It is clear from the definition that the class $\Delta_{\mathcal{Y}}$ of distinguished right triangles is closed under isomorphisms. We need to check the four axioms of Definition 6.1. Throughout the proof we will simply write $[f]$ to denote the image of f under the canonical map $\gamma_{\mathcal{Y}} \colon \mathcal{A} \to \mathrm{St}_{\mathcal{Y}}(\mathcal{A})$, rather than the more cumbersome $[f]_{\mathcal{Y}}$.

(R-Tr1) Let $A \xrightarrow{[f]} B$ be a given morphism in $\mathrm{St}_{\mathcal{Y}}(\mathcal{A})$. It is to be shown that there is a distinguished right triangle $A \xrightarrow{[f]} B \to C \to \Sigma A$. But for the representative morphism $f \in \mathcal{A}$, it follows from the definitions that we have a strict right triangle $A \xrightarrow{[f]} B \to C(f) \to \Sigma A$.

Next, let A be a given object in $\mathrm{St}_{\mathcal{Y}}(\mathcal{A})$. It is to be shown that the right triangle $0 \to A \xrightarrow{[1_A]} A \to 0$ is in $\Delta_{\mathcal{Y}}$. This follows from performing the cone construction on the morphism $0 \to A$. Assuming the convention that the suspension sequence assigned to 0 is just $0 \rightarrowtail 0 \twoheadrightarrow 0$, we see this immediately. (For a more general suspension sequence, $0 \rightarrowtail Y \twoheadrightarrow X$, note that $Y \twoheadrightarrow X$ is necessarily an isomorphism with $Y \cong X \in \mathcal{X} \cap \mathcal{Y}$. The pushout construction leads to a strict right triangle taking the form $0 \to A \to A \bigoplus Y \to X$. Since $Y \cong X$ equals 0 in $\mathrm{St}_{\mathcal{Y}}(\mathcal{A})$ we have the desired right triangle.)

(R-Tr2) We must show that if $A \xrightarrow{[f]} B \to C \to \Sigma A$ is a distinguished right triangle then $B \to C \to \Sigma A \xrightarrow{-\Sigma[f]} \Sigma B$ is also a distinguished right triangle. We note that it is enough to consider the case of being given a strict right triangle. This follows from the observation that if

$$\begin{array}{ccc} A & \xrightarrow{[f]} & B \\ {\scriptstyle [\alpha]}\downarrow & & \downarrow{\scriptstyle [\beta]} \\ A' & \xrightarrow[{[f']}]{} & B' \end{array}$$

is a commutative square with $[\alpha]$ and $[\beta]$ isomorphisms, then of course

$$\begin{array}{ccc} \Sigma A & \xrightarrow{-\Sigma[f]} & \Sigma B \\ {\scriptstyle \Sigma[\alpha]}\downarrow & & \downarrow{\scriptstyle \Sigma[\beta]} \\ \Sigma A' & \xrightarrow[{-\Sigma[f']}]{} & \Sigma B' \end{array}$$

is also commutative with $\Sigma[\alpha]$ and $\Sigma[\beta]$ isomorphisms. So suppose we are given a strict right triangle $A \xrightarrow{[f]} B \xrightarrow{[j_f]} C(f) \xrightarrow{[q_f]} \Sigma A$, coming from the pushout diagram:

$$\begin{array}{ccccc} A & \overset{j_A}{\rightarrowtail} & YA & \overset{q_A}{\twoheadrightarrow} & \Sigma A \\ \downarrow{\scriptstyle f} & & \downarrow{\scriptstyle w_f} & & \| \\ B & \overset{j_f}{\rightarrowtail} & C(f) & \overset{q_f}{\twoheadrightarrow} & \Sigma A. \end{array}$$

Referring back to the definition of Σ we also have the following commutative diagram where $h = Y(f)$ and $c = \Sigma(f)$ are the representative morphisms:

$$\begin{array}{ccccc} A & \overset{j_A}{\rightarrowtail} & YA & \overset{q_A}{\twoheadrightarrow} & \Sigma A \\ \downarrow{\scriptstyle f} & & \downarrow{\scriptstyle h} & & \downarrow{\scriptstyle c} \\ B & \overset{j_B}{\rightarrowtail} & YB & \overset{q_B}{\twoheadrightarrow} & \Sigma B. \end{array}$$

Since $\Sigma[f] = [c]$, our job is to show that $B \xrightarrow{[j_f]} C(f) \xrightarrow{[q_f]} \Sigma A \xrightarrow{-[c]} \Sigma B$ is a distinguished right triangle.

We start by using Corollary 1.13 to factor the morphism (f, h, c) of short exact sequences as

$$\begin{array}{ccccc} A & \overset{j_A}{\rightarrowtail} & YA & \overset{q_A}{\twoheadrightarrow} & \Sigma A \\ {\scriptstyle f}\downarrow & & \downarrow{\scriptstyle w_f} & & \| \\ B & \overset{j_f}{\rightarrowtail} & C(f) & \overset{q_f}{\twoheadrightarrow} & \Sigma A \\ \| & & \downarrow{\scriptstyle t} & & \downarrow{\scriptstyle c} \\ B & \overset{j_B}{\rightarrowtail} & YB & \overset{q_B}{\twoheadrightarrow} & \Sigma B. \end{array}$$

We note that the top of the diagram consists again of the previous pushout diagram, and the lower right-hand square is also bicartesian (both pushout and pullback). So it now follows from Proposition 1.18 that

$$C(f) \overset{\begin{bmatrix} t \\ q_f \end{bmatrix}}{\rightarrowtail} YB \bigoplus \Sigma A \overset{[q_B \ -c]}{\twoheadrightarrow} \Sigma B$$

is a short exact sequence (conflation).

Using this we now construct the following commutative diagram whose rows and columns are now known to all be conflations:

$$\begin{array}{ccccc} B & \overset{j_f}{\rightarrowtail} & C(f) & \overset{q_f}{\twoheadrightarrow} & \Sigma A \\ {\scriptstyle j_B}\downarrow & & \downarrow{\scriptstyle \left[\begin{smallmatrix} t \\ q_f \end{smallmatrix}\right]} & & \| \\ YB & \underset{i_1}{\rightarrowtail} & YB \bigoplus \Sigma A & \overset{\pi_2}{\twoheadrightarrow} & \Sigma A \\ {\scriptstyle q_B}\downarrow & & \downarrow{\scriptstyle [q_B\ -c]} & & \\ \Sigma B & = & \Sigma B. & & \end{array}$$

By Proposition 1.12 we conclude that the upper left square must be a pushout. Thus we have constructed the strict right triangle on j_f, as shown in the top row in the following:

$$\begin{array}{ccccccc} B & \xrightarrow{[j_f]} & C(f) & \xrightarrow{\left[\begin{smallmatrix} t \\ q_f \end{smallmatrix}\right]} & YB \bigoplus \Sigma A & \xrightarrow{[q_B\ -c]} & \Sigma B \\ \| & & \| & & \downarrow{\scriptstyle [\pi_2]} & & \| \\ B & \xrightarrow{[j_f]} & C(f) & \xrightarrow{[q_f]} & \Sigma A & \xrightarrow{-[c]} & \Sigma B. \end{array}$$

The two left squares commute "on the nose", meaning they commute even as maps in $\mathcal{A}$. It is easy to see that $[\pi_2]$ is an isomorphism in $\mathrm{St}_{\mathcal{Y}}(\mathcal{A})$, since YB is a zero object. So it is left to show the right square commutes in $\mathrm{St}_{\mathcal{Y}}(\mathcal{A})$. For this we need to see that their difference factors through an object of $\mathcal{Y}$. In fact it is easy to see that $[q_B\ -c] + c\pi_2$ factors as $YB \bigoplus \Sigma A \overset{\pi_1}{\twoheadrightarrow} YB \overset{q_B}{\twoheadrightarrow} \Sigma B$.

(R-Tr3) We must show that "fill-in maps" exists. We first show this for strict right triangles. That is, assume we have two strict right triangles arising from the pushout construction:

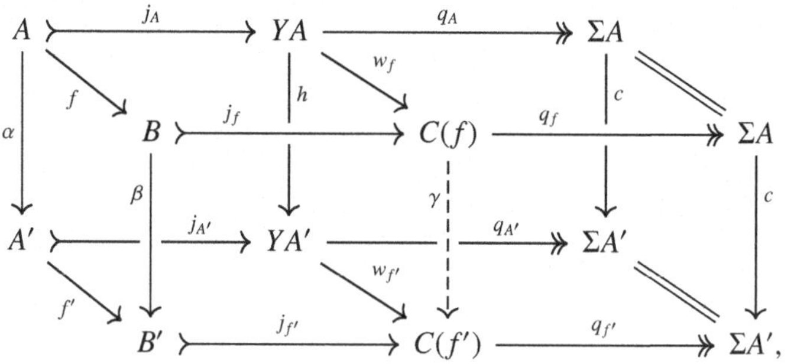

(‡)

along with morphisms α, β such that $[\beta f]_{\mathcal{Y}} = [f'\alpha]_{\mathcal{Y}}$. The back wall of the diagram is a morphism of suspension sequences as in the definition of Σ and where we have set $h = Y(\alpha)$ and $c = \Sigma(\alpha)$. The proof will construct the dashed arrow $C(f) \xrightarrow{[\gamma]} C(f')$ completing the morphism of right triangles in $\mathrm{St}_{\mathcal{Y}}(\mathcal{A})$. Now since $[\beta f] = [f'\alpha]$, we get that $\beta f - f'\alpha$ must factor through some object $Y \in \mathcal{Y}$. But using $\mathrm{Ext}^1_{\mathcal{E}}(\Sigma A, Y) = 0$, it is easy to argue that $\beta f - f'\alpha$ must factor as $A \xrightarrow{j_A} YA \xrightarrow{\psi} B'$ for some map ψ. We claim that $w_{f'}h + j_{f'}\psi$ and $j_{f'}\beta$ are "pushout impostors" of $C(f)$. This would mean $(w_{f'}h + j_{f'}\psi)j_A = j_{f'}\beta f$. Indeed $(w_{f'}h + j_{f'}\psi)j_A = w_{f'}hj_A + j_{f'}\psi j_A = w_{f'}j_{A'}\alpha + j_{f'}\psi j_A = w_{f'}j_{A'}\alpha + j_{f'}(\beta f - f'\alpha) = w_{f'}j_{A'}\alpha + j_{f'}\beta f - j_{f'}f'\alpha = w_{f'}j_{A'}\alpha + j_{f'}\beta f - w_{f'}j_{A'}\alpha = j_{f'}\beta f$.

Since $w_{f'}h + j_{f'}\psi$ and $j_{f'}\beta$ are "pushout impostors" of $C(f)$, there exists a unique map $C(f) \xrightarrow{\gamma} C(f')$ such that

$$\text{(i)}\,\gamma w_f = w_{f'}h + j_{f'}\psi, \qquad \text{(ii)}\,\gamma j_f = j_{f'}\beta.$$

Now one checks that γ is the desired "fill-in" map we seek. Indeed (ii) says that the second square commutes on the nose. We now show that the remaining square also commutes on the nose, meaning $cq_f = q_{f'}\gamma$ (recall $c = \Sigma\alpha$). By the uniqueness property of the pushout $C(f)$, it is enough to show: $cq_fj_f = q_{f'}\gamma j_f$ and $cq_fw_f = q_{f'}\gamma w_f$. But for the first one, we have $cq_fj_f = c0 = 0\beta = (q_{f'}j_{f'})\beta = q_{f'}(j_{f'}\beta) =^{\text{(ii)}} q_{f'}\gamma j_f$. For the second, we also compute: $q_{f'}\gamma w_f =^{\text{(i)}} q_{f'}(w_{f'}h + j_{f'}\psi) = q_{f'}w_{f'}h + q_{f'}j_{f'}\psi = q_{f'}w_{f'}h + 0\psi = q_{A'}h = cq_A = cq_fw_f$.

We have proved (R-Tr3) for strict right triangles. For general distinguished right triangles, the idea of course is that they are isomorphic to strict right triangles, and so the axiom still holds. In more detail, given two right exact triangles, write isomorphisms to some corresponding strict right triangles as indicated next:

$$\begin{array}{ccccccc}
A & \xrightarrow{[f]} & B & \longrightarrow & C(f) & \longrightarrow & \Sigma A \\
\uparrow{\scriptstyle\cong} & & \uparrow{\scriptstyle\cong} & & \uparrow{\scriptstyle\cong} & & \uparrow{\scriptstyle\cong} \\
X & \longrightarrow & Y & \longrightarrow & Z & \longrightarrow & \Sigma X \\
{\scriptstyle[\alpha]}\downarrow & & \downarrow{\scriptstyle[\beta]} & & & & \downarrow{\scriptstyle\Sigma[\alpha]} \\
X' & \longrightarrow & Y' & \longrightarrow & Z' & \longrightarrow & \Sigma X' \\
{\scriptstyle\cong}\downarrow & & \downarrow{\scriptstyle\cong} & & \downarrow{\scriptstyle\cong} & & \downarrow{\scriptstyle\cong} \\
A' & \xrightarrow{[f']} & B' & \longrightarrow & C(f') & \longrightarrow & \Sigma A'.
\end{array}$$

We are assuming the center left square commutes. This gives us a morphism from $[f]$ to $[f']$, and so from what we just proved, a "fill-in" morphism $C(f) \to C(f')$. In the only obvious way this gives rise to a morphism $Z \to Z'$ and it is a "fill-in" morphism as desired.

Having shown that axioms (R-Tr1)–(R-Tr3) hold, we conclude that $(\Sigma, \Delta_{\mathcal{Y}})$ is a right pre-triangulated structure on $\mathrm{St}_{\mathcal{Y}}(\mathcal{A})$. To complete the proof that $\mathrm{St}_{\mathcal{Y}}(\mathcal{A})$ is a right triangulated category we only need to verify the Right Octahedral Axiom, (R-Tr4). But our proof will rely on a uniqueness property for mapping cones, which we will prove in Corollary 6.18. So we will return to the proof of (R-Tr4) afterwards. □

Exercise 6.3.1 Show that, in Diagram (‡) of the proof of Theorem 6.16, we have $[\gamma w_f]_{\mathcal{Y}} = [w_{f'}h]_{\mathcal{Y}}$. Hence the entire diagram commutes in $\mathrm{St}_{\mathcal{Y}}(\mathcal{A})$.

6.3.1 Suspension Connecting Isomorphisms and a Weak Uniqueness Property for Mapping Cones

We now return to clarify the statements made in Remark 6.15. We will use the following observation.

Lemma 6.17 *Consider a commutative square in $\mathcal{A}$ as shown, which is to be thought of as a morphism from f to f':*

$$\begin{array}{ccc} A & \xrightarrow{f} & B \\ {\scriptstyle\alpha}\downarrow & & \downarrow{\scriptstyle\beta} \\ A' & \xrightarrow[f']{} & B'. \end{array}$$

Then for any morphism $\mathbb{S}_A \to \mathbb{S}_{A'}$ of (any choice of) suspension sequences extending $\alpha\colon A \to A'$, there is a unique morphism $\gamma\colon C_{\mathbb{S}_A}(f) \to C_{\mathbb{S}_{A'}}(f')$ making all squares of Diagram (‡) in the proof of Theorem 6.16 commute. In particular, γ is the canonical "fill-in" morphism

$$\begin{array}{ccccccc} A & \xrightarrow{f} & B & \xrightarrow{j_f} & C_{\mathbb{S}_A}(f) & \xrightarrow{q_f} & \Sigma A \\ {\scriptstyle\alpha}\downarrow & & \downarrow{\scriptstyle\beta} & & \downarrow{\scriptstyle\gamma} & & \downarrow{\scriptstyle\Sigma\alpha} \\ A' & \xrightarrow{f'} & B' & \xrightarrow{j_{f'}} & C_{\mathbb{S}_{A'}}(f') & \xrightarrow{q_{f'}} & \Sigma A' \end{array}$$

with respect to the given morphism $\mathbb{S}_A \to \mathbb{S}_{A'}$ of suspension sequences.

Proof Look back to the proof of (R-Tr3) of Theorem 6.16. The proof shows that if $[\beta f]_{\mathcal{Y}} = [f'\alpha]_{\mathcal{Y}}$, then a morphism γ exists completing a morphism of strict right triangles (relative to the choices $\mathbb{S}_A$ and $\mathbb{S}_{A'}$). The uniqueness property of γ relates to a map ψ needed in the proof. However, the reader should go back and read the proof assuming $\beta f = f'\alpha$. That is, look what happens when the left square literally commutes in $\mathcal{A}$, not just in $\mathrm{St}_{\mathcal{Y}}(\mathcal{A})$. We see that we can take $\psi = 0$ in this case. Doing so, we follow the proof to see that γ makes all squares of Diagram (‡) commute in $\mathcal{A}$, and the universal property of pushout guarantees that it is the *unique* map with this property, depending only on the map h which in turn determines the morphism of suspension sequences. □

Corollary 6.18 *Suppose that an assignment $\mathbb{S}\colon A \mapsto \mathbb{S}_A$ of suspension sequences $\mathbb{S}_A\colon A \rightarrowtail YA \twoheadrightarrow \Sigma A$ has been made for each object $A \in \mathcal{A}$. Let $f\colon A \to B$ be a morphism in $\mathcal{A}$. Let $C_{\mathbb{S}_A}(f)$ denote the cone on f and let $C_{\mathbb{S}'_A}(f)$ denote the cone on f coming from another choice of suspension sequence $\mathbb{S}'_A\colon A \rightarrowtail Y'A \twoheadrightarrow \Sigma' A$. Then for any morphism $\mathbb{S}'_A \to \mathbb{S}_A$ of suspension sequences extending the identity $1_A\colon A \to A$, there is a unique morphism $\gamma\colon C_{\mathbb{S}'_A}(f) \to C_{\mathbb{S}_A}(f)$ making Diagram (‡) commute in the proof of Theorem 6.16 (with $\alpha = 1_A, \beta = 1_B$). In particular, γ is the canonical "fill-in" morphism*

$$\begin{array}{ccccccc} A & \xrightarrow{f} & B & \xrightarrow{j'_f} & C_{\mathbb{S}'_A}(f) & \xrightarrow{q'_f} & \Sigma' A \\ \downarrow{\scriptstyle 1_A} & & \downarrow{\scriptstyle 1_B} & & \downarrow{\scriptstyle \exists!\,\gamma} & & \downarrow{\scriptstyle \Sigma(1_A)} \\ A & \xrightarrow{f} & B & \xrightarrow{j_f} & C_{\mathbb{S}_A}(f) & \xrightarrow{q_f} & \Sigma A \end{array}$$

with respect to the given morphism $\mathbb{S}'_A \to \mathbb{S}_A$ of suspension sequences. Moreover, γ is a $\mathcal{Y}$-stable equivalence. Therefore,

$$A \xrightarrow{[f]} B \xrightarrow{[j'_f]} C_{\mathbb{S}'_A}(f) \xrightarrow{\xi^{\mathbb{S}'_A}_{\mathbb{S}_A} \circ [q'_f]} \Sigma A$$

is a distinguished triangle, for it is isomorphic in $\mathrm{St}_{\mathcal{Y}}(\mathcal{A})$ to the strict right triangle on f. Here, $\xi^{\mathbb{S}'_A}_{\mathbb{S}_A}\colon \Sigma' A \to \Sigma A$ is the canonical ***suspension connecting isomorphism*** *defined by $\xi^{\mathbb{S}'_A}_{\mathbb{S}_A} := [\Sigma(1_A)]_{\mathcal{Y}}$ as in Remark 6.15.*

Proof We take $\alpha = 1_A$ and $\beta = 1_B$ in Lemma 6.17. Upon applying the canonical functor $\gamma_{\mathcal{Y}}\colon \mathcal{A} \to \mathrm{St}_{\mathcal{Y}}(\mathcal{A})$ we get a morphism of strict right triangles, but they are relative to two different right pretriangulated structures. Indeed we already have the right pretriangulated structure, $(\mathrm{St}_{\mathcal{Y}}(\mathcal{A}), \Sigma, \Delta_{\mathcal{Y}})$, coming from the given assignment of suspension sequences. Replacing $\mathbb{S}_A$ with $\mathbb{S}'_A$ just leads to another right pretriangulated structure $(\mathrm{St}_{\mathcal{Y}}(\mathcal{A}), \Sigma', \Delta'_{\mathcal{Y}})$. Now one easily checks that the proof of the Five Lemma (Corollary 6.4) still holds when applied to a morphism of right triangles between these two different structures. The reason is because the canonical suspension connecting morphism $\xi^{\mathbb{S}'}_{\mathbb{S}} := \{\xi^{\mathbb{S}'_A}_{\mathbb{S}_A} := [\Sigma(1_A)]\colon \Sigma' A \to \Sigma A\}$ is also an isomorphism. □

Corollary 6.18 and its proof also immediately give us the following.

Corollary 6.19 *Suppose that $\mathbb{S}\colon A \mapsto \mathbb{S}_A$ and $\mathbb{S}'\colon A \mapsto \mathbb{S}'_A$ are any two assignments of suspension sequences for the objects of $\mathcal{A}$. Then the identity functor on* $\mathrm{St}_{\mathcal{Y}}(\mathcal{A})$ *is an isomorphism from the right pretriangulated structure* $(\mathrm{St}_{\mathcal{Y}}(\mathcal{A}), \Sigma', \Delta'_{\mathcal{Y}})$, *to the right pretriangulated structure* $(\mathrm{St}_{\mathcal{Y}}(\mathcal{A}), \Sigma, \Delta_{\mathcal{Y}})$. *In particular, the canonical suspension connecting isomorphism*

$$\xi_{\mathbb{S}}^{\mathbb{S}'} := \left\{\xi_{\mathbb{S}_A}^{\mathbb{S}'_A} : \Sigma' A \xrightarrow{[\Sigma(1_A)]} \Sigma A\right\}$$

is the natural isomorphism $\Sigma' \cong_{\xi_{\mathbb{S}}^{\mathbb{S}'}} \Sigma$ *needed to satisfy Definition 6.6.*

The practical consequence of these corollaries is that we may pushout a morphism $f\colon A \to B$ along any suspension sequence for A to obtain isomorphic distinguished right triangles. One just needs to realize that, technically, we must compose with the canonical suspension connecting isomorphism $\Sigma' A \cong_{\xi_{\mathbb{S}}^{\mathbb{S}'}} \Sigma A$.

There is one more simple point to make before we proceed to the proof of axiom (R-Tr4). Having verified axioms (R-Tr1)–(R-Tr3), the following is immediate by the Five Lemma (Corollary 6.4 and Remark 6.5).

Remark 6.20 Any distinguished right triangle $A \xrightarrow{[f]} B \xrightarrow{[g]} C \xrightarrow{[h]} \Sigma A$ in $\mathrm{St}_{\mathcal{Y}}(\mathcal{A})$ is isomorphic to the strict right triangle $A \xrightarrow{[t]} B \xrightarrow{[j_t]} C(t) \xrightarrow{[q_t]} \Sigma A$, on any morphism $t : A \to B$ representing the equivalence class $[f]_{\mathcal{Y}}$.

6.3.2 Proof of Axiom (R-Tr4)

Now we return to complete the proof of Theorem 6.16.

(Proof of the Right Octahedral Axiom) Consider a given composition $[h] = [g] \circ [f]$ of two morphisms in $\mathrm{St}_{\mathcal{Y}}(\mathcal{A})$ along with three distinguished right triangles on these three morphisms as in the hypotheses of (R-Tr4). The reader should convince themselves that these three given right triangles may be replaced by any isomorphic right triangles (having the same bases $[f]$, $[g]$, and $[h]$). In particular, Corollary 6.18 and Remark 6.20 imply that we may reduce the proof of (R-Tr4) to the case where we are given an honest composition $h = gf$, in $\mathcal{A}$, and strict right triangles on those three built from any choice of suspension sequences we wish.

So we start by taking the suspension sequence of A and then taking two consecutive pushouts as shown:

$$\begin{array}{ccccc} A & \overset{j_A}{\rightarrowtail} & YA & \overset{q_A}{\twoheadrightarrow} & \Sigma A \\ {\scriptstyle f}\downarrow & & \downarrow{\scriptstyle w_f} & & \| \\ B & \overset{j_f}{\rightarrowtail} & C(f) & \overset{q_f}{\twoheadrightarrow} & \Sigma A \\ {\scriptstyle g}\downarrow & & \downarrow{\scriptstyle s} & & \| \\ C & \overset{j_h}{\rightarrowtail} & C(h) & \overset{q_h}{\twoheadrightarrow} & \Sigma A. \end{array} \qquad (\star)$$

By Lemma 1.7, the outer rectangle formed by the two pushout squares is itself a pushout, of h. Thus s satisfies $w_h = sw_f$ and j_h, q_h is the correct notation for these morphisms. We have the two strict right triangles

$$A \xrightarrow{[f]} B \xrightarrow{[j_f]} C(f) \xrightarrow{[q_f]} \Sigma A$$

and

$$A \xrightarrow{[h]} C \xrightarrow{[j_h]} C(h) \xrightarrow{[q_h]} \Sigma A.$$

Take the suspension sequence $C(f) \overset{j_{C(f)}}{\rightarrowtail} YC(f) \overset{q_{C(f)}}{\twoheadrightarrow} \Sigma C(f)$. Then the defined composite $j_B := j_{C(f)} \circ j_f$ is an admissible monomorphism and, by Corollary 1.14(2), there is a bicartesian (meaning both pullback and pushout) diagram with exact rows and columns as shown:

$$\begin{array}{ccccc} B & \overset{j_f}{\rightarrowtail} & C(f) & \overset{q_f}{\twoheadrightarrow} & \Sigma A \\ \| & & \downarrow{\scriptstyle j_{C(f)}} & & \downarrow{\scriptstyle j'} \\ B & \overset{j_B}{\rightarrowtail} & YC(f) & \overset{q_B}{\twoheadrightarrow} & \Sigma B \\ & & \downarrow{\scriptstyle q_{C(f)}} & & \downarrow{\scriptstyle q'} \\ & & \Sigma C(f) & = & \Sigma C(f). \end{array} \qquad (\star\star)$$

Our choice of notation for the middle row suggests that it is an (alternate) suspension sequence for B. This is indeed the case since $YC(f) \in \mathcal{Y}$ and ΣB is displayed as an extension of two objects in $\mathcal{X}$, hence $\Sigma B \in \mathcal{X}$ too. Using this suspension sequence we construct a pushout diagram as shown:

$$\begin{array}{ccccc} B & \overset{j_B}{\rightarrowtail} & YC(f) & \overset{q_B}{\twoheadrightarrow} & \Sigma B \\ \downarrow{\scriptstyle g} & & \downarrow{\scriptstyle w_g} & & \| \\ C & \overset{j_g}{\rightarrowtail} & C(g) & \overset{q_g}{\twoheadrightarrow} & \Sigma B. \end{array} \qquad (\star\star\star)$$

This produces a strict right triangle on g (relative to the new suspension sequence for B), but again, Corollary 6.18 allows us to use it to replace the given strict triangle on g.

The following left square is a previously constructed pushout, but the right square also commutes by the calculation that follows the diagrams:

$$\begin{array}{ccc} A & \xrightarrow{j_A} & YA \\ {\scriptstyle h}\downarrow & & \downarrow{\scriptstyle w_h} \\ C & \xrightarrow[j_h]{} & C(h) \end{array} \qquad \begin{array}{ccc} A & \xrightarrow{j_A} & YA \\ {\scriptstyle h}\downarrow & & \downarrow{\scriptstyle w_g \circ j_{C(f)} \circ w_f} \\ C & \xrightarrow[j_g]{} & C(g) \end{array}$$

$$\begin{aligned} w_g \circ j_{C(f)} \circ w_f \circ j_A &= w_g \circ j_{C(f)} \circ j_f \circ f, \quad \text{by Diagram } (\star) \\ &= w_g \circ j_B \circ f, \quad \text{by Diagram } (\star\star) \\ &= j_g \circ g \circ f, \quad \text{by Diagram } (\star\star\star) \\ &= j_g \circ h. \end{aligned}$$

So by the universal property of the pushout $C(h)$ there exists a unique morphism $C(h) \xrightarrow{j_s} C(g)$ such that

$$\text{(i)}\ \ j_s \circ j_h = j_g \quad \text{and} \quad \text{(ii)}\ \ j_s \circ w_h = w_g \circ j_{C(f)} \circ w_f.$$

(The choice of notation j_s will be explained ahead.) Since Diagram $(\star)$ told us that $sw_f = w_h$, Equation (ii) implies $j_s \circ s \circ w_f = w_g \circ j_{C(f)} \circ w_f$. Hence

$$\text{(iii)}\ \ (j_s \circ s - w_g \circ j_{C(f)}) \circ w_f = 0.$$

But also,

$$\begin{aligned} (j_s \circ s - w_g \circ j_{C(f)}) \circ j_f &= j_s \circ s \circ j_f - w_g \circ j_{C(f)} \circ j_f \\ &= j_s \circ j_h \circ g - w_g \circ j_B, \quad \text{by Diagrams } (\star) \text{ and } (\star\star) \\ &= j_g \circ g - j_g \circ g, \quad \text{by (i) and Diagram } (\star\star\star) \\ &= 0. \end{aligned}$$

Therefore,

$$\text{(iv)}\ \ (j_s \circ s - w_g \circ j_{C(f)}) \circ j_f = 0.$$

Now again, the following left square is a pushout, and the right square is an "impostor", as it trivially commutes:

$$\begin{array}{ccc} A & \xrightarrow{j_A} & YA \\ {\scriptstyle f}\downarrow & & \downarrow{\scriptstyle w_f} \\ B & \xrightarrow[j_f]{} & C(f), \end{array} \qquad \begin{array}{ccc} A & \xrightarrow{j_A} & YA \\ {\scriptstyle f}\downarrow & & \downarrow{\scriptstyle 0} \\ B & \xrightarrow[0]{} & C(g). \end{array}$$

So along with Equations (iii) and (iv), the universal property of the pushout $C(f)$ implies $j_s \circ s - w_g \circ j_{C(f)} = 0$. Hence,

$$\text{(v)}\ \ j_s \circ s = w_g \circ j_{C(f)}.$$

Now we have shown, by Equation (v) and Diagram (⋆), that the following commutes and the left square is a pushout:

$$\begin{array}{ccccc} B & \overset{j_f}{\rightarrowtail} & C(f) & \overset{j_{C(f)}}{\rightarrowtail} & YC(f) \\ \downarrow{\scriptstyle g} & & \downarrow{\scriptstyle s} & & \downarrow{\scriptstyle w_g} \\ C & \overset{j_h}{\rightarrowtail} & C(h) & \overset{j_s}{\rightarrowtail} & C(g). \end{array}$$

The outer rectangle is also a pushout by Equation (i) and Diagram (⋆ ⋆ ⋆). It follows from Lemma 1.7 that the right square is also a pushout. Hence by Lemma 1.10 we have a commutative diagram with exact rows:

$$\begin{array}{ccccc} C(f) & \overset{j_{C(f)}}{\rightarrowtail} & YC(f) & \overset{q_{C(f)}}{\twoheadrightarrow} & \Sigma C(f) \\ \downarrow{\scriptstyle s} & & \downarrow{\scriptstyle w_g} & & \| \\ C(h) & \overset{j_s}{\rightarrowtail} & C(g) & \overset{q_s}{\twoheadrightarrow} & \Sigma C(f). \end{array} \qquad (\ddagger\ddagger)$$

We now note that this explains our choice of notation j_s, q_s, for this provides us with a strict right triangle

$$C(f) \xrightarrow{[s]} C(h) \xrightarrow{[j_s]} C(g) \xrightarrow{[q_s]} \Sigma C(f).$$

We claim that this is the desired triangle, so it remains to show that the following diagram commutes, up to the $\sim_{\mathcal{Y}}$ relation:

$$\begin{array}{ccccccc} A & \xrightarrow{f} & B & \xrightarrow{j_f} & C(f) & \xrightarrow{q_f} & \Sigma A \\ \| & & {\scriptstyle g}\downarrow & & \downarrow{\scriptstyle s} & & \| \\ A & \xrightarrow{h} & C & \xrightarrow{j_h} & C(h) & \xrightarrow{q_h} & \Sigma A \\ & & {\scriptstyle j_g}\downarrow & & \downarrow{\scriptstyle j_s} & & \downarrow{\scriptstyle \Sigma f} \\ & & C(g) & = & C(g) & \xrightarrow{q_g} & \Sigma B \\ & & {\scriptstyle q_g}\downarrow & & \downarrow{\scriptstyle q_s} & & \\ & & \Sigma B & \xrightarrow{\Sigma(j_f)} & \Sigma C(f). & & \end{array}$$

The top left square commutes on the nose, as we chose the representative to be $h = gf$ in $\mathcal{A}$. The top middle and top right squares commute on the nose, by Diagram (⋆). The center square commutes on the nose, by Equation (i).

Let us show that the center right square commutes. To start, we see that the following commutative diagram shows that we may pick the representative for $Y(f)$ to be $j_{C(f)} \circ w_f$ and the representative Σf to be j':

$$
\begin{array}{ccccc}
A & \overset{j_A}{\rightarrowtail} & YA & \overset{q_A}{\twoheadrightarrow} & \Sigma A \\
{\scriptstyle f}\downarrow & & \downarrow{\scriptstyle w_f} & & \| \\
B & \overset{j_f}{\rightarrowtail} & C(f) & \overset{q_f}{\twoheadrightarrow} & \Sigma A \\
\| & & \downarrow{\scriptstyle j_{C(f)}} & & \downarrow{\scriptstyle j'} \\
B & \overset{j_B}{\rightarrowtail} & YC(f) & \overset{q_B}{\twoheadrightarrow} & \Sigma B.
\end{array}
$$

Next, we will show

$$\text{(vi)}\ (\Sigma f \circ q_h - q_g \circ j_s) \circ w_h = 0 \quad \text{and} \quad \text{(vii)}\ (\Sigma f \circ q_h - q_g \circ j_s) \circ j_h = 0.$$

Showing (vi), with the choice $\Sigma f = j'$, we have

$$
\begin{aligned}
j' \circ q_h \circ w_h - q_g \circ j_s \circ w_h &= j' \circ q_A - q_g \circ j_s \circ w_h, \quad \text{by Diagram } (\star) \\
&= q_B \circ j_{C(f)} \circ w_f - q_g \circ (j_s \circ w_h), \quad \text{by previous diagram} \\
&= q_B \circ j_{C(f)} \circ w_f - q_g \circ w_g \circ j_{C(f)} \circ w_f, \quad \text{by Equation (ii)} \\
&= q_B \circ j_{C(f)} \circ w_f - q_B \circ j_{C(f)} \circ w_f, \quad \text{by Diagram } (\star\star\star) \\
&= 0.
\end{aligned}
$$

Showing (vii), we use Equation (i) and look back at Diagrams ($\star$) and ($\star\star\star$) to get

$$j' \circ q_h \circ j_h - q_g(\circ j_s \circ j_h) = j' \circ q_h \circ j_h - q_g \circ j_g = 0 - 0 = 0.$$

Since we have the pushout square and "impostor" shown in the following

$$
\begin{array}{ccc}
A & \overset{j_A}{\longrightarrow} & YA \\
{\scriptstyle h}\downarrow & & \downarrow{\scriptstyle w_h} \\
C & \underset{j_h}{\longrightarrow} & C(h),
\end{array}
\qquad\qquad
\begin{array}{ccc}
A & \overset{j_A}{\longrightarrow} & YA \\
{\scriptstyle h}\downarrow & & \downarrow{\scriptstyle 0} \\
C & \underset{0}{\longrightarrow} & \Sigma B,
\end{array}
$$

another uniqueness argument using Equations (vi) and (vii) tells us $\Sigma f \circ q_h = q_g \circ j_s$.

Finally, let us show the bottom center square commutes. To do this we first note that Diagram ($\star\star$) can be reinterpreted to give a morphism of short exact sequences:

$$
\begin{array}{ccccc}
B & \overset{j_B}{\rightarrowtail} & YC(f) & \overset{q_B}{\twoheadrightarrow} & \Sigma B \\
\downarrow{\scriptstyle j_f} & & \| & & \downarrow{\scriptstyle q'} \\
C(f) & \overset{j_{C(f)}}{\rightarrowtail} & YC(f) & \overset{q_{C(f)}}{\twoheadrightarrow} & \Sigma C(f).
\end{array}
$$

Hence we may take q' to be the representative $\Sigma(j_f) := q'$. Now we, again, will use the pushout technique to show $q' \circ q_g = q_s$. This time we use that $C(g)$ is the pushout square in Diagram ($\star\star\star$). So it only remains to show

$$(q' \circ q_g - q_s) \circ w_g = 0 \quad \text{and} \quad (q' \circ q_g - q_s) \circ j_g = 0.$$

For the first we have

$$\begin{aligned} q' \circ (q_g \circ w_g) - q_s \circ w_g &= q' \circ q_B - q_s \circ w_g\,, \quad \text{by Diagram } (\star\star\star) \\ &= q_{C(f)} - q_s \circ w_g\,, \quad \text{by Diagram } (\star\star\star) \\ &= q_{C(f)} - q_{C(f)}\,, \quad \text{by Diagram } (\ddagger\ddagger) \\ &= 0. \end{aligned}$$

For the second we have

$$\begin{aligned} q' \circ (q_g \circ j_g) - q_s \circ j_g &= 0 - q_s \circ j_g\,, \quad \text{by Diagram } (\star\star\star) \\ &= -q_s \circ (j_s \circ j_h)\,, \quad \text{by Equation (i)} \\ &= (-q_s \circ j_s) \circ j_h \\ &= 0\,, \quad \text{by Diagram } (\ddagger\ddagger). \end{aligned}$$

So all squares commute! □

Example 6.21 Assume that $(\mathcal{A}, \mathcal{E})$ is an exact category with enough injectives. Denoting the class of all injectives by $\mathcal{I}$, this means $(\mathcal{A}, \mathcal{I})$ is a complete cotorsion pair. We call $\mathrm{St}_{\mathcal{I}}(\mathcal{A})$ the *injective stable category* of $\mathcal{A}$ and Theorem 6.16 applies: The associated endofunctor $\Sigma\colon \mathrm{St}_{\mathcal{I}}(\mathcal{A}) \to \mathrm{St}_{\mathcal{I}}(\mathcal{A})$ and the corresponding class $\Delta_{\mathcal{I}}$ of distinguished right triangles so defined, make $(\Sigma, \Delta_{\mathcal{I}})$ a right triangulated structure on $\mathrm{St}_{\mathcal{I}}(\mathcal{A})$. Dually, if $(\mathcal{A}, \mathcal{E})$ is an exact category with enough projectives, then we have a left triangulated structure $(\Omega, \nabla_{\mathcal{P}})$ on the *projective stable category* $\mathrm{St}_{\mathcal{P}}(\mathcal{A})$.

Exercise 6.3.2 (Suspension Connecting Isomorphism) Consider the pushout diagram arising from two choices of suspension sequences, $\mathbb{S}_A$ and $\mathbb{S}'_A$, of an object A:

$$\begin{array}{ccccc} A & \overset{j'_A}{\rightarrowtail} & Y'A & \overset{q'_A}{\twoheadrightarrow} & \Sigma' A \\ {\scriptstyle j_A}\downarrow & & \downarrow{\scriptstyle \bar{j}_A} & & \| \\ YA & \overset{\bar{j}'_A}{\rightarrowtail} & P & \overset{\bar{q}'_A}{\rightarrowtail} & \Sigma' A \\ {\scriptstyle q_A}\downarrow & & \downarrow{\scriptstyle \bar{q}_A} & & \\ \Sigma A & = & \Sigma A. & & \end{array}$$

Show that $\bar{q}'_A$ and $\bar{q}_A$ are $\mathcal{Y}$-stable equivalences, and that $\xi_{\mathbb{S}_A}^{\mathbb{S}'_A} = -[\bar{q}_A] \circ [\bar{q}'_A]^{-1}$. Here, $\xi_{\mathbb{S}_A}^{\mathbb{S}'_A}$ is the natural *suspension connecting isomorphism*. Again, it is the morphism in $\mathrm{St}_{\mathcal{Y}}(\mathcal{A})$ defined by $\xi_{\mathbb{S}_A}^{\mathbb{S}'_A} := [\Sigma(1_A)]$, where $\Sigma(1_A) \colon \Sigma' A \to \Sigma A$ is any morphism obtained as in the Dual Fundamental Lemma 3.4. In fact, for the short exact sequence positioned along the top of the pushout diagram, we do not even need $Y'A \in \mathcal{Y}$ to get $[\Sigma(1_A)] = -[\bar{q}_A] \circ [\bar{q}'_A]^{-1}$; we just need $\Sigma' A \in \mathcal{X}$.

6.4 The Suspension and Loop Functor on $\mathrm{Ho}(\mathfrak{M})$

Throughout this section we let $\mathfrak{M} = (\mathcal{Q}, \mathcal{W}, \mathcal{R})$ denote a Hovey triple on a weakly idempotent complete exact category $(\mathcal{A}, \mathcal{E})$. $\mathfrak{M}$ may even just be a Hovey Ext-triple. Our goal is to construct an adjoint autoequivalence (Σ, Ω) of the homotopy category $\mathrm{Ho}(\mathfrak{M})$ which will serve as the suspension and loop functors in the triangulated structure constructed in the next section.

In Section 4.3 we defined $\mathrm{St}_\ell(\mathcal{A})$, the *left stable category of* $\mathfrak{M}$ and the canonical functor $\gamma_\ell \colon \mathcal{A} \to \mathrm{St}_\ell(\mathcal{A})$. Similarly, we have a canonical functor $\gamma_r \colon \mathcal{A} \to \mathrm{St}_r(\mathcal{A})$ onto the *right stable category of* $\mathfrak{M}$.

We follow the discussion at the start of Section 6.3, but specializing the Ext-pair $\mathfrak{C} = (\mathcal{X}, \mathcal{Y})$ to be $\mathfrak{L} = (\mathcal{Q}, \mathcal{R}_{\mathcal{W}})$, the left Ext-pair associated to $\mathfrak{M}$. That is, having made an assignment $\mathbb{S} \colon A \mapsto \mathbb{S}_A$ of **suspension sequences**, $\mathbb{S}_A \colon A \rightarrowtail \tilde{R}_A \twoheadrightarrow \Sigma A$ (so with $\Sigma A \in \mathcal{Q}$, and $\tilde{R}_A \in \mathcal{R}_{\mathcal{W}}$), we obtain an additive functor $\Sigma_\ell \colon \mathcal{A} \xrightarrow{\Sigma_\omega} \mathrm{St}_\omega(\mathcal{A}) \xrightarrow{\bar{\gamma}_\ell} \mathrm{St}_\ell(\mathcal{A})$. Its value on a morphism $f \colon A \to B$ is computed by constructing a morphism of suspension sequences:

$$\begin{array}{ccccc} A & \overset{k_A}{\rightarrowtail} & \tilde{R}_A & \overset{\sigma_A}{\twoheadrightarrow} & \Sigma A \\ \downarrow{\scriptstyle f} & & \downarrow{\scriptstyle \tilde{f}} & & \downarrow{\scriptstyle \Sigma(f)} \\ B & \overset{k_B}{\rightarrowtail} & \tilde{R}_B & \overset{\sigma_B}{\twoheadrightarrow} & \Sigma B. \end{array}$$

So Σ_ℓ is defined by $f \mapsto [\Sigma(f)]_\ell$. It is compatible with $\mathcal{A} \xrightarrow{\gamma_\ell} \mathrm{St}_\ell(\mathcal{A})$ in the sense that it descends via $[f]_\ell \mapsto [\Sigma(f)]_\ell$ to an additive endofunctor

$$\Sigma_\ell^\ell \colon \mathrm{St}_\ell(\mathcal{A}) \to \mathrm{St}_\ell(\mathcal{A}).$$

In other words, we have the equation

$$\Sigma_\ell := \Sigma_\ell^\ell \circ \gamma_\ell. \tag{6.2}$$

The subscripts and superscripts are meant to emphasize the domain (superscript) and codomain (subscript) of the functor. We have introduced the more cumbersome notation as it will at times be convenient to clarify computations appearing in the rest of the chapter. But owing to the compatibilities, one may certainly drop the subscripts and superscripts. Any of the Σ functors may be called the **suspension functor**.

Now let Δ_ℓ denote the corresponding class of distinguished right triangles in $\mathrm{St}_\ell(\mathcal{A})$, as defined in Definition 6.14. Then we get the following result as a special case of Theorem 6.16.

Corollary 6.22 *$(\Sigma^\ell_\ell, \Delta_\ell)$ is a right triangulated structure on the left stable category $\mathrm{St}_\ell(\mathcal{A})$.*

Of course there is also the dual notion. That is, we have the **loop functor**, Ω, which in particular can be made into an endofunctor $\Omega^r_r \colon \mathrm{St}_r(\mathcal{A}) \to \mathrm{St}_r(\mathcal{A})$. Ω is defined by using enough projectives of the right pair, $\mathfrak{R} = (\mathcal{Q}_{\mathcal{W}}, \mathcal{R})$, as indicated:

$$\begin{array}{ccccc} \Omega A & \overset{\iota_A}{\rightarrowtail} & \tilde{Q}_A & \overset{\rho_A}{\twoheadrightarrow} & A \\ {\scriptstyle \Omega(f)}\downarrow & & \downarrow{\scriptstyle \tilde{f}} & & \downarrow{\scriptstyle f} \\ \Omega B & \overset{\iota_B}{\rightarrowtail} & \tilde{Q}_B & \overset{\rho_B}{\twoheadrightarrow} & B. \end{array}$$

The dual of Corollary 6.22 applies to provide a left triangulated structure (Ω^r_r, ∇_r) on the right stable category $\mathrm{St}_r(\mathcal{A})$.

We will not need the statement in the following exercise elsewhere in the book, but it reveals an interesting relationship between the left and right stable categories of $\mathfrak{M}$ and the classical Yoneda Ext groups. See also Theorem 8.18.

Exercise 6.4.1 Let $\mathfrak{M} = (\mathcal{Q}, \mathcal{W}, \mathcal{R})$ be an abelian model structure on $(\mathcal{A}, \mathcal{E})$. Let $QA \twoheadrightarrow A$ be any cofibrant replacement and $B \rightarrowtail RB$ any fibrant replacement. Then there are natural isomorphisms

$$\mathrm{St}_r(\mathcal{A})\big(\Omega^r_r QA, RB\big) \cong \mathrm{Ext}^1_{\mathcal{E}}(QA, RB) \cong \mathrm{St}_\ell(\mathcal{A})\left(QA, \Sigma^\ell_\ell RB\right).$$

The plan for constructing the desired autoequivalence $\Sigma \colon \mathrm{Ho}(\mathfrak{M}) \to \mathrm{Ho}(\mathfrak{M})$ is to show that it is induced by the suspension functor $\Sigma_\ell \colon \mathcal{A} \to \mathrm{St}_\ell(\mathcal{A})$. To do this, recall that the canonical functor $\gamma \colon \mathcal{A} \to \mathrm{Ho}(\mathfrak{M})$ descends to a functor $\bar{\gamma}^\ell \colon \mathrm{St}_\ell(\mathcal{A}) \to \mathrm{Ho}(\mathfrak{M})$ simply because γ identifies left homotopic maps. (See Definition 5.20.) Our goal is to show that the composition

$$\bar{\gamma}^\ell \circ \Sigma_\ell \colon \mathcal{A} \xrightarrow{\Sigma_\ell} \mathrm{St}_\ell(\mathcal{A}) \xrightarrow{\bar{\gamma}^\ell} \mathrm{Ho}(\mathfrak{M})$$

sends weak equivalences to isomorphisms. Then the Localization Theorem 5.21 will apply to give us the desired functor $\Sigma := \mathrm{Ho}\left(\bar{\gamma}^\ell \circ \Sigma_\ell\right)$. This appears as Proposition 6.25, and we now proceed to break the proof down into two lemmas.

Lemma 6.23 *Let f be a trivial cofibration. Then $\gamma(\Sigma(f))$ is an isomorphism. Consequently, $\Sigma(f)$ is a weak equivalence whenever $\mathfrak{M}$ is a Hovey triple (not just a Hovey Ext-triple). In fact, suppose $f\colon A \rightarrowtail B$ is any cofibration with cokernel $Q \in \mathcal{Q}$ and suppose we are given suspension sequences for A and Q as shown:*

$$A \overset{k_A}{\rightarrowtail} \tilde{R}_A \overset{\sigma_A}{\twoheadrightarrow} \Sigma A$$

and

$$Q \overset{k_Q}{\rightarrowtail} \tilde{R}_Q \overset{\sigma_Q}{\twoheadrightarrow} \Sigma Q.$$

Then there exists a 3×3 diagram with exact rows and columns as shown:

$$\begin{array}{ccccc}
\Sigma A & \overset{\Sigma(f)}{\rightarrowtail} & \Sigma B & \overset{\Sigma(c)}{\twoheadrightarrow} & \Sigma Q \\
{\scriptstyle \sigma_A}\uparrow & & \uparrow & & \uparrow{\scriptstyle \sigma_Q} \\
\tilde{R}_A & \rightarrowtail & \tilde{R}_A \bigoplus \tilde{R}_Q & \twoheadrightarrow & \tilde{R}_Q \\
{\scriptstyle k_A}\uparrow & & \uparrow & & \uparrow{\scriptstyle k_Q} \\
A & \overset{f}{\rightarrowtail} & B & \overset{c}{\twoheadrightarrow} & Q.
\end{array}$$

In particular, the middle column is also a suspension sequence, and $\Sigma(f)$ is a cofibration (resp. trivial cofibration) whenever f is a cofibration (resp. trivial cofibration).

Proof The usual "horseshoe lemma" argument from homological algebra carries over. In more detail, note that the admissible monic $A \rightarrowtail \tilde{R}_A$ extends over f since $\mathrm{Ext}^1_{\mathcal{E}}\left(Q, \tilde{R}_A\right) = 0$. This allows for the construction of the commutative diagram making up the bottom two rows. Indeed with α denoting the extension satisfying $k_A = \alpha f$, and setting $t := k_Q \circ c$, we get that $\left[\begin{smallmatrix} \alpha \\ t \end{smallmatrix}\right]\colon B \rightarrowtail \tilde{R}_A \bigoplus \tilde{R}_Q$ is an admissible monic. (To see this, note that it is the composition of $\left[\begin{smallmatrix} \alpha & 0 \\ 0 & 1_{\tilde{R}_Q} \end{smallmatrix}\right]\colon B \bigoplus \tilde{R}_Q \rightarrowtail \tilde{R}_A \bigoplus \tilde{R}_Q$, which is an admissible monic by Exercise 1.5.1, or see [Bühler, 2010, proposition 2.9], with $\left[\begin{smallmatrix} 1_B \\ t \end{smallmatrix}\right]\colon B \rightarrowtail B \bigoplus \tilde{R}Q$, which the reader can verify is also an admissible monic.) The rest follows from the (3×3)-Lemma 1.16.

So in the case that f is a trivial cofibration, then the constructed $\Sigma(f)$ is also a trivial cofibration. Thus $\gamma(\Sigma(f))$ is an isomorphism by Proposition 5.18. Although we have constructed a new suspension sequence for B, any two suspension sequences are canonically isomorphic in $\mathrm{St}_\ell(\mathcal{A})$, and hence also in $\mathrm{Ho}(\mathfrak{M})$ by Lemma 5.19. Proposition 5.23 applies in the case that $\mathfrak{M}$ is a Hovey triple (not just a Hovey Ext-triple). □

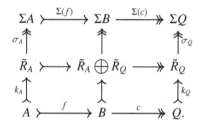

The proof of the next lemma uses the hypothesis that $\mathcal{A}$ is weakly idempotent complete.

Lemma 6.24 *Let f be a trivial fibration. Then $\gamma(\Sigma(f))$ is an isomorphism. Consequently, $\Sigma(f)$ is a weak equivalence whenever $\mathfrak{M}$ is a Hovey triple (not just a Hovey Ext-triple). In fact, if $f\colon A \twoheadrightarrow B$ is a trivial fibration then by altering the suspension sequence for A we can arrange for $\Sigma(f)$ to be a trivial epic (and a trivial fibration if $\mathfrak{M}$ is hereditary).*

Proof Like any morphism, f induces a morphism of suspension sequences:

$$\begin{array}{ccccc} A & \overset{k_A}{\rightarrowtail} & \tilde{R}_A & \overset{\sigma_A}{\twoheadrightarrow} & \Sigma A \\ {\scriptstyle f}\downarrow & & \downarrow{\scriptstyle \tilde{f}} & & \downarrow{\scriptstyle \Sigma(f)} \\ B & \overset{k_B}{\rightarrowtail} & \tilde{R}_B & \overset{\sigma_B}{\twoheadrightarrow} & \Sigma B. \end{array}$$

By the 2-out-of-3 axiom $\tilde{f}$ is a weak equivalence, because the composition

$$\tilde{R}_A \xrightarrow{\tilde{f}} \tilde{R}_A \to 0$$

is a weak equivalence. So we may factor $\tilde{f} = pi$, where $i\colon \tilde{R}_A \rightarrowtail D$ is a trivial cofibration and $p\colon D \twoheadrightarrow \tilde{R}_B$ is a trivial fibration. Then the lower left square in the following diagram commutes and the solid vertical columns are short exact sequences. The ik_A is the composition of two admissible monics and so too is an admissible monic. We let c be its cokernel, so the two solid horizonal rows are also short exact sequences:

$$\begin{array}{ccccc} J & \dashrightarrow & K & \dashrightarrow & L \\ \downarrow & & \downarrow & & \downarrow \\ A & \overset{ik_A}{\rightarrowtail} & D & \overset{c}{\twoheadrightarrow} & C \\ {\scriptstyle f}\downarrow & & \downarrow{\scriptstyle p} & & \downarrow \\ B & \overset{k_B}{\rightarrowtail} & \tilde{R}_B & \overset{\sigma_B}{\twoheadrightarrow} & \Sigma B. \end{array}$$

The dashed arrow $C \twoheadrightarrow \Sigma B$ exists, and makes the lower right corner commute, by the universal property of the cokernel c. By Proposition 1.30, which requires that $\mathcal{A}$ be weakly idempotent complete, we are able to conclude that $C \twoheadrightarrow \Sigma B$ is indeed an admissible epic. (Alternatively, we could argue that $J \rightarrowtail K$ exists and is an admissible monic by weak idempotent completeness.) In any case, the remaining dashed arrows exist so that the diagram is commutative and all rows and columns are short exact sequences by the (3×3)-Lemma 1.16.

We claim that $A \overset{ik_A}{\rightarrowtail} D \overset{c}{\twoheadrightarrow} C$ is a suspension sequence for A. To see this, note first that D is indeed trivially fibrant since it is an extension of two trivially fibrant objects. As for C, we see from Corollary 1.15 that C is an extension of $\operatorname{Cok} i$ and $\operatorname{Cok} k_A$. Since both of these objects are cofibrant, so is C. This completes the proof. Again, although we have constructed a new suspension sequence for A, any two suspension sequences are canonically isomorphic in $\mathrm{St}_\ell(\mathcal{A})$. So the comments made at the end of the proof of Lemma 6.23 also apply here. □

Proposition 6.25 *The composition*

$$\bar{\gamma}^\ell \circ \Sigma_\ell \colon \mathcal{A} \xrightarrow{\Sigma_\ell} \mathrm{St}_\ell(\mathcal{A}) \xrightarrow{\bar{\gamma}^\ell} \mathrm{Ho}(\mathfrak{M})$$

sends weak equivalences to isomorphisms. Therefore, by the Localization Theorem 5.21 it induces an additive endofunctor

$$\Sigma \colon \mathrm{Ho}(\mathfrak{M}) \to \mathrm{Ho}(\mathfrak{M})$$

satisfying $\Sigma \circ \gamma = \bar{\gamma}^\ell \circ \Sigma_\ell$. That is, $\Sigma := \mathrm{Ho}(\bar{\gamma}^\ell \circ \Sigma_\ell)$. Moreover, the following diagram of functors commutes on the nose:

$$\begin{array}{ccc} \mathrm{St}_\ell(\mathcal{A}) & \xrightarrow{\bar{\gamma}^\ell} & \mathrm{Ho}(\mathfrak{M}) \\ {\scriptstyle \Sigma_\ell^\ell}\downarrow & & \downarrow{\scriptstyle \Sigma} \\ \mathrm{St}_\ell(\mathcal{A}) & \xrightarrow{\bar{\gamma}^\ell} & \mathrm{Ho}(\mathfrak{M}). \end{array}$$

Proof Let f be a weak equivalence. By definition, it has a factorization $f = pi$ for some trivial cofibration i and trivial fibration p. We get a commutative diagram with each row a suspension sequence:

$$\begin{array}{ccccc} A & \overset{k_A}{\rightarrowtail} & \tilde{R}_A & \overset{\sigma_A}{\twoheadrightarrow} & \Sigma A \\ {\scriptstyle i}\downarrow & & \downarrow{\scriptstyle \tilde{i}} & & \downarrow{\scriptstyle \Sigma(i)} \\ D & \overset{k_D}{\rightarrowtail} & \tilde{R}_D & \overset{\sigma_D}{\twoheadrightarrow} & \Sigma D \\ {\scriptstyle p}\downarrow & & \downarrow{\scriptstyle \tilde{p}} & & \downarrow{\scriptstyle \Sigma(p)} \\ B & \overset{k_B}{\rightarrowtail} & \tilde{R}_B & \overset{\sigma_B}{\twoheadrightarrow} & \Sigma B. \end{array}$$

The diagram shows $\Sigma_\ell(f) = [\Sigma(p) \circ \Sigma(i)]_\ell$. So we get

$$\left(\bar{\gamma}^\ell \circ \Sigma_\ell\right)(f) = \bar{\gamma}^\ell([\Sigma(p) \circ \Sigma(i)]_\ell) := \gamma(\Sigma(p) \circ \Sigma(i)) = \gamma(\Sigma(p)) \circ \gamma(\Sigma(i)).$$

But $\gamma(\Sigma(p))$ and $\gamma(\Sigma(i))$ are each isomorphisms by Lemmas 6.23 and 6.24. Thus $(\bar{\gamma}^\ell \circ \Sigma_\ell)(f)$ is an isomorphism as desired.

By the Localization Theorem 5.21, with $F = \bar{\gamma}^\ell \circ \Sigma_\ell$, we get the functor $\Sigma := \mathrm{Ho}(F)$ satisfying $\bar{\gamma}^\ell \circ \Sigma_\ell = \Sigma \circ \gamma$. Moreover, taking $* = \ell$ in the Localization Theorem, we see that $\Sigma \circ \bar{\gamma}^\ell = \bar{F}^\ell$. So we will be done once we realize that the unique functor $\bar{F}^\ell$ satisfying $\bar{F}^\ell \circ \gamma_\ell = F$ is just the composition $\bar{F}^\ell = \bar{\gamma}^\ell \circ \Sigma^\ell_\ell$. So we compute

$$\left(\bar{\gamma}^\ell \circ \Sigma^\ell_\ell\right) \circ \gamma_\ell = \bar{\gamma}^\ell \circ \left(\Sigma^\ell_\ell \circ \gamma_\ell\right) = \bar{\gamma}^\ell \circ \Sigma_\ell$$

since we have $\Sigma_\ell = \Sigma^\ell_\ell \circ \gamma_\ell$ as in Equation (6.2) from the beginning of this section. This proves $\Sigma \circ \bar{\gamma}^\ell = \bar{\gamma}^\ell \circ \Sigma^\ell_\ell$ as desired. □

Lemma 6.26 (The Connecting Morphism) *Suppose* $\mathbb{E}\colon A \rightarrowtail B \twoheadrightarrow C$ *and* $\mathbb{S}\colon A \rightarrowtail W \twoheadrightarrow D$ *are short exact sequences in* $(\mathcal{A}, \mathcal{E})$*, where* $W \in \mathcal{W}$*. Construct a pushout diagram as shown:*

$$\begin{array}{ccccc} A & \rightarrowtail & B & \twoheadrightarrow & C \\ \downarrow & & \downarrow & & \| \\ W & \rightarrowtail & P & \overset{w}{\twoheadrightarrow} & C \\ \downarrow & & \downarrow{\scriptstyle v} & & \\ D & = & D. & & \end{array}$$

Regardless of the choices made in constructing the pushout (that is, the object P*, the maps* $W \rightarrowtail P$ *and* $B \rightarrowtail P$*, and the resulting cokernels* $w = \operatorname{cok}(W \rightarrowtail P)$ *and* $v = \operatorname{cok}(B \rightarrowtail P)$*), the composed morphism* $\gamma(v) \circ [\gamma(w)]^{-1} \in$ Ho($\mathfrak{M}$) *is well defined.*

Moreover, the construction is natural in the sense that any morphism of the given diagram built from $\mathbb{E}$ *and* $\mathbb{S}$ *induces a canonical morphism between the corresponding compositions* $\gamma(v) \circ [\gamma(w)]^{-1}$.

Proof Such a diagram exists by Lemma 1.10. Note that w is a weak equivalence (a trivial epic) by Proposition 4.7 and hence $\gamma(w)$ is an isomorphism by Proposition 5.18.

Consider two such pushout squares as shown next, where we set $w = \operatorname{cok} i$, $v = \operatorname{cok} j$ and $w' = \operatorname{cok} i'$, $v' = \operatorname{cok} j'$:

$$\begin{array}{ccc} A & \rightarrowtail & B \\ \downarrow & & \downarrow{\scriptstyle j} \\ W & \underset{i}{\rightarrowtail} & P \end{array} \qquad \begin{array}{ccc} A & \rightarrowtail & B \\ \downarrow & & \downarrow{\scriptstyle j'} \\ W & \underset{i'}{\rightarrowtail} & P'. \end{array}$$

The goal is to show $\gamma(v) \circ [\gamma(w)]^{-1} = \gamma(v') \circ [\gamma(w')]^{-1}$.

By the universal property of P there is a unique isomorphism $\xi\colon P \to P'$ satisfying $i' = \xi i$ and $j' = \xi j$. Moreover, the map w is obtained from the pushout property of P as the *unique* map $P \xrightarrow{w} C$ satisfying $wi = 0$ and $wj = c$, where c is the cokernel of $A \rightarrowtail B$. In the same way, the map w' is the unique map $P' \xrightarrow{w'} C$ satisfying $w'i' = 0$ and $w'j' = c$. But the uniqueness property of w implies $w = w'\xi$, because we have $(w'\xi)i = 0$ and $(w'\xi)j = c$.

Thus we have a commutative diagram with exact rows as follows:

$$\begin{array}{ccccc} W & \overset{i}{\rightarrowtail} & P & \overset{w}{\twoheadrightarrow} & C \\ \| & & \downarrow{\scriptstyle \xi} & & \| \\ W & \overset{i'}{\rightarrowtail} & P' & \overset{w'}{\twoheadrightarrow} & C. \end{array}$$

By symmetry we get a similar commutative diagram satisfying, in particular, the relation $v = v'\xi$. Applying the canonical functor $\gamma\colon \mathcal{A} \to \mathrm{Ho}(\mathfrak{M})$, we get $\gamma(w) = \gamma(w')\gamma(\xi)$ and $\gamma(v) = \gamma(v')\gamma(\xi)$. Thus $[\gamma(w)]^{-1} = [\gamma(\xi)]^{-1}[\gamma(w')]^{-1}$ and $\gamma(v) = \gamma(v')\gamma(\xi)$. Composing these two equations, we get the desired result

$$\gamma(v)[\gamma(w)]^{-1} = \gamma(v')\gamma(\xi)[\gamma(\xi)]^{-1}[\gamma(w')]^{-1} = \gamma(v')[\gamma(w')]^{-1}.$$

We leave it to the reader to verify the last statement by drawing the three-dimensional (3D) diagram needed. To do this, one observes that any morphism of the given diagram (that is, morphisms of short exact sequences $f_{\mathbb{E}}\colon \mathbb{E} \to \mathbb{E}'$ and $f_{\mathbb{S}}\colon \mathbb{S} \to \mathbb{S}'$, agreeing on $f_A\colon A \to A'$), induces a unique morphism $\beta\colon P \to P'$ of their corresponding pushouts. One shows that β will make all squares in sight commute. In particular, one argues that $w'\beta = f_C w$ (and $v'\beta = f_D v$) by applying the *uniqueness* property of pushouts, for these maps agree when composed with either i or j. Applying γ provides us with a commutative square in $\mathrm{Ho}(\mathfrak{M})$:

$$\begin{array}{ccc} C & \xrightarrow{\gamma(v)\circ[\gamma(w)]^{-1}} & D \\ {\scriptstyle \gamma(f_C)}\downarrow & & \downarrow{\scriptstyle \gamma(f_D)} \\ C' & \xrightarrow{\gamma(v')\circ[\gamma(w')]^{-1}} & D'. \end{array}$$

□

Definition 6.27 Given the situation in Lemma 6.26, we define the **connecting morphism for** $\mathbb{E}$ (relative to $\mathbb{S}$) to be the morphism $\gamma(\mathbb{E}) := -\gamma(v) \circ \gamma(w)^{-1}$.

Remark 6.28 Let $\mathbb{S}_A\colon A \rightarrowtail \tilde{R}_A \twoheadrightarrow \Sigma A$ and $\mathbb{S}'_A\colon A \rightarrowtail \tilde{R}'_A \twoheadrightarrow \Sigma' A$ be any two suspension sequences for an object A, relative to the pair $\mathfrak{L} = (\mathcal{Q}, \mathcal{R}_W)$. Let $\xi_{\mathbb{S}_A}^{\mathbb{S}'_A}\colon \Sigma' A \to \Sigma A$ be the corresponding suspension connecting isomorphism of Remark 6.15 and Corollary 6.18, defined by $\xi_{\mathbb{S}_A}^{\mathbb{S}'_A} := [\Sigma(1_A)]_\ell$. Since γ is well defined on $\mathrm{St}_\ell(\mathcal{A})$, we get the analogous **suspension connecting isomorphism** $\gamma(\xi)_{\mathbb{S}_A}^{\mathbb{S}'_A} := \gamma(\Sigma(1_A))$ in $\mathrm{Ho}(\mathfrak{M})$. It follows from Exercise 6.3.2 that $\gamma(\xi)_{\mathbb{S}_A}^{\mathbb{S}'_A} = \gamma(\mathbb{S}'_A)$. That is, it coincides with the connecting morphism for $\mathbb{S}'_A$ (relative to $\mathbb{S}_A$), in the sense of Definition 6.27. In fact, as indicated by the Dual

Fundamental Lemma 3.4 and Exercise 6.3.2, we may take $\mathbb{E}$ to be any short exact sequence $\mathbb{W}_A : A \rightarrowtail W_A \twoheadrightarrow C$ with C cofibrant. Then still, the morphism $\gamma(\mathbb{W}_A): C \to \Sigma A$ coincides with $\gamma(\Sigma(1_A))$.

Proposition 6.29 *The suspension functor* $\Sigma\colon \mathrm{Ho}(\mathfrak{M}) \to \mathrm{Ho}(\mathfrak{M})$ *and the loop functor* $\Omega\colon \mathrm{Ho}(\mathfrak{M}) \to \mathrm{Ho}(\mathfrak{M})$ *are inverse autoequivalences of* Ho($\mathfrak{M}$). *Moreover, for any object A, we can compute* ΣA *(up to isomorphism in* Ho($\mathfrak{M}$)*) by taking the cokernel of any admissible monomorphism* $A \rightarrowtail W$ *with* $W \in \mathcal{W}$. *On the other hand, we can compute* ΩA *by taking the kernel of any admissible epimorphism* $W \twoheadrightarrow A$ *with* $W \in \mathcal{W}$.

Proof Given any object A in $\mathcal{A}$, we may take a loop sequence and follow it by a suspension sequence, and then construct the pushout as shown:

$$\begin{array}{ccccc}
\Omega A & \overset{l_A}{\rightarrowtail} & \tilde{Q}_A & \overset{\rho_A}{\twoheadrightarrow} & A \\
{\scriptstyle k_{\Omega A}}\downarrow & & \downarrow & & \| \\
\tilde{R}_{\Omega A} & \rightarrowtail & P & \overset{w_A}{\twoheadrightarrow} & A \\
{\scriptstyle \sigma_{\Omega A}}\downarrow & & \downarrow{\scriptstyle v_A} & & \\
\Sigma\Omega A & = & \Sigma\Omega A. & &
\end{array}$$

Here, w_A and v_A are both trivial epics, so the connecting morphism $\gamma(v) \circ [\gamma(w)]^{-1}$ of Lemma 6.26 is an isomorphism. We claim that these isomorphisms assemble to provide a natural isomorphism

$$\left\{\gamma(v_A) \circ [\gamma(w_A)]^{-1}\right\}: 1_{\mathrm{Ho}(\mathfrak{M})} \to \Sigma\Omega.$$

To prove this, we will show that any morphism $f\colon A \to B$ in $\mathcal{A}$ induces a commutative diagram:

$$\begin{array}{ccc}
A & \xrightarrow{\gamma(v_A)\circ[\gamma(w_A)]^{-1}} & \Sigma\Omega A \\
{\scriptstyle \gamma(f)}\downarrow & & \downarrow{\scriptstyle \Sigma\Omega(\gamma(f))} \\
B & \xrightarrow{\gamma(v_B)\circ[\gamma(w_B)]^{-1}} & \Sigma\Omega B.
\end{array}$$

This is enough to prove naturality, by Corollary 5.16.

So consider an arbitrary morphism $f\colon A \to B$ in $\mathcal{A}$. It induces commutative diagrams:

$$\begin{array}{ccccc}
\Omega A & \overset{l_A}{\rightarrowtail} & \tilde{Q}_A & \overset{\rho_A}{\twoheadrightarrow} & A \\
\downarrow{\scriptstyle \Omega(f)} & & \downarrow{\scriptstyle \tilde{f}} & & {\scriptstyle f}\downarrow \\
\Omega B & \overset{l_B}{\rightarrowtail} & \tilde{Q}_B & \overset{\rho_B}{\twoheadrightarrow} & B
\end{array}
\quad \text{and} \quad
\begin{array}{ccccc}
\Omega A & \overset{k_{\Omega A}}{\rightarrowtail} & \tilde{R}_{\Omega A} & \overset{\sigma_{\Omega A}}{\twoheadrightarrow} & \Sigma\Omega A \\
\downarrow{\scriptstyle \Omega(f)} & & {\scriptstyle \widetilde{\Omega(f)}}\downarrow & & {\scriptstyle \Sigma(\Omega(f))}\downarrow \\
\Omega B & \overset{k_{\Omega B}}{\rightarrowtail} & \tilde{R}_{\Omega B} & \overset{\sigma_{\Omega B}}{\twoheadrightarrow} & \Sigma\Omega B.
\end{array}$$

We recall that these constructions are unique up to ω-homotopy and that γ identifies homotopic maps. So the naturality of the construction in Lemma 6.26 provides a commutative square in $\mathrm{Ho}(\mathfrak{M})$:

$$\begin{array}{ccc} A & \xrightarrow{\gamma(v_A)\circ[\gamma(w_A)]^{-1}} & \Sigma\Omega A \\ {\scriptstyle\gamma(f)}\downarrow & & \downarrow{\scriptstyle\gamma(\Sigma(\Omega(f)))} \\ B & \xrightarrow{\gamma(v_B)\circ[\gamma(w_B)]^{-1}} & \Sigma\Omega B. \end{array}$$

Using Proposition 6.25 and its dual we see that $\gamma(\Sigma(\Omega(f))) = \left(\bar{\gamma}^{\ell} \circ \Sigma_{\ell}\right)(\Omega(f)) = (\Sigma\circ\gamma)(\Omega(f)) = \Sigma(\gamma(\Omega(f))) = \Sigma(\bar{\gamma}^{r}([\Omega(f)]_r)) = \Sigma\circ(\bar{\gamma}^{r}\circ\Omega_r)(f) = \Sigma\circ(\Omega\circ\gamma)(f) = \Sigma\Omega(\gamma(f))$.

On the other hand, Lemma 6.26 has a dual. The dual arguments, using pullbacks, allow us to conclude a natural isomorphism $\Omega\Sigma \to 1_{\mathrm{Ho}(\mathfrak{M})}$.

The final statement about computing ΣA also follows from Lemma 6.26, and the dual statement for computing ΩA. □

Example 6.30 Let R be a ring and let $\mathrm{Ch}(R)_{\mathcal{P}}$ be the exact category of chain complexes of R-modules, along with some proper class $\mathcal{P}$ of short exact sequences containing every degreewise split short exact sequence. Assume $\mathfrak{M} = (Q, \mathcal{W}, \mathcal{R})$ is an abelian model structure on $\mathrm{Ch}(R)_{\mathcal{P}}$. Assuming that $\mathcal{W}$ contains all contractible complexes, then it follows from Proposition 6.29 that we can take Σ on $\mathrm{Ho}(\mathfrak{M})$ to be the usual suspension functor (shift the complex against arrows, and change differentials to $-d$), and Ω to be its inverse Σ^{-1}. This is made more precise at the end of Section 6.6 and a general version of this is shown in Proposition 10.24.

6.5 Triangulated Structure on Ho($\mathfrak{M}$)

Throughout this section we let $\mathfrak{M} = (Q, \mathcal{W}, \mathcal{R})$ denote a Hovey triple on a weakly idempotent complete exact category $(\mathcal{A}, \mathcal{E})$. Again, $\mathfrak{M}$ may even just be a Hovey Ext-triple. We will show that the homotopy category $\mathrm{Ho}(\mathfrak{M})$ is a triangulated category.

We start by defining the class of distinguished triangles. Our definition is based on the right triangulated category $\left(\mathrm{St}_{\ell}(\mathcal{A}), \Sigma_{\ell}^{\ell}, \Delta_{\ell}\right)$ from Corollary 6.22. Recall that a right triangle in $\mathrm{St}_{\ell}(\mathcal{A})$ is, by definition, a diagram of the form

$$A \xrightarrow{[f]_{\ell}} B \xrightarrow{[g]_{\ell}} C \xrightarrow{[h]_{\ell}} \Sigma A.$$

Observe that by applying the canonical functor $\bar{\gamma}^{\ell} \colon \mathrm{St}_{\ell}(\mathcal{A}) \to \mathrm{Ho}(\mathfrak{M})$ (see Definition 5.20), we get a right triangle in $\mathrm{Ho}(\mathfrak{M})$

$$A \xrightarrow{\gamma(f)} B \xrightarrow{\gamma(g)} C \xrightarrow{\gamma(h)} \Sigma A.$$

Here we have used that, technically, $\Sigma A = \Sigma_\ell^\ell A$ in the first diagram and by Proposition 6.25 we have the compatibility $\Sigma \circ \bar{\gamma}^\ell = \bar{\gamma}^\ell \circ \Sigma_\ell^\ell$.

Recall that a member of Δ_ℓ is any right triangle that is isomorphic to a strict right triangle, and these are defined as follows: For any morphism $f\colon A \to B$ in $\mathcal{A}$, we obtain

$$A \xrightarrow{[f]_\ell} B \xrightarrow{[k_f]_\ell} C(f) \xrightarrow{[\sigma_f]_\ell} \Sigma A, \tag{†}$$

the *strict right triangle on* f, in $\mathrm{St}_\ell(\mathcal{A})$, by constructing a pushout diagram:

$$\begin{array}{ccccc} A & \overset{k_A}{\rightarrowtail} & \tilde{R}_A & \overset{\sigma_A}{\twoheadrightarrow} & \Sigma A \\ {\scriptstyle f}\downarrow & & \downarrow{\scriptstyle w_f} & & \| \\ B & \underset{k_f}{\rightarrowtail} & C(f) & \underset{\sigma_f}{\twoheadrightarrow} & \Sigma A. \end{array}$$

We now define strict right triangles in Ho($\mathfrak{M}$) by simply applying the canonical functor $\bar{\gamma}^\ell\colon \mathrm{St}_\ell(\mathcal{A}) \to \mathrm{Ho}(\mathfrak{M})$ to such right triangles (†). This will also make $\bar{\gamma}^\ell$ a right exact functor as defined in Definition 6.6, but in a very simple way because the compatibility $\Sigma \circ \bar{\gamma}^\ell = \bar{\gamma}^\ell \circ \Sigma_\ell^\ell$ is an equality, not just an isomorphism.

Definition 6.31 A strict right triangle in Ho($\mathfrak{M}$) is defined by applying the canonical functor $\bar{\gamma}^\ell\colon \mathrm{St}_\ell(\mathcal{A}) \to \mathrm{Ho}(\mathfrak{M})$ to a strict right triangle in $\mathrm{St}_\ell(\mathcal{A})$. That is, by a **strict right triangle on** $f \in \mathcal{A}$, we mean a right triangle

$$A \xrightarrow{\gamma(f)} B \xrightarrow{\gamma(k_f)} C(f) \xrightarrow{\gamma(\sigma_f)} \Sigma A,$$

in Ho($\mathfrak{M}$), constructed via the pushout diagram:

$$\begin{array}{ccccc} A & \overset{k_A}{\rightarrowtail} & \tilde{R}_A & \overset{\sigma_A}{\twoheadrightarrow} & \Sigma A \\ {\scriptstyle f}\downarrow & & \downarrow{\scriptstyle w_f} & & \| \\ B & \underset{k_f}{\rightarrowtail} & C(f) & \underset{\sigma_f}{\twoheadrightarrow} & \Sigma A. \end{array}$$

Any right triangle in Ho($\mathfrak{M}$) that is isomorphic to such a strict right triangle is called a **distinguished right triangle**. We let Δ denote the class of all of them.

Lemma 6.32 *The canonical functor* $\bar{\gamma}^\ell\colon \mathrm{St}_\ell(\mathcal{A}) \to \mathrm{Ho}(\mathfrak{M})$ *preserves right triangles, strict right triangles, and distinguished right triangles.*

Proof We have already discussed how $\bar{\gamma}^\ell$ preserves right triangles and, by definition, strict right triangles. It therefore also preserves distinguished right triangles because $\bar{\gamma}^\ell$ preserves isomorphisms. Of course, this is simply because $\bar{\gamma}^\ell$ is a functor; again, see Lemma 5.19 and Definition 5.20. □

Lemma 6.33 *Consider a commutative square in $\mathcal{A}$ as shown, which is to be thought of as a morphism from f to f':*

$$\begin{array}{ccc} A & \xrightarrow{f} & B \\ {\scriptstyle\alpha}\downarrow & & \downarrow{\scriptstyle\beta} \\ A' & \xrightarrow[f']{} & B'. \end{array}$$

If α and β are each trivial fibrations, or trivial cofibrations, then given any suspension sequences

$$\mathbb{S}_A : A \overset{k_A}{\rightarrowtail} \tilde{R}_A \overset{\sigma_A}{\twoheadrightarrow} \Sigma A \quad \text{and} \quad \mathbb{S}_{A'} : A' \overset{k_{A'}}{\rightarrowtail} \tilde{R}_{A'} \overset{\sigma_{A'}}{\twoheadrightarrow} \Sigma A',$$

there exists a "fill-in" morphism $c\colon C_{\mathbb{S}_A}(f) \to C_{\mathbb{S}_{A'}}(f')$ as shown and such that $\gamma(c)$ is an isomorphism in $\mathrm{Ho}(\mathfrak{M})$*:*

$$\begin{array}{ccccccc} A & \xrightarrow{f} & B & \xrightarrow{k_f} & C_{\mathbb{S}_A}(f) & \xrightarrow{\sigma_f} & \Sigma A \\ {\scriptstyle\alpha}\downarrow & & \downarrow{\scriptstyle\beta} & & \downarrow{\scriptstyle c} & & \downarrow{\scriptstyle\Sigma(\alpha)} \\ A' & \xrightarrow{f'} & B' & \xrightarrow{k_{f'}} & C_{\mathbb{S}_{A'}}(f') & \xrightarrow{\sigma_{f'}} & \Sigma A'. \end{array}$$

To be clear, this is a commutative diagram in $\mathcal{A}$ and we note that c is necessarily a weak equivalence when $\mathfrak{M}$ is a Hovey triple (not just a Hovey Ext-triple).

Proof Using Lemma 6.24 we will prove the statement for the case that α and β are trivial fibrations. The proof for trivial cofibrations is similar, using Lemma 6.23. Indeed by Lemma 6.24, we can construct a new suspension sequence for A,

$$A \overset{k'_A}{\rightarrowtail} \tilde{R}'_A \overset{\sigma'_A}{\twoheadrightarrow} \Sigma' A,$$

and arrange for $\Sigma(\alpha)$ to be a trivial epic as shown:

$$\begin{array}{ccccc} A & \overset{k'_A}{\rightarrowtail} & \tilde{R}'_A & \overset{\sigma'_A}{\twoheadrightarrow} & \Sigma' A \\ {\scriptstyle\alpha}\downarrow & & \downarrow{\scriptstyle\tilde{\alpha}} & & \downarrow{\scriptstyle\Sigma(\alpha)} \\ A' & \overset{k_{A'}}{\rightarrowtail} & \tilde{R}_{A'} & \overset{\sigma_{A'}}{\twoheadrightarrow} & \Sigma A'. \end{array}$$

Of course there also is a morphism from the original suspension sequence for A to this new one, extending 1_A, as shown:

$$\begin{array}{ccccc} A & \stackrel{k_A}{\rightarrowtail} & \tilde{R}_A & \stackrel{\sigma_A}{\twoheadrightarrow} & \Sigma A \\ {\scriptstyle 1_A}\downarrow & & \downarrow{\scriptstyle \tilde{1}_A} & & \downarrow{\scriptstyle \Sigma(1_A)} \\ A & \stackrel{k'_A}{\rightarrowtail} & \tilde{R}'_A & \stackrel{\sigma'_A}{\twoheadrightarrow} & \Sigma' A. \end{array}$$

We know that any such morphism of suspension sequences provides a canonical isomorphism between the two sequences, in $\mathrm{St}_\ell(\mathcal{A})$. In particular, $\Sigma(1_A)$ is a left homotopy equivalence. Stacking the two diagrams and performing the pushout construction as in the definition of strict right triangles, we get unique morphisms c_1 and c_2 so that all squares commute in the large 3D diagram that follows; see Lemma 6.17 and Corollary 6.18:

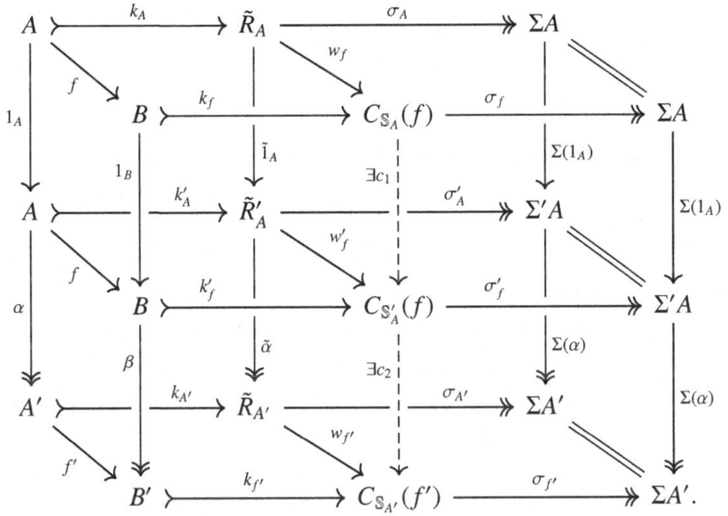

This gives us in particular the following commutative diagram in $\mathcal{A}$:

$$\begin{array}{ccccccc} A & \xrightarrow{f} & B & \stackrel{k_f}{\rightarrowtail} & C_{\mathbb{S}_A}(f) & \stackrel{\sigma_f}{\twoheadrightarrow} & \Sigma A \\ {\scriptstyle 1_A}\downarrow & & \downarrow{\scriptstyle 1_B} & & \downarrow{\scriptstyle c_1} & & \downarrow{\scriptstyle \Sigma(1_A)} \\ A & \xrightarrow{f} & B & \stackrel{k'_f}{\rightarrowtail} & C_{\mathbb{S}'_A}(f) & \stackrel{\sigma'_f}{\twoheadrightarrow} & \Sigma' A \\ {\scriptstyle \alpha}\downarrow & & \downarrow{\scriptstyle \beta} & & \downarrow{\scriptstyle c_2} & & \downarrow{\scriptstyle \Sigma(\alpha)} \\ A' & \xrightarrow{f'} & B' & \stackrel{k_{f'}}{\rightarrowtail} & C_{\mathbb{S}_{A'}}(f') & \stackrel{\sigma_{f'}}{\twoheadrightarrow} & \Sigma A'. \end{array}$$

Since β and $\Sigma(\alpha)$ are admissible epics it follows from the Five Lemma (Exercise 1.4.2) that c_2 is also an admissible epic. But since β and $\Sigma(\alpha)$ also have trivial kernels it then follows from the (3×3)-Lemma 1.16 that c_2 is a trivial epic. In particular c_2 is a weak equivalence and therefore $\gamma(c_2)$ is an isomorphism in $\mathrm{Ho}(\mathfrak{M})$, by Proposition 5.18. As for c_1, applying the canonical functor $\gamma_\ell \colon \mathcal{A} \to \mathrm{St}_\ell(\mathcal{A})$, the top two rows become a morphism of strict right triangles, just constructed with different suspensions sequences. So by Corollary 6.18, $[c_1]_\ell \colon C_{\mathbb{S}_A}(f) \to C_{\mathbb{S}'_A}(f)$ is an isomorphism in $\mathrm{St}_\ell(\mathcal{A})$. Therefore, $\bar{\gamma}^\ell([c_1]_\ell) = \gamma(c_1)$ is an isomorphism. It follows that the composition $c := c_2 \circ c_1$ is the desired "fill-in" morphism. Indeed $\gamma(c)$ is an isomorphism and Proposition 5.23 guarantees that it is a weak equivalence in the case that $\mathfrak{M}$ is a Hovey triple (not just a Hovey Ext-triple). □

Theorem 6.34 *Let* $\Sigma \colon \mathrm{Ho}(\mathfrak{M}) \to \mathrm{Ho}(\mathfrak{M})$ *denote the suspension functor and* Δ *the class of all distinguished right triangles associated to* $\mathfrak{M}$. *Then* (Σ, Δ) *is a triangulated structure on* $\mathrm{Ho}(\mathfrak{M})$ *and* $\bar{\gamma}^\ell \colon \mathrm{St}_\ell(\mathcal{A}) \to \mathrm{Ho}(\mathfrak{M})$ *is a right triangulated functor.*

Proof We already saw in Proposition 6.29 that Σ is an autoequivalence of $\mathrm{Ho}(\mathfrak{M})$. Also, it is clear by the comments made at the beginning of this section, up through Lemma 6.32, that $\bar{\gamma}^\ell \colon \mathrm{St}_\ell(\mathcal{A}) \to \mathrm{Ho}(\mathfrak{M})$ will be a right exact functor. But of course it remains to show that (Σ, Δ) is a right triangulated structure on $\mathrm{Ho}(\mathfrak{M})$, by verifying the axioms (R-Tr1)–(R-Tr4).

(R-Tr1) We first need to check that for any morphism $f \colon A \to B$ in $\mathrm{Ho}(\mathfrak{M})$, there is some right triangle $A \xrightarrow{f} B \to C \to \Sigma A$ in Δ. By definition, we have that $f = {}_A[h]_B$ is represented by some morphism $h \colon RQA \to RQB$ in $\mathcal{A}$, and by Proposition 5.15 we have

$$f = \gamma(p_B) \circ [\gamma(j_{QB})]^{-1} \circ \gamma(h) \circ \gamma(j_{QA}) \circ [\gamma(p_A)]^{-1}.$$

We form a strict right triangle on h,

$$RQA \xrightarrow{[h]_\ell} RQB \xrightarrow{[k_h]_\ell} C(h) \xrightarrow{[\sigma_h]_\ell} \Sigma RQA,$$

in the category $\mathrm{St}_\ell(\mathcal{A})$. Applying $\bar{\gamma}^\ell$, it becomes a strict right triangle in $\mathrm{Ho}(\mathfrak{M})$ sitting in the bottom row of the following isomorphism of right triangles:

$$\begin{array}{ccccccc}
A & \xrightarrow{f} & B & \xrightarrow{\gamma(k_h)\circ\beta} & C(h) & \xrightarrow{[\Sigma(\alpha)]^{-1}\circ\gamma(\sigma_h)} & \Sigma A \\
\downarrow{\scriptstyle \alpha} & & \downarrow{\scriptstyle \beta} & & \| & & \downarrow{\scriptstyle \Sigma(\alpha)} \\
RQA & \xrightarrow{\gamma(h)} & RQB & \xrightarrow{\gamma(k_h)} & C(h) & \xrightarrow{\gamma(\sigma_h)} & \Sigma RQA.
\end{array}$$

Here $\alpha = \gamma(j_{QA}) \circ [\gamma(p_A)]^{-1}$ and $\beta = \gamma(j_{QB}) \circ [\gamma(p_B)]^{-1}$. This proves the top row is a distinguished right triangle in Ho($\mathfrak{M}$), as desired.

Second, we need to check that for any object A in Ho($\mathfrak{M}$), the right triangle $0 \to A \xrightarrow{1_A} A \to 0$ is a distinguished right triangle. But this is certainly the case in the category $\mathrm{St}_\ell(\mathcal{A})$, so this follows at once from Lemma 6.32.

(R-Tr2) (Right Rotations) For the same formal reason as in the proof of Theorem 6.16, it is enough to assume we are given a strict right triangle

$$A \xrightarrow{\gamma(f)} B \xrightarrow{\gamma(k_f)} C(f) \xrightarrow{\gamma(\sigma_f)} \Sigma A.$$

We must show that the right rotation $B \xrightarrow{\gamma(k_f)} C(f) \xrightarrow{\gamma(\sigma_f)} \Sigma A \xrightarrow{-\Sigma(\gamma(f))} \Sigma B$ is in Δ. By definition, the given right triangle is the image under $\bar{\gamma}^\ell : \mathrm{St}_\ell(\mathcal{A}) \to$ Ho($\mathfrak{M}$) of a strict right triangle $A \xrightarrow{[f]_\ell} B \xrightarrow{[k_f]_\ell} C(f) \xrightarrow{[\sigma_f]_\ell} \Sigma A$. By Theorem 6.16, its right rotation $B \xrightarrow{[k_f]_\ell} C(f) \xrightarrow{[\sigma_f]_\ell} \Sigma A \xrightarrow{-\Sigma^\ell_\ell([f]_\ell)} \Sigma B$ is a distinguished right triangle in $\mathrm{St}_\ell(\mathcal{A})$. Applying $\bar{\gamma}^\ell$ to this triangle gives us what we need. Indeed $\bar{\gamma}^\ell\left(-\Sigma^\ell_\ell([f]_\ell)\right) = -\bar{\gamma}^\ell\left(\Sigma^\ell_\ell([f]_\ell)\right) = -\left(\bar{\gamma}^\ell \circ \Sigma^\ell_\ell\right)([f]_\ell) = -(\Sigma \circ \bar{\gamma}^\ell)([f]_\ell) = -\Sigma\left(\bar{\gamma}^\ell([f]_\ell)\right) = -\Sigma(\gamma(f))$ where we have used Proposition 6.25, and Lemma 6.32 tells us the right triangle is in the class Δ of distinguished right triangles.

(R-Tr3) (Right Fill-Ins) Let us suppose that we are given two strict right triangles as shown in the following rows, along with the morphisms α and β, in Ho($\mathfrak{M}$), making the left square commute:

$$\begin{array}{ccccccc}
A & \xrightarrow{\gamma(f)} & B & \xrightarrow{\gamma(k_f)} & C(f) & \xrightarrow{\gamma(\sigma_f)} & \Sigma A \\
\downarrow{\scriptstyle\alpha} & & \downarrow{\scriptstyle\beta} & & \downarrow & & \downarrow{\scriptstyle\Sigma(\alpha)} \\
A' & \xrightarrow{\gamma(f')} & B' & \xrightarrow{\gamma(k_{f'})} & C(f') & \xrightarrow{\gamma(\sigma_{f'})} & \Sigma A'.
\end{array}$$

Our job is to prove the existence of a fill-in map as indicated by the dashed arrow. The rows are, of course, images under $\gamma\colon \mathcal{A} \xrightarrow{\gamma_\ell} \mathrm{St}_\ell(\mathcal{A}) \xrightarrow{\bar{\gamma}^\ell}$ Ho($\mathfrak{M}$) of actual morphisms in $\mathcal{A}$. However, by Proposition 5.15, the vertical morphisms really take the form

$${}_A[\alpha]_{A'} = \gamma(p_{A'}) \circ [\gamma(j_{QA'})]^{-1} \circ \gamma(\alpha) \circ \gamma(j_{QA}) \circ [\gamma(p_A)]^{-1}$$

for some representative morphism $\alpha\colon RQA \to RQA'$ in $\mathcal{A}$. Similarly,

$${}_B[\beta]_{B'} = \gamma(p_{B'}) \circ [\gamma(j_{QB'})]^{-1} \circ \gamma(\beta) \circ \gamma(j_{QB}) \circ [\gamma(p_B)]^{-1}$$

for some representative morphism $\beta\colon RQB \to RQB'$. So in fact the commutative square on the left side of the diagram induces a ladder of commutative squares as shown on the left side of the following diagram:

$$
\begin{array}{ccccccc}
A & \xrightarrow{f} & B & \xrightarrow{k_f} & C(f) & \xrightarrow{\sigma_f} & \Sigma A \\
{\scriptstyle p_A}\uparrow & & \uparrow{\scriptstyle p_B} & & \uparrow & & \uparrow{\scriptstyle \Sigma(p_A)} \\
QA & \xrightarrow{Q(f)} & QB & \longrightarrow & C(Q(f)) & \longrightarrow & \Sigma QA \\
{\scriptstyle j_{QA}}\downarrow & & \downarrow{\scriptstyle j_{QB}} & & \downarrow & & \downarrow{\scriptstyle \Sigma(j_{QA})} \\
RQA & \xrightarrow{R(Q(f))} & RQB & \longrightarrow & C(R(Q(f))) & \longrightarrow & \Sigma RQA \\
{\scriptstyle \alpha}\downarrow & & \downarrow{\scriptstyle \beta} & & \downarrow & & \downarrow{\scriptstyle \Sigma(\alpha)} \\
RQA' & \xrightarrow{R(Q(f'))} & RQB' & \longrightarrow & C(R(Q(f'))) & \longrightarrow & \Sigma RQA' \\
{\scriptstyle j_{QA'}}\uparrow & & \uparrow{\scriptstyle j_{QB'}} & & \uparrow & & \uparrow{\scriptstyle \Sigma(j_{QA'})} \\
QA' & \xrightarrow{Q(f')} & QB' & \longrightarrow & C(Q(f')) & \longrightarrow & \Sigma QA' \\
{\scriptstyle p_{A'}}\downarrow & & \downarrow{\scriptstyle p_{B'}} & & \downarrow & & \downarrow{\scriptstyle \Sigma(p_{A'})} \\
A' & \xrightarrow[f']{} & B' & \xrightarrow[k_{f'}]{} & C(f') & \xrightarrow[\sigma_{f'}]{} & \Sigma A'.
\end{array}
$$

Actually, only four of the squares in the left ladder literally commute as morphisms in $\mathcal{A}$. The middle square only commutes up to a homotopy

$$\beta \circ R(Q(f)) \sim R(Q(f')) \circ \alpha.$$

But this means it is a commutative square in the category $\mathrm{St}_\ell(\mathcal{A})$. So by the (R-Tr3) Axiom already proved for this category in Corollary 6.22, there exists a fill-in morphism $C(R(Q(f))) \to C(R(Q(F')))$ as in the middle of the diagram. As for the four other dashed arrows, these exist and their images under γ are isomorphisms by Lemma 6.33. So, applying $\gamma\colon \mathcal{A} \to \mathrm{Ho}(\mathfrak{M})$ to the entire diagram provides us with a "fill-in" morphism as required. Let us just point out that a fill-in morphism obtained this way is compatible with Σ. Indeed the very definition of Σ is $\Sigma := \mathrm{Ho}\left(\bar{\gamma}^\ell \circ \Sigma_\ell\right)$, so by the Localization Theorem 5.21 we have that $\Sigma({}_A[\alpha]_{A'})$ coincides with the composition

$$\left(\bar{\gamma}^\ell \circ \Sigma_\ell\right)(p_{A'}) \circ \left[\left(\bar{\gamma}^\ell \circ \Sigma_\ell\right)(j_{QA'})\right]^{-1} \circ \left(\bar{\gamma}^\ell \circ \Sigma_\ell\right)(\alpha) \circ \left(\bar{\gamma}^\ell \circ \Sigma_\ell\right)(j_{QA}) \circ \left[\left(\bar{\gamma}^\ell \circ \Sigma_\ell\right)(p_A)\right]^{-1}$$

$$= \gamma(\Sigma(p_{A'})) \circ [\gamma(\Sigma(j_{QA'}))]^{-1} \circ \gamma(\Sigma(\alpha)) \circ \gamma(\Sigma(j_{QA})) \circ [\gamma(\Sigma(p_A))]^{-1}.$$

At this point we have shown $\mathrm{Ho}(\mathfrak{M})$ to be a pretriangulated category in the sense that it satisfies axioms (Tr1)–(Tr3). These three axioms are enough to prove the following lemma, which we will use to prove Verdier's Octahedral Axiom.

Lemma 6.35 *Any distinguished triangle* $A \xrightarrow{f} B \xrightarrow{f'} C \xrightarrow{f''} \Sigma A$ *in* Ho($\mathfrak{M}$) *is isomorphic to the strict right triangle on* α, *where* $f = {}_A[\alpha]_B$ *is the homotopy class represented by the* $\mathcal{A}$-*morphism* $\alpha\colon RQA \to RQB$.

To prove Lemma 6.35, start by forming the strict right triangle

$$RQA \xrightarrow{\gamma(\alpha)} RQB \xrightarrow{\gamma(k_\alpha)} C(\alpha) \xrightarrow{\gamma(\sigma_\alpha)} \Sigma RQA$$

on the representative morphism α. We have $\gamma(\alpha) = [\alpha]$ and by Proposition 5.15 we have $f = \gamma(p_B) \circ [\gamma(j_{QB})]^{-1} \circ \gamma(\alpha) \circ \gamma(j_{QA}) \circ [\gamma(p_A)]^{-1}$. So using the isomorphisms $\gamma(j_{QA}) \circ [\gamma(p_A)]^{-1}$ and $\gamma(j_{QB}) \circ [\gamma(p_B)]^{-1}$, we get the commutative square in Ho($\mathfrak{M}$) shown on the left here:

$$\begin{array}{ccccccc} A & \xrightarrow{f} & B & \xrightarrow{f'} & C & \xrightarrow{f''} & \Sigma A \\ \downarrow & & \downarrow & & & & \\ RQA & \xrightarrow{\gamma(\alpha)} & RQB & \xrightarrow{\gamma(k_\alpha)} & C(\alpha) & \xrightarrow{\gamma(\sigma_\alpha)} & \Sigma RQA. \end{array}$$

It follows from the Five Lemma (Corollary 6.4) that this extends to an isomorphism of right triangles. This completes the proof of Lemma 6.35.

(R-Tr4) (Right Octahedrals) Now with Lemma 6.35 we return to complete the proof of Theorem 6.34. It only remains to prove Verdier's Octahedral Axiom. So consider a given composition $h = g \circ f$ of two morphisms $A \xrightarrow{f} B$ and $B \xrightarrow{g} C$ in Ho($\mathfrak{M}$). Moreover, let us be given three distinguished triangles on these morphisms as follows:

$$A \xrightarrow{f} B \xrightarrow{f'} D \xrightarrow{f''} \Sigma A, \quad B \xrightarrow{g} C \xrightarrow{g'} E \xrightarrow{g''} \Sigma B$$

and

$$A \xrightarrow{h=gf} C \xrightarrow{h'} F \xrightarrow{h''} \Sigma A.$$

Using Proposition 5.15 we have

$$f = \gamma(p_B) \circ [\gamma(j_{QB})]^{-1} \circ \gamma(\alpha) \circ \gamma(j_{QA}) \circ [\gamma(p_A)]^{-1}$$

for some representative morphism $\alpha\colon RQA \to RQB$ and similarly,

$$g = \gamma(p_C) \circ [\gamma(j_{QC})]^{-1} \circ \gamma(\beta) \circ \gamma(j_{QB}) \circ [\gamma(p_B)]^{-1}$$

for some representative morphism $\beta\colon RQB \to RQC$. Thus $\delta := \beta \circ \alpha$ is a representative morphism for the composition h. Therefore we may use

Lemma 6.35 to replace the three distinguished triangles with the following three strict right triangles:

$$RQA \xrightarrow{\gamma(\alpha)} RQB \xrightarrow{\gamma(k_\alpha)} C(\alpha) \xrightarrow{\gamma(\sigma_\alpha)} \Sigma RQA,$$

$$RQB \xrightarrow{\gamma(\beta)} RQC \xrightarrow{\gamma(k_\beta)} C(\beta) \xrightarrow{\gamma(\sigma_\beta)} \Sigma RQB$$

and

$$RQA \xrightarrow{\gamma(\delta)} RQC \xrightarrow{\gamma(k_\delta)} C(\delta) \xrightarrow{\gamma(\sigma_\delta)} \Sigma RQA.$$

But recall that, by definition, these are each the image under $\bar{\gamma}^\ell \colon \mathrm{St}_\ell(\mathcal{A}) \to \mathrm{Ho}(\mathfrak{M})$ of strict right triangles in $\mathrm{St}_\ell(\mathcal{A})$ as indicated in the first two rows and second column of the following diagram:

$$\begin{array}{ccccccc}
RQA & \xrightarrow{[\alpha]} & RQB & \xrightarrow{[k_\alpha]} & C(\alpha) & \xrightarrow{[\sigma_\alpha]} & \Sigma RQA \\
\| & & {\scriptstyle [\beta]}\downarrow & & \downarrow{\scriptstyle \exists!\,[s]} & & \| \\
RQA & \xrightarrow{[\delta]} & RQC & \xrightarrow{[k_\delta]} & C(\delta) & \xrightarrow{[\sigma_\delta]} & \Sigma RQA \\
 & & {\scriptstyle [k_\beta]}\downarrow & & \downarrow{\scriptstyle \exists!\,[t]} & & \downarrow{\scriptstyle \Sigma[\alpha]} \\
 & & C(\beta) & = & C(\beta) & \xrightarrow{[\sigma_\beta]} & \Sigma RQB \\
 & & {\scriptstyle [\sigma_\beta]}\downarrow & & \downarrow & & \\
 & & \Sigma RQB & \xrightarrow{\Sigma[k_\alpha]} & \Sigma C(\alpha). & &
\end{array}$$

(These equivalence classes represent morphisms in $\mathrm{St}_\ell(\mathcal{A})$.) The point is that we now may apply the (R-Tr4) Axiom in the right triangulated category $\mathrm{St}_\ell(\mathcal{A})$. This provides $\mathcal{A}$-morphisms s and t such that the entire diagram commutes in $\mathrm{St}_\ell(\mathcal{A})$ and such that the second to last column is a distinguished right triangle in $\mathrm{St}_\ell(\mathcal{A})$. Applying $\bar{\gamma}^\ell \colon \mathrm{St}_\ell(\mathcal{A}) \to \mathrm{Ho}(\mathfrak{M})$ we get a commutative diagram in $\mathrm{Ho}(\mathfrak{M})$ and the second to last column is a distinguished (right) triangle in $\mathrm{Ho}(\mathfrak{M})$ by Lemma 6.32. Since the remaining right triangles are isomorphic to the ones we were first given, and since $\bar{\gamma}^\ell$ is compatible with the two endofunctors Σ, we conclude that the (R-Tr4) Axiom passes to $\mathrm{Ho}(\mathfrak{M})$. □

6.6 General Suspension Functors and Cofiber Sequences

We continue with the same hypotheses on $\mathfrak{M} = (Q, \mathcal{W}, \mathcal{R})$ and $(\mathcal{A}, \mathcal{E})$ as in Section 6.5. We just showed there that $\mathrm{Ho}(\mathfrak{M})$ is a triangulated category.

In doing so, we defined a strict right triangle by pushout along a suspension sequence: For a given object A, this is an assigned short exact sequence $\mathbb{S}_A\colon A \rightarrowtail \tilde{R}_A \twoheadrightarrow \Sigma A$ where $\tilde{R}_A \in \mathcal{R}_W$ and $\Sigma A \in Q$. But Proposition 6.29 says that we can compute ΣA, up to isomorphism in $\mathrm{Ho}(\mathfrak{M})$, by taking any short exact sequence $\mathbb{W}\colon A \rightarrowtail W \twoheadrightarrow C$ with $W \in \mathcal{W}$. Indeed the connecting morphism $\gamma(\mathbb{W})$ of Definition 6.27 is then a $\mathrm{Ho}(\mathfrak{M})$-isomorphism, $\gamma(\mathbb{W})\colon C \cong \Sigma A$. It raises a question that begs to be answered: Do we obtain distinguished triangles via pushout along such general suspension sequences? We now show that in fact we do. After making this clear, we will then describe the triangulated structure on $\mathrm{Ho}(\mathfrak{M})$ in terms of general suspension functors.

So now, given any admissible monic $w\colon A \rightarrowtail W$ with $W \in \mathcal{W}$, we will refer to any short exact sequence $\mathbb{W}\colon A \rightarrowtail W \twoheadrightarrow \mathrm{Cok}(w)$ as a *general suspension sequence (for A)*. We again will call the associated connecting morphism $\gamma(\mathbb{W})\colon \mathrm{Cok}(w) \to \Sigma A$ of Definition 6.27 the *suspension connecting isomorphism (for $\mathbb{W}$)*. It generalizes the previous notion given the same name; see Remark 6.28.

Lemma 6.36 *The pushout of any morphism $f\colon A \to B$*

$$\begin{array}{ccccc} A & \overset{w}{\rightarrowtail} & W & \overset{t}{\twoheadrightarrow} & \mathrm{Cok}(w) \\ \downarrow{\scriptstyle f} & & \downarrow & & \| \\ B & \rightarrowtail & P & \twoheadrightarrow & \mathrm{Cok}(w), \end{array}$$

along any general suspension sequence $\mathbb{W}\colon A \rightarrowtail W \twoheadrightarrow \mathrm{Cok}(w)$, with $\mathrm{Cok}(w) \in Q$ (so $w\colon A \rightarrowtail W$ a cofibration), yields a distinguished triangle

$$A \xrightarrow{\gamma(f)} B \to P \to \mathrm{Cok}(w) \cong \Sigma A$$

in Δ, where $\cong$ denotes the suspension connecting isomorphism $\gamma(\mathbb{W})$.

Proof First we assume $f\colon A \rightarrowtail B$ is an admissible monic, and we will prove the result for this special case first. Take a short exact sequence $W \rightarrowtail RW \twoheadrightarrow Q_W$ with $Q_W \in Q_W$ and $RW \in \mathcal{R}_W$. By Corollary 1.14(2), we may take the pushout of $RW \leftarrowtail W \twoheadrightarrow \mathrm{Cok}(w)$ to construct the back wall of the following 3-D commutative diagram. We denote this pushout by $\Sigma' A$, for we see that $A \rightarrowtail RW \twoheadrightarrow \Sigma' A$ is an (alternate) strict suspension sequence. We have denoted the map by $\Sigma(1_A)$ in accordance with the Dual Fundamental Lemma 3.4 which only requires $\mathrm{Cok}(w) \in Q$ and $RW \in \mathcal{R}_W$ as we have here. So then we go on to construct the following diagram by taking pushouts of both $B \leftarrowtail A \rightarrowtail W$ and $B \leftarrowtail A \rightarrowtail RW$ and using the universal property of pushout to construct the morphism $P \dashrightarrow C(f)$. One verifies that the right front face of the diagram commutes by the uniqueness portion of the universal property of the pushout P:

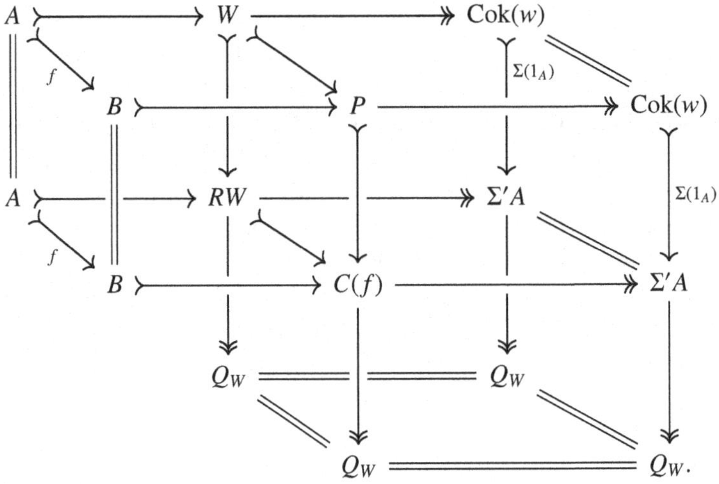

Moreover, we note that we must have $P/W \cong \mathrm{Cok}\, f$ and $C(f)/RW \cong \mathrm{Cok}\, f$. Hence $C(f)$ must be the pushout of $RW \leftarrowtail W \rightarrowtail P$ by Proposition 1.12. This is why we have written $C(f)/P \cong Q_W$ in the diagram. Now by applying γ to the commutative diagram

$$\begin{array}{ccccccc} A & \overset{f}{\rightarrowtail} & B & \rightarrowtail & P & \twoheadrightarrow & \mathrm{Cok}(w) \\ \| & & \| & & \downarrow & & \downarrow{\scriptstyle \Sigma(1_A)} \\ A & \overset{f}{\rightarrowtail} & B & \rightarrowtail & C(f) & \twoheadrightarrow & \Sigma' A, \end{array}$$

we note that the bottom row is a strict triangle in $\mathrm{Ho}(\mathfrak{M})$, but relative to the newly constructed suspension sequence for A. Moreover, the vertical arrows become isomorphisms because $Q_W \in \mathcal{W}$. In fact, the image of $\mathrm{Cok}(w) \rightarrowtail \Sigma' A$ under γ is exactly $\gamma(\mathbb{W})\colon\ \mathrm{Cok}(w) \xrightarrow{\gamma(\Sigma(1_A))} \Sigma' A$, the suspension connecting isomorphism from $\mathbb{W}$ to the newly constructed suspension sequence; see Remark 6.28. The Dual Fundamental Lemma 3.4 and Remark 6.28 even imply that, relative to the initially assigned suspension sequence, $\gamma(\mathbb{W})$ equals the composite $\mathrm{Cok}(w) \xrightarrow{\gamma(\Sigma(1_A))} \Sigma' A \xrightarrow{\gamma(\Sigma(1_A))} \Sigma A$. All together we have a commutative diagram:

$$
\begin{array}{ccccccccc}
A & \xrightarrow{\gamma(f)} & B & \longrightarrow & P & \longrightarrow & \mathrm{Cok}(w) & \xrightarrow[\cong]{\gamma(\mathbb{W})} & \Sigma A \\
\| & & \| & & \downarrow & & \cong\downarrow \gamma(\Sigma(1_A)) & & \| \\
A & \xrightarrow{\gamma(f)} & B & \longrightarrow & C(f) & \longrightarrow & \Sigma' A & \xrightarrow[\cong]{\gamma(\Sigma(1_A))} & \Sigma A.
\end{array}
$$

The bottom row is a distinguished triangle in $\mathrm{Ho}(\mathfrak{M})$, by Corollary 6.18. Since the vertical maps are isomorphisms, the top row is also in Δ.

Now let's prove the general case of the result for a general morphism $f: A \to B$. We begin by factoring $f = pi$ where p is a trivial fibration and i is a cofibration. We construct the commutative diagram with exact rows

$$
\begin{array}{ccccc}
A & \overset{w}{\rightarrowtail} & W & \overset{t}{\twoheadrightarrow} & \mathrm{Cok}(w) \\
{\scriptstyle i}\downarrow & & {\scriptstyle i'}\downarrow & & \| \\
X & \rightarrowtail & P(i) & \twoheadrightarrow & \mathrm{Cok}(w) \\
{\scriptstyle p}\downarrow & & {\scriptstyle p'}\downarrow & & \| \\
B & \rightarrowtail & P & \twoheadrightarrow & \mathrm{Cok}(w)
\end{array}
$$

by taking successive pushouts, and using Corollary 1.14(2), we have $\mathrm{Ker}\, p' = \mathrm{Ker}\, p$. In particular, p' is also a trivial fibration. So by applying γ to the (three left squares of the) commutative diagram

$$
\begin{array}{ccccccccc}
A & \overset{i}{\rightarrowtail} & X & \rightarrowtail & P(i) & \twoheadrightarrow & \mathrm{Cok}(w) & \xrightarrow[\cong]{\gamma(\mathbb{W})} & \Sigma A \\
\| & & \downarrow{\scriptstyle p} & & \downarrow{\scriptstyle p'} & & \| & & \| \\
A & \xrightarrow{f} & B & \rightarrowtail & P & \twoheadrightarrow & \mathrm{Cok}(w) & \xrightarrow[\cong]{\gamma(\mathbb{W})} & \Sigma A,
\end{array}
$$

we obtain an isomorphism of right triangles. Since i is an admissible monic, the top triangle is in Δ by what we already proved previously. Therefore the bottom row triangle is also in Δ, proving the assertion. □

Lemma 6.37 *The pushout of any morphism* $f: A \to B$

$$
\begin{array}{ccccc}
A & \overset{w}{\rightarrowtail} & W & \overset{t}{\twoheadrightarrow} & \mathrm{Cok}(w) \\
\downarrow{\scriptstyle f} & & \downarrow & & \| \\
B & \rightarrowtail & P & \twoheadrightarrow & \mathrm{Cok}(w),
\end{array}
$$

along any general suspension sequence $\mathbb{W}: A \rightarrowtail W \twoheadrightarrow \mathrm{Cok}(w)$, *yields a distinguished triangle*

$$A \xrightarrow{\gamma(f)} B \to P \to \mathrm{Cok}(w) \cong \Sigma A$$

in Δ, *where* $\cong$ *denotes the suspension connecting isomorphism* $\gamma(\mathbb{W})$.

Proof Setting $C := \mathrm{Cok}(w)$, we start by constructing the pullback diagram

$$\begin{array}{ccccc} & & R_C & = & R_C \\ & & \downarrow & & \downarrow \\ A & \overset{v}{\rightarrowtail} & V & \twoheadrightarrow & QC \\ \| & & \downarrow & & \downarrow{\scriptstyle p_C} \\ A & \overset{w}{\rightarrowtail} & W & \twoheadrightarrow & C, \end{array}$$

where the right-hand column is a cofibrant approximation sequence of C. Note that $V \in \mathcal{W}$, and v is a cofibration. So the middle row, which we denote by $\mathbb{V}$, is a general suspension sequence of the type in the previous Lemma 6.36. Using arguments similar in spirit to those in the proof of that lemma, one obtains a commutative diagram

$$\begin{array}{ccccccccc} A & \xrightarrow{f} & B & \rightarrowtail & P' & \twoheadrightarrow & QC & \xrightarrow[\cong]{\gamma(\mathbb{V})} & \Sigma A \\ \| & & \| & & \downarrow & & \downarrow{\scriptstyle p_C} & & \| \\ A & \xrightarrow{f} & B & \rightarrowtail & P & \twoheadrightarrow & C & \xrightarrow[\cong]{\gamma(\mathbb{W})} & \Sigma A, \end{array} \qquad (6.3)$$

which becomes an isomorphism of right triangles after applying γ. Note that the commutativity, $\gamma(1_{\Sigma A}) \circ \gamma(\mathbb{V}) = \gamma(\mathbb{W}) \circ \gamma(p_C)$, follows from the naturality of connecting morphisms as in Lemma 6.26. The top row is in Δ by Lemma 6.36, so the bottom row is too. We leave the remaining details of constructing the isomorphism of triangles in Diagram (6.3) to Exercise 6.6.1. □

Exercise 6.6.1 Complete the proof of Lemma 6.37 by showing that the morphism $P' \twoheadrightarrow P$ in Diagram (6.3) is indeed a trivial epic.

We now state the first main result of this section.

Theorem 6.38 *Given any morphism* $f: A \to B$ *in* $\mathcal{A}$, *the pushout*

$$\begin{array}{ccccc} A & \rightarrowtail & W & \twoheadrightarrow & \mathrm{Cok}(w) \\ \downarrow{\scriptstyle f} & & \downarrow & & \| \\ B & \rightarrowtail & C_{\mathbb{W}}(f) & \twoheadrightarrow & \mathrm{Cok}(w) \end{array} \qquad (\dagger\dagger)$$

along any general suspension sequence $\mathbb{W}\colon A \rightarrowtail W \twoheadrightarrow \mathrm{Cok}(w)$, *meaning* $W \in \mathcal{W}$, *yields a distinguished triangle*

$$A \xrightarrow{\gamma(f)} B \rightarrow C_{\mathbb{W}}(f) \rightarrow \mathrm{Cok}(w) \cong \Sigma A \tag{6.4}$$

in $\mathrm{Ho}(\mathfrak{M})$, *where* $\cong$ *denotes the suspension connecting isomorphism* $\gamma(\mathbb{W})$. *Regardless of the choice of suspension sequence used for A, distinguished triangles constructed this way are unique up to isomorphism. In particular, the class* Δ *of distinguished triangles is precisely the class of all right triangles that are isomorphic to one as in* (6.4).

Proof Lemma 6.37 shows that right triangles constructed this way are in Δ. It follows from the Five Lemma (Corollary 6.4) that any two such distinguished right triangles are isomorphic. Of course any strict triangle in the sense of Definition 6.31 is a particular instance of a right triangle of the form as in (6.4). It follows that the isomorphic closure of all such right triangles is nothing more than Δ. □

The Triangulated Structure Relative to General Suspension Functors In applications one might have a preferred candidate for their suspension functor, or one may simply wish to use a different suspension functor. For example, given a chain complex X, there is the usual suspension, ΣX, obtained by shifting indices against the arrows and changing the sign of the differential to $-d$. But depending on the model, and the complex X, the usual ΣX need not sit in a strict suspension sequence $X \rightarrowtail \tilde{R}_X \twoheadrightarrow \Sigma X$ (with $\tilde{R}_X$ trivially fibrant and ΣX cofibrant). However, there is a functorial assignment of short exact sequences $\mathbb{W}_X\colon X \rightarrowtail W_X \twoheadrightarrow \Sigma X$, where W_X is contractible; see Lemma 10.16. In a typical model structure on chain complexes, contractible complexes are trivial and the following lemma would apply.

Lemma 6.39 *Suppose* $\mathbb{W}\colon A \mapsto \mathbb{W}_A$ *is an assignment of general suspension sequences with the property that for each* $\mathcal{A}$*-morphism* $f\colon A \rightarrow B$, *there exists some* $\mathcal{A}$*-morphism* $\hat{\Sigma}(f)$ *fitting into a morphism of short exact sequences:*

$$\begin{array}{lccccc}
\mathbb{W}_A: & A & \overset{k_A}{\rightarrowtail} & W_A & \overset{\sigma_A}{\twoheadrightarrow} & \hat{\Sigma}A \\
& \downarrow{\scriptstyle f} & & \downarrow & & \downarrow{\scriptstyle \hat{\Sigma}(f)} \\
\mathbb{W}_B: & B & \underset{k_B}{\rightarrowtail} & W_B & \underset{\sigma_B}{\twoheadrightarrow} & \hat{\Sigma}B.
\end{array}$$

Then the rule $f \mapsto \gamma(\hat{\Sigma}(f))$ *is well defined and it determines a functor which we will denote by* $\gamma\hat{\Sigma}\colon \mathcal{A} \rightarrow \mathrm{Ho}(\mathfrak{M})$. *The functor* $\gamma\hat{\Sigma}$ *induces an autoequivalence*

$\hat{\Sigma}\colon \mathrm{Ho}(\mathfrak{M}) \to \mathrm{Ho}(\mathfrak{M})$ and $\hat{\Sigma}$ is the unique functor satisfying $\hat{\Sigma} \circ \gamma = \gamma\hat{\Sigma}$. Moreover, there is a natural equivalence

$$\{\gamma(\mathbb{W}_A)\}\colon \hat{\Sigma} \xrightarrow{\gamma(\mathbb{W})} \Sigma$$

to the initially assinged suspension functor, where each $\gamma(\mathbb{W}_A)\colon \hat{\Sigma}A \to \Sigma A$ is the suspension connecting isomorphism. Any such assignment $\mathbb{W}\colon A \mapsto \mathbb{W}_A$ will be called a **general suspension functor** *for $\mathfrak{M}$. Abusing language, we also call $\hat{\Sigma}$ a general suspension functor for $\mathfrak{M}$.*

Note that our originally assigned suspension functor Σ (from Section 6.4) is a particular instance of a general suspension functor in the aforementioned sense.

Proof of Lemma 6.39 Suppose we are given a morphism $f\colon A \to B$ of $\mathcal{A}$. By hypothesis there is some $\mathcal{A}$-morphism $\hat{\Sigma}(f)\colon \hat{\Sigma}A \to \hat{\Sigma}B$ fitting into a commutative diagram of short exact sequences:

$$\begin{array}{lccccc}
\mathbb{W}_A\colon & A & \overset{k_A}{\rightarrowtail} & W_A & \overset{\sigma_A}{\twoheadrightarrow} & \hat{\Sigma}A \\
& \downarrow{\scriptstyle f} & & \downarrow & & \downarrow{\scriptstyle \hat{\Sigma}(f)} \\
\mathbb{W}_B\colon & B & \overset{k_B}{\rightarrowtail} & W_B & \overset{\sigma_B}{\twoheadrightarrow} & \hat{\Sigma}B.
\end{array}$$

By the naturality of the connecting morphism of Lemma 6.26 and Definition 6.27 we obtain a commutative square in $\mathrm{Ho}(\mathfrak{M})$:

$$\begin{array}{ccc}
\hat{\Sigma}A & \xrightarrow{\gamma(\mathbb{W}_A)} & \Sigma A \\
{\scriptstyle \gamma(\hat{\Sigma}(f))}\downarrow & & \downarrow{\scriptstyle \gamma(\Sigma_\ell(f))} \\
\hat{\Sigma}B & \xrightarrow{\gamma(\mathbb{W}_B)} & \Sigma B.
\end{array} \tag{6.5}$$

By Proposition 6.25 we have $\gamma(\Sigma_\ell(f)) = \left(\bar{\gamma}^\ell \circ \Sigma_\ell\right)(f) = (\Sigma \circ \gamma)(f) = \Sigma(\gamma(f))$. Such a commutative diagram exists no matter what our choice is for the morphism $\mathbb{W}_A \to \mathbb{W}_B$ of short exact sequences. In particular, no matter what map $\hat{\Sigma}(f)$ is used we have

$$\gamma\left(\hat{\Sigma}(f)\right) = \gamma(\mathbb{W}_B)^{-1} \circ \Sigma(\gamma(f)) \circ \gamma(\mathbb{W}_A).$$

It follows that the rule $f \mapsto \gamma\left(\hat{\Sigma}(f)\right)$ is a well-defined functor $\gamma\hat{\Sigma}\colon \mathcal{A} \to \mathrm{Ho}(\mathfrak{M})$. If f is a weak equivalence then $\Sigma(\gamma(f))$ is an isomorphism, and so $\gamma\hat{\Sigma}(f)$ is also an isomorphism. It follows from the Localization Theorem 5.21 that $\gamma\hat{\Sigma}$ descends to an endofunctor on $\mathrm{Ho}(\mathfrak{M})$. That is, there is a unique functor, which we will denote by $\hat{\Sigma}\colon \mathrm{Ho}(\mathfrak{M}) \to \mathrm{Ho}(\mathfrak{M})$, such that $\gamma\hat{\Sigma} = \hat{\Sigma} \circ \gamma$. The commutative diagrams shown in (6.5) provide a natural equivalence

$$\gamma(\mathbb{W})\colon \gamma\hat{\Sigma} \xrightarrow{\{\gamma(\mathbb{W}_A)\}} \bar{\gamma}^\ell \circ \Sigma_\ell,$$

which by the universal properties means we have a natural equivalence

$$\gamma(\mathbb{W})\colon \hat{\Sigma} \circ \gamma \xrightarrow{\{\gamma(\mathbb{W}_A)\}} \Sigma \circ \gamma.$$

By Corollary 5.16 this is equivalent to a natural equivalence

$$\gamma(\mathbb{W}) = \{\gamma(\mathbb{W}_A)\}\colon \hat{\Sigma} \cong \Sigma$$

of the Ho($\mathfrak{M}$)-endofunctors. In particular, $\hat{\Sigma}$ is also an autoequivalence of Ho($\mathfrak{M}$). □

By a $\hat{\Sigma}$-*triangle* we mean a diagram in Ho($\mathfrak{M}$) of the form $A \to B \to C \to \hat{\Sigma}A$. Next we generalize our previous notion of strict triangles to strict $\hat{\Sigma}$-triangles.

Definition 6.40 Let $\hat{\Sigma}$ be a general suspension functor arising from an assignment $\mathbb{W}\colon A \mapsto \mathbb{W}_A$ of general suspension sequences for $\mathfrak{M}$. For each $\mathcal{A}$-morphism $f\colon A \to B$, we define the **cone on f** (relative to $\mathbb{W}$), denoted again by $C(f)$, or $C_{\mathbb{W}}(f)$ if necessary, via the pushout:

$$\begin{array}{ccccc} A & \overset{k_A}{\rightarrowtail} & W_A & \overset{\sigma_A}{\twoheadrightarrow} & \hat{\Sigma}A \\ \downarrow{\scriptstyle f} & & \downarrow & & \| \\ B & \rightarrowtail & C(f) & \twoheadrightarrow & \hat{\Sigma}A. \end{array} \qquad (\dagger\dagger)$$

Any resulting diagram $A \xrightarrow{f} B \to C(f) \to \hat{\Sigma}A$ will be refered to as a **cofiber sequence** of f (relative to $\mathbb{W}$). Its image under γ is called the **strict $\hat{\Sigma}$-triangle on f**.

We already showed in Theorem 6.38 that, upon composing with the connecting isomorphism $\gamma(\mathbb{W}_A)\colon \hat{\Sigma}A \to \Sigma A$, any strict $\hat{\Sigma}$-triangle

$$A \xrightarrow{\gamma(f)} B \to C(f) \to \hat{\Sigma}A \cong \Sigma A \qquad (\ddagger)$$

is a distinguished triangle. That is, it is in the class Δ defined in Definition 6.31. The proof of the following lemma is straightforward and is left as an exercise.

Lemma 6.41 *Let $\Delta_{\mathbb{W}}$ be the class of all $\hat{\Sigma}$-triangles $X \to Y \to Z \to \hat{\Sigma}X$ such that composing the third morphism with the suspension connecting isomorphism $\gamma(\mathbb{W}_X)\colon \hat{\Sigma}X \to \Sigma X$ produces a distinguished triangle in Δ. Then $\Delta_{\mathbb{W}}$ coincides with the class of all $\hat{\Sigma}$-triangles that are isomorphic to some strict $\hat{\Sigma}$-triangle $A \xrightarrow{\gamma(f)} B \to C_{\mathbb{W}}(f) \to \hat{\Sigma}A$.*

Exercise 6.6.2 Use Theorem 6.38 and the Five Lemma (Corollary 6.4) to prove Lemma 6.41.

The following corollary to Theorem 6.38 shows how the triangulated structure on $\mathrm{Ho}(\mathfrak{M})$ may be described entirely in terms of a general suspension functor.

Corollary 6.42 *Assume $\hat{\Sigma}$ is a general suspension functor arising from an assignment $\mathbb{W}: A \mapsto \mathbb{W}_A$ of general suspension sequences for $\mathfrak{M}$. Let $\Delta_{\mathbb{W}}$ be the class of all $\hat{\Sigma}$-triangles that are isomorphic to some strict $\hat{\Sigma}$-triangle*

$$A \xrightarrow{\gamma(f)} B \rightarrow C_{\mathbb{W}}(f) \rightarrow \hat{\Sigma}A.$$

Then $(\hat{\Sigma}, \Delta_{\mathbb{W}})$ is a triangulated structure on $\mathrm{Ho}(\mathfrak{M})$ and the identity functor on $\mathrm{Ho}(\mathfrak{M})$ is a triangulated isomorphism from $\left(\mathrm{Ho}(\mathfrak{M}), \hat{\Sigma}, \Delta_{\mathbb{W}}\right)$ to $(\mathrm{Ho}(\mathfrak{M}), \Sigma, \Delta)$. Indeed the natural equivalence $\gamma(\mathbb{W}) := \{\gamma(\mathbb{W}_A)\}: \hat{\Sigma} \rightarrow \Sigma$, where each component $\gamma(\mathbb{W}_A): \hat{\Sigma}A \rightarrow \Sigma A$ is the suspension connecting isomorphism, is the one needed to show $1_{\mathrm{Ho}(\mathfrak{M})}$ is a triangulated functor.

Proof Using the characterization of $\Delta_{\mathbb{W}}$ given in Lemma 6.41 it is easy to check that $\left(\mathrm{Ho}(\mathfrak{M}), \hat{\Sigma}, \Delta_{\mathbb{W}}\right)$ is a triangulated category. Indeed each of the axioms (Tr1)–(Tr4) is inherited from the corresponding one for $(\mathrm{Ho}(\mathfrak{M}), \Sigma, \Delta)$ by translating back and forth with the natural equivalence $\gamma(\mathbb{W})$. Then it is clear that $(1_{\mathrm{Ho}(\mathfrak{M})}, \gamma(\mathbb{W}))$ is a triangulated isomorphism. □

6.7 Triangulated Localization Theorem

Throughout this section we continue to let $\mathfrak{M} = (\mathcal{Q}, \mathcal{W}, \mathcal{R})$ denote a Hovey triple on a weakly idempotent complete exact category $(\mathcal{A}, \mathcal{E})$; again $\mathfrak{M}$ may even just be a Hovey Ext-triple. We have shown that $\mathrm{Ho}(\mathfrak{M})$ is both a triangulated category and the localization with respect to the class of weak equivalences. We now show that $\mathrm{Ho}(\mathfrak{M})$ is the *triangulated localization* of $(\mathcal{A}, \mathcal{E})$ with respect to $\mathcal{W}$, because it satisfies a suitable universal property.

To explain, we recall the connecting morphism of Definition 6.27. Given any short exact sequence $\mathbb{E}: \ A \overset{f}{\rightarrowtail} B \overset{g}{\twoheadrightarrow} C$ in $(\mathcal{A}, \mathcal{E})$, and any general suspension sequence $\mathbb{W}_A: \ A \overset{k_A}{\rightarrowtail} W_A \overset{\sigma_A}{\twoheadrightarrow} \Sigma A$ for A, it provides a well-defined morphism $\gamma(\mathbb{E}): C \rightarrow \Sigma A$ defined by

$$\gamma(\mathbb{E}) := -\gamma(\sigma_f) \circ [\gamma(\bar{g})]^{-1} \in \mathrm{Ho}(\mathfrak{M})$$

as indicated by the pushout diagram:

$$\begin{array}{ccccc} A & \overset{f}{\rightarrowtail} & B & \overset{g}{\twoheadrightarrow} & C \\ \downarrow{\scriptstyle k_A} & & \downarrow{\scriptstyle k_f} & & \| \\ W_A & \rightarrowtail & C(f) & \overset{\bar{g}}{\twoheadrightarrow} & C \\ \downarrow{\scriptstyle \sigma_A} & & \downarrow{\scriptstyle \sigma_f} & & \\ \Sigma A & = & \Sigma A. & & \end{array}$$

We call $\gamma(\mathbb{E})$ the *connecting morphism* for $\mathbb{E}$ (relative to $\mathbb{W}_A$). We have the following.

Proposition 6.43 *The canonical functor $\gamma\colon \mathcal{A} \to \mathrm{Ho}(\mathfrak{M})$ is an exact functor in the sense that it takes any short exact sequence*

$$\mathbb{E}\colon\ A \overset{f}{\rightarrowtail} B \overset{g}{\twoheadrightarrow} C$$

to a distinguished triangle

$$A \xrightarrow{\gamma(f)} B \xrightarrow{\gamma(g)} C \xrightarrow{-\gamma(\mathbb{E})} \Sigma A,$$

where $\gamma(\mathbb{E})$ is the connecting morphism for $\mathbb{E}$. Moreover, any morphism $(a, b, c)\colon \mathbb{E} \to \mathbb{E}'$ of short exact sequences induces a corresponding morphism of distinguished triangles; in particular, $\gamma(\mathbb{E}') \circ \gamma(c) = \Sigma(\gamma(a)) \circ \gamma(\mathbb{E})$.

Throughout the remainder of this section we assume that Σ is a general suspension functor for $\mathfrak{M}$, arising from an assignment $\mathbb{W}\colon A \mapsto \mathbb{W}_A$ of general suspension sequences as in Lemma 6.39. Again, the originally assigned suspension functor Σ from Section 6.4 is a special case. The only property we really need is $\gamma(\Sigma(f)) = \Sigma(\gamma(f))$ for all $\mathcal{A}$-morphisms f.

Proof of Propositioin 6.43 The proof is given by the following isomorphism of triangles:

$$\begin{array}{ccccccc} A & \xrightarrow{\gamma(f)} & B & \xrightarrow{\gamma(g)} & C & \xrightarrow{-\gamma(\mathbb{E})} & \Sigma A \\ \| & & \| & & \downarrow{\scriptstyle \gamma(\bar{g})^{-1}} & & \| \\ A & \underset{\gamma(f)}{\longrightarrow} & B & \underset{\gamma(k_f)}{\longrightarrow} & C(f) & \underset{\gamma(\sigma_f)}{\longrightarrow} & \Sigma A. \end{array}$$

The bottom row is in Δ (by definition for the assigned suspension sequence, and by Theorem 6.38 for a general suspension sequence).

Now given any morphism $(a, b, c)\colon \mathbb{E} \to \mathbb{E}'$ of short exact sequences, one argues with Lemma 6.26 that we have $\gamma(\mathbb{E}') \circ \gamma(c) = \Sigma(\gamma(a)) \circ \gamma(\mathbb{E})$. It amounts to applying γ to the entire morphism of pushout diagrams, and using the relation $\gamma(\Sigma(f)) = \Sigma(\gamma(f))$ of Lemma 6.39. □

In general, if $\mathcal{T} = (\mathcal{T}, \Sigma_\mathcal{T}, \Delta_\mathcal{T})$ is any triangulated category, let us say that an additive functor $F\colon \mathcal{A} \to \mathcal{T}$ is **exact** if to each short exact sequence, such as $\mathbb{E} \in \mathcal{E}$ given earlier, there is assigned (functorially!) a $\mathcal{T}$-morphism $F(\mathbb{E})\colon FC \to \Sigma_\mathcal{T} FA$ in such a way that

$$FA \xrightarrow{F(f)} FB \xrightarrow{F(g)} FC \xrightarrow{-F(\mathbb{E})} \Sigma_\mathcal{T} FA$$

is a distinguished triangle in $\Delta_\mathcal{T}$. The functoriality means that the pair $(F, \Sigma_\mathcal{T})$ respects morphisms $(a, b, c)\colon \mathbb{E} \to \mathbb{E}'$ of short exact sequences. This amounts to the functor F extending to a functor $F\colon \mathcal{E} \to \Delta_\mathcal{T}$, from the category of all short exact sequences to the category of distinguished $\mathcal{T}$-triangles.

We can now show that $\mathrm{Ho}(\mathfrak{M})$ is the *triangulated localization* of $(\mathcal{A}, \mathcal{E})$ with respect to the class of trivial objects $\mathcal{W}$. Compare this to how the stable category of Section 4.3, $\mathrm{St}(\mathcal{A}) := \mathrm{St}_\mathcal{W}(\mathcal{A})$, is the *additive* localization of $\mathcal{A}$ with respect to $\mathcal{W}$.

Theorem 6.44 (Triangulated Localization Theorem) *The canonical functor $\gamma\colon \mathcal{A} \to \mathrm{Ho}(\mathfrak{M})$ is universally initial among all exact functors $F\colon \mathcal{A} \to \mathcal{T}$ to triangulated categories $\mathcal{T}$ satisfying $F(\mathcal{W}) = 0$. That is, for any other exact functor $F\colon \mathcal{A} \to \mathcal{T}$ with $F(\mathcal{W}) = 0$, there exists a unique functor* $\mathrm{Ho}(F)$ *such that $F = \mathrm{Ho}(F) \circ \gamma$. Moreover,* $\mathrm{Ho}(F)$ *is necessarily a triangulated functor.*

Proof Say $F\colon \mathcal{A} \to \mathcal{T}$ is a given exact functor with $\mathcal{T}$ triangulated and $F(\mathcal{W}) = 0$. Then given any trivial cofibration $A \rightarrowtail B \twoheadrightarrow W$ we have an exact triangle

$$FA \to FB \to FW \to \Sigma_\mathcal{T} FA.$$

But since $FW = 0$ and $\mathcal{T}$ is triangulated, it follows that $FA \to FB$ must be an isomorphism (Exercise 6.1.5(2)). So F sends trivial cofibrations to isomorphisms. Similarly, it sends trivial fibrations to isomorphisms. Therefore *any* weak equivalence, which must factor as a trivial cofibration followed by a trivial fibration, must also be sent to an isomorphism. We thus have proved that F sends all weak equivalences to isomorphisms, and so by the Localization Theorem 5.21 we get that F factors uniquely through γ as $F = \mathrm{Ho}(F) \circ \gamma$. Since F is additive, the functor $\mathrm{Ho}(F)$ is also additive.

But we wish to show that $\mathrm{Ho}(F)$ is a triangulated functor. Referring to the Definition 6.6, we first need a candidate for the needed natural equivalence $\mathrm{Ho}(F) \circ \Sigma \cong \Sigma_\mathcal{T} \circ \mathrm{Ho}(F)$. By Corollary 5.16 it is equivalent to find a natural equivalence $\mathrm{Ho}(F) \circ \Sigma \circ \gamma \cong \Sigma_\mathcal{T} \circ \mathrm{Ho}(F) \circ \gamma = \Sigma_\mathcal{T} \circ F$. Now recall that by

Lemma 6.39, $\Sigma \circ \gamma = \gamma\Sigma \colon \mathcal{A} \to \mathrm{Ho}(\mathfrak{M})$ is the functor given by $f \mapsto \gamma(\Sigma(f))$. So since $F = \mathrm{Ho}(F) \circ \gamma$, it follows that $\mathrm{Ho}(F) \circ \Sigma \circ \gamma$ is the functor acting by $f \mapsto F(\Sigma(f))$. So to find the desired natural equivalence, we note that a morphism of general suspension sequences

$$\begin{array}{llcccc} \mathbb{W}_A\colon & A & \overset{k_A}{\rightarrowtail} & W_A & \overset{\sigma_A}{\twoheadrightarrow} & \Sigma A \\ & \downarrow{\scriptstyle f} & & \downarrow & & \downarrow{\scriptstyle \Sigma(f)} \\ \mathbb{W}_B\colon & B & \overset{k_B}{\rightarrowtail} & W_B & \overset{\sigma_B}{\twoheadrightarrow} & \Sigma B \end{array}$$

gives rise to a morphism of distinguished triangles in $\Delta_{\mathcal{T}}$:

$$\begin{array}{ccccccc} FA & \longrightarrow & 0 & \longrightarrow & F\Sigma A & \xrightarrow{-F(\mathbb{W}_A)} & \Sigma_{\mathcal{T}} FA \\ \downarrow{\scriptstyle F(f)} & & \downarrow & & \downarrow{\scriptstyle F(\Sigma(f))} & & \downarrow{\scriptstyle \Sigma_{\mathcal{T}}(F(f))} \\ FB & \longrightarrow & 0 & \longrightarrow & F\Sigma B & \xrightarrow[-F(\mathbb{W}_B)]{} & \Sigma_{\mathcal{T}} FB. \end{array}$$

It follows that $F(\mathbb{W}) := \{F(\mathbb{W}_A)\colon F\Sigma A \to \Sigma_{\mathcal{T}} FA\}$ defines a natural equivalence $\mathrm{Ho}(F) \circ \Sigma \circ \gamma \cong \Sigma_{\mathcal{T}} \circ F$. This provides the desired natural equivalence

$$\mathrm{Ho}(F) \circ \Sigma \xrightarrow{F(\mathbb{W}):=\{F(\mathbb{W}_A)\}} \Sigma_{\mathcal{T}} \circ \mathrm{Ho}(F).$$

Now we will show that $(\mathrm{Ho}(F), F(\mathbb{W}))\colon (\mathrm{Ho}(\mathfrak{M}), \Sigma, \Delta_{\mathbb{W}}) \to (\mathcal{T}, \Sigma_{\mathcal{T}}, \Delta_{\mathcal{T}})$ is a triangulated functor. So let us be given a strict Σ-triangle

$$A \xrightarrow{\gamma(f)} B \xrightarrow{\gamma(k_f)} C(f) \xrightarrow{\gamma(\sigma_f)} \Sigma A,$$

which we recall arises from a pushout diagram:

$$\begin{array}{llcccc} \mathbb{W}_A\colon & A & \overset{k_A}{\rightarrowtail} & W_A & \overset{\sigma_A}{\twoheadrightarrow} & \Sigma A \\ & \downarrow{\scriptstyle f} & & \downarrow & & \| \\ \mathbb{P}_f\colon & B & \overset{k_f}{\rightarrowtail} & C(f) & \overset{\sigma_f}{\twoheadrightarrow} & \Sigma A. \end{array}$$

By hypothesis, this gives rise to a morphism of distinguished triangles in $\Delta_{\mathcal{T}}$:

$$\begin{array}{ccccccc} FA & \longrightarrow & 0 & \longrightarrow & F\Sigma A & \xrightarrow{-F(\mathbb{W}_A)} & \Sigma_{\mathcal{T}} FA \\ \downarrow{\scriptstyle F(f)} & & \downarrow & & \| & & \downarrow{\scriptstyle \Sigma_{\mathcal{T}}(F(f))} \\ FB & \xrightarrow[F(k_f)]{} & FC(f) & \xrightarrow[F(\sigma_f)]{} & F\Sigma A & \xrightarrow[-F(\mathbb{P}_f)]{} & \Sigma_{\mathcal{T}} FB. \end{array}$$

In particular, the right square commutes and the bottom row is a distinguished triangle in $\Delta_{\mathcal{T}}$. Because $\mathrm{Ho}(F) \circ \gamma = F$, our goal is to show that

$$FA \xrightarrow{F(f)} FB \xrightarrow{F(k_f)} FC(f) \xrightarrow{F(\sigma_f)} F\Sigma A \cong_{F(\mathbb{W}_A)} \Sigma_{\mathcal{T}} FA$$

is a distinguished triangle in $\Delta_{\mathcal{T}}$. But, we have an isomorphism of $\mathcal{T}$-triangles, and the top row has already been shown to be in $\Delta_{\mathcal{T}}$:

$$\begin{array}{ccccccc} FB & \xrightarrow{F(k_f)} & FC(f) & \xrightarrow{F(\sigma_f)} & F\Sigma A & \xrightarrow{-F(\mathbb{P}_f)} & \Sigma_{\mathcal{T}} FB \\ \| & & \| & & \downarrow{\scriptstyle F(\mathbb{W}_A)} & & \| \\ FB & \xrightarrow[F(k_f)]{} & FC(f) & \xrightarrow[F(\mathbb{W}_A)\circ F(\sigma_f)]{} & \Sigma_{\mathcal{T}} FA & \xrightarrow[-\Sigma_{\mathcal{T}} F(f)]{} & \Sigma_{\mathcal{T}} FB. \end{array}$$

We conclude that the bottom row is also in $\Delta_{\mathcal{T}}$. Therefore,

$$FA \xrightarrow{F(f)} FB \xrightarrow{F(k_f)} FC(f) \xrightarrow{F(\mathbb{W}_A)\circ F(\sigma_f)} \Sigma_{\mathcal{T}} FA$$

is also in $\Delta_{\mathcal{T}}$, by the converse of Axiom (Tr-2) (see Lemma 6.7). This completes the proof that $(\mathrm{Ho}(F), F(\mathbb{W}))$ is a triangulated functor. □

7

Derived Functors and Abelian Monoidal Model Structures

In this chapter we study left and right derived functors of additive functors $F: \mathcal{A} \to \mathcal{B}$, where $\mathcal{A}$ has a (closed) abelian model structure. If $\mathcal{B}$ too has an abelian model structure, there is the fundamental notion of left and right Quillen functors. This is a special type of adjoint pair

$$\mathcal{A} \underset{G}{\overset{F}{\rightleftarrows}} \mathcal{B}$$

which induces an adjunction between their homotopy categories

$$\mathrm{Ho}(\mathfrak{M}_{\mathcal{A}}) \underset{\mathbf{R}G}{\overset{\mathbf{L}F}{\rightleftarrows}} \mathrm{Ho}(\mathfrak{M}_{\mathcal{B}}).$$

Here, $\mathbf{L}F$ is the (total) left derived functor of F and $\mathbf{R}G$ is the right derived functor of G. We will see that these are triangulated functors and we will characterize Quillen equivalences, which are adjoint pairs (F, G) for which $(\mathbf{L}F, \mathbf{R}G)$ becomes an equivalence of categories.

The end of the chapter turns to the basics of abelian monoidal model structures. The main result is that a tensor product on $\mathcal{A}$, compatible with the model structure, will descend to a well-behaved tensor product on the homotopy category. We give Hovey's criteria for abelian monoidal model structures which provides a powerful way to construct tensor triangulated categories.

Throughout this chapter $\mathfrak{M} = \mathfrak{M}_{\mathcal{A}} = (Q, \mathcal{W}, \mathcal{R})$ *will denote a Hovey triple on a weakly idempotent complete exact category* $(\mathcal{A}, \mathcal{E}) = (\mathcal{A}, \mathcal{E}_{\mathcal{A}})$.

7.1 Derived Functors

The goal of this section is to define derived functors and prove a theorem guaranteeing their existence. A crucial first step to this is a result known as Ken

Brown's Lemma. To describe it, we will need to work with a second exact category, which we will denote by $(\mathcal{B}, \mathcal{E}_{\mathcal{B}})$. So at times we might denote $(\mathcal{A}, \mathcal{E})$ by $(\mathcal{A}, \mathcal{E}_{\mathcal{A}})$ for added clarity.

Generally speaking, let us say that by a class of *trivial objects* in $(\mathcal{B}, \mathcal{E}_{\mathcal{B}})$ we will mean a class of objects $\mathcal{T}$, containing 0, satisfying the 2 out of 3 property on short exact sequences. Then by a **$\mathcal{T}$-weak equivalence** we mean a morphism of the form $f = pi$ where i is an admissible monic with trivial cokernel and p is an admissible epic with trivial kernel. Of course, if $(\mathcal{Q}_{\mathcal{B}}, \mathcal{W}_{\mathcal{B}}, \mathcal{R}_{\mathcal{B}})$ is a Hovey triple on $(\mathcal{B}, \mathcal{E}_{\mathcal{B}})$ then $\mathcal{W}_{\mathcal{B}}$ is a class of trivial objects in this general sense too. As the following exercise implies, the class of all isomorphisms in any additive category can be construed as a class of weak equivalences.

Exercise 7.1.1 Let $(\mathcal{B}, \mathcal{E}_{\mathcal{B}})$ be any exact category and let 0 denote the class of all zero objects. Show that an isomorphism in $\mathcal{B}$ is precisely a 0-weak equivalence.

Lemma 7.1 (Ken Brown's Lemma) *Let $F\colon \mathcal{A} \to \mathcal{B}$ be an additive functor and assume $(\mathcal{B}, \mathcal{E}_{\mathcal{B}})$ is an exact category that comes with a class of trivial objects $\mathcal{W}_{\mathcal{B}}$. If F maps all trivial cofibrations between cofibrant objects to $\mathcal{W}_{\mathcal{B}}$-weak equivalences, then F takes all weak equivalences between cofibrant objects to $\mathcal{W}_{\mathcal{B}}$-weak equivalences. Dually, if F maps all trivial fibrations between fibrant objects to $\mathcal{W}_{\mathcal{B}}$-weak equivalences, then F maps all weak equivalences between fibrant objects to $\mathcal{W}_{\mathcal{B}}$-weak equivalences.*

Proof As shown in the proof of Proposition 2.13, any morphism $f\colon A \to B$ of $\mathcal{A}$ factors as

$$A \xrightarrow{\left[\begin{smallmatrix} 1_A \\ 0 \end{smallmatrix}\right]} A \bigoplus B \xrightarrow{[f\ 1_B]} B$$

and there is a short exact sequence

$$A \rightarrowtail^{\left[\begin{smallmatrix} 1_A \\ -f \end{smallmatrix}\right]} A \bigoplus B \overset{[f\ 1_B]}{\twoheadrightarrow} B.$$

Now assume A and B are cofibrant and $f\colon A \to B$ is a weak equivalence in $\mathfrak{M}$. Using enough injectives of $(\mathcal{Q}, \mathcal{R}_{\mathcal{W}})$, write a short exact sequence $A \rightarrowtail R \twoheadrightarrow Q$ with Q cofibrant and R trivially fibrant. Note then that R is in fact (trivially) cofibrant too since it is an extension of two objects in $\mathcal{Q}$. Construct the following pushout diagram:

$$\begin{array}{ccccc}
A & \overset{\left[\begin{smallmatrix}1_A\\ -f\end{smallmatrix}\right]}{\rightarrowtail} & A \bigoplus B & \overset{[f\ 1_B]}{\twoheadrightarrow} & B \\
\downarrow & & \downarrow j & & \| \\
R & \rightarrowtail & C & \overset{p}{\twoheadrightarrow} & B \\
\downarrow & & \downarrow & & \\
Q & = & Q. & &
\end{array}$$

Let us write $j_A := j \circ \left[\begin{smallmatrix}1_A\\ 0\end{smallmatrix}\right]$. Since the composite of two cofibrations is again a cofibration, j_A is a cofibration. But $f = pj_A$, and so by the 2 out of 3 Axiom in $\mathfrak{M}$, j_A must be a *trivial* cofibration, and it is between cofibrant objects. By hypothesis, $F(j_A)$ is a $\mathcal{W}_{\mathcal{B}}$-weak equivalence. So we can write $F(j_A) = \beta\alpha$ where α is an admissible monic with $\operatorname{Cok}\alpha \in \mathcal{W}_{\mathcal{B}}$ and β is an admissible epic with $\operatorname{Ker}\beta \in \mathcal{W}_{\mathcal{B}}$.

Next, note that the middle row of the previous diagram must be a split exact sequence. Certainly F preserves split exact sequences, so $F(p)$ is a split epimorphism in $\mathcal{B}$, with kernel $F(R)$. In particular, $F(p)$ is an admissible epimorphism, and we have $F(R) \in \mathcal{W}_{\mathcal{B}}$. Indeed the hypothesis again implies $F(0 \rightarrowtail R) = 0 \rightarrowtail F(R)$ is a $\mathcal{W}_{\mathcal{B}}$-weak equivalence. We get that $F(f) = F(p) \circ F(j_A) = (F(p)\beta)\alpha$. But $F(p)\beta$ is an admissible epimorphism and its kernel is an extension of R and $\operatorname{Ker}\beta$, by Corollary 1.15. So $\operatorname{Ker}(F(p)\beta) \in \mathcal{W}_{\mathcal{B}}$, and we have shown directly that $F(f) = (F(p)\beta)\alpha$ satisfies the definition of a $\mathcal{W}_{\mathcal{B}}$-weak equivalence. □

Exercise 7.1.2 The proof of Ken Brown's Lemma works even when the model structure on $(\mathcal{A}, \mathcal{E}_{\mathcal{A}})$ is not closed. Find a shorter proof that works whenever the model structure is closed. (*Hint*: Let $f: A \to B$ be a weak equivalence between cofibrant objects and factor it as $f = pi$, where i is a trivial cofibration and p is a trivial fibration. Show that $F(p)$ is an admissible (split) epic with kernel in $\mathcal{W}_{\mathcal{B}}$.)

Recall the universal property enjoyed by the canonical localization functor $\gamma: \mathcal{A} \to \operatorname{Ho}(\mathfrak{M})$. Given any functor $F: \mathcal{A} \to \mathcal{B}$ sending weak equivalences to isomorphisms, it uniquely factors through $\operatorname{Ho}(\mathfrak{M})$ as $\operatorname{Ho}(F) \circ \gamma = F$. So here F induces a functor $\operatorname{Ho}(F): \operatorname{Ho}(\mathfrak{M}) \to \mathcal{B}$ which is unique with respect to satisfying this equality. A left derived functor, for any given functor $F: \mathcal{A} \to \mathcal{B}$, is a more general idea: It is a functor $LF: \operatorname{Ho}(\mathfrak{M}) \to \mathcal{B}$ but with only a natural transformation $t: LF \circ \gamma \to F$ satisfying a universal property. We now make this precise.

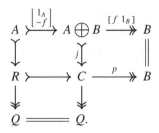

Definition 7.2 Let $\mathcal{B}$ be any additive category and let $F\colon \mathcal{A} \to \mathcal{B}$ be an additive functor. A **left derived functor** of F is an additive functor $LF\colon \mathrm{Ho}(\mathfrak{M}) \to \mathcal{B}$ along with a natural transformation $LF \circ \gamma \xrightarrow{t} F$ that is universal from the left in the following sense: For any other additive $G\colon \mathrm{Ho}(\mathfrak{M}) \to \mathcal{B}$ and natural transformation $G \circ \gamma \xrightarrow{s} F$, there exists a unique natural transformation $G \circ \gamma \xrightarrow{u} LF \circ \gamma$ such that s factors through t as shown in the following diagram:

$$\begin{array}{ccc} G \circ \gamma & & \\ {\scriptstyle \exists! u}\downarrow & \searrow^{s} & \\ LF \circ \gamma & \xrightarrow[t]{} & F. \end{array}$$

On the other hand, a **right derived functor** of F is an additive functor $RF\colon \mathrm{Ho}(\mathfrak{M}) \to \mathcal{B}$ along with a natural transformation $F \xrightarrow{t} RF \circ \gamma$ that is universal to the right: For any other additive $G\colon \mathrm{Ho}(\mathfrak{M}) \to \mathcal{B}$ and natural transformation $F \xrightarrow{s} G \circ \gamma$, there exists a unique natural transformation $RF \circ \gamma \xrightarrow{u} G \circ \gamma$ such that s factors through t as shown in the following diagram:

$$\begin{array}{ccc} & & G \circ \gamma \\ & \nearrow^{s} & \uparrow{\scriptstyle \exists! u} \\ F & \xrightarrow[t]{} & RF \circ \gamma. \end{array}$$

Remark 7.3 The definition of a left derived functor F is usually stated so that the last sentence asserts the existence of a unique natural transformation $G \xrightarrow{u} LF$ such that s equals the composite natural transformation

$$G \circ \gamma \xrightarrow{u_\gamma} LF \circ \gamma \xrightarrow{t} F.$$

However, it was shown in Corollary 5.16 that such a u is exactly the same thing as a natural transformation $G \circ \gamma \xrightarrow{u} LF \circ \gamma$.

Note the similarity of Definition 7.2 to well-known universal properties such as the kernel of a morphism.

Exercise 7.1.3 Show that left and right derived functors, if they exist, are unique up to a canonical natural equivalence.

In particular, if $L'F$ is any other left derived functor of F, then we have a canonical $\mathcal{B}$-isomorphism $\alpha_A\colon LF(A) \cong L'F(A)$ for each object A.

Exercise 7.1.4 Assume $F\colon \mathcal{A} \to \mathcal{B}$ is an additive functor sending weak equivalences to isomorphisms. Show that the canonical localization functor $\mathrm{Ho}(F)$, along with the identity natural transformation $1_F\colon \mathrm{Ho}(F) \circ \gamma \to F$, is the left derived functor of F. In the same way, it is also the right derived functor for F.

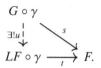

We now prove the main result of this section. It provides a condition guaranteeing the existence of left derived functors.

Theorem 7.4 *Let $\mathcal{B}$ be any additive category and let $F: \mathcal{A} \to \mathcal{B}$ be an additive functor that takes trivial cofibrations between cofibrant objects to isomorphisms. Then the left derived functor of F exists and may be taken to be the unique functor $LF: \mathrm{Ho}(\mathfrak{M}) \to \mathcal{B}$ satisfying $FQ = LF \circ \gamma$. In particular, we may take $LF(A) = F(QA)$ for all objects $A \in \mathcal{A}$, and $LF(A) = F(A)$ whenever A is cofibrant.*

Proof The hypothesis implies that F "kills" all trivially cofibrant objects. Therefore, it descends to a functor

$$F: \mathrm{St}_*(\mathcal{A}) \to \mathcal{B},$$

where $* \in \{r, \omega\}$. That is, $\mathrm{St}_*(\mathcal{A})$ may be taken to be either $\mathrm{St}_r(\mathcal{A})$, the right stable category, or $\mathrm{St}_\omega(\mathcal{A})$, the ω-stable category. Moreover, the codomain of Q may be taken to be $\mathrm{St}_*(\mathcal{A})$ for $* \in \{r, \omega\}$. See Exercise 4.4.1. Consequently, we have a well-defined composition

$$FQ: \mathcal{A} \xrightarrow{Q} \mathrm{St}_*(\mathcal{A}) \xrightarrow{F} \mathcal{B},$$

which simply acts by $f \mapsto F(Q(f))$ where $Q(f)$ is any morphism making the square commute:

$$\begin{array}{ccc} QA & \xrightarrow{p_A} & A \\ {\scriptstyle Q(f)}\downarrow & & \downarrow{\scriptstyle f} \\ QB & \xrightarrow{p_B} & B. \end{array} \qquad (**)$$

This composite functor FQ takes all weak equivalences to isomorphisms. Indeed Lemma 5.5 says precisely that $Q(f)$ is a weak equivalence iff f is a weak equivalence, and, the hypothesis combined with Ken Brown's Lemma 7.1 provides the rest. (Considering $\mathcal{B}$ as an exact category with the split exact structure, and taking $\mathcal{W}_{\mathcal{B}} = 0$, we see that a $\mathcal{W}_{\mathcal{B}}$-weak equivalence is equivalent to an isomorphism – see Exercise 7.1.1. So by Ken Brown's Lemma 7.1 we get that $F: \mathcal{A} \to \mathcal{B}$ takes *all* weak equivalences between cofibrant objects to isomorphisms.)

Therefore the Localization Theorem 5.21 applies to FQ. We obtain a unique functor $LF: \mathrm{Ho}(\mathfrak{M}) \to \mathcal{B}$ satisfying $LF \circ \gamma = FQ$. Moreover, applying F to all squares of the form $(**)$ shows that the $F(p_A)$ assemble to a natural transformation $\{F(p_A)\}_A$ which we denote by $F(p): LF \circ \gamma \to F$.

We are to show that $(LF, F(p))$ is a left derived functor of F. So suppose that $G: \mathrm{Ho}(\mathfrak{M}) \to \mathcal{B}$ is a functor and $G \circ \gamma \xrightarrow{s} F$ is a natural transformation. We must find a natural transformation $G \circ \gamma \xrightarrow{u} LF \circ \gamma$ such that $s = F(p) \circ u$ and show that u is unique with respect to this property. The u we seek is

$$\left\{u_A := s_{QA} \circ [G(\gamma(p_A))]^{-1}\right\}.$$

Each such u_A is indeed a morphism from $G\gamma(A) \to LF(\gamma(A)) = FQ(A)$ because

$$G\gamma(A) \xleftarrow{G(\gamma(p_A))} G\gamma(QA) \xrightarrow{s_{QA}} F(QA).$$

Given any morphism $f \colon A \to B$ in $\mathcal{A}$, we need to check commutativity of the outer rectangle:

$$\begin{array}{ccccc} G\gamma(A) & \xrightarrow{[G\gamma(p_A)]^{-1}} & G\gamma(QA) & \xrightarrow{s_{QA}} & F(QA) \\ \downarrow{\scriptstyle G\gamma(f)} & & \downarrow{\scriptstyle G\gamma(Q(f))} & & \downarrow{\scriptstyle FQ(f)} \\ G\gamma(B) & \xrightarrow[{[G\gamma(p_B)]^{-1}}]{} & G\gamma(QB) & \xrightarrow[s_{QB}]{} & F(QB). \end{array}$$

But the left square clearly commutes because it is just $G \circ \gamma$ applied to the squares of Diagram (∗∗). Also the right square commutes because we are given that $G \circ \gamma \xrightarrow{s} F$ is a natural transformation and it is just applied to $QA \xrightarrow{Q(f)} QB$.

It is only left to show that the only natural transformation $G \circ \gamma \xrightarrow{u} LF \circ \gamma$ satisfying $s = F(p) \circ u$ is the one we have previously defined. For this, let us suppose that u is such a natural transformation. Then for each object $A \in \mathcal{A}$, applying the composite $F(p) \circ u$ to the morphism $QA \xrightarrow{p_A} A$ yields the following commutative diagram:

$$\begin{array}{ccccc} G(QA) & \xrightarrow{u_{QA}} & LF(QA) & \xrightarrow{F(p_{QA})} & F(QA) \\ \downarrow{\scriptstyle G\gamma(p_A)} & & \downarrow{\scriptstyle LF(\gamma(p_A))} & & \downarrow{\scriptstyle F(p_A)} \\ G(A) & \xrightarrow[u_A]{} & LF(A) & \xrightarrow[F(p_A)]{} & F(A). \end{array}$$

But $p_{QA} = 1_{QA}$, so $F(p_{QA}) = F(1_{QA}) = 1_{FQA}$. So if we are to have $s = F(p) \circ u$, then the top composite forces $u_{QA} = s_{QA}$. Moreover,

$$(LF(QA) \xrightarrow{LF(\gamma(p_A))} LF(A)) = (F(QA) \xrightarrow{FQ(p_A)} F(QA)) = 1_{F(QA)}.$$

Therefore, commutativity of the left square forces $s_{QA} = u_A \circ G\gamma(p_A)$. Since $\gamma(p_A)$ is an isomorphism we conclude $u_A = s_{QA} \circ [G\gamma(p_A)]^{-1}$. This is exactly our definition of u so we have proved the uniqueness claim. □

Remark 7.5 There is an important point to be made here concerning the fact that LF "may be taken to be the unique functor" satisfying $FQ = LF \circ \gamma$. Indeed we have shown that there is only one functor satisfying this property and it will serve as LF, but one should realize that there may still be other choices for LF. For example, take F to be the canonical functor $\gamma\colon \mathcal{A} \to \mathrm{Ho}(\mathfrak{M})$. It is easy to see that (LF, t) may be taken to be the identity functor $1_{\mathrm{Ho}(\mathfrak{M})}$ along with the identity transformation $\gamma \xrightarrow{1} \gamma$. (This is a special case of Exercise 7.1.4.) On the other hand, Theorem 7.4 applies and asserts that we may take $L(\gamma)$ to be the unique functor LF satisfying $\gamma \circ Q = LF \circ \gamma$. The functor $I \circ \mathrm{Ho}(Q)$ described before Proposition 5.25 is the one satisfying this property. Of course there must be a canonical isomorphism between the two functors and, as shown in the proof of Proposition 5.25, it is given by the natural isomorphism $\{\gamma(p_A)\}\colon I \circ \mathrm{Ho}(Q) \to 1_{\mathrm{Ho}(\mathfrak{M})}$, where $p_A\colon QA \to A$ is the fixed cofibrant approximation.

7.2 Quillen Functors and Derived Adjunctions

Throughout this section, we let $(\mathcal{A}, \mathcal{E}_{\mathcal{A}})$ and $(\mathcal{B}, \mathcal{E}_{\mathcal{B}})$ each be weakly idempotent complete exact categories along with Hovey triples $\mathfrak{M}_{\mathcal{A}}$ and $\mathfrak{M}_{\mathcal{B}}$. We will define Quillen adjunctions and show that they induce adjunctions of homotopy categories.

Assume throughout that we have an adjunction (F, G, η, ϵ) from $\mathcal{A}$ to $\mathcal{B}$:

$$\mathcal{A} \underset{G}{\overset{F}{\rightleftarrows}} \mathcal{B}.$$

So $F\colon \mathcal{A} \to \mathcal{B}$ is the left adjoint functor and $G\colon \mathcal{B} \to \mathcal{A}$ is its right adjoint. Like any adjunction, recall that it comes with its *unit* natural transformation $\eta\colon 1_{\mathcal{A}} \to GF$ and *counit* natural transformation $\epsilon\colon FG \to 1_{\mathcal{B}}$. We have adjunction isomorphisms, natural in A and B,

$$\mathrm{Hom}_{\mathcal{B}}(FA, B) \underset{\flat}{\overset{\sharp}{\rightleftarrows}} \mathrm{Hom}_{\mathcal{A}}(A, GB),$$

which work from left to right via composition as follows:

$$FA \xrightarrow{s} B \quad \rightsquigarrow \quad \begin{array}{ccc} GFA & \xrightarrow{G(s)} & GB. \\ {\scriptstyle \eta_A}\uparrow & \nearrow_{s^\sharp} & \\ A & & \end{array}$$

Inversely, from right to left the function works by forming the following composition:

$$A \xrightarrow{t} GB \quad \rightsquigarrow \quad \begin{array}{ccc} FA & \xrightarrow{F(t)} & FGB \\ & \searrow^{t^\flat} & \downarrow \epsilon_B \\ & & B. \end{array}$$

For brevity we will sometimes denote the adjunction simply by (F, G), knowing that η and ϵ are implicitly defined.

Definition 7.6 We say that (F, G) is a **Quillen adjunction** if F preserves cofibrations and G preserves fibrations. In this case we also say that F is a **left Quillen functor** and that G is a **right Quillen functor**.

Proposition 7.7 *The following are equivalent for an adjunction* (F, G).

(1) (F, G) *is a Quillen adjunction.*
(2) F *preserves cofibrations and trivial cofibrations.*
(3) G *preserves fibrations and trivial fibrations.*

Proof Let us prove (3) $\implies$ (1). So let $f\colon A \rightarrowtail B$ be a cofibration in $\mathfrak{M}_{\mathcal{A}}$. We must show that $F(f)$ is a cofibration in $\mathfrak{M}_{\mathcal{B}}$. By Corollary 2.15 we only need to show that $F(f)$ satisfies the LLP with respect to all trivial fibrations in $\mathfrak{M}_{\mathcal{B}}$. So let $g\colon X \twoheadrightarrow Y$ be such a trivial fibration. We need to find a lift in the following commutative square on the right. But the adjunction provides the corresponding square on the left:

$$\begin{array}{ccc} A & \longrightarrow & GX \\ f \downarrow & \nearrow^{t} & \downarrow G(g) \\ B & \longrightarrow & GY, \end{array} \qquad \begin{array}{ccc} FA & \longrightarrow & X \\ F(f) \downarrow & \nearrow^{t^\flat} & \downarrow g \\ FB & \longrightarrow & Y. \end{array}$$

By hypothesis, a lift t does exist for the left square. The adjunction brings it back over to a lift $t^\flat$ for the right square. This proves (3) $\implies$ (1) and the reverse argument will also show (1) $\implies$ (3). Similarly, (1) $\iff$ (2). □

The following corollary is now immediate from Ken Brown's Lemma 7.1.

Corollary 7.8 (Corollary to Ken Brown's Lemma) *A left Quillen functor F maps all weak equivalences between cofibrant objects to weak equivalences. Dually, a right Quillen functor G maps all weak equivalences between fibrant objects to weak equivalences.*

The goal now is to see that a Quillen adjunction induces an adjunction between the corresponding homotopy categories. This brings us to the fundamental notion of total derived functors.

Definition 7.9 Suppose $F\colon \mathcal{A} \to \mathcal{B}$ is just any additive functor (not necessarily a left adjoint) between our exact categories possessing abelian model structures $\mathfrak{M}_{\mathcal{A}}$ and $\mathfrak{M}_{\mathcal{B}}$. A **total left derived functor** for F is a functor $\mathbf{L}F\colon \mathrm{Ho}(\mathfrak{M}_{\mathcal{A}}) \to \mathrm{Ho}(\mathfrak{M}_{\mathcal{B}})$ which is a left derived functor for the composite $\mathcal{A} \xrightarrow{F} \mathcal{B} \xrightarrow{\gamma_{\mathcal{B}}} \mathrm{Ho}(\mathfrak{M}_{\mathcal{B}})$. That is, $\mathbf{L}F := L(\gamma_{\mathcal{B}} \circ F)$, and we may picture this as a square

$$\begin{array}{ccc} \mathcal{A} & \xrightarrow{F} & \mathcal{B} \\ {\scriptstyle \gamma_{\mathcal{A}}}\downarrow & \nearrow\!\!\!\!\nearrow & \downarrow{\scriptstyle \gamma_{\mathcal{B}}} \\ \mathrm{Ho}(\mathfrak{M}_{\mathcal{A}}) & \xrightarrow[\mathbf{L}F]{} & \mathrm{Ho}(\mathfrak{M}_{\mathcal{B}}), \end{array}$$

which is not necessarily commutative but instead a universal square with the arrow $\Longrightarrow$ indicating a natural transformation. Note that if $\mathbf{L}F$ exists, then by Theorem 7.4 it may be computed for a given object A by $\mathbf{L}F(A) = F(QA)$; in particular we can take $\mathbf{L}F(A) = F(A)$ whenever A is cofibrant.

On the other hand, a **total right derived functor** for F is a functor $\mathbf{R}F\colon \mathrm{Ho}(\mathfrak{M}_{\mathcal{A}}) \to \mathrm{Ho}(\mathfrak{M}_{\mathcal{B}})$ which is a right derived functor for the composite $\mathcal{A} \xrightarrow{F} \mathcal{B} \xrightarrow{\gamma_{\mathcal{B}}} \mathrm{Ho}(\mathfrak{M}_{\mathcal{B}})$. That is, $\mathbf{R}F := R(\gamma_{\mathcal{B}} \circ F)$, and we may picture this as a universal transformation square:

$$\begin{array}{ccc} \mathcal{A} & \xrightarrow{F} & \mathcal{B} \\ {\scriptstyle \gamma_{\mathcal{A}}}\downarrow & \swarrow\!\!\!\!\swarrow & \downarrow{\scriptstyle \gamma_{\mathcal{B}}} \\ \mathrm{Ho}(\mathfrak{M}_{\mathcal{A}}) & \xrightarrow[\mathbf{R}F]{} & \mathrm{Ho}(\mathfrak{M}_{\mathcal{B}}). \end{array}$$

If $\mathbf{R}F$ exists then it may be computed for a given object A by $\mathbf{R}F(A) = F(RA)$; in particular we can take $\mathbf{R}F(A) = F(A)$ whenever A is fibrant.

Observe that a left Quillen functor F descends to a functor $F\colon \mathrm{St}_r(\mathcal{A}) \to \mathrm{St}_r(\mathcal{B})$ on the right stable categories, for F preserves trivially cofibrant objects. Dually, a right Quillen functor G determines a functor $G\colon \mathrm{St}_\ell(\mathcal{B}) \to \mathrm{St}_\ell(\mathcal{A})$ on the left stable categories. But more importantly, we have the following crucial lemma.

Lemma 7.10 *Assume (F, G) is a Quillen adjunction. The natural isomorphism*

$$\mathrm{Hom}_{\mathcal{B}}(FA, B) \underset{\flat}{\overset{\sharp}{\rightleftarrows}} \mathrm{Hom}_{\mathcal{A}}(A, GB)$$

respects the homotopy relation $\sim\, = \,\sim_\omega$ whenever the source is cofibrant and the target is fibrant. That is, for any cofibrant object $A \in \mathcal{A}$ and fibrant object $B \in \mathcal{B}$ we have a natural isomorphism at the level of stable category hom-sets

$$\underline{\mathrm{Hom}}_{\mathcal{B}}(FA,B) \underset{\flat}{\overset{\sharp}{\rightleftarrows}} \underline{\mathrm{Hom}}_{\mathcal{A}}(A,GB). \qquad (\sharp/\flat)$$

Proof Assume A is cofibrant and B is fibrant. The natural isomorphism

$$\mathrm{Hom}_{\mathcal{B}}(FA,B) \cong \mathrm{Hom}_{\mathcal{A}}(A,GB)$$

works from left to right by the action $s \mapsto s^{\sharp}$ described at the beginning of this section. Let us show $s_1 \sim s_2$ implies $s_1^{\sharp} \sim s_2^{\sharp}$. Looking at the definition of $\sharp$, the additivity of G shows it is enough to show $s \sim 0$ implies $s^{\sharp} \sim 0$. Now to write $s \sim 0$ means, by Lemma 4.11, that $FA \xrightarrow{s} B$ factors through a trivially bifibrant object of $\mathcal{B}$. But G preserves trivially fibrant objects, so $GFA \xrightarrow{G(s)} GB$ factors through a trivially fibrant object of $\mathcal{A}$. Hence the composition $s^{\sharp}\colon A \xrightarrow{\eta_A} GFA \xrightarrow{G(s)} GB$ also factors through a trivially fibrant object of $\mathcal{A}$. The morphism $s^{\sharp}$ has a cofibrant source and a fibrant target, so $s^{\sharp} \sim 0$ by Lemma 4.11. In the same way we can show that the reverse action satisfies $t \sim 0$ implies $t^{\flat} \sim 0$. This proves that the adjunction isomorphism descends to $\underline{\mathrm{Hom}}_{\mathcal{B}}(FA,B) \cong \underline{\mathrm{Hom}}_{\mathcal{A}}(A,GB)$. Naturality in both variables is inherited. □

Theorem 7.11 *Assume (F,G) is a Quillen adjunction from $\mathfrak{M}_{\mathcal{A}}$ to $\mathfrak{M}_{\mathcal{B}}$. Then the total derived functors* $\mathbf{L}F$ *and* $\mathbf{R}G$ *exist and form an adjunction*

$$\mathrm{Ho}(\mathfrak{M}_{\mathcal{A}}) \underset{\mathbf{R}G}{\overset{\mathbf{L}F}{\rightleftarrows}} \mathrm{Ho}(\mathfrak{M}_{\mathcal{B}}).$$

That is, $\mathbf{L}F\colon \mathrm{Ho}(\mathfrak{M}_{\mathcal{A}}) \to \mathrm{Ho}(\mathfrak{M}_{\mathcal{B}})$ *is a left adjoint functor and* $\mathbf{R}G\colon \mathrm{Ho}(\mathfrak{M}_{\mathcal{B}}) \to \mathrm{Ho}(\mathfrak{M}_{\mathcal{A}})$ *is its right adjoint. Moreover, we have the following.*

(1) *The unit of the adjunction, denoted* $\boldsymbol{\eta}$, *is the family of all composites*

$$\boldsymbol{\eta}_A := A \xrightarrow{[\gamma(p_A)]^{-1}} QA \xrightarrow{\gamma(j^{\sharp}_{F(QA)})} GRF(QA) = (\mathbf{R}G)(\mathbf{L}F)(A),$$

where $j^{\sharp}_{F(QA)}$ *is the adjunct of* $j_{F(QA)}\colon F(QA) \rightarrowtail RF(QA)$.

(2) *The counit of the adjunction, denoted* $\boldsymbol{\epsilon}$, *is the family of all composites*

$$\boldsymbol{\epsilon}_B := (\mathbf{L}F)(\mathbf{R}G)(B) = FQG(RB) \xrightarrow{\gamma(p^{\flat}_{G(RB)})} RB \xrightarrow{[\gamma(j_B)]^{-1}} B,$$

where $p^{\flat}_{G(RB)}$ *is the adjunct of* $p_{G(RB)}\colon QG(RB) \twoheadrightarrow G(RB)$.

We call $(\mathbf{L}F, \mathbf{R}G, \boldsymbol{\eta}, \boldsymbol{\epsilon})$ *the* ***derived adjunction*** *of* (F,G,η,ϵ).

Proof To show that $\mathbf{L}F := L(\gamma_{\mathcal{B}} \circ F)$ exists, we use Theorem 7.4. It asks that we show $\gamma_{\mathcal{B}} \circ F$ takes all trivial cofibrations between cofibrant objects to isomorphisms. This translates to showing that F takes all trivial cofibrations between cofibrant objects to weak equivalences. But F preserves *all* trivial cofibrations, by Proposition 7.7. So $\mathbf{L}F$ exists and by Theorem 7.4 it may be taken to be the unique functor $\mathbf{L}F\colon \mathrm{Ho}(\mathfrak{M}_{\mathcal{A}}) \to \mathrm{Ho}(\mathfrak{M}_{\mathcal{B}})$ satisfying $(\gamma_{\mathcal{B}} \circ F) \circ Q = \mathbf{L}F \circ \gamma_{\mathcal{A}}$, where Q is the cofibrant replacement functor. In particular we have $\mathbf{L}F(A) = F(QA)$ on objects $A \in \mathcal{A}$. On the other hand, $\mathbf{R}G := R(\gamma_{\mathcal{B}} \circ G)$ exists and $\mathbf{R}G(B) = G(RB)$ on objects $B \in \mathcal{B}$.

We need to show there is an isomorphism

$$\mathrm{Ho}(\mathfrak{M}_{\mathcal{B}})(\mathbf{L}F(A), B) \cong \mathrm{Ho}(\mathfrak{M}_{\mathcal{A}})(A, \mathbf{R}G(B)),$$

natural in both variables A and B. This naturality in A (resp. B) is meant to be with respect to morphisms in $\mathrm{Ho}(\mathfrak{M}_{\mathcal{A}})$ (resp. $\mathrm{Ho}(\mathfrak{M}_{\mathcal{B}})$). But by Corollary 5.16, the problem reduces to only needing to show the naturality with respect to morphisms in $\mathcal{A}$ (resp. B). Thus we are to show there is an isomorphism,

$$\mathrm{Ho}(\mathfrak{M}_{\mathcal{B}})(F(QA), B) \cong \mathrm{Ho}(\mathfrak{M}_{\mathcal{A}})(A, G(RB)),$$

natural in $A \in \mathcal{A}$ and $B \in \mathcal{B}$. The standard natural isomorphism of Corollary 5.17, $\mathrm{Ho}(\mathfrak{M})(A, B) \cong \underline{\mathrm{Hom}}(QA, RB)$, reduces this further to showing an isomorphism

$$\underline{\mathrm{Hom}}_{\mathcal{B}}(F(QA), RB) \cong \underline{\mathrm{Hom}}_{\mathcal{A}}(QA, G(RB)),$$

natural in $A \in \mathcal{A}$ and $B \in \mathcal{B}$. But this follows from Lemma 7.10.

Let us turn to describing the unit and counit of the derived adjunction. As with any adjunction, we compute the unit by setting $B = \mathbf{L}F(A)$ and tracking how the identity morphism $1_{\mathbf{L}F(A)}$ passes through the natural isomorphism

$$\mathrm{Ho}(\mathfrak{M}_{\mathcal{B}})(\mathbf{L}F(A), \mathbf{L}F(A)) \cong \mathrm{Ho}(\mathfrak{M}_{\mathcal{A}})(A, (\mathbf{R}G)(\mathbf{L}F)(A)).$$

To do this, we must use not only Lemma 7.10, but the explicit description of the isomorphism of Corollary 5.17. It is described in detail after that corollary, and it may help the reader to review this now before proceeding.

So passing from left to right, and using $\mathbf{L}F(A) := F(QA)$, we have

$$\mathrm{Ho}(\mathfrak{M}_{\mathcal{B}})(F(QA), F(QA)) := \underline{\mathrm{Hom}}_{\mathcal{B}}(RF(QA), RF(QA)),$$

which identifies the identity morphisms by $1_{\mathbf{L}F(A)} := {}_{F(QA)}[1_{RF(QA)}]_{F(QA)}$. We continue by using the natural isomorphism

$$\mathrm{Ho}(\mathfrak{M}_{\mathcal{B}})(F(QA), F(QA)) \xrightarrow{[\gamma(j_{F(QA)})]_*} \mathrm{Ho}(\mathfrak{M}_{\mathcal{B}})(F(QA), RF(QA)),$$

the fact that $\gamma(j_{F(QA)}) = {}_{F(QA)}[\tau_{F(QA)}]_{RF(QA)}$ by Lemma 5.12, and also the fact that

$${}_{F(QA)}[\tau_{F(QA)}]_{RF(QA)} = {}_{F(QA)}[1_{RF(QA)}]_{RF(QA)}$$

by Theorem 5.6 (since $F(QA)$ is indeed cofibrant). Consequently, $[\gamma(j_{F(QA)})]_*$ sends ${}_{F(QA)}[1_{RF(QA)}]_{F(QA)}$ to ${}_{F(QA)}[1_{RF(QA)}]_{RF(QA)}$, (note it is just a change of codomain).

Now since we have a cofibrant source object and a fibrant target object, we use the natural isomorphism $\mathrm{Ho}(\mathfrak{M})(A, B) \cong \underline{\mathrm{Hom}}(A, B)$ of Corollary 5.17. As described in the paragraph after that corollary, its general action is given by ${}_A[h]_B \mapsto [p_B \circ h \circ j_A]$.

Continuing the isomorphisms, this means

$$\mathrm{Ho}(\mathfrak{M}_{\mathcal{B}})(F(QA), RF(QA)) \xrightarrow{\cong} \underline{\mathrm{Hom}}_{\mathcal{B}}(F(QA), RF(QA))$$

maps ${}_{F(QA)}[1_{RF(QA)}]_{RF(QA)} \mapsto [p_{RF(QA)} \circ 1_{RF(QA)} \circ j_{F(QA)}] = [j_{F(QA)}]$, where the last equality holds because we have set $p_C = 1_C$ whenever C is cofibrant; indeed $C = RF(QA)$ is cofibrant.

We are now at the point where we may apply Lemma 7.10. The isomorphism

$$\underline{\mathrm{Hom}}_{\mathcal{B}}(F(QA), RF(QA)) \cong \underline{\mathrm{Hom}}_{\mathcal{A}}(QA, GRF(QA))$$

takes $[j_{F(QA)}]$ to $\left[j^{\#}_{F(QA)}\right]$. We note here that $(\mathbf{R}G)(\mathbf{L}F)(A) := GRF(QA)$, and so the train of isomorphisms continues

$$\underline{\mathrm{Hom}}_{\mathcal{A}}(QA, GRF(QA)) \cong \mathrm{Ho}(\mathfrak{M}_{\mathcal{A}})(QA, (\mathbf{R}G)(\mathbf{L}F)(A))$$

via the action $\left[j^{\#}_{F(QA)}\right] \mapsto \gamma\left(j^{\#}_{F(QA)}\right)$. Finally, the isomorphism $\gamma(p_A)$ induces a natural isomorphism $[\gamma(p_A)]^*$ with inverse

$$\mathrm{Ho}(\mathfrak{M}_{\mathcal{A}})(QA, (\mathbf{R}G)(\mathbf{L}F)(A)) \xrightarrow{[\gamma(p_A)^{-1}]^*} \mathrm{Ho}(\mathfrak{M}_{\mathcal{A}})(A, (\mathbf{R}G)(\mathbf{L}F)(A)).$$

So we get $\left[\gamma(p_A)^{-1}\right]^*\left(\gamma\left(j^{\#}_{F(QA)}\right)\right) = \gamma\left(j^{\#}_{F(QA)}\right) \circ \gamma(p_A)^{-1}$. So we have proved that the unit of the derived adjunction is exactly how η is defined. □

Exercise 7.2.1 Let R be a commutative ring with identity. Consider the following standard model structures on $\mathrm{Ch}(R)$, the category of chain complexes of R-modules. (See the Introduction and Main Examples for details.)

- Let $\left(dg\widetilde{\mathcal{P}}, \mathcal{E}\right)$ be the DG-projective cotorsion pair, which determines the standard projective model structure $\mathrm{Ch}(R)_{proj} = \left(dg\widetilde{\mathcal{P}}, \mathcal{E}, All\right)$ on $\mathrm{Ch}(R)$.
- Let $\left(\mathcal{E}, dg\widetilde{\mathcal{I}}\right)$ be the DG-injective cotorsion pair, which corresponds to the standard injective model structure $\mathrm{Ch}(R)_{inj} = \left(All, \mathcal{E}, dg\widetilde{\mathcal{I}}\right)$ on $\mathrm{Ch}(R)$.
- Let $\mathrm{Ch}(R)_{flat} = \left(dg\widetilde{\mathcal{F}}, \mathcal{E}, dw\widetilde{\mathcal{C}}\right)$ be the standard flat model structure on $\mathrm{Ch}(R)$, where $dg\widetilde{\mathcal{F}}$ is the class of all DG-flat complexes and $dw\widetilde{\mathcal{C}}$ is the class of all complexes that are degreewise cotorsion.

In each case, the associated homotopy category is equivalent to the usual derived category, $\mathcal{D}(R)$.

(1) Show that for any given R-module M, the functor $-\otimes_R M\colon \mathrm{Ch}(R)_{proj} \to \mathrm{Ch}(R)_{inj}$ is a left Quillen functor with right adjoint $\mathrm{Hom}_R(M,-)$. More generally, for any chain complex X of R-modules, show that the usual chain complex tensor product $-\otimes_R X\colon \mathrm{Ch}(R)_{proj} \to \mathrm{Ch}(R)_{inj}$ is a left Quillen functor with right adjoint $\mathit{Hom}_R(X,-)$, the usual Hom complex; see the definitions in Example 7.6.4. Conclude we have a left adjoint functor $\mathbf{L}(-\otimes_R X)\colon \mathcal{D}(R) \to \mathcal{D}(R)$ with right adjoint given by $\mathbf{R}\mathit{Hom}_R(X,-)\colon \mathcal{D}(R) \to \mathcal{D}(R)$.
(2) Show that for any R-module M, the usual left derived functors $\mathrm{Tor}_n^R(-,M)$ of $-\otimes_R M$ can be recovered by taking the nth homology of the total left derived functor $\mathbf{L}(-\otimes_R M)$. Similarly, we can recover $\mathrm{Ext}_R^n(M,-)$ by taking the nth cohomology of the total right derived functor $\mathbf{R}\,\mathrm{Hom}_R(M,-)$.
(3) Show that everything mentioned previously still works if we replace $\mathrm{Ch}(R)_{proj}$ with the flat model structure $\mathrm{Ch}(R)_{flat}$.

Exercise 7.2.2 Continuing Exercise 7.2.1, note that for an arbitrary complex X, the functor $-\otimes_R X\colon \mathrm{Ch}(R)_{proj} \to \mathrm{Ch}(R)_{proj}$ need not be a left Quillen functor. Nevertheless, use Theorem 7.4 to argue directly from Definition 7.9 that $\mathbf{L}(-\otimes_R X)$ exists. The same is true if we replace $\mathrm{Ch}(R)_{proj}$ with $\mathrm{Ch}(R)_{flat}$.

7.3 Total Derived Functors Are Triangulated

We just showed that a Quillen adjunction between abelian model structures descends to a total derived adjunction between their homotopy categories. Knowing that these homotopy categories are always triangulated it is natural to ask if total derived functors are triangulated functors. So the goal of this section is to prove the following theorem.

Theorem 7.12 *Assume (F,G) is a Quillen adjunction from $\mathfrak{M}_{\mathcal{A}}$ to $\mathfrak{M}_{\mathcal{B}}$. Then the total derived functors $\mathbf{L}F\colon \mathrm{Ho}(\mathfrak{M}_{\mathcal{A}}) \to \mathrm{Ho}(\mathfrak{M}_{\mathcal{B}})$ and $\mathbf{R}G\colon \mathrm{Ho}(\mathfrak{M}_{\mathcal{B}}) \to \mathrm{Ho}(\mathfrak{M}_{\mathcal{A}})$ are triangulated functors.*

The proof will come after and make use of some easy lemmas. First, let us say that a class of distinguished triangles C in a triangulated category $(\mathcal{T},\Sigma,\Delta)$ is *dense* if each triangle in Δ is isomorphic to one in C.

Lemma 7.13 *Let $F\colon \mathcal{T} \to \mathcal{T}'$ be an additive functor between triangulated categories $(\mathcal{T},\Sigma,\Delta)$ and $(\mathcal{T}',\Sigma',\Delta')$. Assume we have a natural equivalence $\xi\colon F\Sigma \cong \Sigma' F$ such that for all right triangles $A \xrightarrow{f} B \xrightarrow{g} C \xrightarrow{h} \Sigma A$ in some dense class $C \subseteq \Delta$, the right triangle $FA \xrightarrow{F(f)} FB \xrightarrow{F(g)} FC \xrightarrow{\xi_A \circ F(h)} \Sigma' FA$ is in Δ'. Then (F,ξ) is a triangulated functor.*

Proof This is straightforward and left for the reader to verify. Refer to Definition 6.6. □

Lemma 7.14 *Let $\mathfrak{M} = (\mathcal{Q}, \mathcal{W}, \mathcal{R})$ be an abelian model structure. Then the class C of all strict triangles on morphisms $f\colon A \to B$ between cofibrant objects is dense in the class Δ of all distinguished triangles in* $\mathrm{Ho}(\mathfrak{M})$.

Proof By definition, Δ consists of all the (right) triangles that are isomorphic to a strict triangle on some $\mathcal{A}$-morphism. But each of these is isomorphic to one in C as indicated:

$$\begin{array}{ccccccc}
QA & \xrightarrow{\gamma(Q(f))} & QB & \xrightarrow{\gamma(k_{Q(f)})} & C(Q(f)) & \xrightarrow{\gamma(\sigma_{Q(f)})} & \Sigma QA \\
{\scriptstyle \gamma(p_A)}\downarrow & & {\scriptstyle \gamma(p_B)}\downarrow & & \exists \, \vdots & & \downarrow{\scriptstyle \Sigma(p_A)} \\
A & \xrightarrow[\gamma(f)]{} & B & \xrightarrow[\gamma(k_f)]{} & C(f) & \xrightarrow[\gamma(\sigma_f)]{} & \Sigma A.
\end{array}$$

The dashed fill-in morphism exists, and by the Five Lemma (Corollary 6.4) it provides an isomorphism or triangles, proving that C is dense in Δ. □

Proof of Theorem 7.12 Recall the definition of a triangulated functor from Definition 6.6. We first need to see that $\mathbf{L}F$ commutes with suspensions in the sense that there is a natural isomorphism of functors $\mathbf{L}F \circ \Sigma^{\mathcal{A}} \cong \Sigma^{\mathcal{B}} \circ \mathbf{L}F$. By Corollary 5.16 it is equivalent to find a natural isomorphism

$$\left(\mathbf{L}F \circ \Sigma^{\mathcal{A}}\right) \circ \gamma_{\mathcal{A}} \cong \left(\Sigma^{\mathcal{B}} \circ \mathbf{L}F\right) \circ \gamma_{\mathcal{A}}.$$

Using Proposition 6.25 and the definition for $\mathbf{L}F$ given in the proof of Theorem 7.11, this reduces to wanting a natural isomorphism

$$\gamma_{\mathcal{B}} \circ F \circ Q \circ \Sigma^{\mathcal{A}}_{\ell} \cong \gamma_{\mathcal{B}} \circ \Sigma^{\mathcal{B}}_{\ell} \circ F \circ Q.$$

The composition on the right-hand side is well defined, as shown in the proof of Theorem 7.4. On the other hand, noting that $\Sigma^{\mathcal{A}}_{\ell}\colon \mathcal{A} \to \mathrm{St}_{\ell}(\mathcal{A})$ takes values in the full subcategory of cofibrant objects, we seek a natural isomorphism

$$\gamma_{\mathcal{B}} \circ F \circ \Sigma^{\mathcal{A}}_{\ell} \cong \gamma_{\mathcal{B}} \circ \Sigma^{\mathcal{B}}_{\ell} \circ F \circ Q.$$

For any given object A, we take its cofibrant replacement and construct the commutative diagram:

$$\begin{array}{ccccc}
QA & \xrightarrow{k_{QA}} & \tilde{R}_{QA} & \xrightarrow{\sigma_{QA}} & \Sigma QA \\
{\scriptstyle p_A}\downarrow & & \vdots & & \vdots\,{\scriptstyle \Sigma(p_A)} \\
A & \xrightarrow{k_A} & \tilde{R}_A & \xrightarrow{\sigma_A} & \Sigma A.
\end{array}$$

By Lemma 6.24 we have that $\Sigma(p_A)$ is a weak equivalence. Since it is between cofibrant objects, Ken Brown's Corollary 7.8 tells us that $F\Sigma(p_A)\colon F\Sigma QA \to$

$F\Sigma A$ is a weak equivalence. We note that the rows of the diagram are cofibrations, so applying F will keep a commutative diagram with exact rows. Moreover, the new top row of said diagram will be a general suspension sequence of FQA since $F(\tilde{R}_{QA})$ must be trivially cofibrant. Denote this general suspension sequence by $\mathbb{W}_{FQA}$. Then by Lemma 6.26 we have the suspension connecting isomorphism $\gamma_{\mathcal{B}}(\mathbb{W}_{FQA})\colon F\Sigma QA \to \Sigma FQA$. By composition, this provides a natural $\mathrm{Ho}(\mathfrak{M}_{\mathcal{B}})$-isomorphism

$$\xi_A := F\Sigma A \xrightarrow{\gamma_{\mathcal{B}}(F\Sigma(p_A))^{-1}} F\Sigma QA \xrightarrow{\gamma_{\mathcal{B}}(\mathbb{W}_{FQA})} \Sigma FQA.$$

This is our definition of the natural isomorphism

$$\xi\colon \mathbf{L}F \circ \Sigma^{\mathcal{A}} \cong \Sigma^{\mathcal{B}} \circ \mathbf{L}F,$$

but we should verify the naturality of the construction. To check the naturality of the isomorphism $\gamma_{\mathcal{B}}(F\Sigma(p_A))^{-1}$, note that for any $\mathcal{A}$-morphism $f\colon A \to B$, we have $\Sigma(f) \circ \Sigma(p_A) \sim_{\omega} \Sigma(p_B) \circ \Sigma(Q(f))$ by the Fundamental Lemmas 3.3 and 3.4. Since F preserves trivially cofibrant objects, we get that $F(\Sigma(f)) \circ F(\Sigma(p_A)) \sim_r F(\Sigma(p_B)) \circ F(\Sigma(Q(f)))$. Since $\gamma_{\mathcal{B}}$ identifies r-homotopic maps we have the desired naturality of the isomorphism $\gamma_{\mathcal{B}}(F\Sigma(p_A))^{-1}$. Now to check the naturality of $\gamma_{\mathcal{B}}(\mathbb{W}_{FQA})$, we note that for any map $Q(f)\colon QA \to QB$ such that $p_B \circ Q(f) = f \circ p_A$, we have a corresponding morphism of short exact sequences:

$$\begin{array}{ccccc}
FQA & \xrightarrow{F(k_{QA})} & F\tilde{R}_{QA} & \xrightarrow{F(\sigma_{QA})} & F\Sigma QA \\
\downarrow{\scriptstyle F(Q(f))} & & \downarrow & & \downarrow{\scriptstyle F(\Sigma(Q(f)))} \\
FQB & \xrightarrow{F(k_{QB})} & F\tilde{R}_{QB} & \xrightarrow{F(\sigma_{QB})} & F\Sigma QB.
\end{array}$$

The naturality of Lemma 6.26 provides the rest: $\gamma(\Sigma(F(Q(f)))) \circ \gamma(\mathbb{W}_{FQA}) = \gamma(\mathbb{W}_{FQB}) \circ \gamma(F(\Sigma(Q(f))))$.

Now let C be the class of all strict triangles on morphisms $f\colon A \to B$ between cofibrant objects of $\mathfrak{M}_{\mathcal{A}}$. By Lemma 7.14 it is dense in the class of all distinguished triangles in $\mathrm{Ho}(\mathfrak{M}_{\mathcal{A}})$. So it will follow from Lemma 7.13 that the pair $(\mathbf{L}F, \xi)$ is a triangulated functor once we show that it sends all triangles in C to distinguished triangles in $\mathrm{Ho}(\mathfrak{M}_{\mathcal{B}})$. So assume that A and B are cofibrant objects of $\mathfrak{M}_{\mathcal{A}}$ and let $f\colon A \to B$ be an $\mathcal{A}$-morphism. An arbitrary triangle in C

takes the form of a strict right triangle $A \xrightarrow{\gamma(f)} B \xrightarrow{\gamma(k_f)} C(f) \xrightarrow{\gamma(\sigma_f)} \Sigma A$ arising from the pushout:

$$\begin{array}{ccccc} A & \overset{k_A}{\rightarrowtail} & \tilde{R}_A & \overset{\sigma_A}{\twoheadrightarrow} & \Sigma A \\ {\scriptstyle f}\downarrow & & \downarrow & & \| \\ B & \underset{k_f}{\rightarrowtail} & C(f) & \underset{\sigma_f}{\twoheadrightarrow} & \Sigma A. \end{array}$$

Being a left Quillen functor, F preserves cofibrations and pushouts (its a left adjoint), so we obtain another pushout diagram with exact rows:

$$\begin{array}{ccccc} FA & \overset{F(k_A)}{\rightarrowtail} & F\tilde{R}_A & \overset{F(\sigma_A)}{\twoheadrightarrow} & F\Sigma A \\ {\scriptstyle F(f)}\downarrow & & \downarrow & & \| \\ FB & \underset{F(k_f)}{\rightarrowtail} & FC(f) & \underset{F(\sigma_f)}{\twoheadrightarrow} & F\Sigma A. \end{array} \tag{7.1}$$

As we have already observed, the top row is a general suspension sequence, $\mathbb{W}_{FA}$. On the other hand, using the definition of $\mathbf{L}F(\gamma_{\mathcal{A}}(f)) = \gamma_{\mathcal{B}}(F(Q(f))$ given in the proof of Theorem 7.11, and the fact that all of the objects A, B, $C(f)$, and ΣA are cofibrant, we will be done if we can show that

$$FA \xrightarrow{\gamma(F(f))} FB \xrightarrow{\gamma(F(k_f))} FC(f) \xrightarrow{\gamma(F(\sigma_f))} F\Sigma A \cong_{\xi_A} \Sigma FA$$

is a distinguished triangle in $\mathrm{Ho}(\mathfrak{M}_{\mathcal{B}})$. Because of Theorem 6.38, the pushout in Diagram (7.1) does indeed show this to be a distinguished triangle in $\Delta_{\mathcal{B}}$, as long as ξ_A coincides with $\gamma(\mathbb{W}_{FA})$ whenever A is a cofibrant object. But this is easy to see because $p_A = 1_A$ implies that $\gamma_{\mathcal{B}}(F\Sigma(p_A))^{-1} = 1_{F\Sigma A}$ in this case. Likewise, $\gamma(\mathbb{W}_{FQA}) = \gamma(\mathbb{W}_{FA})$ when A is cofibrant. This completes the proof that the pair $(\mathbf{L}F, \{\xi_A\})$ is a triangulated functor. □

7.4 Quillen Equivalence

We continue to let $(\mathcal{A}, \mathcal{E}_{\mathcal{A}})$ and $(\mathcal{B}, \mathcal{E}_{\mathcal{B}})$ each be weakly idempotent complete exact categories along with Hovey triples $\mathfrak{M}_{\mathcal{A}}$ and $\mathfrak{M}_{\mathcal{B}}$. If (F, G) is a Quillen adjunction from $\mathfrak{M}_{\mathcal{A}}$ to $\mathfrak{M}_{\mathcal{B}}$, then, as shown in Section 7.2, the total derived functors $(\mathbf{L}F, \mathbf{R}G)$ form an adjunction from $\mathrm{Ho}(\mathfrak{M}_{\mathcal{A}})$ to $\mathrm{Ho}(\mathfrak{M}_{\mathcal{B}})$. The proof boiled down to the fact that the adjunction (F, G) descends to a natural isomorphism between the hom-sets (in the stable categories)

$$\underline{\mathrm{Hom}}_{\mathcal{B}}(FA, B) \underset{\flat}{\overset{\sharp}{\rightleftarrows}} \underline{\mathrm{Hom}}_{\mathcal{A}}(A, GB) \tag{$\sharp/\flat$}$$

for cofibrant A and fibrant B. Recall that if $f \sim g$, then f is a weak equivalence if and only if g is a weak equivalence; see Exercise 5.4.2. So any particular element of one of the aforementioned hom-sets is either a weak equivalence or not a weak equivalence. A special thing happens when the adjunction mappings $\sharp$ and $\flat$ preserve weak equivalences. In this case, we say that (F, G) is a *Quillen equivalence* and it turns out that $(\mathbf{L}F, \mathbf{R}G)$ is an equivalence of categories.

Definition 7.15 A Quillen adjunction (F, G) from $\mathfrak{M}_{\mathcal{A}}$ to $\mathfrak{M}_{\mathcal{B}}$ is a called a **Quillen equivalence** if for all cofibrant $A \in \mathfrak{M}_{\mathcal{A}}$ and fibrant $B \in \mathfrak{M}_{\mathcal{B}}$, the adjunction bijection of Diagram ($\sharp/\flat$) preserves weak equivalences.

Lemma 7.16 *Assume (F, G) is a Quillen adjunction from $\mathfrak{M}_{\mathcal{A}}$ to $\mathfrak{M}_{\mathcal{B}}$.*

(1) *The following are equivalent.*
   (a) *For all cofibrant $A \in \mathfrak{M}_{\mathcal{A}}$ and fibrant $B \in \mathfrak{M}_{\mathcal{B}}$, if $w\colon FA \to B$ is a weak equivalence in $\mathfrak{M}_{\mathcal{B}}$ then $w^\sharp\colon A \to GB$ is a weak equivalence in $\mathfrak{M}_{\mathcal{A}}$.*
   (b) *For all cofibrant $A \in \mathfrak{M}_{\mathcal{A}}$, the morphism $j_{FA}^\sharp$ is a weak equivalence in $\mathfrak{M}_{\mathcal{A}}$. That is, $A \xrightarrow{\eta_A} GFA \xrightarrow{G(j_{FA})} GRFA$ is a weak equivalence.*
   (c) *The unit η of the derived adjunction $(\mathbf{L}F, \mathbf{R}G, \eta, \epsilon)$ is a natural isomorphism $\eta \to (\mathbf{R}G)(\mathbf{L}F)$.*

(2) *Dually, the following are equivalent.*
   (a) *For all cofibrant $A \in \mathfrak{M}_{\mathcal{A}}$ and fibrant $B \in \mathfrak{M}_{\mathcal{B}}$, if $w\colon A \to GB$ is a weak equivalence in $\mathfrak{M}_{\mathcal{A}}$ then $w^\flat\colon FA \to B$ is a weak equivalence in $\mathfrak{M}_{\mathcal{B}}$.*
   (b) *For all fibrant $B \in \mathfrak{M}_{\mathcal{B}}$, the morphism $p_{GB}^\flat$ is a weak equivalence in $\mathfrak{M}_{\mathcal{B}}$. That is, $FQGB \xrightarrow{F(p_{GB})} FGB \xrightarrow{\epsilon_B} B$ is a weak equivalence.*
   (c) *The counit ϵ of the derived adjunction $(\mathbf{L}F, \mathbf{R}G, \eta, \epsilon)$ is a natural isomorphism $(\mathbf{L}F)(\mathbf{R}G) \to \epsilon$.*

Proof We will prove the first set. First, we will construct a diagram that will be used to prove (a) and (b) are equivalent. Let A be a cofibrant object of $\mathfrak{M}_{\mathcal{A}}$, B be a fibrant object of $\mathfrak{M}_{\mathcal{B}}$, and let $w\colon FA \to B$ be a weak equivalence of $\mathfrak{M}_{\mathcal{B}}$. To show (a) and (b) are equivalent we will show $w^\sharp$ is a weak equivalence if and only if $j_{FA}^\sharp$ is a weak equivalence. Take a fibrant replacement of w as shown here:

$$\begin{array}{ccc} RFA & \xrightarrow{R(w)} & B \\ {\scriptstyle j_{FA}}\uparrow & & \| \\ FA & \xrightarrow{w} & B. \end{array}$$

Note that $R(w)$ is a weak equivalence. Apply the functor G and form the following commutative diagram in $\mathcal{A}$:

$$\begin{array}{ccccc} A & \xrightarrow{j^{\#}_{FA}} & GRFA & \xrightarrow{G(R(w))} & GB \\ \| & & \uparrow{\scriptstyle G(j_{FA})} & & \| \\ A & \xrightarrow[\eta_A]{} & GFA & \xrightarrow[G(w)]{} & GB. \end{array}$$

Note that $G(R(w))$ is weak equivalence by Ken Brown's Corollary 7.8. Moreover, the top row composite equals the bottom row composite, which is exactly $w^{\#}$. We infer from the 2 out of 3 Axiom that $w^{\#}$ is a weak equivalence if and only if $j^{\#}_{FA}$ is a weak equivalence. This proves (a) and (b) are equivalent.

Now we show (b) is equivalent to (c). By Theorem 7.11, for any object A, the component of the unit at A is

$$\eta_A := A \xrightarrow{[\gamma(p_A)]^{-1}} QA \xrightarrow{\gamma(j^{\#}_{F(QA)})} GRF(QA) = (\mathbf{R}G)(\mathbf{L}F)(A),$$

where $j^{\#}_{F(QA)}$ is the adjunct of $j_{F(QA)}\colon F(QA) \rightarrowtail RF(QA)$. Since p_A is a weak equivalence, Proposition 5.23 tells us that this composite is an isomorphism in $\mathrm{Ho}(\mathfrak{M}_{\mathcal{A}})$ if and only if $j^{\#}_{F(QA)}$ is a weak equivalence. It follows that the natural transformation $\eta \to (\mathbf{R}G)(\mathbf{L}F)$ is an isomorphism if and only if $j^{\#}_{F(C)}$ is a weak equivalence for all cofibrant objects C. □

It is standard category theory that a right (resp. left) adjoint is fully faithful, and so exhibits a reflective subcategory (resp. coreflective subcategory), if and only if the counit (resp. unit) is an isomorphism. For example, see Mac Lane [1998, section IV.3, theorem 1]. Lemma 7.16 gives useful conditions for applying this to $\mathbf{R}G$ or $\mathbf{L}F$. Most importantly, an adjunction is an (adjoint) equivalence if and only if both the unit and counit are isomorphisms. See Mac Lane [1998, section IV.4, theorem 1]. So the following can be used to decide if $(\mathbf{L}F, \mathbf{R}G)$ is an equivalence of categories.

Theorem 7.17 *Assume (F, G) is a Quillen adjunction from $\mathfrak{M}_{\mathcal{A}}$ to $\mathfrak{M}_{\mathcal{B}}$. The following are equivalent.*

(1) *(F, G) is a Quillen equivalence. That is, for all cofibrant $A \in \mathfrak{M}_{\mathcal{A}}$ and fibrant $B \in \mathfrak{M}_{\mathcal{B}}$, a morphism $w\colon FA \to B$ is a weak equivalence in $\mathfrak{M}_{\mathcal{B}}$ if and only if $w^{\#}\colon A \to GB$ is a weak equivalence in $\mathfrak{M}_{\mathcal{A}}$.*

(2) *For all cofibrant $A \in \mathfrak{M}_{\mathcal{A}}$, the morphism*

$$j^{\#}_{FA}\colon A \xrightarrow{\eta_A} GFA \xrightarrow{G(j_{FA})} GRFA$$

is a weak equivalence in $\mathfrak{M}_{\mathcal{A}}$*, and, for all fibrant* $B \in \mathfrak{M}_{\mathcal{B}}$*, the morphism*

$$p^{\flat}_{GB}\colon FQGB \xrightarrow{F(p_{GB})} FGB \xrightarrow{\epsilon_B} B$$

is a weak equivalence in $\mathfrak{M}_{\mathcal{B}}$.

(3) *The derived adjunction* $(\mathbf{L}F, \mathbf{R}G, \eta, \epsilon)$ *is an equivalence of categories.*

Example 7.18 Theorem 7.17(2) can be used to show that the Gorenstein injective model structure (see Example 5 of the Introduction and Main Examples) is Quillen equivalent to a model structure on chain complexes. See Exercise 8.8.2.

7.5 Abelian Monoidal Model Structures

In this section we will assume that the reader is familiar with the idea of a monoidal category as described in say [Mac Lane, 1998]. Nevertheless, we begin this section by briefly recalling the definition of a closed symmetric monoidal category. The ultimate goal here is to prove the fundamental result that the homotopy category of an abelian monoidal model structure is itself a closed symmetric monoidal category.

Again, $(\mathcal{A}, \mathcal{E})$ is assumed to be a weakly idempotent complete exact category. But we now assume throughout that the underlying additive category $\mathcal{A}$ is also closed symmetric monoidal. By this we mean we have a bifunctor

$$-\otimes -\colon \mathcal{A} \times \mathcal{A} \to \mathcal{A},$$

called the *tensor product*, which is additive in each variable and comes along with some natural isomorphisms. The first of these is a natural equivalence $\alpha\colon (-\otimes -)\otimes - \cong -\otimes(-\otimes -)$ called the *associator*. It means that for each triple of objects (A, B, C), there is an isomorphism $\alpha_{A,B,C}\colon (A\otimes B)\otimes C \to A\otimes(B\otimes C)$, natural in all of three variables. Next, there is a *unit* object I, which comes with a natural equivalence $\lambda_A\colon I \otimes A \to A$ called the *left unitor* as well as a natural equivalence $\rho_A\colon A \otimes I \to A$ called the *right unitor*. For simplicity we will also assume that our tensor product is commutative. That is, there is another natural equivalence $\sigma_{A,B}\colon A \otimes B \to B \otimes A$ satisfying $\sigma_{B,A} \circ \sigma_{A,B} = 1_{A\otimes B}$. In order for all this data to be called a *symmetric monoidal category*, several coherence diagrams, which will not be written down here, are assumed to commute; see Mac Lane [1998, chapter VII]. Finally, we assume it to be a *closed* symmetric monoidal category which means that the functor $-\otimes B$ has a right adjoint, for all objects B. Here we will denote the right adjoint of

$- \otimes B$ by $\mathcal{H}om(B, -)$. It is then a standard fact that this determines a bifunctor $\mathcal{H}om(-,-)\colon \mathcal{A}^{\mathrm{op}} \times \mathcal{A} \to \mathcal{A}$ in such a way that the adjunction bijection

$$\mathrm{Hom}_{\mathcal{A}}(A \otimes B, C) \cong \mathrm{Hom}_{\mathcal{A}}(A, \mathcal{H}om(B, C)) \qquad (7.2)$$

is natural in all three variables A, B, and C; see Mac Lane [1998, section IV.7, theorem 3].

The situation of having functors $- \otimes -$ and $\mathcal{H}om(-,-)$ as previously is a main example of what is sometimes called a *two variable adjunction*, or an *adjunction with parameter*. The next exercise makes a basic point about tensor products.

Exercise 7.5.1 Let R be a commutative ring with identity. Show that the tensor product bifunctor $- \otimes_R -\colon R\text{-Mod} \times R\text{-Mod} \to R\text{-Mod}$ is *not* a left adjoint by finding a colimit it doesn't preserve.

We denote a closed symmetric monoidal category simply by $(\mathcal{A}, \otimes, \mathcal{H}om)$. Having fixed such a structure $(\mathcal{A}, \otimes, \mathcal{H}om)$ on our additive category $\mathcal{A}$, we now bring back our usual setting of an abelian model structure $\mathfrak{M} = (\mathcal{Q}, \mathcal{W}, \mathcal{R})$ on $(\mathcal{A}, \mathcal{E})$. Our immediate goal is to define what it means to say that all this data determines an *abelian monoidal model structure*. Of course, the idea is that $\mathfrak{M}$ should be compatible in an appropriate sense, again relative to $\mathcal{E}$, with the tensor product $\otimes$. For a general model category, the standard way to do this is in terms of pushout-products (and the dual pullback-homs) as we now define. However, in the next section we will show that these more cumbersome notions can be replaced by much simpler criteria in our case of abelian model structures.

Definition 7.19 Let $f\colon A \to B$ and $g\colon X \to Y$ be any pair of $\mathcal{A}$-morphisms.

(1) Because the tensor product $\otimes$ is a bifunctor, we obtain a commutative square:

$$\begin{array}{ccc} A \otimes X & \xrightarrow{1_A \otimes g} & A \otimes Y \\ {\scriptstyle f\otimes 1_X}\downarrow & & \downarrow{\scriptstyle f\otimes 1_Y} \\ B \otimes X & \xrightarrow{1_B \otimes g} & B \otimes Y. \end{array}$$

Assuming the pushout of $1_A \otimes g$ and $f \otimes 1_Y$ exists, we obtain the pushout square

$$\begin{array}{ccc} A \otimes X & \xrightarrow{1_A \otimes g} & A \otimes Y \\ {\scriptstyle f\otimes 1_X}\downarrow & & \downarrow{\scriptstyle f'} \\ B \otimes X & \xrightarrow[g']{} & {}_fP_g \end{array} \qquad (\dagger)$$

and a unique morphism $f \boxtimes g \colon {}_fP_g \to B \otimes Y$ satisfying the equations

$$1_B \otimes g = (f \boxtimes g) \circ g' \text{ and } f \otimes 1_Y = (f \boxtimes g) \circ f'. \qquad (\ddagger)$$

The morphism $f \boxtimes g \colon {}_fP_g \to B \otimes Y$ is called the **pushout-product** of $f \colon A \to B$ and $g \colon X \to Y$.

(2) There is a sort of dual notion based on the bifunctor $\mathcal{H}om$ coming from the closed structure. We obtain a commutative square:

$$\begin{array}{ccc} \mathcal{H}om(B,X) & \xrightarrow{\mathcal{H}om(1_B,g)} & \mathcal{H}om(B,Y) \\ {\scriptstyle \mathcal{H}om(f,1_X)}\big\downarrow & & \big\downarrow{\scriptstyle \mathcal{H}om(f,1_Y)} \\ \mathcal{H}om(A,X) & \xrightarrow{\mathcal{H}om(1_A,g)} & \mathcal{H}om(A,Y). \end{array}$$

Assuming the pullback of $\mathcal{H}om(1_A, g)$ and $\mathcal{H}om(f, 1_Y)$ exists, we obtain the pullback square

$$\begin{array}{ccc} {}_fD_g & \xrightarrow{\bar{g}} & \mathcal{H}om(B,Y) \\ {\scriptstyle \bar{f}}\big\downarrow & & \big\downarrow{\scriptstyle \mathcal{H}om(f,1_Y)} \\ \mathcal{H}om(A,X) & \xrightarrow{\mathcal{H}om(1_A,g)} & \mathcal{H}om(A,Y) \end{array}$$

and a unique morphism $\langle f, g\rangle \colon \mathcal{H}om(B,X) \to {}_fD_g$ satisfying the equations

$$\mathcal{H}om(1_B, g) = \bar{g} \circ \langle f, g\rangle \text{ and } \mathcal{H}om(f, 1_X) = \bar{f} \circ \langle f, g\rangle.$$

The morphism $\langle f, g\rangle \colon \mathcal{H}om(B,X) \to {}_fD_g$ is called the **pullback-hom** of $f \colon A \to B$ and $g \colon X \to Y$.

Consider the following compatibility axioms on $(\mathcal{A}, \otimes, \mathcal{H}om, \mathcal{E}, \mathfrak{M})$ based on the pushout-product $f \boxtimes g$ and pullback-hom $\langle f, g\rangle$.

($\boxtimes$ **1**) The pushout-product $f \boxtimes g \colon {}_fP_g \to B \otimes Y$ is a cofibration whenever $f \colon A \rightarrowtail B$ and $g \colon X \rightarrowtail Y$ are each cofibrations.

($\boxtimes$ **2**) The pushout-product $f \boxtimes g \colon {}_fP_g \to B \otimes Y$ is a trivial cofibration whenever $f \colon A \rightarrowtail B$ and $g \colon X \rightarrowtail Y$ are each cofibrations with at least one of them trivial.

($\diamond$ **1**) The pullback-hom $\langle f, g\rangle \colon \mathcal{H}om(B,X) \to {}_fD_g$ is a fibration whenever $f \colon A \rightarrowtail B$ is a cofibration and $g \colon X \rightarrowtail Y$ is a fibration.

($\diamond$ **2**) The pullback-hom $\langle f, g\rangle \colon \mathcal{H}om(B,X) \to {}_fD_g$ is a trivial fibration whenever $f \colon A \rightarrowtail B$ is a cofibration and $g \colon X \rightarrowtail Y$ is a fibration with at least one of them trivial.

Lemma 7.20 *Let $(\mathcal{A}, \otimes, \mathcal{H}om)$ be a closed symmetric monoidal structure on $\mathcal{A}$ and $\mathfrak{M} = (\mathcal{Q}, \mathcal{W}, \mathcal{R})$ be an abelian model structure relative to $(\mathcal{A}, \mathcal{E})$. Then the following conditions are equivalent.*

(1) *Axioms* ($\boxtimes$ 1) *and* ($\boxtimes$ **2**) *are satisfied. In this case we say that*

$$- \otimes - \colon \mathcal{A} \times \mathcal{A} \to \mathcal{A}$$

is a ***left Quillen bifunctor****.*

(2) *Axioms* ($\diamond$ **1**) *and* ($\diamond$ **2**) *are satisfied. In this case we say that*

$$\mathcal{H}om(-,-) \colon \mathcal{A}^{\text{op}} \times \mathcal{A} \to \mathcal{A}$$

is a ***right Quillen bifunctor****.*

When these statements hold, we also refer to the data $(\mathcal{A}, \otimes, \mathcal{H}om, \mathcal{E}, \mathfrak{M})$ as a ***two variable Quillen adjunction****, or simply say that $\otimes$ is a* ***Quillen bifunctor****.*

Proof Assume (1) is true. To prove axiom ($\diamond$ **1**), let $f \colon A \rightarrowtail B$ be a cofibration and let $g \colon X \twoheadrightarrow Y$ be a fibration. We need to show that the pullback-hom $\langle f, g \rangle \colon \mathcal{H}om(B, X) \to {}_fD_g$ is a fibration. Using Corollary 2.15, it is enough to show that $\langle f, g \rangle$ satisfies the RLP with respect to all trivial cofibrations. That is, given any trivial cofibration $i \colon M \rightarrowtail N$, we seek a lift as indicated in the following commutative diagram on the left:

$$\begin{array}{ccc} M & \longrightarrow & \mathcal{H}om(B,X) \\ {\scriptstyle i}\downarrow & \overset{?}{\nearrow} & \downarrow{\scriptstyle \langle f,g\rangle} \\ N & \longrightarrow & {}_fD_g, \end{array} \qquad\qquad \begin{array}{ccc} {}_iP_f & \longrightarrow & X \\ {\scriptstyle i\boxtimes f}\downarrow & \overset{\exists!}{\nearrow} & \downarrow{\scriptstyle g} \\ N \otimes B & \longrightarrow & Y. \end{array}$$

Note the similarity here to the argument used to prove Proposition 7.7. The question is whether or not a more intricate version of that argument holds, enabling us to replace the lifting problem on the left with the one represented by the square on the right. Note that the lifting problem on the right is solvable because $i \boxtimes f \colon {}_iP_f \to N \otimes B$ must be a trivial cofibration by axiom ($\boxtimes$ **2**). In fact this idea works; using the adjointness bijection of Equation (7.2), it is an exercise to show that the two lifting problems are equivalent. In fact, let $\vec{\mathcal{A}}$ denote the arrow category, whose objects are the arrows of $\mathcal{A}$ and whose morphisms are commutative squares. It is an exercise to check that there is a bijection

$$\vec{\mathcal{A}}(i \boxtimes f, g) \cong \vec{\mathcal{A}}(i, \langle f, g \rangle)$$

that is natural in i, f, and g. From this it follows that we also have (1) implies axiom ($\diamond$ **2**), proving (1) implies (2). On the other hand, (2) implies (1) by similar arguments. $\square$

Products of abelian model structures, and opposites of abelian model structure, are constructed in the obvious way. Everything is done componentwise for products, and the role of fibration and cofibration swap for opposite model structures. More precisely, the opposite model structure of $\mathfrak{M} = (\mathcal{Q}, \mathcal{W}, \mathcal{R})$ is given by the Hovey triple $\mathfrak{M}^{\mathrm{op}} = (\mathcal{R}, \mathcal{W}, \mathcal{Q})$ on the opposite exact category $(\mathcal{A}^{\mathrm{op}}, \mathcal{E}^{\mathrm{op}})$. See Exercises 5.3.2 and 5.5.3. In this way we have the bifunctors

$$- \otimes - \colon \mathfrak{M} \times \mathfrak{M} \to \mathfrak{M}$$

and

$$\mathcal{H}om(-,-) \colon \mathfrak{M}^{\mathrm{op}} \times \mathfrak{M} \to \mathfrak{M}$$

between abelian model structures. The next lemma tells us that when these bifunctors satisfy the conditions of Lemma 7.20, then in particular they are Quillen functors in each variable.

Lemma 7.21 *Let $(\mathcal{A}, \otimes, \mathcal{H}om, \mathcal{E}, \mathfrak{M})$ be a two variable Quillen adjunction. Then each of the following hold.*

(1) *For each cofibrant object $Q \in \mathcal{Q}$, the functor $- \otimes Q \colon \mathfrak{M} \to \mathfrak{M}$ is a left Quillen functor with right adjoint $\mathcal{H}om(Q,-) \colon \mathfrak{M} \to \mathfrak{M}$.*
(2) *For each fibrant object $R \in \mathcal{R}$, the functor $\mathcal{H}om(-,R) \colon \mathfrak{M} \to \mathfrak{M}^{\mathrm{op}}$ is a left Quillen functor with right adjoint $\mathcal{H}om(-,R) \colon \mathfrak{M}^{\mathrm{op}} \to \mathfrak{M}$.*

Proof Let $Q \in \mathcal{Q}$ be a given cofibrant object. We want to show that if we are given a cofibration $f \colon A \rightarrowtail B$, then $f \otimes 1_Q \colon A \otimes Q \rightarrowtail B \otimes Q$ is also a cofibration, which is trivial whenever f is trivial. Take g to be the cofibration $g \colon 0 \rightarrowtail Q$. Then considering the pushout-product construction from Diagram (†) of Definition 7.19, the pushout square becomes:

$$\begin{array}{ccc} A \otimes 0 & \xrightarrow{1_A \otimes g} & A \otimes Q \\ {\scriptstyle f \otimes 0}\downarrow & & \downarrow{\scriptstyle f'} \\ B \otimes 0 & \xrightarrow[g']{} & {}_fP_g. \end{array}$$

Therefore, ${}_fP_g = A \otimes Q$ and $f' = 1_{A \otimes Q}$, and so the identities in Equation (‡) tell us that $f \boxtimes g \colon {}_fP_g \to B \otimes Q$ is precisely $f \otimes 1_Q \colon A \otimes Q \rightarrowtail B \otimes Q$. So axioms ($\boxtimes$ **1**) and ($\boxtimes$ **2**) tell us that $f \otimes 1_Q \colon A \otimes Q \rightarrowtail B \otimes Q$ is a cofibration whenever f is a cofibration and that it is trivial whenever f is such. By Proposition 7.7 this proves that $(- \otimes Q, \mathcal{H}om(Q,-))$ is a Quillen adjunction.

The second statement is similar. Taking g to be the trivial morphism $R \twoheadrightarrow 0$ in Definition 7.19, an argument similar to the earlier one, but using axioms (◇ **1**) and (◇ **2**), shows that $\mathcal{H}om(-, R)$ sends (trivial) $\mathfrak{M}$-cofibrations to (trivial) $\mathfrak{M}$-fibrations. Since the role of cofibrations and fibrations swap in the opposite model structure, this shows that $\mathcal{H}om(-, R)\colon \mathfrak{M} \to \mathfrak{M}^{\text{op}}$ is a left Quillen functor, while $\mathcal{H}om(-, R)\colon \mathfrak{M}^{\text{op}} \to \mathfrak{M}$ is a right Quillen functor. Note that $(\mathcal{H}om(-, R), \mathcal{H}om(-, R))$ is indeed an adjoint pair because, by the adjoint associativity bijection of Equation (7.2), we have

$$\begin{aligned}\operatorname{Hom}_{\mathcal{A}^{\text{op}}}(\mathcal{H}om(A, R), B) &:= \operatorname{Hom}_{\mathcal{A}}(B, \mathcal{H}om(A, R)) \cong \operatorname{Hom}_{\mathcal{A}}(B \otimes A, R)\\ &\cong \operatorname{Hom}_{\mathcal{A}}(A \otimes B, R) \cong \operatorname{Hom}_{\mathcal{A}}(A, \mathcal{H}om(B, R)),\end{aligned}$$

where the second-to-last isomorphism utilizes the natural symmetry equivalence $\sigma_{A,B}\colon A \otimes B \to B \otimes A$. □

Finally, we can state the definition of an abelian monoidal model category.

Definition 7.22 Let $(\mathcal{A}, \otimes, \mathcal{H}om)$ be a closed symmetric monoidal structure on $\mathcal{A}$ and let $\mathfrak{M} = (\mathcal{Q}, \mathcal{W}, \mathcal{R})$ be an abelian model structure relative to $(\mathcal{A}, \mathcal{E})$. Then all of $(\mathcal{A}, \otimes, \mathcal{H}om, \mathcal{E}, \mathfrak{M})$ is said to be an **abelian monoidal model structure** if the following hold.

(1) The tensor product $- \otimes -\colon \mathcal{A} \times \mathcal{A} \to \mathcal{A}$ is a left Quillen bifunctor. Equivalently, $\mathcal{H}om(-, -)\colon \mathcal{A}^{\text{op}} \times \mathcal{A} \to \mathcal{A}$ is a right Quillen bifunctor.
(2) For all cofibrant objects $C \in \mathcal{Q}$, the map $C \otimes QI \xrightarrow{1_C \otimes p_I} C \otimes I \cong C$ is a weak equivalence, where $R_I \overset{i_I}{\rightarrowtail} QI \overset{p_I}{\twoheadrightarrow} I$ is a cofibrant approximation sequence for the tensor unit I.

We can now proceed to prove the main result about monoidal model structures. The point is that the total left derived functor of the bifunctor $- \otimes -$ exists, the total right derived functor of the bifunctor $\mathcal{H}om(-, -)$ exists, and the resulting bifunctors determine a closed symmetric monoidal structure on the homotopy category, $\operatorname{Ho}(\mathfrak{M})$.

Theorem 7.23 *Let $(\mathcal{A}, \otimes, \mathcal{H}om, \mathcal{E}, \mathfrak{M})$ be an abelian monoidal model structure. Then the following hold.*

(1) *The total left derived tensor product*

$$- \otimes^{\mathbf{L}} - := L(\gamma \circ \otimes)\colon \operatorname{Ho}(\mathfrak{M}) \times \operatorname{Ho}(\mathfrak{M}) \to \operatorname{Ho}(\mathfrak{M})$$

exists. It may be taken to be the unique bifunctor $- \otimes^{\mathbf{L}} -$ satisfying

$$\gamma(-) \otimes^{\mathbf{L}} \gamma(-) = \gamma(Q(-) \otimes Q(-)).$$

In particular, we may take $A \otimes^{\mathbf{L}} X = QA \otimes QX$ *for all objects* $A, X \in \mathcal{A}$, *and so* $A \otimes^{\mathbf{L}} X = A \otimes X$ *whenever* A *and* X *are cofibrant.*

(2) *The total right derived hom functor*

$$\mathbf{R}\mathcal{H}om(-,-) := R(\gamma \circ \mathcal{H}om) \colon \mathrm{Ho}(\mathfrak{M})^{\mathrm{op}} \times \mathrm{Ho}(\mathfrak{M}) \to \mathrm{Ho}(\mathfrak{M})$$

exists. It may be taken to be the unique bifunctor $\mathbf{R}\mathcal{H}om(-,-)$ *satisfying*

$$\mathbf{R}\mathcal{H}om(\gamma(-), \gamma(-)) = \gamma\mathcal{H}om(Q(-), R(-)).$$

In particular, we may take $\mathbf{R}\mathcal{H}om(A, X) = \mathcal{H}om(QA, RX)$ *for all objects* $A, X \in \mathcal{A}$, *and so* $\mathbf{R}\mathcal{H}om(A, X) = \mathcal{H}om(A, X)$ *whenever* A *is cofibrant and* X *is fibrant.*

(3) $\left(\mathrm{Ho}(\mathfrak{M}), \otimes^{\mathbf{L}}, \mathbf{R}\mathcal{H}om\right)$ *is a closed symmetric monoidal category.*

Proof We will prove the existence of $- \otimes^{\mathbf{L}} - := L(\gamma \circ (- \otimes -))$, by using Theorem 7.4, and then it will follow from that same theorem that $- \otimes^{\mathbf{L}} -$ may be computed as described. To verify the hypotheses of Theorem 7.4 we will show that $- \otimes -$ sends trivial cofibrations between cofibrant objects to trivial cofibrations. Since the product model structure $\mathfrak{M} \times \mathfrak{M}$ on $\mathcal{A} \times \mathcal{A}$ is defined componentwise, a trivial cofibration between cofibrant objects in $\mathfrak{M} \times \mathfrak{M}$ is nothing more than a pair of trivial cofbration sequences

$$A \overset{f}{\rightarrowtail} B \overset{p}{\twoheadrightarrow} P \quad \text{and} \quad X \overset{g}{\rightarrowtail} Y \overset{q}{\twoheadrightarrow} Q,$$

with $A, B, X, Y \in \mathcal{Q}$, and $P, Q \in \mathcal{Q}_{\mathcal{W}}$. Note that the composite $A \otimes X \xrightarrow{f \otimes g} B \otimes Y$ coincides with the composition

$$A \otimes X \overset{f \otimes 1_X}{\rightarrowtail} B \otimes X \overset{1_B \otimes g}{\rightarrowtail} B \otimes Y.$$

Since B and X are cofibrant, the functors $B \otimes -$ and $- \otimes X$ are each left Quillen functors, by Lemma 7.21. Therefore, $f \otimes 1_X$ and $1_B \otimes g$ are each trivial cofibrations and thus the composite $f \otimes g$ is too by Corollary 4.9. This proves statement (1).

On the other hand, we use the dual of Theorem 7.4 to show that the total right derived hom functor $\mathbf{R}\mathcal{H}om(-,-) := R(\gamma \circ \mathcal{H}om(-,-))$ exists. And again, this will imply that $\mathbf{R}\mathcal{H}om(-,-)$ can be computed as described, for the role of fibration and cofibration swap in the first variable of the functor $\mathcal{H}om(-,-) \colon \mathfrak{M}^{\mathrm{op}} \times \mathfrak{M} \to \mathfrak{M}$. So this time we will show that $\mathcal{H}om(-,-)$ sends trivial fibrations between fibrant objects to trivial fibrations. From definitions, a trivial fibration between fibrant objects of $\mathfrak{M}^{\mathrm{op}} \times \mathfrak{M}$ is equivalent to a pair

$$A \overset{f}{\rightarrowtail} B \overset{p}{\twoheadrightarrow} Q \quad \text{and} \quad R \overset{k}{\rightarrowtail} X \overset{q}{\twoheadrightarrow} Y,$$

where f is a trivial cofibration between cofibrant objects of $\mathfrak{M}$, and q is a trivial fibration between fibrant objects of $\mathfrak{M}$. So these are short exact sequences in $\mathcal{E}$ with $A, B \in \mathcal{Q}$, and $Q \in \mathcal{Q}_W$, and $X, Y \in \mathcal{R}$, and $R \in \mathcal{R}_W$. We want to show that

$$\mathcal{H}om(f, q)\colon \mathcal{H}om(B, X) \to \mathcal{H}om(A, Y)$$

is a trivial $\mathfrak{M}$-fibration. This time we note that this is the composition

$$\mathcal{H}om(B, X) \xrightarrow{\mathcal{H}om(f, 1_X)} \mathcal{H}om(A, X) \xrightarrow{\mathcal{H}om(1_A, q)} \mathcal{H}om(A, Y).$$

By Lemma 7.21, the map $\mathcal{H}om(A, q) := \mathcal{H}om(1_A, q)$ is a trivial $\mathfrak{M}$-fibration, and also $\mathcal{H}om(f, X) := \mathcal{H}om(f, 1_X)$ is a trivial $\mathfrak{M}$-fibration. So again, the composite $\mathcal{H}om(f, q)$ is a trivial $\mathfrak{M}$-fibration as desired.

Recall that $(\mathcal{A}, \otimes, \mathcal{H}om)$ comes with an associator $\alpha\colon (- \otimes -) \otimes - \cong - \otimes (- \otimes -)$, a tensor unit object, I, and a natural equivalence σ expressing the symmetry $\sigma_{A,B}\colon A \otimes B \cong B \otimes A$, all satisfying several coherence diagrams. We can use this data to obtain the corresponding data needed for $- \otimes^{\mathbf{L}} -$. For example, $\otimes^{\mathbf{L}}$ inherits a natural equivalence $\sigma^{\mathbf{L}}_{A,B}\colon A \otimes^{\mathbf{L}} B \cong B \otimes^{\mathbf{L}} A$ by using $\sigma_{QA,QB}\colon QA \otimes QB \cong QB \otimes QA$. Also, $\left(- \otimes^{\mathbf{L}} -\right) \otimes^{\mathbf{L}} -$ is the total left derived functor of $(- \otimes -) \otimes -$ while $- \otimes^{\mathbf{L}} \left(- \otimes^{\mathbf{L}} -\right)$ is the total left derived functor of $- \otimes (- \otimes -)$. So an associator natural equivalence $\alpha^{\mathbf{L}}\colon \left(- \otimes^{\mathbf{L}} -\right) \otimes^{\mathbf{L}} - \cong - \otimes^{\mathbf{L}} \left(- \otimes^{\mathbf{L}} -\right)$ is induced by $\alpha_{QA,QB,QC}\colon (QA \otimes QB) \otimes QC \cong QA \otimes (QB \otimes QC)$. Finally, we show that the unit I, or technically $\gamma(I)$, also serves as the unit for $\otimes^{\mathbf{L}}$. Indeed for all objects A, using the right unitor equivalence, $\rho_A\colon A \otimes I \to A$, we have a natural composite

$$A \otimes^{\mathbf{L}} I = QA \otimes QI \xrightarrow{1_{QA} \otimes p_I} QA \otimes I \xrightarrow{\rho_{QA}} QA \xrightarrow{p_A} A$$

of weak equivalences by the second condition of Definition 7.22. The fact that the coherence diagrams commute follows from the fact that they commute in the ground category $\mathcal{A}$.

Finally, to complete the proof that $\left(\mathrm{Ho}(\mathfrak{M}), \otimes^{\mathbf{L}}, \mathbf{R}\mathcal{H}om\right)$ is a closed symmetric monoidal category, we only need to see that $\mathbf{R}\mathcal{H}om(B, -)$ is the right adjoint of $- \otimes^{\mathbf{L}} B$, for each object B. But now this is straightforward, for on one hand we have

$$\mathrm{Ho}(\mathfrak{M})\left(A \otimes^{\mathbf{L}} B, C\right) = \mathrm{Ho}(\mathfrak{M})(QA \otimes QB, C) \cong \underline{\mathrm{Hom}}(QA \otimes QB, RC)$$

by the standard natural isomorphism of Corollary 5.17. On the other hand, we similarly have

$$\mathrm{Ho}(\mathfrak{M})(A, \mathbf{R}\mathcal{H}om(B, C)) = \mathrm{Ho}(\mathfrak{M})(A, \mathcal{H}om(QB, RC)) \cong \underline{\mathrm{Hom}}(QA, \mathcal{H}om(QB, RC)).$$

Since $(- \otimes QB, \mathcal{H}om(QB, -))$ is a Quillen adjunction by Lemma 7.21, we have a natural isomorphism $\underline{\mathrm{Hom}}(QA \otimes QB, RC) \cong \underline{\mathrm{Hom}}(QA, \mathcal{H}om(QB, RC))$ by Lemma 7.10. This completes the proof. □

Theorem 7.23 is just the tip of the iceberg when it comes to the usefulness of abelian monoidal model categories. For one, since we now know that the category $\mathrm{Ho}(\mathfrak{M})$ possesses both a closed symmetric monoidal structure as well as a triangulated structure, there are some natural compatibilities between the two structures that one might inquire about. This leads to the idea of a *tensor triangulated category*. At the very least, this is a triangulated category along with a tensor product that determines a triangulated functor in each variable; see Balmer [2005, definition 1.1]. But there is slew of other compatibilities between the triangulated structure and the derived tensor product that one may impose and which do in fact hold for homotopy categories of stable model categories. To verify all of these would lead us quite far astray so we will have to instead refer the reader to Hovey et al. [1997] and May [2001]. However, by quoting relevant results from this chapter, the reader can easily verify that $\mathrm{Ho}(\mathfrak{M})$ is a tensor triangulated category as asked in the following exercise.

Exercise 7.5.2 ($\mathrm{Ho}(\mathfrak{M})$ is Tensor Triangulated) Let $(\mathcal{A}, \otimes, \mathcal{H}om, \mathcal{E}, \mathfrak{M})$ be an abelian monoidal model structure. By referring to proven results in this chapter, show that the functor $X \otimes^{\mathbf{L}} - \cong - \otimes^{\mathbf{L}} X$ is a triangulated functor, for each object X. That is, we have a distinguished triangle

$$A \otimes^{\mathbf{L}} X \to B \otimes^{\mathbf{L}} X \to C \otimes^{\mathbf{L}} X \to \Sigma A \otimes^{\mathbf{L}} X \cong \Sigma\left(A \otimes^{\mathbf{L}} X\right)$$

whenever $A \to B \to C \to \Sigma A$ is a distinguished triangle.

7.6 Hovey's Criteria for Abelian Monoidal Model Structures

A theme throughout this book is that abelian model categories are easier than general model categories because most of the concepts can be expressed solely in terms of objects rather than (the more difficult) morphisms. Most notably, an abelian model structure is equivalent to a triple of classes of objects as opposed to a triple of classes of morphisms. Keeping with this theme, we will show in this section that we may replace the notion of a two-variable Quillen adjunction (i.e. a Quillen bifunctor), with a much simpler set of criteria, based only on the classes of cofibrant and trivially cofibrant objects. It is perhaps not surprising that the notion of purity (in the sense of $\otimes$-exactness) plays a role in this.

So again, we let $(\mathcal{A}, \otimes, \mathcal{H}om)$ be a closed symmetric monoidal category on $(\mathcal{A}, \mathcal{E})$. Note that since each $A \otimes - \cong - \otimes A$ is a left adjoint, these functors preserve all colimits that might exist. In particular $A \otimes -$ preserves cokernels, though one should be careful; there is no reason for it to preserve $\mathcal{E}$-admissible epics. The following notion of purity helps to rectify such incompatibilities and indeed cofibration sequences appearing in practice are often $\otimes$-pure exact in this sense.

Definition 7.24 Let $(\mathcal{A}, \otimes, \mathcal{H}om)$ be a closed symmetric monoidal category on $(\mathcal{A}, \mathcal{E})$.

(1) A short exact sequence $A \rightarrowtail B \twoheadrightarrow C$ is called **⊗-pure** if for all objects $X \in \mathcal{A}$, the sequence $X \otimes A \rightarrowtail X \otimes B \twoheadrightarrow X \otimes C$ is again in $\mathcal{E}$.
(2) An object $F \in \mathcal{A}$ is called **⊗-flat** if for all short exact sequences $A \rightarrowtail B \twoheadrightarrow C$ in $\mathcal{E}$, the sequence $F \otimes A \rightarrowtail F \otimes B \twoheadrightarrow F \otimes C$ is again in $\mathcal{E}$.

Note that 0 is always flat, and split exact sequences are always pure.

Exercise 7.6.1 Let $A \rightarrowtail F \twoheadrightarrow C$ be a $\otimes$-pure sequence. Show that if F is $\otimes$-flat, then both A and C are also $\otimes$-flat.

Exercise 7.6.2 Let $(\mathcal{A}, \otimes, \mathcal{H}om)$ be a closed symmetric monoidal category on $(\mathcal{A}, \mathcal{E})$. Show that the class $\mathcal{E}_\otimes$, of all $\otimes$-pure exact sequences, determines an exact structure on $\mathcal{A}$. That is, $(\mathcal{A}, \mathcal{E}_\otimes)$ is an exact category. *Note*: This is a special case of Exercise 1.5.4!

Bringing back our abelian model structure $\mathfrak{M} = (\mathcal{Q}, \mathcal{W}, \mathcal{R})$ on $(\mathcal{A}, \mathcal{E})$, the following compatibility axioms ($\otimes$**0**)–($\otimes$**3**) may be imposed on all of the data $(\mathcal{A}, \otimes, \mathcal{H}om, \mathcal{E}, \mathfrak{M})$. These conditions were suggested by Hovey and we will refer to them as *Hovey's monoidal criteria* for abelian model structures. Indeed we will proceed to show that these conditions imply that $(\mathcal{A}, \otimes, \mathcal{H}om, \mathcal{E}, \mathfrak{M})$ is an abelian monoidal model structure.

($\otimes$**0**) Every cofibration is $\otimes$-pure. That is, if $A \rightarrowtail B \twoheadrightarrow Q$ is a short exact sequence with $Q \in \mathcal{Q}$, then $X \otimes A \rightarrowtail X \otimes B \twoheadrightarrow X \otimes Q$ is also a short exact sequence in $\mathcal{E}$, for any object $X \in \mathcal{A}$.

($\otimes$**1**) The class $\mathcal{Q}$ of cofibrant objects is closed under tensor products. That is, if $P \in \mathcal{Q}$ and $Q \in \mathcal{Q}$, then $P \otimes Q \in \mathcal{Q}$.

($\otimes$**2**) If $P \in \mathcal{Q}$ and $Q \in \mathcal{Q}$, and $P \in \mathcal{W}$ or $Q \in \mathcal{W}$, then $P \otimes Q \in \mathcal{Q}_{\mathcal{W}}$.

($\otimes$**3**) $I \in \mathcal{Q}$. That is, the tensor unit I is cofibrant.

Let us make some simple observations regarding these conditions. First, note that conditions ($\otimes$**1**) and ($\otimes$**3**) together say that $\otimes$ restricts to a symmetric monoidal structure on the full subcategory $\mathcal{Q}$. Therefore, ($\otimes$ **2**) may be viewed as saying that $\mathcal{Q}_W$ is a $\otimes$-*ideal* in $\mathcal{Q}$. Second, note that it is immediate from conditions ($\otimes$ **0**)–($\otimes$ **2**) that $-\otimes Q$ is a left Quillen functor whenever Q is cofibrant. Finally, note the similarity of condition ($\otimes$ **1**) to condition ($\boxtimes$ **1**) from Section 7.5, and likewise the similarity of ($\otimes$ **2**) to ($\boxtimes$ **2**). We now show that they are in fact equivalent when assuming axiom ($\otimes$ **0**).

Proposition 7.25 *Assume axiom* ($\otimes$ **0**) *holds. Then we have the following.*

(1) *Axiom* ($\otimes$ **1**) *is equivalent to axiom* ($\boxtimes$ **1**).
(2) *Axiom* ($\otimes$ **2**) *is equivalent to axiom* ($\boxtimes$ **2**).

Consequently, $(\mathcal{A}, \otimes, \mathcal{H}om, \mathcal{E}, \mathfrak{M})$ *is a two-variable Quillen adjunction (i.e.* $\otimes$ *is a Quillen bifunctor) whenever axioms* ($\otimes$ **0**)–($\otimes$ **2**) *hold.*

Proof Assume P and Q are each cofibrant and set $f: 0 \rightarrowtail P$ and $g: 0 \rightarrowtail Q$. Then the pushout-product construction of Definition 7.19 applies and yields a trivial pushout square, with ${}_fP_g = 0$. Therefore the canonical morphism is just $f \boxtimes g: 0 \rightarrowtail P \otimes Q$. So it is clear that condition ($\boxtimes$ **1**) implies condition ($\otimes$ **1**) and condition ($\boxtimes$ **2**) implies condition ($\otimes$ **2**).

To prove the converse, let

$$A \overset{f}{\rightarrowtail} B \overset{p}{\twoheadrightarrow} P \quad \text{and} \quad X \overset{g}{\rightarrowtail} Y \overset{q}{\twoheadrightarrow} Q$$

be two given short exact sequences with P and Q cofibrant. By our assumption that axiom ($\otimes$ **0**) holds, Lemma 1.10 provides a pushout diagram

$$\begin{array}{ccccc}
A \otimes X & \overset{1_A \otimes g}{\rightarrowtail} & A \otimes Y & \overset{1_A \otimes q}{\twoheadrightarrow} & A \otimes Q \\
{\scriptstyle f \otimes 1_X} \downarrow & & \downarrow {\scriptstyle f'} & & \| \\
B \otimes X & \underset{g'}{\rightarrowtail} & {}_fP_g & \twoheadrightarrow & A \otimes Q \\
{\scriptstyle g \otimes 1_X} \downarrow & & \downarrow & & \\
P \otimes X & = & P \otimes X & &
\end{array}$$

satisfying the equations in (‡) of Definition 7.19. So we may continue to construct the following commutative diagram with exact rows:

$$\begin{array}{ccccc}
A \otimes X & \overset{1_A \otimes g}{\rightarrowtail} & A \otimes Y & \overset{1_A \otimes q}{\twoheadrightarrow} & A \otimes Q \\
{\scriptstyle f \otimes 1_X} \downarrow & & \downarrow {\scriptstyle f'} & & \| \\
B \otimes X & \overset{g'}{\rightarrowtail} & {}_fP_g & \twoheadrightarrow & A \otimes Q \\
\| & & \downarrow {\scriptstyle \exists !\, f \boxtimes g} & & \downarrow {\scriptstyle f \otimes 1_Q} \\
B \otimes X & \overset{1_B \otimes g}{\rightarrowtail} & B \otimes Y & \overset{1_B \otimes q}{\twoheadrightarrow} & B \otimes Q
\end{array}$$

and where $f \otimes 1_Y = (f \boxtimes g) \circ f'$. The lower right-hand square commutes by the uniqueness portion of the pushout ${}_fP_g$, for the two paths from ${}_fP_g$ to $B \otimes Q$ become equal upon composing each with f' and g'. It now follows from Proposition 1.12(2) that the lower right-hand square is a pullback. But then Corollary 1.14 assures us that $f \boxtimes g$ is in fact an admissible monic and that it sits in another commutative diagram with exact rows:

$$\begin{array}{ccccc} {}_fP_g & \xrightarrow{f \boxtimes g} & B \otimes Y & \xrightarrow{p \otimes q} & P \otimes Q \\ \downarrow & & \downarrow{\scriptstyle 1_B \otimes q} & & \| \\ A \otimes Q & \xrightarrow{f \otimes 1_Q} & B \otimes Q & \xrightarrow{p \otimes 1_Q} & P \otimes Q. \end{array}$$

Now the short exact sequence in the top row makes it clear that condition ($\otimes$ **1**) implies condition ($\boxtimes$ **1**) and condition ($\otimes$ **2**) implies condition ($\boxtimes$ **2**).

The final statement of the proposition is now a consequence of Lemma 7.20. □

The following result shows that we may use Hovey's criteria to check if $\mathfrak{M} = (Q, \mathcal{W}, \mathcal{R})$ is a monoidal abelian model structure. Although it does not characterize Definition 7.22, it is quite practical for constructing abelian monoidal models.

Theorem 7.26 (Hovey's Criteria for Abelian Monoidal Model Structures) *Let $(\mathcal{A}, \otimes, \mathcal{H}om)$ be a closed symmetric monoidal category and let $\mathfrak{M} = (Q, \mathcal{W}, \mathcal{R})$ be an abelian model structure relative to $(\mathcal{A}, \mathcal{E})$. If axioms ($\otimes$ **0**)–($\otimes$ **3**) each hold, then $(\mathcal{A}, \otimes, \mathcal{H}om, \mathcal{E}, \mathfrak{M})$ is an abelian monoidal model structure.*

Consequently, $\left(\mathrm{Ho}(\mathfrak{M}), \otimes^{\mathbf{L}}, \mathbf{R}\mathcal{H}om\right)$ is a tensor triangulated category by Theorem 7.23 and Exercise 7.5.2.

Proof Conditions ($\otimes$ **0**)–($\otimes$ **2**) together imply that $(\mathcal{A}, \otimes, \mathcal{H}om, \mathcal{E}, \mathfrak{M})$ is a two variable Quillen adjunction (i.e. $\otimes$ is a Quillen bifunctor), by Proposition 7.25. Condition ($\otimes$ **3**) states that the unit I is cofibrant, and in this case the second condition of Definition 7.22 becomes trivial. □

Given a closed symmetric monoidal structure $(\mathcal{A}, \otimes, \mathcal{H}om)$ on an exact category $(\mathcal{A}, \mathcal{E})$, we say that $(\mathcal{A}, \mathcal{E})$ has **enough $\otimes$-flats** if for each object $X \in \mathcal{A}$, there is an admissible epic $F \twoheadrightarrow X$ with F a $\otimes$-flat object. The next exercise gives conditions for which axiom ($\otimes$ **0**) is automatic if each cofibrant object is $\otimes$-flat.

Exercise 7.6.3 Let $(\mathcal{A}, \otimes, \mathcal{H}om)$ be a closed symmetric monoidal structure on an abelian category $\mathcal{A}$ along with a proper class $\mathcal{P}$ of short exact sequences. Assume $(\mathcal{A}, \mathcal{P})$ has enough $\otimes$-flats. Let $A \rightarrowtail B \twoheadrightarrow C$ be a short exact sequence in $\mathcal{P}$ with C a $\otimes$-flat object. Show that this must be a $\otimes$-pure exact sequence. In particular, axiom ($\otimes$ **0**) holds when $(\mathcal{A}, \mathcal{P})$ has enough $\otimes$-flats and $\mathfrak{M}$ is an abelian model structure in which each cofibrant object is $\otimes$-flat.

Exercise 7.6.4 (Chain Complexes) Let R be a commutative ring with identity and $(\mathrm{Ch}(R), \otimes_R, Hom_R)$ be the standard closed symmetric monoidal category of chain complexes of R-modules. So here $X \otimes_R Y$ is the total chain complex tensor product whose degree n component is given by the direct sum

$$\bigoplus_{i+j=n} \left(X_i \otimes_R Y_j\right),$$

while Hom_R is the total Hom complex as in Definition 10.2. Prove each of the following statements.

(1) A chain complex X is $\otimes_R$-flat in the sense of Definition 7.24 if and only if each X_n is a flat R-module.
(2) A short exact sequence $0 \to W \to X \to Y \to 0$ of chain complexes is $\otimes_R$-pure in the sense of Definition 7.24 if and only if each $0 \to W_n \to X_n \to Y_n \to 0$ is a pure exact sequence of R-modules.

Exercise 7.6.5 (Chain Complexes) Assume R is a commutative ring and consider the standard projective model structure on $\mathrm{Ch}(R)$ given by the Hovey triple $\mathfrak{M}_{proj} = (dg\widetilde{\mathcal{P}}, \widetilde{\mathcal{E}}, All)$. Here $\widetilde{\mathcal{E}}$ is the class of all exact (acyclic) chain complexes of R-modules, and $dg\widetilde{\mathcal{P}}$ is the class of all DG-projective chain complexes. A chain complex $P \in dg\widetilde{\mathcal{P}}$ if and only if each P_n is a projective R-module and $P \otimes_R E$ remains acyclic for each $E \in \widetilde{\mathcal{E}}$. Show that conditions ($\otimes$ **0**) – ($\otimes$ **3**) hold. Thus $\mathfrak{M}_{proj}$ is an abelian monoidal model structure for $\mathcal{D}(R)$, the derived category of R.

More examples of abelian monoidal model structures on chain complexes appear at the end of Section 10.9. In particular, see Exercises 10.9.6 and 10.9.9, and Example 10.51.

8

Hereditary Model Structures

An abelian model structure $\mathfrak{M} = (Q, \mathcal{W}, \mathcal{R})$ is called *hereditary* if its two associated cotorsion pairs are each hereditary. In practice, almost every abelian model structure we see is hereditary. In this chapter we will see that the homotopy category of such an $\mathfrak{M}$ is equivalent to the stable category of a Frobenius category. Triangulated categories with this property are sometimes called *algebraic triangulated categories*. Theorem 8.16 provides a powerful method for constructing hereditary abelian model structures. It may be viewed as a handy addition to Hovey's Correspondence for hereditary cotorsion pairs. The end of this chapter focuses on adjunctions between homotopy categories and conditions guaranteeing that the homotopy category of an hereditary model structure is compactly generated.

Throughout this chapter, we assume $(\mathcal{A}, \mathcal{E})$ is a weakly idempotent complete exact category, and we let $\mathfrak{M} = (Q, \mathcal{W}, \mathcal{R})$ denote a Hovey triple on $(\mathcal{A}, \mathcal{E})$. Equivalently, $\mathfrak{M}$ is an abelian model structure on $(\mathcal{A}, \mathcal{E})$.

8.1 Projective, Injective, and Frobenius Model Structures

Let us state the main definition that is the bedrock of this chapter.

Definition 8.1 $\mathfrak{M} = (Q, \mathcal{W}, \mathcal{R})$ is called **hereditary** if each of the associated cotorsion pairs, $\mathfrak{L} = (Q, \mathcal{R}_{\mathcal{W}})$ and $\mathfrak{R} = (Q_{\mathcal{W}}, \mathcal{R})$, are hereditary in the sense of Section 2.4.

The following special cases of hereditary abelian model structures frequently appear in practice.

- Assume $(\mathcal{A}, \mathcal{E})$ has enough projectives. We say that $\mathfrak{M}$ is a **projective model structure** if every object is fibrant. That is, $\mathfrak{M} = (Q, \mathcal{W}, \mathcal{A})$. Note that this data is equivalent to a projective cotorsion pair $\mathfrak{M} = (\mathcal{Q}, \mathcal{W})$ in the sense of Definition 2.20.

- Assume $(\mathcal{A}, \mathcal{E})$ has enough injectives. We say that $\mathfrak{M}$ is an **injective model structure** if every object is cofibrant. That is, $\mathfrak{M} = (\mathcal{A}, \mathcal{W}, \mathcal{R})$. Note that this data is equivalent to an injective cotorsion pair $\mathfrak{M} = (\mathcal{W}, \mathcal{R})$ in the sense of Definition 2.20.
- $(\mathcal{A}, \mathcal{E})$ is called a **Frobenius category** if it has enough projectives and injectives and if an object is injective if and only if it is projective. Note that this data is equivalent to a Hovey triple $\mathfrak{M} = (\mathcal{A}, \omega, \mathcal{A})$, where ω denotes the class of **projective–injective** objects.

In Section 10.4 we will prove a much more general version of the result in the following exercise.

Exercise 8.1.1 Let R be a ring and let $\text{Ch}(R)_{dw}$ denote the exact category of all chain complexes of R-modules, along with the degreewise split short exact sequences. Show that $\text{Ch}(R)_{dw}$ is a Frobenius category. *Hint:* The projective–injective complexes are the contractible complexes, which are those complexes isomorphic to direct sums of "disk" complexes of the form

$$D^n(M) \equiv \cdots \to 0 \to M \xrightarrow{1_M} M \to 0 \cdots ,$$

where M is concentrated in degrees n and $n-1$.

It is interesting to note that, for Frobenius categories, not only do the projective and injective objects coincide, but the notions of projective and injective cotorsion pairs also coincide; see Exercise 8.1.3.

The homotopy category of a Frobenius category recovers Happel's triangulated *stable category*. The next theorem makes this clear and Happel's construction is described in the proof.

Theorem 8.2 *Let $(\mathcal{A}, \mathcal{E})$ be a Frobenius category with ω the class of projective–injective objects. Then $\mathfrak{M} = (\mathcal{A}, \omega, \mathcal{A})$ is an hereditary abelian model structure with homotopy category* $\text{Ho}(\mathfrak{M}) = \text{St}_\omega(\mathcal{A})$. *Each of the following hold.*

(1) *The class of all weak equivalences associated to $\mathfrak{M}$ coincides with the class of all ω-stable equivalences.*
(2) *The canonical functor to the homotopy category, $\gamma \colon \mathcal{A} \to \text{Ho}(\mathfrak{M})$, agrees exactly with the canonical functor to the stable category, $\gamma_\omega \colon \mathcal{A} \to \text{St}_\omega(\mathcal{A})$. They each act by $f \mapsto [f]_\omega$.*

(3) *The triangulated category* $\mathrm{Ho}(\mathfrak{M}) = (\mathrm{Ho}(\mathfrak{M}), \Sigma, \Delta)$ *of Theorem 6.34 coincides exactly with the classical triangulated structure on* $\mathrm{St}_\omega(\mathcal{A})$ *due to Happel.*

Proof It is clear that $\mathfrak{M} = (\mathcal{A}, \omega, \mathcal{A})$ is an hereditary abelian model structure. It is also clear by the very definition of the homotopy category that $\mathrm{Ho}(\mathfrak{M}) = \mathrm{St}_\omega(\mathcal{A})$, and that the canonical functor γ acts by $f \mapsto [f]_\omega$. It follows from Corollary 5.4 that the class of all $\mathfrak{M}$-weak equivalences coincides with the class of all ω-stable equivalences.

Set $\mathrm{St}(\mathcal{A}) := \mathrm{St}_\omega(\mathcal{A})$. The suspension functor $\Sigma\colon \mathrm{St}(\mathcal{A}) \to \mathrm{St}(\mathcal{A})$ of Section 6.4 boils down to the simple action $[f]_\omega \mapsto [\Sigma(f)]_\omega$ where we use a short exact sequence $A \rightarrowtail W \twoheadrightarrow \Sigma A$ with $W \in \omega$. By Proposition 6.29, it is an autoequivalence with inverse the loop functor $\Omega\colon \mathrm{St}(\mathcal{A}) \to \mathrm{St}(\mathcal{A})$. Following Definition 6.31, a *distinguished right triangle* is any right triangle isomorphic in $\mathrm{St}(\mathcal{A})$ to a *strict right triangle* on a morphism $f \in \mathcal{A}$. The latter is formed, for each morphism $f\colon A \to B$ of $\mathcal{A}$, via the pushout construction:

$$\begin{array}{ccccc} A & \overset{k_A}{\rightarrowtail} & W & \overset{\sigma_A}{\twoheadrightarrow} & \Sigma A \\ \downarrow{\scriptstyle f} & & \downarrow{\scriptstyle w_f} & & \| \\ B & \underset{k_f}{\rightarrowtail} & C(f) & \underset{\sigma_f}{\twoheadrightarrow} & \Sigma A, \end{array} \qquad (\dagger)$$

where W is some projective–injective object. Then

$$A \xrightarrow{[f]} B \xrightarrow{[k_f]} C(f) \xrightarrow{[\sigma_f]} \Sigma A$$

is a strict right triangle in $\mathrm{St}(\mathcal{A})$. This data coincides exactly with Happel's definitions, and by Theorem 6.34 it determines a triangulated structure on $\mathrm{Ho}(\mathfrak{M}) = \mathrm{St}(\mathcal{A})$. □

The following easy proposition is a fundamental fact about any hereditary abelian model structure $\mathfrak{M}$.

Proposition 8.3 *Let* $\mathfrak{M} = (Q, \mathcal{W}, \mathcal{R})$ *be an hereditary abelian model structure. Then the following statements hold.*

(1) *The full subcategory* $Q \cap \mathcal{R}$ *of bifibrant objects, along with its inherited exact structure (consisting of all short exact sequences with all three terms in* $Q \cap \mathcal{R}$*), is a Frobenius category. The projective–injective objects are precisely those in the core* $\omega = Q \cap \mathcal{W} \cap \mathcal{R}$*.* $\mathfrak{M}$ *restricts to the canonical model structure* $(Q \cap \mathcal{R}, \omega, Q \cap \mathcal{R})$ *associated to the Frobenius category* $Q \cap \mathcal{R}$*.*
(2) *The full subcategory* $\mathcal{R}$ *of fibrant objects, along with its inherited exact structure, is an exact category with enough projectives.* $\mathfrak{M}$ *restricts to a projective cotorsion pair* $(Q \cap \mathcal{R}, \mathcal{R}_{\mathcal{W}})$ *in* $\mathcal{R}$*.*

(3) *The full subcategory $\mathcal{Q}$ of cofibrant objects, along with its inherited exact structure, is an exact category with enough injectives. $\mathfrak{M}$ restricts to an injective cotorsion pair $(\mathcal{Q}_{\mathcal{W}}, \mathcal{Q} \cap \mathcal{R})$ in $\mathcal{Q}$.*

Proof First, any strictly full subcategory $\mathcal{S} \subseteq \mathcal{A}$ that is closed under $\mathcal{E}$-extensions naturally inherits an exact structure from $(\mathcal{A}, \mathcal{E})$. It consists of the ambient short exact sequences but with all three terms in $\mathcal{S}$; see Exercise 1.4.3. Since $\mathcal{Q} \cap \mathcal{R}$ is closed under extensions it inherits such an exact structure. Second, we see that if $A \rightarrowtail B \twoheadrightarrow C$ is a short exact sequence with $A, B, C \in \mathcal{Q} \cap \mathcal{R}$, then it will split if either A or C is in ω, since this would imply $\mathrm{Ext}^1_{\mathcal{E}}(C, A) = 0$. So everything in ω is both injective and projective with respect to the inherited exact structure on $\mathcal{Q} \cap \mathcal{R}$. On the other hand, say $I \in \mathcal{Q} \cap \mathcal{R}$ is injective with respect to the exact structure. Using the complete cotorsion pair $(\mathcal{Q}, \mathcal{R}_{\mathcal{W}})$, write a short exact sequence $I \rightarrowtail W \twoheadrightarrow Q$ with $W \in \mathcal{W} \cap \mathcal{R}$ and $Q \in \mathcal{Q}$. Since the model structure is hereditary, $Q \in \mathcal{Q} \cap \mathcal{R}$. Also note that $W \in \omega$. So we have an admissible short exact sequence, which must split by the hypothesis on I. We infer $I \in \omega$, since ω is closed under direct summands. Hence all injective objects of $\mathcal{Q} \cap \mathcal{R}$ are in ω. Performing the same argument as previously for an *arbitrary* $A \in \mathcal{Q} \cap \mathcal{R}$ (in place of I) shows that $\mathcal{Q} \cap \mathcal{R}$ has enough injectives. Finally, a similar argument with the cotorsion pair $(\mathcal{Q}_{\mathcal{W}}, \mathcal{R})$ shows that all projectives are in ω, and that $\mathcal{Q} \cap \mathcal{R}$ has enough projectives.

We leave it to the reader to similarly verify statements (2) and (3). □

Example 8.4 Consider the flat model structure $\mathrm{Ch}(R)_{flat} = \left(dg\widetilde{\mathcal{F}}, \widetilde{\mathcal{E}}, dw\widetilde{\mathcal{C}}\right)$ from Example 8 of the Introduction and Main Examples. This is an hereditary model structure. Consequently, the class $dg\widetilde{\mathcal{F}} \cap dw\widetilde{\mathcal{C}}$, of all DG-flat complexes with cotorsion components, is a Frobenius category. Its projective–injective objects are precisely the contractible complexes with flat–cotorsion components.

Exercise 8.1.2 Prove part (2) or (3) of Proposition 8.3.

Exercise 8.1.3 Show that the following conditions are equivalent for a complete cotorsion pair $(\mathcal{U}, \mathcal{V})$ on a Frobenius category $(\mathcal{A}, \mathcal{E})$.

(1) $(\mathcal{U}, \mathcal{V})$ is a projective cotorsion pair.
(2) $(\mathcal{U}, \mathcal{V})$ is an injective cotorsion pair.
(3) $\mathcal{U}$ is thick. (Satisfies the 2 out of 3 property on short exact sequences.)
(4) $\mathcal{V}$ is thick.

We say that such an $\mathfrak{M} = (\mathcal{U}, \mathcal{V})$ is a *localizing cotorsion pair* on $(\mathcal{A}, \mathcal{E})$. Note that it induces both an injective model structure $\mathfrak{M}^i := (\mathcal{A}, \mathcal{U}, \mathcal{V})$ and a projective model structure $\mathfrak{M}^p := (\mathcal{U}, \mathcal{V}, \mathcal{A})$.

8.2 The Homotopy Category of an Hereditary Model Structure

Throughout this section we let $\mathfrak{M} = (\mathcal{Q}, \mathcal{W}, \mathcal{R})$ be an hereditary abelian model structure. In the previous section we saw that $\mathcal{Q} \cap \mathcal{R}$, the full subcategory of bifibrant objects, is a Frobenius category. We will now show that the equivalence $\mathrm{Ho}(R) \circ \mathrm{Ho}(Q)\colon \mathrm{Ho}(\mathfrak{M}) \to \mathrm{St}_\omega(\mathcal{Q} \cap \mathcal{R})$ of Proposition 5.25 is a triangulated equivalence to the stable category of $\mathcal{Q} \cap \mathcal{R}$. In particular, this shows that the homotopy category of any hereditary abelian model structure is triangle equivalent to the stable category of a Frobenius category.

The following lemma will be the key, for it shows that (co)fibrant approximations respect the construction of cofiber sequences.

Lemma 8.5 *Assume $\mathfrak{M} = (\mathcal{Q}, \mathcal{W}, \mathcal{R})$ is hereditary. Then cofibrant, fibrant, and bifibrant replacements may be assigned to diagram sequences as follows.*

(1) *Short exact sequences have cofibrant, fibrant, and bifibrant replacements. (In particular, as guaranteed by the Generalized Horseshoe Lemma 2.18.)*
(2) *Suspension sequences have cofibrant, fibrant, and bifibrant replacements. (In particular, as guaranteed by the Generalized Horseshoe Lemma 2.18.)*
(3) *Cofiber sequences have cofibrant, fibrant, and bifibrant replacements which are again cofiber sequences. More specifically, given any cofiber sequence for a morphism f, we can construct cofiber sequences for $Q(f)$ and $R(Q(f))$ along with commutative diagrams in $\mathcal{A}$ such that all the vertical morphisms are the weak equivalences obtained by bifibrant replacment:*

$$
\begin{array}{ccccccc}
RQA & \xrightarrow{R(Q(f))} & RQB & \overset{k_{R(Q(f))}}{\rightarrowtail} & C(R(Q(f))) & \overset{\sigma_{R(Q(f))}}{\twoheadrightarrow} & RQ\Sigma A \\
\uparrow{\scriptstyle j_{QA}} & & \uparrow{\scriptstyle j_B} & & \uparrow{\scriptstyle j_{C(Q(f))}} & & \uparrow{\scriptstyle j_{Q\Sigma A}} \\
QA & \xrightarrow{Q(f)} & QB & \overset{k_{Q(f)}}{\rightarrowtail} & C(Q(f)) & \overset{\sigma_{Q(f)}}{\twoheadrightarrow} & Q\Sigma A \\
{\scriptstyle p_A}\downarrow & & \downarrow{\scriptstyle p_B} & & \downarrow{\scriptstyle p_{C(f)}} & & \downarrow{\scriptstyle p_{\Sigma A}} \\
A & \xrightarrow{f} & B & \overset{k_f}{\rightarrowtail} & C(f) & \overset{\sigma_f}{\twoheadrightarrow} & \Sigma A.
\end{array}
$$

In particular, the first, second, and fourth columns may be any originally assigned bifibrant replacements, and we have $RQ\Sigma A \cong \Sigma RQA$ and $Q\Sigma A \cong \Sigma QA$. It follows that we may take $Q(k_f) = k_{Q(f)}$, $R(Q(k_f)) = k_{R(Q(f))}$, $Q(\sigma_f) = \sigma_{Q(f)}$, and $R(Q(\sigma_f)) = \sigma_{R(Q(f))}$.

Proof Since the left cotorsion pair, $\mathfrak{L} = (\mathcal{Q}, \mathcal{R}_W)$, is an hereditary cotorsion pair, the Generalized Horseshoe Lemma 2.18 immediately implies that short exact sequences extend to short exact sequences of cofibrant replacements. (Note that the two outer $\mathcal{Q}$-approximation sequences may be any given cofibrant replacements. Only the middle one is newly constructed.) The right cotorsion pair, $\mathfrak{R} = (\mathcal{Q}_{\mathcal{W}}, \mathcal{R})$, implies the same thing for fibrant replacements. This proves the first statement (1).

Statement (2) follows from (1). In particular, take any general suspension sequence $A \rightarrowtail W_A \twoheadrightarrow \Sigma A$ of an object A (so $W_A \in \mathcal{W}$). Then we may construct the commutative diagram with exact rows:

$$\begin{array}{lccccc}
R(Q(\mathbb{W}_A)): & RQA & \overset{R(Q(k_A))}{\rightarrowtail} & RQW_A & \overset{R(Q(\sigma_A))}{\twoheadrightarrow} & RQ\Sigma A \\
 & \uparrow{\scriptstyle j_{QA}} & & \uparrow{\scriptstyle j} & & \uparrow{\scriptstyle j_{Q\Sigma A}} \\
Q(\mathbb{W}_A): & QA & \overset{Q(k_A)}{\rightarrowtail} & QW_A & \overset{Q(\sigma_A)}{\twoheadrightarrow} & Q\Sigma A \\
 & \downarrow{\scriptstyle p_A} & & \downarrow{\scriptstyle p} & & \downarrow{\scriptstyle p_{\Sigma A}} \\
\mathbb{W}_A: & A & \overset{k_A}{\rightarrowtail} & W_A & \overset{\sigma_A}{\twoheadrightarrow} & \Sigma A,
\end{array}$$

and it is the case that we can do this from any given cofibrant and fibrant approximation sequences (the solid vertical arrows). We see that $Q(\mathbb{W}_A)$, the sequence of cofibrant replacements making up the middle row, is indeed a general suspension sequence of QA. Similarly, the sequence $R(Q(\mathbb{W}_A))$ of bifibrant replacements is a general suspension sequence of RQA. (One may note that if $\mathbb{W}_A$ is a strict suspension sequence, that is $W_A \in \mathcal{R}_{\mathcal{W}}$, then $Q(\mathbb{W}_A)$ and $R(Q(\mathbb{W}_A))$ are also strict suspension sequences.)

We have shown (1) and (2), and we now prove (3). So assume we have a cofiber sequence arising from a pushout along a general suspension sequence:

$$\begin{array}{lccccc}
\mathbb{W}_A: & A & \overset{k_A}{\rightarrowtail} & W_A & \overset{\sigma_A}{\twoheadrightarrow} & \Sigma A \\
 & \downarrow{\scriptstyle f} & & \downarrow{\scriptstyle w_f} & & \| \\
\mathbb{P}_f: & B & \overset{k_f}{\rightarrowtail} & C(f) & \overset{\sigma_f}{\twoheadrightarrow} & \Sigma A.
\end{array}$$

We first extend the suspension sequence $\mathbb{W}_A$ to a commutative diagram with exact rows and columns, and center row $Q(\mathbb{W}_A)$, as described earlier. We also extend f, in the usual way, to a commutative diagram of cofibrant replacements:

$$\begin{array}{ccccc}
R_A & \overset{i_A}{\rightarrowtail} & QA & \overset{p_A}{\twoheadrightarrow} & A \\
\downarrow & & \downarrow{\scriptstyle Q(f)} & & \downarrow{\scriptstyle f} \\
R_B & \overset{i_A}{\rightarrowtail} & QB & \overset{p_B}{\twoheadrightarrow} & B.
\end{array}$$

We then pushout the former diagram along each of the three morphisms. One checks that the three pushouts are connected by unique morphisms making all squares in sight commute. (Similar arguments already appeared in the proof of Lemma 6.17 where Diagram (‡) in the proof of Theorem 6.16 was referenced, and again in the proof of Lemma 6.26.) The front face of the pushout diagram is the following commutative diagram:

$$\begin{array}{ccccc}
R_B & \rightarrowtail & C & \twoheadrightarrow & R_{\Sigma A} \\
{\scriptstyle i_B}\downarrow & & \downarrow{\scriptstyle \exists !} & & \downarrow{\scriptstyle i_{\Sigma A}} \\
QB & \xrightarrow{k_{Q(f)}} & C(Q(f)) & \xrightarrow{\sigma_{Q(f)}} & Q\Sigma A \\
{\scriptstyle p_B}\downarrow & & \downarrow{\scriptstyle \exists !} & & \downarrow{\scriptstyle p_{\Sigma A}} \\
B & \xrightarrow{k_f} & C(f) & \xrightarrow{\sigma_f} & \Sigma A,
\end{array}$$

where $C(Q(f))$ is the cone relative to $Q(\mathbb{W}_A)$ and $C(f)$ is the cone relative to $\mathbb{W}_A$. We note that each row of the diagram is a short exact sequence and we already know that the outer two columns are as well. Moreover, by the universal property of the pushout C, the two dashed arrows making up the middle column must compose to 0. Therefore, the (3×3)-Lemma 1.16 applies and we may conclude that the middle column is a short exact sequence too. In particular, $C(Q(f))$, the cone on $Q(f)$, is a cofibrant replacement of $C(f)$. Now the lower half of the entire pushout diagram is shown in the following, and it clearly displays a cofiber sequence for $Q(f)$ which is in fact a "cofibrant replacement" of the cofiber sequence for f:

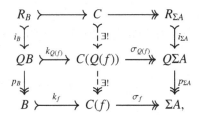

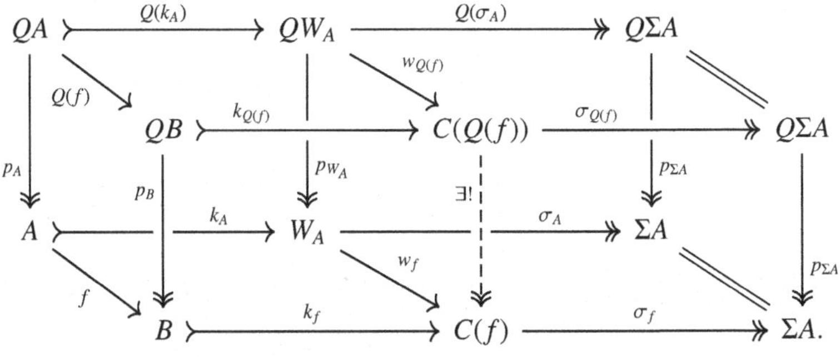

In particular, we have constructed a cofiber sequence for $Q(f)$ that is weakly equivalent to the given cofiber sequence for f. In a similar way we can construct fibrant replacements of cofiber sequences. In particular, we can proceed the same way to construct a cofiber sequence for $R(Q(f))$ that is a "bifibrant replacement" of the cofiber sequence we just constructed for $Q(f)$. □

The functors Ho(Q) and Ho(R) were defined in Section 5.5, and Proposition 5.25 showed that we have category equivalences

$$\mathrm{Ho}(\mathfrak{M}) \underset{I}{\overset{\mathrm{Ho}(Q)}{\rightleftarrows}} \mathrm{Ho}(Q) \underset{I}{\overset{\mathrm{Ho}(R)}{\rightleftarrows}} \mathrm{St}_\omega(\mathcal{Q} \cap \mathcal{R}).$$

We are now ready to state the main result.

Theorem 8.6 *Assume $\mathfrak{M} = (\mathcal{Q}, \mathcal{W}, \mathcal{R})$ is hereditary. The composition*

$$\mathrm{Ho}(R) \circ \mathrm{Ho}(Q) \colon \mathrm{Ho}(\mathfrak{M}) \to \mathrm{St}(\mathcal{Q} \cap \mathcal{R})$$

is a triangulated equivalence to the stable category of the Frobenius category $\mathcal{Q} \cap \mathcal{R}$. In particular, the homotopy category of any hereditary abelian model structure is triangle equivalent to the stable category of a Frobenius category.

Proof Our proof will allow for Σ to be a general suspension functor, arising from an assignment $\mathbb{W} \colon A \mapsto \mathbb{W}_A$, in the sense of Lemma 6.39. We also assume that this assignment is such that $\Sigma B \in \mathcal{Q} \cap \mathcal{R}$ whenever $B \in \mathcal{Q} \cap \mathcal{R}$. This is the case for a standard assignment $\mathbb{S} \colon A \mapsto \mathbb{S}_A$ of strict suspension sequences as in Section 6.4. Indeed if $B \in \mathcal{Q} \cap \mathcal{R}$ is a bifibrant object, then we compute ΣB, in Ho($\mathfrak{M}$), by taking the assigned suspension sequence

$$\mathbb{S}_B \colon B \rightarrowtail \tilde{R}_B \twoheadrightarrow \Sigma B,$$

which has $\tilde{R}_B \in \mathcal{R}_{\mathcal{W}}$ and $\Sigma B \in \mathcal{Q}$. Since $B, \tilde{R}_B \in \mathcal{R}$, the hereditary hypothesis implies $\Sigma B \in \mathcal{Q} \cap \mathcal{R}$. Moreover, $\tilde{R}_B \in \omega$, the class of projective–injective objects in the Frobenius category $\mathcal{Q} \cap \mathcal{R}$. So the suspension functor $\Sigma_{\mathrm{St}(\mathcal{Q} \cap \mathcal{R})}$ is nothing more than $\Sigma_{\mathrm{Ho}(\mathfrak{M})}$ but with its domain restricted to $\mathrm{St}(\mathcal{Q} \cap \mathcal{R})$.

Now, to show that the functor $\mathrm{Ho}(R) \circ \mathrm{Ho}(Q)$ is triangulated (Definition 6.6) we need a candidate for the required natural equivalence

$$\mathrm{Ho}(R) \circ \mathrm{Ho}(Q) \circ \Sigma \cong \Sigma \circ \mathrm{Ho}(R) \circ \mathrm{Ho}(Q).$$

By Corollary 5.16, it is enough to have a natural equivalence

$$\mathrm{Ho}(R) \circ \mathrm{Ho}(Q) \circ \Sigma \circ \gamma \cong \Sigma \circ \mathrm{Ho}(R) \circ \mathrm{Ho}(Q) \circ \gamma. \tag{8.1}$$

So let $f \colon A \to B$ be any $\mathcal{A}$-morphism. By tracing through the definitions, the functor $\mathrm{Ho}(R) \circ \mathrm{Ho}(Q) \circ \Sigma \circ \gamma$ sends f to the homotopy class

$$[R(Q(\Sigma(f)))] \colon RQ\Sigma A \to RQ\Sigma B.$$

On the other hand, the functor $\Sigma \circ \mathrm{Ho}(R) \circ \mathrm{Ho}(Q) \circ \gamma$ sends f to

$$[\Sigma(R(Q(f)))] \colon \Sigma RQA \to \Sigma RQB.$$

But the Generalized Horseshoe Lemma 2.18 can be seen to be functorial; see Exercise 2.4.2. In particular, f induces a morphism $\mathbb{W}_A \to \mathbb{W}_B$ of suspension sequences, and in turn the construction of Lemma 8.5 leads to a corresponding morphism of their bifibrant replacement sequences:

$$\begin{array}{llcll} R(Q(\mathbb{W}_A)): & RQA & \xrightarrow{R(Q(k_A))} & RQW_A & \xrightarrow{R(Q(\sigma_A))} RQ\Sigma A \\ & \downarrow{\scriptstyle R(Q(f))} & & \downarrow & \qquad \downarrow{\scriptstyle R(Q(\Sigma(f)))} \\ R(Q(\mathbb{W}_B)): & RQB & \xrightarrow{R(Q(k_A))} & RQW_B & \xrightarrow{R(Q(\sigma_B))} RQ\Sigma B. \end{array}$$

Now for each A, by comparing the two suspension sequences for RQA, that is $R(Q(\mathbb{W}_A))$ and $\mathbb{W}_{RQA}$, Lemma 6.26 yields the suspension connecting isomorphism $\gamma(R(Q(\mathbb{W}_A)))\colon RQ\Sigma A \to \Sigma RQA$. Moreover, the naturality in that lemma means that the previous morphism of suspension sequences induces a commutative square:

$$\begin{array}{ccc} RQ\Sigma A & \xrightarrow{\gamma(R(Q(\mathbb{W}_A)))} & \Sigma RQA \\ {\scriptstyle [R(Q(\Sigma(f)))]}\downarrow & & \downarrow{\scriptstyle [\Sigma(R(Q(f)))]} \\ RQ\Sigma B & \xrightarrow{\gamma(R(Q(\mathbb{W}_B)))} & \Sigma RQB. \end{array}$$

In this way, the collection

$$\gamma(R(Q(\mathbb{W}))) := \{\gamma(R(Q(\mathbb{W}_A)))\colon RQ\Sigma A \to \Sigma RQA\}$$

defines the desired natural equivalence that we indicated earlier in (8.1).

The fact that $\gamma(R(Q(\mathbb{W})))$ makes $\mathrm{Ho}(R) \circ \mathrm{Ho}(Q)\colon \mathrm{Ho}(\mathfrak{M}) \to \mathrm{St}(\mathcal{Q} \cap \mathcal{R})$ a triangulated functor now follows from everything shown in Lemma 8.5. In particular, the diagram in the statement of the lemma makes it clear that any strict triangle $A \xrightarrow{\gamma(f)} B \xrightarrow{\gamma(k_f)} C(f) \xrightarrow{\gamma(\sigma_f)} \Sigma A$ is mapped by $\mathrm{Ho}(R) \circ \mathrm{Ho}(Q)$ to (the γ-image of) the cofiber sequence shown along the top row of the diagram in Lemma 8.5(3). This cofiber sequence is relative to the suspension sequence $R(Q(\mathbb{W}_A))$. But upon composing the last morphism in the sequence with the suspension connecting isomorphism, $\gamma(R(Q(\mathbb{W})))$, we obtain a distinguished triangle in $\mathrm{St}(\mathcal{Q} \cap \mathcal{R})$ relative to $\mathbb{W}_{RQA}$. This is guaranteed by Theorem 6.38. (We even have that the restriction of the assignment $\mathbb{W}\colon A \mapsto \mathbb{W}_A$ to $\mathcal{Q} \cap \mathcal{R}$ becomes an assignment of *strict* suspension sequences.) □

8.3 The Homotopy Category of Projective and Injective Models

In this section we look at the special case of hereditary model structures that are projective or injective. Such model structures naturally occur in algebra and often their properties can be expressed more simply than those of general model structures. For example, in this section we simplify our description of coproducts and products in the homotopy category of a projective (or injective) model structure.

Since the notions of projective and injective model structures are dual to one another, we only need to develop the theory for one, and we choose to work with projective model structures. So, *throughout this section we assume that* $(\mathcal{A}, \mathcal{E})$ *has enough projectives and that* $\mathfrak{M} = (Q, \mathcal{W})$ *is a projective cotorsion pair.* Recall that this means $\mathfrak{M}$ is a complete cotorsion pair with $\mathcal{W}$ thick and such that $Q \cap \mathcal{W} = \mathcal{P}$, the class of all projective objects in the exact category $(\mathcal{A}, \mathcal{E})$. Equivalently, $\mathfrak{M} = (Q, \mathcal{W}, \mathcal{A})$ is a projective model structure.

For any collection of objects $\{X_i\}_{i \in I}$ in $\mathcal{A}$, if the coproduct $\bigoplus_{i \in I} X_i$ exists, then there is a canonical monomorphism of abelian groups

$$\operatorname{Ext}^1_{\mathcal{E}}\left(\bigoplus_{i \in I} X_i, Y\right) \to \prod_{i \in I} \operatorname{Ext}^1_{\mathcal{E}}(X_i, Y),$$

by Lemma 1.24. It follows immediately that Q is closed under direct sums whenever $\mathcal{A}$ has coproducts. Likewise, coproducts of projective objects are again projective; see Lemma 1.26.

Definition 8.7 The **projective stable category** of $(\mathcal{A}, \mathcal{E})$ is defined via Proposition 3.1 to be the category $\mathrm{St}_{\mathcal{P}}(\mathcal{A})$. Note then that we denote its morphism sets by $\underline{\mathrm{Hom}}_{\mathcal{P}}(A, B)$, and the kernel of the canonical functor $\mathcal{A} \xrightarrow{\gamma} \mathrm{St}_{\mathcal{P}}(\mathcal{A})$ is precisely the class of all projectives $\mathcal{P}$. Moreover, if $\mathcal{A}$ has coproducts then they produce coproducts in $\mathrm{St}_{\mathcal{P}}(\mathcal{A})$ by Proposition 3.2.

$\mathrm{St}_{\mathcal{P}}(\mathcal{A})$ has the structure of a left triangulated category; see Example 6.21. The following statement summarizes first properties of the homotopy category of a projective model structure.

Corollary 8.8 *Assume* $(\mathcal{A}, \mathcal{E})$ *has enough projectives and let* $\mathfrak{M} = (Q, \mathcal{W})$ *be a projective cotorsion pair. Then the following hold.*

(1) *The full subcategory* Q*, along with the class of all short exact sequences with all three terms in* Q*, is a Frobenius category. Its class of projective–injective objects is precisely* $\mathcal{P}$*. Cofibrant replacement induces a triangulated equivalence* $\mathrm{Ho}(Q)\colon \mathrm{Ho}(\mathfrak{M}) \to \mathrm{St}(Q)$ *with inverse the inclusion* $I\colon \mathrm{St}(Q) \to \mathrm{Ho}(\mathfrak{M})$.

(2) *We have group isomorphisms* $\mathrm{Ho}(\mathfrak{M})(A, B) \cong \underline{\mathrm{Hom}}_{\mathcal{P}}(QA, B)$*, natural in both* A *and* B.

(3) $\mathrm{St}(\mathcal{Q}) \xrightarrow{I} \mathrm{St}_{\mathcal{P}}(\mathcal{A})$ *is a coreflective subcategory. Indeed it has a right adjoint, the* $\mathcal{Q}$*-approximation functor* $\mathrm{St}_{\mathcal{P}}(\mathcal{A}) \xrightarrow{Q} \mathrm{St}(\mathcal{Q})$*, defined via Definition 4.15. It is a left triangulated functor.*

Proof Statement (1) is a special case of Proposition 8.3(1) and Theorem 8.6. Statement (2) is a special case of Corollary 5.17. Also, (3) is a special case of Proposition 3.7. The fact that Q is a left triangulated functor follows from Lemma 6.32 combined with Theorem 8.6. □

Corollary 8.9 (Coproducts and Products for Projective Model Structures) *Assume* $(\mathcal{A}, \mathcal{E})$ *has enough projectives and let* $\mathfrak{M} = (\mathcal{Q}, \mathcal{W})$ *be a projective cotorsion pair. Coproducts and products exist in* $\mathrm{Ho}(\mathfrak{M})$ *as follows.*

(1) *If coproducts exist in* $\mathcal{A}$*, then coproducts also exist in* $\mathrm{Ho}(\mathfrak{M})$*. In fact, the coproduct of a collection of objects* $\{A_i\}_{i\in I}$ *in* $\mathrm{Ho}(\mathfrak{M})$ *may be obtained by taking the coproduct of* $\{QA_i\}_{i\in I}$ *in* $\mathrm{St}(\mathcal{Q})$*. More precisely, the coproduct is*

$$\left(\bigoplus_{i\in I} QA_i, \{[\eta_{QA_i}]\}_{i\in I}\right),$$

where $QA_i \xrightarrow{\eta_{QA_i}} \bigoplus_{i\in I} QA_i$ *is the canonical injection into the coproduct* $\bigoplus_{i\in I} QA_i$ *taken in* $\mathcal{A}$*.*

Moreover, assume $\mathcal{W}$ *and the class of all short exact sequences are each closed under coproducts. Then we have* $Q\left(\bigoplus_{i\in I} A_i\right) = \bigoplus_{i\in I} QA_i$ *and*

$$\left(\bigoplus_{i\in I} A_i, \{\gamma(\eta_i) = Q[\eta_i]\}_{i\in I}\right)$$

is a coproduct of $\{A_i\}_{i\in I}$ *in* $\mathrm{Ho}(\mathfrak{M})$*, whenever* $A_i \xrightarrow{\eta_i} \bigoplus_{i\in I} A_i$ *denotes the canonical injections into a coproduct of* $\{A_i\}_{i\in I}$ *in* $\mathcal{A}$*.*

(2) *If products exist in* $\mathcal{A}$ *then products also exist in* $\mathrm{Ho}(\mathfrak{M})$*. In fact, for a collection of objects* $\{A_i\}_{i\in I}$ *we may compute their product as follows: Let* $\prod_{i\in I} A_i \xrightarrow{\pi_i} A_i$ *denote the canonical projections out of their product taken in* $\mathcal{A}$*. Then*

$$\left(\prod_{i\in I} A_i, \{\gamma(\pi_i) = Q[\pi_i]\}_{i\in I}\right)$$

is a product of $\{A_i\}_{i\in I}$*, in* $\mathrm{Ho}(\mathfrak{M})$*.*

Proof For a projective or injective model structure, we have that the RQ-representation of the homotopy category, $\mathrm{Ho}_{RQ}(\mathfrak{M})$, coincides exactly with the QR-representation, $\mathrm{Ho}_{QR}(\mathfrak{M})$. Moreover, the isomorphism

$$\mathrm{Ho}_{RQ}(\tau)\colon \mathrm{Ho}_{RQ}(\mathfrak{M}) \to \mathrm{Ho}_{QR}(\mathfrak{M})$$

of Corollary 5.24 is exactly the identity functor. This follows from what was shown in Corollary 5.8 and Theorem 5.6(2). So then the statements are deduced from Propositions 5.28 and 5.30. □

Of course everything can be dualized for injective cotorsion pairs $\mathfrak{M} = (\mathcal{W}, \mathcal{R})$. The reader can easily formulate the dual of Corollary 8.8. For easy reference we state the dual of Corollary 8.9.

Corollary 8.10 (Coproducts and Products for Injective Model Structures) *Assume $(\mathcal{A}, \mathcal{E})$ has enough injectives and let $\mathfrak{M} = (\mathcal{W}, \mathcal{R})$ be an injective cotorsion pair. Coproducts and products exist in* $\mathrm{Ho}(\mathfrak{M})$ *as follows.*

(1) *If products exist in $\mathcal{A}$ then products also exist in* $\mathrm{Ho}(\mathfrak{M})$. *In fact, the product of a collection of objects $\{A_i\}_{i\in I}$ in* $\mathrm{Ho}(\mathfrak{M})$ *may be obtained by taking the product of $\{RA_i\}_{i\in I}$ in* $\mathrm{St}(\mathcal{R})$. *More precisely, the product is*

$$\left(\prod_{i\in I} RA_i, \{[\pi_{RA_i}]\}_{i\in I}\right),$$

where $\prod_{i\in I} RA_i \xrightarrow{\pi_{RA_i}} RA_i$ is the canonical projection out of the product $\prod_{i\in I} RA_i$ taken in $\mathcal{A}$.

Moreover, assume $\mathcal{W}$ and the class of all short exact sequences are each closed under products. Then we have $R\left(\prod_{i\in I} A_i\right) = \prod_{i\in I} RA_i$ and

$$\left(\prod_{i\in I} A_i, \{\gamma(\eta_i) = R[\eta_i]\}_{i\in I}\right)$$

is a product of $\{A_i\}_{i\in I}$ in $\mathrm{Ho}(\mathfrak{M})$, *whenever $\prod_{i\in I} A_i \xrightarrow{\pi_i} A_i$ denotes the canonical projections out of a product of $\{A_i\}_{i\in I}$ in $\mathcal{A}$.*

(2) *If coproducts exist in $\mathcal{A}$ then coproducts also exist in* $\mathrm{Ho}(\mathfrak{M})$. *In fact, for a collection of objects $\{A_i\}_{i\in I}$ we may compute their coproduct as follows: Let $A_i \xrightarrow{\eta_i} \bigoplus_{i\in I} A_i$ denote the canonical injections into their coproduct taken in $\mathcal{A}$. Then*

$$\left(\bigoplus_{i\in I} A_i, \{\gamma(\eta_i) = R[\eta_i]\}_{i\in I}\right)$$

is a coproduct of $\{A_i\}_{i\in I}$, in $\mathrm{Ho}(\mathfrak{M})$.

8.4 How to Construct a Hovey Triple from Two Cotorsion Pairs

The goal of this section is to prove Theorem 8.16, a very useful tool for constructing hereditary abelian model structures. It constructs a Hovey triple from two complete hereditary cotorsion pairs, $(\mathcal{Q}, \widetilde{\mathcal{R}})$ and $(\widetilde{\mathcal{Q}}, \mathcal{R})$, satisfying a couple of simple conditions.

Assume throughout this section that $(\mathcal{Q}, \widetilde{\mathcal{R}})$ and $(\widetilde{\mathcal{Q}}, \mathcal{R})$ are two complete cotorsion pairs. Until explicitly stated otherwise later, these cotorsion pairs need not be hereditary. We also allow for $(\mathcal{A}, \mathcal{E})$ to be any exact category in this section. (Weak idempotent completeness is not needed to construct the Hovey triple, only for the Hovey triple to be equivalent to an abelian model structure.) We want to know what precisely is needed in order for $(\mathcal{Q}, \widetilde{\mathcal{R}})$ and $(\widetilde{\mathcal{Q}}, \mathcal{R})$ to be the left and right cotorsion pairs of a Hovey triple $(\mathcal{Q}, \mathcal{W}, \mathcal{R})$. Certainly, we will then have $\widetilde{\mathcal{Q}} = \mathcal{Q} \cap \mathcal{W}$ and $\widetilde{\mathcal{R}} = \mathcal{W} \cap \mathcal{R}$. So we must require $\widetilde{\mathcal{Q}} \subseteq \mathcal{Q}$, or equivalently, $\widetilde{\mathcal{R}} \subseteq \mathcal{R}$. This can be stated more succinctly, and symmetrically, by simply stating that $(\widetilde{\mathcal{Q}}, \widetilde{\mathcal{R}})$ is an Ext-pair. That is, $\mathrm{Ext}^1_{\mathcal{E}}(\widetilde{\mathcal{Q}}, \widetilde{\mathcal{R}}) = 0$; see Lemma 2.1.

Since $\widetilde{\mathcal{Q}}$ will turn out to be the class of trivially cofibrant objects, and $\widetilde{\mathcal{R}}$ the class of trivially fibrant objects, it follows from Lemma 4.2 that the class $\mathcal{W}$, of all trivial objects, must coincide with the classes $\mathcal{W}^f$ and $\mathcal{W}^i$ defined as follows.

Definition 8.11 Let $(\mathcal{Q}, \widetilde{\mathcal{R}})$ and $(\widetilde{\mathcal{Q}}, \mathcal{R})$ be two complete cotorsion pairs such that $(\widetilde{\mathcal{Q}}, \widetilde{\mathcal{R}})$ is an Ext-pair. We define the following classes:

$$\mathcal{W}^f := \left\{X \in \mathcal{A} \mid \exists \text{ a short exact sequence } X \rightarrowtail \tilde{R} \twoheadrightarrow \tilde{Q} \text{ with } \tilde{R} \in \widetilde{\mathcal{R}}, \tilde{Q} \in \widetilde{\mathcal{Q}}\right\},$$
$$\mathcal{W}^i := \left\{X \in \mathcal{A} \mid \exists \text{ a short exact sequence } \tilde{R} \rightarrowtail \tilde{Q} \twoheadrightarrow X \text{ with } \tilde{R} \in \widetilde{\mathcal{R}}, \tilde{Q} \in \widetilde{\mathcal{Q}}\right\}.$$

Note that the statement $\mathcal{W}^f = \mathcal{W}^i$ is equivalent to the assertion that $0 \rightarrowtail X$ is an $\mathfrak{M}$-weak equivalence if and only if $X \twoheadrightarrow 0$ is an $\mathfrak{M}$-weak equivalence for the desired Hovey triple $\mathfrak{M} = (\mathcal{Q}, \mathcal{W}, \mathcal{R})$. Our goal now is to prove Theorem 8.15, which says that we do indeed have such a triple whenever $\mathcal{W}^f = \mathcal{W}^i$. The proof will build off of the following series of lemmas.

Lemma 8.12 $\mathcal{W}^f = \mathcal{W}^i \implies \mathcal{Q} \cap \widetilde{\mathcal{R}} = \widetilde{\mathcal{Q}} \cap \mathcal{R} \implies$ *Both* $\mathcal{W}^i$ *and* $\mathcal{W}^f$ *each contain both* $\widetilde{\mathcal{Q}}$ *and* $\widetilde{\mathcal{R}}$.

Exercise 8.4.1 Prove Lemma 8.12

Lemma 8.13 *Assume* $\mathcal{Q} \cap \widetilde{\mathcal{R}} = \widetilde{\mathcal{Q}} \cap \mathcal{R}$. *If* $R \rightarrowtail Y \twoheadrightarrow W$ *is exact with* $R \in \widetilde{\mathcal{R}}$ *and* $W \in \mathcal{W}^i$, *then* $Y \in \mathcal{W}^i$ *too. Dually, if* $W \rightarrowtail Y \twoheadrightarrow Q$ *is exact with* $Q \in \widetilde{\mathcal{Q}}$ *and* $W \in \mathcal{W}^f$, *then* $Y \in \mathcal{W}^f$ *too.*

Proof By Lemma 8.12, $R \in \mathcal{W}^i$. So if $W \in \mathcal{W}^i$ too, there are short exact sequences $\widetilde{R} \rightarrowtail \widetilde{Q} \twoheadrightarrow R$ and $\widetilde{R}' \rightarrowtail \widetilde{Q}' \twoheadrightarrow W$ with $\widetilde{R}, \widetilde{R}' \in \widetilde{\mathcal{R}}$ and $\widetilde{Q}, \widetilde{Q}' \in \widetilde{\mathcal{Q}}$. Since we have $R \in \widetilde{\mathcal{R}} \subseteq \mathcal{R}$ we have $\mathrm{Ext}^1_{\mathcal{E}}(\widetilde{Q}', R) = 0$. This means there exists a lift as shown:

$$
\begin{array}{ccccc}
 & & & & \widetilde{Q}' \\
 & & & \swarrow & \downarrow \\
R & \rightarrowtail & Y & \twoheadrightarrow & W.
\end{array}
$$

This lift allows for the construction, analogous to the usual Horseshoe Lemma in homological algebra, of a commutative diagram as follows. The bottom-middle vertical arrow is automatically an admissible epic by the Five Lemma, see Exercise 1.4.2; the top row is then automatically exact by the (3×3)-Lemma 1.16:

$$
\begin{array}{ccccc}
\widetilde{R} & \rightarrowtail & \widetilde{R}'' & \twoheadrightarrow & \widetilde{R}' \\
\downarrow & & \downarrow & & \downarrow \\
\widetilde{Q} & \rightarrowtail & \widetilde{Q} \bigoplus \widetilde{Q}' & \twoheadrightarrow & \widetilde{Q}' \\
\downarrow & & \downarrow & & \downarrow \\
R & \rightarrowtail & Y & \twoheadrightarrow & W.
\end{array}
$$

Since any class that is part of a cotorsion pair is closed under extensions we now have $\widetilde{Q} \bigoplus \widetilde{Q}' \in \widetilde{\mathcal{Q}}$ and $\widetilde{R}'' \in \widetilde{\mathcal{R}}$, and so we have proved $Y \in \mathcal{W}^i$. □

Lemma 8.14 *If $R \rightarrowtail W \twoheadrightarrow B$ is exact with $R \in \widetilde{\mathcal{R}}$ and $W \in \mathcal{W}^i$, then $B \in \mathcal{W}^i$ too. Dually, if $B \rightarrowtail W \twoheadrightarrow Q$ is exact with $Q \in \widetilde{\mathcal{Q}}$ and $W \in \mathcal{W}^f$, then $B \in \mathcal{W}^f$ too.*

Proof Assuming $W \in \mathcal{W}^i$, write a short exact sequence $\tilde{R} \rightarrowtail \tilde{Q} \twoheadrightarrow W$ where $\tilde{R} \in \widetilde{\mathcal{R}}$ and $\tilde{Q} \in \widetilde{\mathcal{Q}}$. We take a pullback and from Corollary 1.14(1) we get the following commutative diagram with exact rows and columns and whose lower left corner is a bicartesian (pushpull) square:

$$
\begin{array}{ccccc}
\tilde{R} & = & \tilde{R} & & \\
\downarrow & & \downarrow & & \\
P & \rightarrowtail & \tilde{Q} & \twoheadrightarrow & B \\
\downarrow & & \downarrow & & \| \\
R & \rightarrowtail & W & \twoheadrightarrow & B.
\end{array}
$$

Since $\widetilde{\mathcal{R}}$ is closed under extensions we are done. □

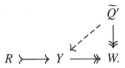

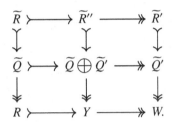

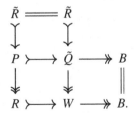

Theorem 8.15 *Let $(\mathcal{Q}, \widetilde{\mathcal{R}})$ and $(\widetilde{\mathcal{Q}}, \mathcal{R})$ be two complete cotorsion pairs such that $(\widetilde{\mathcal{Q}}, \widetilde{\mathcal{R}})$ is an Ext-pair, and $\mathcal{W}^f = \mathcal{W}^i$. Then $(\mathcal{Q}, \mathcal{W}, \mathcal{R})$ is a Hovey triple,*

where $\mathcal{W} := \mathcal{W}^f = \mathcal{W}^i$. *In particular, we have an abelian model structure* $\mathfrak{M} = (\mathcal{Q}, \mathcal{W}, \mathcal{R})$ *whenever* $\mathcal{A}$ *is weakly idempotent complete.*

Proof We must show that $\mathcal{W}$ satisfies the 2 out of 3 property on short exact sequences and that $\widetilde{\mathcal{Q}} = \mathcal{Q} \cap \mathcal{W}$ and $\widetilde{\mathcal{R}} = \mathcal{W} \cap \mathcal{R}$. (That $\mathcal{W}$ is closed under direct summands is then automatic, by Proposition 4.4.)

We first prove $\widetilde{\mathcal{Q}} = \mathcal{Q} \cap \mathcal{W}$ and $\widetilde{\mathcal{R}} = \mathcal{W} \cap \mathcal{R}$. Since the proof of each is similar, we will only show $\mathcal{W} \cap \mathcal{R} = \widetilde{\mathcal{R}}$. Here, we clearly have $\mathcal{W} \cap \mathcal{R} \supseteq \widetilde{\mathcal{R}}$, so we just need to show $\mathcal{W} \cap \mathcal{R} \subseteq \widetilde{\mathcal{R}}$. Let $W \in \mathcal{W} \cap \mathcal{R}$. Since $W \in \mathcal{W}$, we may write a short exact sequence $W \rightarrowtail R \twoheadrightarrow Q$ where $R \in \widetilde{\mathcal{R}}$ and $Q \in \widetilde{\mathcal{Q}}$. It is now rather immediate that $W \in \widetilde{\mathcal{R}}$, for it must be a direct summand of R.

(*$\mathcal{W}$ is Closed under Extensions*) Now suppose $W \rightarrowtail B \twoheadrightarrow W'$ is exact with $W, W' \in \mathcal{W}$. We need to prove that $B \in \mathcal{W}$ too. Since $W \in \mathcal{W}$, we may write a short exact sequence $W \rightarrowtail R \twoheadrightarrow Q$ where $R \in \widetilde{\mathcal{R}}$ and $Q \in \widetilde{\mathcal{Q}}$. Now form the pushout diagram shown:

$$\begin{array}{ccccc} W & \rightarrowtail & B & \twoheadrightarrow & W' \\ \downarrow & & \downarrow & & \| \\ R & \rightarrowtail & P & \twoheadrightarrow & W' \\ \downarrow & & \downarrow & & \\ Q & = & Q. & & \end{array}$$

Note the second row is the type of sequence from Lemma 8.13. Hence $P \in \mathcal{W}$. This makes the second column the type of sequence from Lemma 8.14 (dual). Hence $B \in \mathcal{W}$, proving $\mathcal{W}$ is closed under extensions.

(*$\mathcal{W}$ Has 2 out of 3 Property*) Now suppose $A \rightarrowtail B \twoheadrightarrow C$ is a short exact sequence. To finish proving the 2 out of 3 property we must show that $A, B \in \mathcal{W}$ implies $C \in \mathcal{W}$, and $B, C \in \mathcal{W}$ implies $A \in \mathcal{W}$. The two are similar, so we will only prove the first statement. So suppose $A, B \in \mathcal{W}$. We again may write a short exact sequence $A \rightarrowtail R \twoheadrightarrow Q$ where $R \in \widetilde{\mathcal{R}}$ and $Q \in \widetilde{\mathcal{Q}}$. We again form the pushout diagram shown:

$$\begin{array}{ccccc} A & \rightarrowtail & B & \twoheadrightarrow & C \\ \downarrow & & \downarrow & & \| \\ R & \rightarrowtail & P & \twoheadrightarrow & C \\ \downarrow & & \downarrow & & \\ Q & = & Q. & & \end{array}$$

By Lemma 8.13 (dual), we get $P \in \mathcal{W}$. The second row is now seen to be of the type of sequence from Lemma 8.14. Hence $C \in \mathcal{W}$. □

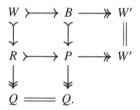

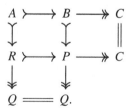

The following exercise shows how Theorem 8.15 generalizes to Ext-pairs.

Exercise 8.4.2 Assume $(\mathcal{Q}, \widetilde{\mathcal{R}})$ and $(\widetilde{\mathcal{Q}}, \mathcal{R})$ are each complete Ext-pairs with $\widetilde{\mathcal{Q}} \subseteq \mathcal{Q}$ and $\widetilde{\mathcal{R}} \subseteq \mathcal{R}$ and such that $\mathcal{W}^f = \mathcal{W}^i$. Generalize the proof of Theorem 8.15 to show that we still get a Hovey triple $(\mathcal{Q}, \mathcal{W}, \mathcal{R})$, not just a Hovey Ext-triple, where $\mathcal{W} := \mathcal{W}^f = \mathcal{W}^i$. Show that the class $\mathcal{R}_{\mathcal{W}} := \mathcal{W} \cap \mathcal{R}$, of all trivially fibrant objects, is precisely the class of all direct summands of objects in $\widetilde{\mathcal{R}}$. Similarly the class $\mathcal{Q}_{\mathcal{W}} := \mathcal{Q} \cap \mathcal{W}$, of all trivially cofibrant objects, is precisely the class of all direct summands of objects in $\widetilde{\mathcal{Q}}$.

The next theorem shows that when the two cotorsion pairs of Theorem 8.15 are hereditary, the condition $\mathcal{W}^f = \mathcal{W}^i$ is equivalent to the much more useful condition that $\widetilde{\mathcal{Q}} \cap \mathcal{R} = \mathcal{Q} \cap \widetilde{\mathcal{R}}$. That is, that the two cotorsion pairs have the same *core*.

Theorem 8.16 (How to Construct a Hovey Triple from Two Cotorsion Pairs) *Let $(\mathcal{Q}, \widetilde{\mathcal{R}})$ and $(\widetilde{\mathcal{Q}}, \mathcal{R})$ be two complete hereditary cotorsion pairs satisfying the following conditions.*

(1) $(\widetilde{\mathcal{Q}}, \widetilde{\mathcal{R}})$ *is an Ext-pair. That is,* $\mathrm{Ext}^1_{\mathcal{E}}(\widetilde{\mathcal{Q}}, \widetilde{\mathcal{R}}) = 0$.

(2) $\widetilde{\mathcal{Q}} \cap \mathcal{R} = \mathcal{Q} \cap \widetilde{\mathcal{R}}$. *That is, the two cotorsion pairs share the same core.*

Then there is a Hovey triple $(\mathcal{Q}, \mathcal{W}, \mathcal{R})$, where $\mathcal{W} := \mathcal{W}^f = \mathcal{W}^i$. In particular, we have an hereditary abelian model structure $\mathfrak{M} = (\mathcal{Q}, \mathcal{W}, \mathcal{R})$ whenever $\mathcal{A}$ is weakly idempotent complete.

Proof By Theorem 8.15 we only need to show $\mathcal{W}^f = \mathcal{W}^i$. So say $X \in \mathcal{W}^f$, that is, there exists a short exact sequence $X \rightarrowtail \tilde{R} \twoheadrightarrow \tilde{Q}$ where $\tilde{R} \in \widetilde{\mathcal{R}}$ and $\tilde{Q} \in \widetilde{\mathcal{Q}}$. Since $(\mathcal{Q}, \widetilde{\mathcal{R}})$ has enough projectives we can find a short exact sequence $R' \rightarrowtail Q' \twoheadrightarrow \tilde{R}$ where $R' \in \widetilde{\mathcal{R}}$ and $Q' \in \mathcal{Q}$. We take a pullback and from Corollary 1.14(1) we get the following commutative diagram with exact rows and columns and whose lower left corner is a bicartesian (pushpull) square:

$$
\begin{array}{ccccc}
R' & = & R' & & \\
\downarrow & & \downarrow & & \\
P & \rightarrowtail & Q' & \twoheadrightarrow & \tilde{Q} \\
\downarrow & & \downarrow & & \| \\
X & \rightarrowtail & \tilde{R} & \twoheadrightarrow & \tilde{Q}.
\end{array}
$$

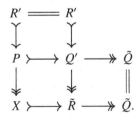

Since $\widetilde{\mathcal{R}}$ is closed under extensions, by the assumption (2) $\widetilde{\mathcal{Q}} \cap \mathcal{R} = \mathcal{Q} \cap \widetilde{\mathcal{R}}$, we deduce that $Q' \in \widetilde{\mathcal{Q}} \cap \mathcal{R}$. Now since $Q', \tilde{Q} \in \widetilde{\mathcal{Q}}$ and the cotorsion pairs are hereditary we conclude $P \in \widetilde{Q}$. Now the left vertical column shows that $X \in \mathcal{W}^i$, proving $\mathcal{W}^f \subseteq \mathcal{W}^i$. A similar argument shows $\mathcal{W}^i \subseteq \mathcal{W}^f$. □

Exercise 8.4.3 Let $\left(Q, \widetilde{\mathcal{R}}\right)$ and $\left(\widetilde{Q}, \mathcal{R}\right)$ be two complete hereditary cotorsion pairs. Show that the following are equivalent.

(1) There is an (hereditary) Hovey triple $\mathfrak{M} = (Q, \mathcal{W}, \mathcal{R})$.
(2) $\widetilde{\mathcal{R}} \subseteq \mathcal{R}$, and $\widetilde{\mathcal{R}}$ is projectively resolving in $\mathcal{R}$ when viewed as an exact category along with its inherited exact structure.
(3) $\widetilde{Q} \subseteq Q$, and $\widetilde{Q}$ is injectively coresolving in Q when viewed as an exact category along with its inherited exact structure.
(4) $\left(\widetilde{Q}, \widetilde{\mathcal{R}}\right)$ is an Ext-pair and $\widetilde{Q} \cap \mathcal{R} = Q \cap \widetilde{\mathcal{R}}$.

8.5 The Ext Functors for Hereditary Model Structures

Throughout this section we let $\mathfrak{M} = (Q, \mathcal{W}, \mathcal{R})$ be an hereditary abelian model structure. The goal here is to relate morphism sets in the homotopy category, $\mathrm{Ho}(\mathfrak{M})$, to the classical Yoneda Ext groups. The main result is Theorem 8.18. It will be used in the next section to characterize weak generators for $\mathrm{Ho}(\mathfrak{M})$.

We recall from Section 4.3 that $\mathrm{St}_\ell(\mathcal{A})$, the left stable category of $\mathfrak{M}$, is the stable category obtained by identifying two maps whose difference factors through an object of $\mathcal{R}_\mathcal{W}$. On the other hand, the right stable category, $\mathrm{St}_r(\mathcal{A})$, identifies maps whose difference factors through an object of $Q_\mathcal{W}$

Proposition 8.17 *Let $\mathfrak{M} = (Q, \mathcal{W}, \mathcal{R})$ be an hereditary abelian model structure.*

(1) *For each cofibrant $C \in Q$ and $n \geq 1$, the covariant Yoneda Ext functor $\mathrm{Ext}^n_\mathcal{E}(C, -) \colon \mathcal{A} \to \boldsymbol{Ab}$ descends to an additive functor $\mathrm{Ext}^n_\mathcal{E}(C, -) \colon \mathrm{St}_\ell(\mathcal{A}) \to \boldsymbol{Ab}$ by factoring through $\gamma_\ell \colon \mathcal{A} \to \mathrm{St}_\ell(\mathcal{A})$.*
(2) *For each fibrant $F \in \mathcal{R}$ and $n \geq 1$, the contravariant Yoneda Ext functor $\mathrm{Ext}^n_\mathcal{E}(-, F) \colon \mathcal{A} \to \boldsymbol{Ab}$ descends to an additive functor $\mathrm{Ext}^n_\mathcal{E}(-, F) \colon \mathrm{St}_r(\mathcal{A}) \to \boldsymbol{Ab}$ by factoring through $\gamma_r \colon \mathcal{A} \to \mathrm{St}_r(\mathcal{A})$.*

Proof We will only prove (1), as statement (2) is dual. We need to show that $\mathrm{Ext}^n_\mathcal{E}(C, f) = \mathrm{Ext}^n_\mathcal{E}(C, g)$ whenever $[f]_\ell = [g]_\ell$. That is, we shall show that the two group homomorphisms $\mathrm{Ext}^n_\mathcal{E}(C, A) \to \mathrm{Ext}^n_\mathcal{E}(C, B)$, induced by $f, g \colon A \to B$, are equal whenever their difference $g - f$ factors through an object of $\mathcal{R}_\mathcal{W} := \mathcal{W} \cap \mathcal{R}$.

Recall the Yoneda description of $\mathrm{Ext}^n_{\mathcal{E}}(C,-)$ from Section 1.6. For a given object A, the elements of $\mathrm{Ext}^n_{\mathcal{E}}(C,A)$ are equivalence classes of exact n-sequences of the form

$$\mathbb{E}\colon\ A \rightarrowtail Y_n \to \cdots \to Y_2 \to Y_1 \twoheadrightarrow C.$$

The equivalence relation is *generated* by a relation $\sim$, where $\mathbb{E}' \sim \mathbb{E}$ means there exists some commutative diagram of the following form:

$$\begin{array}{lccccccccccc}
\mathbb{E}': & A & \rightarrowtail & Y'_n & \longrightarrow & Y'_{n-1} & \longrightarrow & \cdots & \longrightarrow & Y'_1 & \twoheadrightarrow & C \\
 & \| & & \downarrow & & \downarrow & & & & \downarrow & & \| \\
\mathbb{E}: & A & \rightarrowtail & Y_n & \longrightarrow & Y_{n-1} & \longrightarrow & \cdots & \longrightarrow & Y_1 & \twoheadrightarrow & C.
\end{array}$$

Since $\mathrm{Ext}^n_{\mathcal{E}}(C,-)\colon \mathcal{A} \to \mathbf{Ab}$ is well known to be an additive functor, we will only show $\mathrm{Ext}^n_{\mathcal{E}}(C,f) = 0$ whenever $f\colon A \to B$ factors through an object of $\mathcal{R}_{\mathcal{W}}$. Our proof will rely on the fact that we can replace any representative

$$\mathbb{E}\colon\ A \xrightarrow{k} Y_n \xrightarrow{f_n} Y_{n-1} \xrightarrow{f_{n-1}} \cdots \xrightarrow{f_3} Y_2 \xrightarrow{f_2} Y_1 \xrightarrow{f_1} C$$

of $\mathrm{Ext}^n_{\mathcal{E}}(C,A)$, with an equivalent n-sequence

$$\mathbb{E}'\colon\ A \xrightarrow{k'} P_n \xrightarrow{t_n} Q_{n-1} \xrightarrow{t_{n-1}} \cdots \xrightarrow{t_3} Q_2 \xrightarrow{t_2} Q_1 \xrightarrow{t_1} C$$

such that for each $i = 1,2,\ldots,n-1$, the object $L_i := \mathrm{Ker}\, t_i \in \mathcal{Q}$. So first we prove this.

We start on the right end of $\mathbb{E}$, considering the exact 2-sequence

$$\mathrm{Ker}\, f_2 \rightarrowtail Y_2 \xrightarrow{f_2} Y_1 \xrightarrow{f_1} C.$$

Since $(\mathcal{Q},\mathcal{R}_{\mathcal{W}})$ has enough projectives, there is an epimorphism $p\colon Q_1 \twoheadrightarrow Y_1$ with $Q_1 \in \mathcal{Q}$. Using Corollary 1.14(1), one constructs the pullback diagram:

$$\begin{array}{ccccc}
\mathrm{Ker}\, t_1 & \rightarrowtail & Q_1 & \xrightarrow{t_1} & C \\
\downarrow & & \downarrow p & & \| \\
\mathrm{Ker}\, f_1 & \rightarrowtail & Y_1 & \xrightarrow{f_1} & C.
\end{array}$$

Then we take another pullback:

$$\begin{array}{ccccc}
\mathrm{Ker}\, f_2 & \rightarrowtail & P_2 & \twoheadrightarrow & \mathrm{Ker}\, t_1 \\
\| & & \downarrow p' & & \downarrow \\
\mathrm{Ker}\, f_2 & \rightarrowtail & Y_2 & \twoheadrightarrow & \mathrm{Ker}\, f_1.
\end{array}$$

Composing the two pullback diagrams results in a morphism of exact 2-sequences:

$$\begin{array}{ccccccc}
\operatorname{Ker} f_2 & \rightarrowtail & P_2 & \xrightarrow{f_2'} & Q_1 & \xrightarrow{t_1} & C \\
\| & & \downarrow{\scriptstyle p'} & & \downarrow{\scriptstyle p} & & \| \\
\operatorname{Ker} f_2 & \rightarrowtail & Y_2 & \xrightarrow{f_2} & Y_1 & \xrightarrow{f_1} & C.
\end{array}$$

But then $L_1 := \operatorname{Ker} t_1$ is also cofibrant because the model structure is hereditary. This morphism of exact 2-sequences can be extended to a morphism of exact n-sequences, by "pasting" the top row, at $\operatorname{Ker} f_2$, together with the rest of $\mathbb{E}$, yielding the (Yoneda equivalent) exact n-sequence shown:

$$A \overset{k}{\rightarrowtail} Y_n \xrightarrow{f_n} Y_{n-1} \xrightarrow{f_{n-1}} \cdots \xrightarrow{f_4} Y_3$$

$$\xrightarrow{f_3'} P_2 \xrightarrow{f_2'} Q_1 \xrightarrow{t_1} C.$$

Next, we focus on the portion of this new n-sequence shown here:

$$\operatorname{Ker} f_3' \rightarrowtail Y_3 \xrightarrow{f_3'} P_2 \xrightarrow{f_2'} L_1.$$

Recalling that L_1 is cofibrant, we repeat the same procedure, obtaining another Yoneda equivalent exact n-sequence

$$A \overset{k}{\rightarrowtail} Y_n \xrightarrow{f_n} Y_{n-1} \xrightarrow{f_{n-1}} \cdots \xrightarrow{f_5} Y_4$$

$$\xrightarrow{f_4'} P_3 \xrightarrow{f_3''} Q_2 \xrightarrow{t_2} Q_1 \xrightarrow{t_1} C$$

and with this one also having $L_2 := \operatorname{Ker} t_2$ cofibrant.

In this way, we can continue the process, from right to left, finally obtaining the desired (Yoneda equivalent) exact n-sequence

$$\mathbb{E}'\colon\ A \overset{k'}{\rightarrowtail} P_n \xrightarrow{f_n''} Q_{n-1} \xrightarrow{t_{n-1}} Q_{n-2} \xrightarrow{t_{n-2}} \cdots$$

$$\cdots \xrightarrow{t_3} Q_2 \xrightarrow{t_2} Q_1 \xrightarrow{t_1} C$$

having $L_i := \operatorname{Ker} t_i$ cofibrant for each $i = 1, 2, \ldots, n-1$.

So now, finally, we are able to show $\operatorname{Ext}^n_{\mathcal{E}}(C, f) = 0$ whenever $f\colon A \to B$ factors as $f = \beta\alpha$ through an object $W \in \mathcal{R}_W$. In this case, since $\operatorname{Ext}^1_{\mathcal{E}}(L_{n-1}, W) = 0$, the morphism $A \overset{\alpha}{\to} W$ extends over $A \overset{k'}{\to} P_n$. The Homotopy Lemma 1.21

now applies and implies that the pushed-out n-sequence in the bottom row of the following diagram represents 0 in $\mathrm{Ext}^n_{\mathcal{E}}(C, B)$:

$$\begin{array}{ccccccccc} A & \stackrel{k'}{\rightarrowtail} & P_n & \xrightarrow{t_n} & Q_{n-1} & \xrightarrow{t_{n-1}} & \cdots & \stackrel{t_1}{\twoheadrightarrow} & C \\ {\scriptstyle f}\downarrow & & \downarrow & & \| & & & & \| \\ B & \rightarrowtail & P & \longrightarrow & Q_{n-1} & \xrightarrow{t_{n-1}} & \cdots & \stackrel{t_1}{\twoheadrightarrow} & C. \end{array}$$

□

Exercise 8.5.1 Show that for a given cofibrant C, the Ext groups $\mathrm{Ext}^n_{\mathcal{E}}(C, RB)$ are well defined, up to canonical isomorphism, regardless of the choice of fibrant replacement $B \rightarrowtail RB$ used. Similarly, for a given fibrant F, the Ext groups $\mathrm{Ext}^n_{\mathcal{E}}(QA, F)$ are well defined with respect to cofibrant replacements $QA \twoheadrightarrow A$.

Theorem 8.18 *Let $\mathfrak{M} = (\mathcal{Q}, \mathcal{W}, \mathcal{R})$ be an hereditary abelian model structure on $(\mathcal{A}, \mathcal{E})$. Then for each $n \geq 1$ there are natural isomorphisms*

$$\mathrm{Ho}(\mathfrak{M})(\Omega^n A, B) \cong \mathrm{Ext}^n_{\mathcal{E}}(QA, RB) \cong \mathrm{Ho}(\mathfrak{M})(A, \Sigma^n B).$$

Note that the previous isomorphisms may be viewed as extensions of the $n = 0$ isomorphism

$$\mathrm{Ho}(\mathfrak{M})(A, B) \cong \underline{\mathrm{Hom}}(QA, RB).$$

Exercise 8.5.2 expands upon this perspective.

Proof We will first show $\mathrm{Ho}(\mathfrak{M})(\Omega A, B) \cong \mathrm{Ext}^1_{\mathcal{E}}(QA, RB)$. As shown in Proposition 6.29, given an object A we may compute ΩA by taking a short exact sequence $\Omega A \rightarrowtail W_A \twoheadrightarrow A$ with $W_A \in \mathcal{W}$. (A potentially different ΩA resulting from a different short exact sequence will be isomorphic, in $\mathrm{Ho}(\mathfrak{M})$). From the Generalized Horseshoe Lemma 2.18, we can find a cofibrant replacement sequence as in the top row here:

$$\begin{array}{ccccc} Q\Omega A & \rightarrowtail & QW_A & \twoheadrightarrow & QA \\ \downarrow & & \downarrow & & \downarrow \\ \Omega A & \rightarrowtail & W_A & \twoheadrightarrow & A. \end{array}$$

Since $\mathcal{W}$ is closed under extensions we see that $QW_A \in \mathcal{Q}_{\mathcal{W}}$, the class of trivially cofibrant objects. Now given the other object B, we apply $\mathrm{Hom}_{\mathcal{A}}(-, RB)$ to the top row and it gives us a homomorphism

$$\delta\colon \mathrm{Hom}_{\mathcal{A}}(Q\Omega A, RB) \to \mathrm{Ext}^1_{\mathcal{E}}(QA, RB).$$

In the Yoneda Ext description, δ is defined via pushout, as indicated in Diagram (8.2):

$$\begin{array}{rccccc} & Q\Omega A & \rightarrowtail & QW_A & \twoheadrightarrow & QA \\ & {\scriptstyle f}\downarrow & & \downarrow & & \| \\ \delta(f): & RB & \rightarrowtail & P & \twoheadrightarrow & QA. \end{array} \tag{8.2}$$

We can prove directly that δ is onto. Indeed given any short exact sequence as in the bottom row in the following, we use that $\mathrm{Ext}^1_{\mathcal{E}}(QW_A, RB) = 0$ to construct a morphism of short exact sequences as shown:

$$\begin{array}{ccccc} Q\Omega A & \rightarrowtail & QW_A & \twoheadrightarrow & QA \\ {\scriptstyle t}\downarrow & & \downarrow & & \| \\ RB & \rightarrowtail & D & \twoheadrightarrow & QA. \end{array}$$

By Proposition 1.12(1) the left square is a pushout, proving that the bottom row is $\delta(t)$ as desired. Next, we claim that

$$\operatorname{Ker}\delta = \{ f \in \mathrm{Hom}_{\mathcal{A}}(Q\Omega A, RB) \mid f \sim 0 \}. \tag{8.3}$$

To prove Equation (8.3), suppose $\delta(f) = 0$. It means that there exists a lift $QA \to P$ (or *section*) in the pushout (Diagram (8.2)) making the lower right triangle commute. But by the Homotopy Lemma 1.21, this is equivalent to a morphism $QW_A \to RB$ making the upper left triangle of Diagram (8.2) commute. This proves f factors through an object of $\mathcal{Q}_{\mathcal{W}}$. But since the source object $Q\Omega A$ is cofibrant and the target object RB is fibrant, we conclude by Lemma 4.11 that f actually factors through an object of ω. This proves the containment ($\subseteq$). To prove the reverse containment ($\supseteq$), suppose f factors as $Q\Omega A \xrightarrow{\alpha} W \xrightarrow{\beta} RB$ where $W \in \omega$. Then applying $\mathrm{Hom}_{\mathcal{A}}(-, W)$ to the top row of Diagram (8.2), and using $\mathrm{Ext}^1_{\mathcal{E}}(QA, W) = 0$, we see that α extends through $Q\Omega A \to QW_A$. Composing the new map with β, the Homotopy Lemma now allows us to conclude $\delta(f)$ splits, so $f \in \operatorname{Ker}\delta$. This completes the proof of Equation (8.3).

Thus δ descends to an isomorphism

$$\underline{\mathrm{Hom}}(Q\Omega A, RB) \xrightarrow{\bar{\delta}} \mathrm{Ext}^1_{\mathcal{E}}(QA, RB).$$

The result for $n = 1$ now follows by composing with the natural isomorphism

$$\mathrm{Ho}(\mathfrak{M})(\Omega A, B) \cong \underline{\mathrm{Hom}}(Q\Omega A, RB).$$

For $n > 1$, we may use an inductive dimension shifting argument. For example, from what we just proved we have $\mathrm{Ho}(\mathfrak{M})(\Omega^2 A, B) \cong \mathrm{Ext}^1_{\mathcal{E}}(Q\Omega A, RB)$. But applying $\mathrm{Hom}_{\mathcal{A}}(-, RB)$ to the short exact sequence $Q\Omega A \rightarrowtail QW_A \twoheadrightarrow QA$ we deduce $\mathrm{Ext}^1_{\mathcal{E}}(Q\Omega A, RB) \cong \mathrm{Ext}^2_{\mathcal{E}}(QA, RB)$. Note that this dimension-shifting argument relies on the fact that $(\mathcal{Q}_W, \mathcal{R})$ is an hereditary cotorsion pair because we need $\mathrm{Ext}^i_{\mathcal{E}}(QW_A, RB) = 0$ for all $i \geq 1$.

We have shown $\mathrm{Ho}(\mathfrak{M})(\Omega^n A, B) \cong \mathrm{Ext}^n_{\mathcal{E}}(QA, RB)$ and a dual argument will prove $\mathrm{Ho}(\mathfrak{M})(A, \Sigma^n B) \cong \mathrm{Ext}^n_{\mathcal{E}}(QA, RB)$. Of course this also must be automatic since Σ and Ω are inverse autoequivalences on $\mathrm{Ho}(\mathfrak{M})$. □

Example 8.19 Let R be a ring and let $K(R)$ be the chain homotopy category of all chain complexes of R-modules. As we will make clear in Section 10.4, Theorem 8.18 recovers the well-known statement

$$K(R)(\Sigma^{-n} X, Y) \cong \mathrm{Ext}^n_{dw}(X, Y) \cong K(R)(X, \Sigma^n Y).$$

Here, Σ is the usual suspension of chain complexes obtained by shifting indices, and $\mathrm{Ext}^n_{dw}(X, Y)$ is the Yoneda Ext group of all (equivalence classes of) n-fold *degreewise* split exact sequences of chain complexes.

Exercise 8.5.2 Following up on Proposition 8.17 and Exercise 8.5.1, we may define

$$\underline{\mathrm{Ext}}^n_{\mathrm{Ho}(\mathfrak{M})}(A, B) := \mathrm{Ext}^n_{\mathcal{E}}(RQA, RQB),$$

for $n \geq 1$. (Note the analogy to how we defined the morphism sets in Definition 5.11.) Show that for each $n \geq 1$, $\underline{\mathrm{Ext}}^n_{\mathrm{Ho}(\mathfrak{M})}(-, -)$ defines a functor with the following properties.

(1) $\underline{\mathrm{Ext}}^n_{\mathrm{Ho}(\mathfrak{M})}(A, -) \colon \mathcal{A} \to \mathbf{Ab}$ is a covariant additive functor. (With respect to the choices involved, it is well defined up to a canonical isomorphism.) Short exact sequences in $(\mathcal{A}, \mathcal{E})$ are sent to long exact cohomology sequences of $\underline{\mathrm{Ext}}^n_{\mathrm{Ho}(\mathfrak{M})}$ groups. Moreover, $\underline{\mathrm{Ext}}^n_{\mathrm{Ho}(\mathfrak{M})}(A, -)$ descends via $\gamma \colon \mathcal{A} \to \mathrm{Ho}(\mathfrak{M})$ to a well-defined functor on $\mathrm{Ho}(\mathfrak{M})$.

(2) $\underline{\mathrm{Ext}}^n_{\mathrm{Ho}(\mathfrak{M})}(-, B) \colon \mathcal{A} \to \mathbf{Ab}$ is a contravariant additive functor and short exact sequences in $(\mathcal{A}, \mathcal{E})$ are sent to long exact cohomology sequences of $\underline{\mathrm{Ext}}^n_{\mathrm{Ho}(\mathfrak{M})}$ groups. Moreover, $\underline{\mathrm{Ext}}^n_{\mathrm{Ho}(\mathfrak{M})}(-, B)$ descends via $\gamma \colon \mathcal{A} \to \mathrm{Ho}(\mathfrak{M})$ to a well-defined functor on $\mathrm{Ho}(\mathfrak{M})$.

(3) We have isomorphisms for each $n \geq 1$, natural in both A and B:

$$\underline{\mathrm{Ext}}^n_{\mathrm{Ho}(\mathfrak{M})}(A, B) \cong \mathrm{Ext}^n_{\mathcal{E}}(QA, RB).$$

Consequently, we have natural isomorphisms for each $n \geq 1$:

$$\mathrm{Ho}(\mathfrak{M})(\Omega^n A, B) \cong \underline{\mathrm{Ext}}^n_{\mathrm{Ho}(\mathfrak{M})}(A, B) \cong \mathrm{Ho}(\mathfrak{M})(A, \Sigma^n B).$$

8.6 Weak Generators for Homotopy Categories

We continue to assume throughout this section that $\mathfrak{M} = (\mathcal{Q}, \mathcal{W}, \mathcal{R})$ is hereditary. We wish to make a link between, on one hand, the notion of weak generators for the homotopy category $\mathrm{Ho}(\mathfrak{M})$, and on the other hand, the notion of cogenerating sets for the cotorsion pairs associated to $\mathfrak{M} = (\mathcal{Q}, \mathcal{W}, \mathcal{R})$. For the latter, recall that a set $\mathcal{S}$ of objects *cogenerates* a cotorsion pair $(\mathcal{X}, \mathcal{Y})$ in $(\mathcal{A}, \mathcal{E})$ if $\mathcal{S}^{\perp} = \mathcal{Y}$. That is, $Y \in \mathcal{Y}$ if and only if $\mathrm{Ext}^1_{\mathcal{E}}(S, Y) = 0$ for all $S \in \mathcal{S}$. On the other hand, we know that $\mathrm{Ho}(\mathfrak{M})$ is a triangulated category, and we have the following general notion.

Definition 8.20 Let $\mathcal{T}$ be a triangulated category and $\mathcal{S}$ a set of objects. We say that $\mathcal{S}$ is a set of **weak generators** for $\mathcal{T}$ if $X = 0$ in $\mathcal{T}$ if and only if $\mathcal{T}(\Sigma^n S, X) = 0$ for all $n \in \mathbb{Z}$ and $S \in \mathcal{S}$.

Theorem 8.21 *If a set S cogenerates $(\mathcal{Q}, \mathcal{R}_{\mathcal{W}})$ then $\mathcal{S}$ also serves as a set of weak generators for* $\mathrm{Ho}(\mathfrak{M})$. *More generally, let $\mathcal{S}$ be any set of objects in $\mathcal{A}$ and suppose we have a set of their cofibrant replacements, $Q(\mathcal{S}) = \{QS\}$, one for each object $S \in \mathcal{S}$. If $Q(\mathcal{S})$ cogenerates $(\mathcal{Q}, \mathcal{R}_{\mathcal{W}})$, then $\mathcal{S}$ is a set of weak generators for* $\mathrm{Ho}(\mathfrak{M})$.

Proof We prove the more general statement. So suppose $Q(\mathcal{S})$ cogenerates the cotorsion pair $(\mathcal{Q}, \mathcal{R}_{\mathcal{W}})$. To show $\mathcal{S}$ is a set of weak generators for $\mathrm{Ho}(\mathfrak{M})$ we assume $\mathrm{Ho}(\mathfrak{M})(\Sigma^n S, X) = 0$ for all $n \in \mathbb{Z}$ and $S \in \mathcal{S}$. We wish to show that $X = 0$ in $\mathrm{Ho}(\mathfrak{M})$, but this just means we wish to show $X \in \mathcal{W}$. Using Theorem 8.18 we have, for each $S \in \mathcal{S}$,

$$0 = \mathrm{Ho}(\mathfrak{M})\left(\Sigma^{-1} S, X\right) = \mathrm{Ho}(\mathfrak{M})(\Omega S, X) \cong \mathrm{Ext}^1_{\mathcal{E}}(QS, RX).$$

Hence $RX \in Q(\mathcal{S})^{\perp} = \mathcal{R}_{\mathcal{W}}$. In particular, $RX \in \mathcal{W}$. But by the 2 out of 3 property on $\mathcal{W}$, this happens if and only if $X \in \mathcal{W}$. So this proves $\mathcal{S}$ is a set of weak generators for $\mathrm{Ho}(\mathfrak{M})$. □

One would like to have a converse to the theorem. We have the following.

Theorem 8.22 *Suppose $\mathcal{U}$ is a set that cogenerates $(\mathcal{Q}_{\mathcal{W}}, \mathcal{R})$ and $\mathcal{S}$ is a set of weak generators for* $\mathrm{Ho}(\mathfrak{M})$. *Then there is a set which cogenerates the other cotorsion pair, $(\mathcal{Q}, \mathcal{R}_{\mathcal{W}})$. The set is $\mathcal{U} \cup Q\Sigma^n(\mathcal{S})$, where $Q\Sigma^n(\mathcal{S}) = \{Q(\Sigma^n S)\}$, and here S ranges through $\mathcal{S}$ and $n \in \mathbb{Z}$.*

Proof Assume $\mathcal{U}^{\perp}$, the right Ext-orthogonal of $\mathcal{U}$, satisfies $\mathcal{U}^{\perp} = \mathcal{R}$, and that $\mathcal{S}$ is a set of weak generators for $\mathrm{Ho}(\mathfrak{M})$. Again, it means $X \in \mathcal{W}$ whenever

$\mathrm{Ho}(\mathfrak{M})(\Sigma^n S, X) = 0$ for all $n \in \mathbb{Z}$ and $S \in \mathcal{S}$. For each $S \in \mathcal{S}$ we can certainly find a two-sided exact resolution

$$\cdots \to W_2 \to W_1 \to W_0 \to W^0 \to W^1 \to \cdots$$

of S, where each $W_i, W^i \in \mathcal{W}$. We construct it so that $S = \mathrm{Ker}\left(W^0 \to W^1\right)$, and we note that all the cycles appearing in the complex represent loops and suspensions of S, by Proposition 6.29. Precisely, $\Sigma^n S = \mathrm{Ker}\left(W^n \to W^{n+1}\right)$ for $n \geq 0$ and $\Sigma^{-n} S := \Omega^n S = \mathrm{Im}\,(W_n \to W_{n-1})$ for $n > 0$.

We then take cofibrant replacements $Q(\Sigma^n S)$ and let $Q\Sigma^n(\mathcal{S})$ denote the set of all such objects $\{Q(\Sigma^n S)\}$ as S ranges through $\mathcal{S}$ and $n \in \mathbb{Z}$. We wish to show $Q\Sigma^n(\mathcal{S}) \cup \mathcal{U}$ cogenerates $(\mathcal{Q}, \mathcal{R}_{\mathcal{W}})$. That is, we wish to show $X \in \mathcal{W} \cap \mathcal{R}$ if and only if $\mathrm{Ext}^1_{\mathcal{E}}(A, X) = 0$ for all $A \in \mathcal{U} \cup Q\Sigma^n(\mathcal{S})$.

To prove the "if" part, we start by letting A range through $\mathcal{U}$ to first conclude that $X \in \mathcal{U}^{\perp} = \mathcal{R}$. So $RX = X$, and then letting A range through $Q\Sigma^n(\mathcal{S})$ and using Theorem 8.18 we get

$$0 = \mathrm{Ext}^1_{\mathcal{E}}\left(Q\left(\Sigma^{n+1} S\right), RX\right) \cong \mathrm{Ho}(\mathfrak{M})\left(\Omega\left(\Sigma^{n+1} S\right), X\right) \cong \mathrm{Ho}(\mathfrak{M})(\Sigma^n S, X)$$

for each $n \in \mathbb{Z}$ and $S \in \mathcal{S}$. We conclude $X \in \mathcal{W}$, as $\mathcal{S}$ is a set of weak generators for $\mathrm{Ho}(\mathfrak{M})$ by hypothesis. Thus $X \in \mathcal{W} \cap \mathcal{R}$, proving the "if" part. The "only if" part can be proved similarly. □

If $(\mathcal{A}, \mathcal{E})$ is an exact category with enough projectives, note that the canonical projective cotorsion pair is vacuously cogenererated by the empty set $\mathcal{U} = \phi$. Thus we get the last sentence of the following corollary.

Corollary 8.23 *Assume $(\mathcal{A}, \mathcal{E})$ has enough projectives and let $\mathfrak{M} = (\mathcal{Q}, \mathcal{W})$ be a projective cotorsion pair. If $\mathcal{S}$ is a set that cogenerates $(\mathcal{Q}, \mathcal{W})$, then $\mathcal{S}$ is a set of weak generators for* $\mathrm{Ho}(\mathfrak{M})$. *More generally, $\mathcal{S}$ is a set of weak generators for* $\mathrm{Ho}(\mathfrak{M})$ *whenever a set of cofibrant replacements, $Q(\mathcal{S})$, cogenerates $(\mathcal{Q}, \mathcal{W})$. Conversely, suppose $\mathcal{S}$ is a set of weak generators for* $\mathrm{Ho}(\mathfrak{M})$. *Then there is a set, precisely $Q\Sigma^n(\mathcal{S})$, which cogenerates the cotorsion pair $(\mathcal{Q}, \mathcal{W})$.*

8.7 Adjunctions and Recollement of Homotopy Categories

Some important work in homological algebra has utilized adjunctions and recollement of triangulated categories to pass nice properties of those categories between one another. Loosely, a recollement is an "attachment" of two triangulated categories described in terms of adjunctions satisfying strong exactness conditions.

Recall that the homotopy category of an abelian model category is always triangulated (Theorem 6.34). Moreover, we saw in Chapter 7 that given a

Quillen adjunction (F, G), the total derived functors $\mathbf{L}F$ and $\mathbf{R}G$ are triangulated and $(\mathbf{L}F, \mathbf{R}G)$ is an adjunction between the homotopy categories. In this section we will describe recollements of triangulated categories by way of abelian model structures and derived Quillen adjunctions. We continue to assume throughout that $(\mathcal{A}, \mathcal{E})$ is a weakly idempotent complete exact category, so Hovey triples are synonymous with abelian model structures.

Definition 8.24 Let $\mathcal{T}' \xrightarrow{F} \mathcal{T} \xrightarrow{G} \mathcal{T}''$ be a sequence of triangulated functors between triangulated categories. We say it is a **localization sequence** when there exists right adjoints F_ρ and G_ρ giving a diagram of functors as follows, with the listed properties:

$$\mathcal{T}' \underset{F_\rho}{\overset{F}{\rightleftarrows}} \mathcal{T} \underset{G_\rho}{\overset{G}{\rightleftarrows}} \mathcal{T}''.$$

(1) The right adjoint F_ρ of F satisfies $F_\rho \circ F \cong \mathrm{id}_{\mathcal{T}'}$.
(2) The right adjoint G_ρ of G satisfies $G \circ G_\rho \cong \mathrm{id}_{\mathcal{T}''}$.
(3) For any object $X \in \mathcal{T}$, we have $GX = 0$ iff $X \cong FX'$ for some $X' \in \mathcal{T}'$.

The notion of a **colocalization sequence** is the dual. That is, there must exist left adjoints F_λ and G_λ with the analogous properties.

Note the similarity in the previous definitions to the notion of a split exact sequence, but for adjunctions. It is true that if $\mathcal{T}' \xrightarrow{F} \mathcal{T} \xrightarrow{G} \mathcal{T}''$ is a localization sequence then $\mathcal{T}'' \xrightarrow{G_\rho} \mathcal{T} \xrightarrow{F_\rho} \mathcal{T}'$ is a colocalization sequence and if $\mathcal{T}' \xrightarrow{F} \mathcal{T} \xrightarrow{G} \mathcal{T}''$ is a colocalization sequence then $\mathcal{T}'' \xrightarrow{G_\lambda} \mathcal{T} \xrightarrow{F_\lambda} \mathcal{T}'$ is a localization sequence. This brings us to the definition of a recollement where the sequence of functors $\mathcal{T}' \xrightarrow{F} \mathcal{T} \xrightarrow{G} \mathcal{T}''$ is both a localization sequence and a colocalization sequence.

Definition 8.25 Let $\mathcal{T}' \xrightarrow{F} \mathcal{T} \xrightarrow{G} \mathcal{T}''$ be a sequence of triangulated functors between triangulated categories. We say $\mathcal{T}' \xrightarrow{F} \mathcal{T} \xrightarrow{G} \mathcal{T}''$ induces a **recollement** if it is both a localization sequence and a colocalization sequence as shown in the picture:

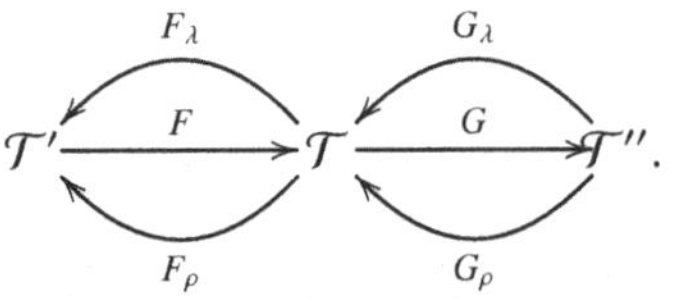

So the idea is that a recollement is a colocalization sequence "glued" with a localization sequence.

Theorems 8.30 and 8.32 automatically produce a recollement from three interrelated Hovey triples. As a first step we prove Proposition 8.29. It constructs (co)localization sequences from two Hovey triples satisfying a simple class containment condition.

Our first lemma is very easy, but it is quite useful in general for obtaining adjunctions between homotopy categories. In what follow, we let $\widetilde{Q}$ and $\widetilde{Q'}$ denote classes of trivially cofibrant objects $\widetilde{Q} := Q \cap \mathcal{W}$ and $\widetilde{Q'} := Q' \cap \mathcal{W}'$.

Lemma 8.26 (Derived Identity Adjunctions) *Assume $(\mathcal{A}, \mathcal{E})$ has two abelian model structures, $\mathfrak{M} = (Q, \mathcal{W}, \mathcal{R})$ and $\mathfrak{M}' = (Q', \mathcal{W}', \mathcal{R}')$. The following conditions are equivalent.*

(1) $Q \subseteq Q'$ *and* $\mathcal{R}' \subseteq \mathcal{R}$.
(2) $Q \subseteq Q'$ *and* $\widetilde{Q} \subseteq \widetilde{Q'}$.
(3) $\mathcal{R}' \subseteq \mathcal{R}$ *and* $\widetilde{\mathcal{R}'} \subseteq \widetilde{\mathcal{R}}$.
(4) *The identity adjunction* $\mathfrak{M} \underset{\text{Id}}{\overset{\text{Id}}{\rightleftarrows}} \mathfrak{M}'$ *is a Quillen adjunction.*

In this case, the identity adjunction descends to an adjunction of triangulated functors

$$\text{Ho}(\mathfrak{M}) \underset{\mathbf{R}(\text{Id})}{\overset{\mathbf{L}(\text{Id})}{\rightleftarrows}} \text{Ho}(\mathfrak{M}')$$

and we may take $\mathbf{L}(\text{Id})$ *to be (abusing notation) the unique functor* $Q\colon \text{Ho}(\mathfrak{M}) \to \text{Ho}(\mathfrak{M}')$ *making the diagram commute:*

$$\begin{array}{ccc} \mathcal{A} & \xrightarrow{Q} & \text{St}_\omega(\mathcal{A}) \\ {\scriptstyle\gamma}\downarrow & & \downarrow{\scriptstyle\gamma'} \\ \text{Ho}(\mathfrak{M}) & \xrightarrow{Q} & \text{Ho}(\mathfrak{M}'). \end{array}$$

Here, $\mathcal{A} \xrightarrow{Q} \text{St}_\omega(\mathcal{A})$ *is the cofibrant replacement functor. On the other hand,* $\mathfrak{M}'$*-fibrant replacement serves as the total right derived functor* $R' := \mathbf{R}(\text{Id})\colon \text{Ho}(\mathfrak{M}') \to \text{Ho}(\mathfrak{M})$*, and satisfies* $R' \circ \gamma' = \gamma \circ R'$.

Proof The first three conditions are clearly equivalent and immediately translate to the statement that the identity adjunction $\mathfrak{M} \underset{\text{Id}}{\overset{\text{Id}}{\rightleftarrows}} \mathfrak{M}'$ is a Quillen adjunction. So the total derived adjunction exists by Theorem 7.11, and by Definition 7.9 and Theorem 7.4 we see that $\mathbf{L}(\text{Id})\colon \text{Ho}(\mathfrak{M}) \to \text{Ho}(\mathfrak{M}')$ may be taken to be the unique functor satisfying $\mathbf{L}(\text{Id}) \circ \gamma = (\gamma' \circ \text{Id}) \circ Q$ where $\mathcal{A} \xrightarrow{Q} \text{St}_\omega(\mathcal{A})$

is $\mathfrak{M}$-cofibrant replacement. So $Q := \mathbf{L}(\mathrm{Id})$ is the unique functor making the stated diagram commute. Note that the canonical functor $\gamma' : \mathcal{A} \to \mathrm{Ho}(\mathfrak{M}')$ does descend to a functor $\mathrm{St}_\omega(\mathcal{A}) \to \mathrm{Ho}(\mathfrak{M}')$ since $\omega := Q \cap \mathcal{W} \cap \mathcal{R} \subseteq \widetilde{Q}' \subseteq \mathcal{W}'$. On the other hand, $\gamma : \mathcal{A} \to \mathrm{Ho}(\mathfrak{M})$ descends to a functor $\mathrm{St}_{\omega'}(\mathcal{A}) \to \mathrm{Ho}(\mathfrak{M})$ since $\omega' := Q' \cap \mathcal{W}' \cap \mathcal{R}' \subseteq \widetilde{\mathcal{R}}' \subseteq \mathcal{W}$. □

Recall that derived functors are only unique up to a canonical natural equivalence. In some cases the total left or right derived functor may be taken to be a localization functor as is the case in the following lemma.

Lemma 8.27 *Suppose $\mathfrak{M} = (Q, \mathcal{W}, \mathcal{R})$ and $\mathfrak{M}' = (Q', \mathcal{W}', \mathcal{R}')$ are two abelian model structures on $(\mathcal{A}, \mathcal{E})$, and that $\mathcal{W} \subseteq \mathcal{W}'$. Then we have the following.*

(1) *There exists a canonical* **quotient functor** $\bar{\gamma}' : \mathrm{Ho}(\mathfrak{M}) \to \mathrm{Ho}(\mathfrak{M}')$. *It is the unique functor satisfying $\gamma' = \bar{\gamma}' \circ \gamma$.*
(2) *$\bar{\gamma}'$ serves as both the total left derived functor,* $\mathbf{L}(\mathrm{Id})$, *of the identity functor $\mathrm{Id} : \mathfrak{M} \to \mathfrak{M}'$, and, the total right derived functor* $\mathbf{R}(\mathrm{Id})$, *of the identity functor* $\mathrm{Id} : \mathfrak{M} \to \mathfrak{M}'$.

Proof In any abelian model structure, a map is a weak equivalence if and only if it factors as an admissible monic with trivial cokernel followed by an admissible epic with trivial kernel; see Proposition 4.22. So if $\mathcal{W} \subseteq \mathcal{W}'$, then the localization functor $\gamma' : \mathcal{A} \to \mathrm{Ho}(\mathfrak{M}')$ sends weak equivalences in $\mathfrak{M}$ to isomorphisms. So the universal property of γ guaranteed by the Localization Theorem 5.21 provides the unique functor $\bar{\gamma}' : \mathrm{Ho}(\mathfrak{M}) \to \mathrm{Ho}(\mathfrak{M}')$ satisfying $\gamma' = \bar{\gamma}' \circ \gamma$.

By definition, the total left derived functor of the identity functor $\mathrm{Id} : \mathfrak{M} \to \mathfrak{M}'$ is $\mathbf{L}(\mathrm{Id}) := L(\gamma' \circ \mathrm{Id}) = L(\gamma')$, the left derived functor of $\gamma' : \mathcal{A} \to \mathrm{Ho}(\mathfrak{M}')$. It is easy to check that $\bar{\gamma}'$ (along with the identity natural transformation) satisfies the required universal property; see Exercise 7.1.4. In the same way, $\mathbf{R}(\mathrm{Id}) := R(\gamma' \circ \mathrm{Id}) = R(\gamma') = \bar{\gamma}'$. □

We need just one more lemma before we can start putting together our main results.

Lemma 8.28 *Let $\mathfrak{M} = (Q, \mathcal{W}, \mathcal{R})$ and $\mathfrak{M}' = (Q, \mathcal{W}', \mathcal{R}')$ be abelian model structures having the same cofibrant objects and $\mathcal{R}' \subseteq \mathcal{R}$. Then we have $\mathcal{W} \subseteq \mathcal{W}'$. In particular, there exists a unique quotient functor $\bar{\gamma}' : \mathrm{Ho}(\mathfrak{M}) \to \mathrm{Ho}(\mathfrak{M}')$ satisfying $\gamma' = \bar{\gamma}' \circ \gamma$, and we may take* $\mathbf{L}(\mathrm{Id}) = \bar{\gamma}'$ *and* $\mathbf{R}(\mathrm{Id}) = \bar{\gamma}'$.

Proof Since $\mathcal{R}' \subseteq \mathcal{R}$, we have a containment of trivially cofibrant objects $\widetilde{Q} \subseteq \widetilde{Q}'$. Similarly, since the two model structures share the same cofibrant objects we have equality of trivially fibrant objects $\widetilde{\mathcal{R}} = \widetilde{\mathcal{R}}'$. It follows immediately from the characterization of trivial objects given in Lemma 4.2 that $\mathcal{W} \subseteq \mathcal{W}''$. Lemma 8.27 provides for the quotient functor $\bar{\gamma}'$. □

Now we are ready to show how to obtain a (co)localization sequence from two simply related Hovey triples.

Proposition 8.29 *Let $\mathfrak{M} = (\mathcal{Q}, \mathcal{W}, \mathcal{R})$ and $\mathfrak{M}' = (\mathcal{Q}, \mathcal{W}'', \mathcal{R}')$ be hereditary abelian model structures with equal cores $\omega := \mathcal{Q} \cap \mathcal{W} \cap \mathcal{R} = \mathcal{Q} \cap \mathcal{W}'' \cap \mathcal{R}'$, and $\mathcal{R}' \subseteq \mathcal{R}$. Then there exists an hereditary abelian model structure*

$$\mathfrak{M}/\mathfrak{M}' := \left(\widetilde{\mathcal{Q}'}, \mathcal{V}, \mathcal{R}\right),$$

where $\widetilde{\mathcal{Q}'} := \mathcal{Q} \cap \mathcal{W}''$, called the **right localization** *of $\mathfrak{M}$ with respect to $\mathfrak{M}'$. Here*

$$\begin{aligned} \mathcal{V} &= \left\{ V \in \mathcal{A} \mid \exists \text{ a short exact sequence } V \rightarrowtail R' \twoheadrightarrow \tilde{Q} \text{ with } R' \in \mathcal{R}', \tilde{Q} \in \widetilde{\mathcal{Q}} \right\} \\ &= \left\{ V \in \mathcal{A} \mid \exists \text{ a short exact sequence } R' \rightarrowtail \tilde{Q} \twoheadrightarrow V \text{ with } R' \in \mathcal{R}', \tilde{Q} \in \widetilde{\mathcal{Q}} \right\}. \end{aligned}$$

Moreover, there is a colocalization sequence

$$\mathrm{Ho}(\mathfrak{M}') \xrightarrow{\mathbf{R}(\mathrm{Id}) = R'} \mathrm{Ho}(\mathfrak{M}) \xrightarrow{\mathbf{R}(\mathrm{Id}) = \bar{\gamma}^{\backprime}} \mathrm{Ho}(\mathfrak{M}/\mathfrak{M}')$$

with left adjoints $\mathbf{L}(\mathrm{Id}) = \bar{\gamma}'$ and $\widetilde{Q'} := \mathbf{L}(\mathrm{Id})$ as shown in the following:

$$\begin{array}{ccccc} \mathrm{Ho}(\mathfrak{M}') & \underset{R'}{\overset{\bar{\gamma}'}{\leftrightarrows}} & \mathrm{Ho}(\mathfrak{M}) & \underset{\bar{\gamma}^{\backprime}}{\overset{\widetilde{Q'}}{\leftrightarrows}} & \mathrm{Ho}(\mathfrak{M}/\mathfrak{M}') \\ \mathrm{Ho}(R') \circ \mathrm{Ho}(Q) \downarrow\uparrow & & \mathrm{Ho}(R) \circ \mathrm{Ho}(Q) \downarrow\uparrow & & \mathrm{Ho}(R) \circ \mathrm{Ho}(\widetilde{Q'}) \downarrow\uparrow \\ \mathrm{St}_\omega(\mathcal{Q} \cap \mathcal{R}') & \underset{\mathrm{inc}}{\overset{R'}{\leftrightarrows}} & \mathrm{St}_\omega(\mathcal{Q} \cap \mathcal{R}) & \underset{\widetilde{Q'}}{\overset{\mathrm{inc}}{\leftrightarrows}} & \mathrm{St}_\omega(\widetilde{\mathcal{Q}'} \cap \mathcal{R}). \end{array}$$

Here the quotient functors $\bar{\gamma}'$ and $\bar{\gamma}^{\backprime}$ are derived from the canonical localization functors $\gamma' \colon \mathcal{A} \to \mathrm{Ho}(\mathfrak{M}')$ and $\gamma^{\backprime} \colon \mathcal{A} \to \mathrm{Ho}(\mathfrak{M}/\mathfrak{M}')$. The vertical arrows are the standard equivalences passing between the homotopy categories and their equivalent full subcategories of bifibrant objects; see Proposition 5.25 and Theorem 8.6.

In Proposition 8.29 we have followed a convention we use throughout the book: No matter which direction functors are going, we will always write left adjoints on the top or the left, while the right adjoints will be written on the bottom or right.

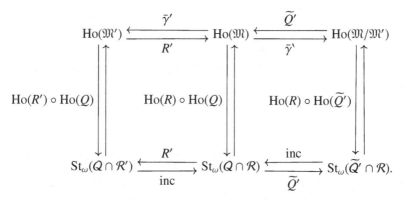

Proof We have the two complete hereditary cotorsion pairs $(\widetilde{Q}, \mathcal{R})$ and $(\widetilde{Q}', \mathcal{R}')$ satisfying $\mathcal{R}' \subseteq \mathcal{R}$ and $\widetilde{Q} \cap \mathcal{R} = \omega = \widetilde{Q}' \cap \mathcal{R}'$. Applying Theorem 8.16 we immediately obtain a unique thick class $\mathcal{V}$, with the two previous descriptions, making $\mathfrak{M}/\mathfrak{M}' := (\widetilde{Q}', \mathcal{V}, \mathcal{R})$ an hereditary abelian model structure.

Lemmas 8.26 and 8.28 apply to both of the identity adjunctions

$$\mathfrak{M}/\mathfrak{M}' \underset{\text{Id}}{\overset{\text{Id}}{\rightleftarrows}} \mathfrak{M} \underset{\text{Id}}{\overset{\text{Id}}{\rightleftarrows}} \mathfrak{M}'$$

and together they yield composable derived adjunctions

$$\text{Ho}(\mathfrak{M}') \underset{R'}{\overset{\bar{\gamma}'}{\leftrightarrows}} \text{Ho}(\mathfrak{M}) \underset{\bar{\gamma}^{\backprime}}{\overset{\widetilde{Q}'}{\leftrightarrows}} \text{Ho}(\mathfrak{M}/\mathfrak{M}').$$

To show this is a colocalization sequence, it remains to show the following.

(1) $\bar{\gamma}' \circ R' \cong 1_{\text{Ho}(\mathfrak{M}')}$.
(2) $\bar{\gamma}^{\backprime} \circ \widetilde{Q}' \cong 1_{\text{Ho}(\mathfrak{M}/\mathfrak{M}')}$.
(3) The essential image of R' equals the kernel of $\bar{\gamma}^{\backprime}$.

To prove (1), we use Corollary 5.16 which reduces the problem to showing

$$\bar{\gamma}' \circ R' \circ \gamma' \cong 1_{\text{Ho}(\mathfrak{M}')} \circ \gamma' = \gamma'.$$

Using Lemma 8.26, the left-hand side can be rearranged as $\bar{\gamma}' \circ (R' \circ \gamma') = \bar{\gamma}' \circ (\gamma \circ R') = (\bar{\gamma}' \circ \gamma) \circ R' = \gamma' \circ R'$. So given any morphism $f: A \to B$ in $\mathcal{A}$, the problem reduces to finding natural commutative squares where the isomorphisms are in $\text{Ho}(\mathfrak{M}')$:

$$\begin{array}{ccc} R'A & \xrightarrow{\cong} & A \\ {\scriptstyle \gamma'(R'(f))}\downarrow & & \downarrow{\scriptstyle \gamma'(f)} \\ R'B & \xrightarrow{\cong} & B. \end{array}$$

But the fibrant replacement functor R' acts by $f \mapsto R'(f)$ where $R'(f)$ is any map making the following diagram commute. Here the rows are exact, $R'A, R'B \in \mathcal{R}'$ and $Q_A, Q_B \in \widetilde{Q}'$:

$$\begin{array}{ccccc} A & \overset{j_A}{\rightarrowtail} & R'A & \overset{q_A}{\twoheadrightarrow} & Q_A \\ \downarrow{\scriptstyle f} & & \downarrow{\scriptstyle R'(f)} & & \\ B & \overset{j_B}{\rightarrowtail} & R'B & \overset{q_B}{\twoheadrightarrow} & Q_B. \end{array}$$

So applying γ' to this diagram proves we have a natural equivalence

$$\{\gamma'(j_A)^{-1}\}: \bar{\gamma}' \circ R' \cong 1_{\text{Ho}(\mathfrak{M}')}.$$

We have proved (1) and a similar type of argument will prove (2).

For (3), note that $\operatorname{Ker}\bar{\gamma}' = \mathcal{V}$, the class of trivial objects in $\mathfrak{M}/\mathfrak{M}'$. This contains the literal image of $R'\colon \mathrm{Ho}(\mathfrak{M}') \to \mathrm{Ho}(\mathfrak{M})$, which is the class $\mathcal{R}' = \mathcal{V} \cap \mathcal{R}$. Clearly the kernel of any additive functor is closed under isomorphims, so the essential image of R' is contained in $\operatorname{Ker}\bar{\gamma}'$. On the other hand, if $V \in \operatorname{Ker}\bar{\gamma}' = \mathcal{V}$, it means there is a short exact sequence $V \rightarrowtail R' \twoheadrightarrow \tilde{Q}$ with $R' \in \mathcal{R}'$ and $\tilde{Q} \in \widetilde{Q}$. In particular, $V \rightarrowtail R'$ is a trivial cofibration in $\mathfrak{M}$, and hence V is isomorphic in $\mathrm{Ho}(\mathfrak{M})$ to $R' \in \mathcal{R}'$. This proves $\mathcal{V}$ coincides with the essential image of $R'\colon \mathrm{Ho}(\mathfrak{M}') \to \mathrm{Ho}(\mathfrak{M})$. □

We are now ready to prove the main theorem constructing a recollement from three interrelated Hovey triples. Since we will now be dealing with three Hovey triples on the same category, we will denote them by $\mathfrak{M}_1 = (\mathcal{Q}_1, \mathcal{W}_1, \mathcal{R}_1)$, $\mathfrak{M}_2 = (\mathcal{Q}_2, \mathcal{W}_2, \mathcal{R}_2)$, and $\mathfrak{M}_3 = (\mathcal{Q}_3, \mathcal{W}_3, \mathcal{R}_3)$. For convenience, we will let $\widetilde{\mathcal{Q}_i} := \mathcal{Q}_i \cap \mathcal{W}_i$ denote the class of trivially cofibrant objects, for each of $i = 1, 2, 3$. We will also denote the cofibrant replacement functors by Q_i and the fibrant replacement functors by R_i. Moreover, by using enough projectives of $\left(\widetilde{\mathcal{Q}_i}, \mathcal{R}_i\right)$, we also get an approximation functor which we will denote by $\widetilde{Q}_i$.

Theorem 8.30 (Right Recollement Theorem) *Assume $(\mathcal{A}, \mathcal{E})$ possesses three hereditary model structures, sharing the same cofibrant objects, and whose cores ω all coincide:*

$$\mathfrak{M}_1 = (\mathcal{Q}, \mathcal{W}_1, \mathcal{R}_1), \quad \mathfrak{M}_2 = (\mathcal{Q}, \mathcal{W}_2, \mathcal{R}_2), \quad \mathfrak{M}_3 = (\mathcal{Q}, \mathcal{W}_3, \mathcal{R}_3).$$

If $\mathcal{W}_3 \cap \mathcal{R}_1 = \mathcal{R}_2$ and $\mathcal{R}_3 \subseteq \mathcal{R}_1$ (or equivalently, $\widetilde{\mathcal{Q}_2} \cap \mathcal{W}_3 = \widetilde{\mathcal{Q}_1}$ and $\mathcal{R}_2 \subseteq \mathcal{W}_3$), then $\mathrm{Ho}(\mathfrak{M}_1/\mathfrak{M}_2) \cong \mathrm{Ho}(\mathfrak{M}_3)$ and $\mathrm{Ho}(\mathfrak{M}_1/\mathfrak{M}_3) \cong \mathrm{Ho}(\mathfrak{M}_2)$. In fact, $\mathfrak{M}_1/\mathfrak{M}_3$ is Quillen equivalent to $\mathfrak{M}_2$, while $\mathfrak{M}_1/\mathfrak{M}_2$ is Quillen equivalent to $\mathfrak{M}_3$, and we even have a recollement:

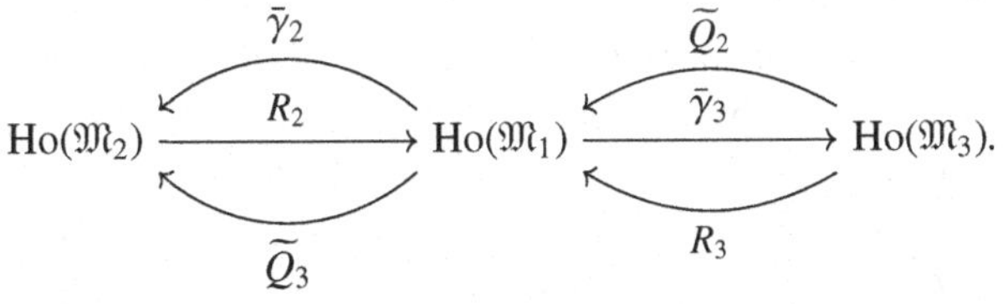

Here (for $i = 2, 3$) the quotient functors $\bar{\gamma}_i$ uniquely satisfy $\bar{\gamma}_i \circ \gamma_1 = \gamma_i$. Similarly, R_i and $\widetilde{Q}_i$ are induced by the approximation functors $R_i, \widetilde{Q}_i\colon \mathcal{A} \to \mathrm{St}_\omega(\mathcal{A})$ and are the unique functors commuting with the canonical localization functors.

Proof First we show that the two conditions are equivalent. That is, $\mathcal{W}_3 \cap \mathcal{R}_1 = \mathcal{R}_2$ and $\mathcal{R}_3 \subseteq \mathcal{R}_1$ if and only if $\widetilde{\mathcal{Q}_2} \cap \mathcal{W}_3 = \widetilde{\mathcal{Q}_1}$ and $\mathcal{R}_2 \subseteq \mathcal{W}_3$. For the "only if" part, the only part that is not clear is $\widetilde{Q}_2 \cap \mathcal{W}_3 \subseteq \widetilde{Q}_1$. So assume $Q \in \widetilde{Q}_2 \cap \mathcal{W}_3$. Use enough projectives of $(\widetilde{Q}_1, \mathcal{R}_1)$ to find a short exact sequence $R_1 \rightarrowtail Q_1 \twoheadrightarrow Q$ with $R_1 \in \mathcal{R}_1$, and $Q_1 \in \widetilde{Q}_1 \subseteq \mathcal{W}_3$. Since $\mathcal{W}_3$ is thick we see that $R_1 \in \mathcal{W}_3 \cap \mathcal{R}_1 = \mathcal{R}_2$. So with $R_1 \in \mathcal{R}_2$ and $Q \in \widetilde{Q}_2$, the short exact sequence must split, making Q a direct summand of Q_1. Hence Q must be in $\widetilde{Q}_1$ finishing the proof of the "only if" part. For the "if" part, the analogous part to show is $\mathcal{W}_3 \cap \mathcal{R}_1 \subseteq \mathcal{R}_2$. This follows by a similar argument: For $W \in \mathcal{W}_3 \cap \mathcal{R}_1$ start by finding a short exact sequence $W \rightarrowtail R_2 \twoheadrightarrow Q_2$ with $R_2 \in \mathcal{R}_2$, and $Q_2 \in \widetilde{Q}_2$. Since $\mathcal{W}_3$ is thick we get $Q_2 \in \widetilde{Q}_2 \cap \mathcal{W}_3 = \widetilde{Q}_1$ and the sequence splits.

Having shown that the two conditions are equivalent, we see at once that they make $(\widetilde{Q}_2, \mathcal{W}_3, \mathcal{R}_1)$ into a Hovey triple. By uniqueness of the thick class in a Hovey triple (see Lemma 4.2) we conclude that $\mathfrak{M}_1/\mathfrak{M}_2 = (\widetilde{Q}_2, \mathcal{W}_3, \mathcal{R}_1)$. It is then easy to see that the identity functor from $\mathfrak{M}_3$ to $\mathfrak{M}_1/\mathfrak{M}_2$ is a right Quillen functor. But since $\mathfrak{M}_3$ and $\mathfrak{M}_1/\mathfrak{M}_2$ both have $\mathcal{W}_3$ as their class of trivial objects, it follows (by Proposition 4.22) that the model structures have the same weak equivalences. So it is immediate by the very definition of Quillen equivalence that the identity adjunction is a Quillen equivalence and thus $\mathrm{Ho}(\mathfrak{M}_3) \cong \mathrm{Ho}(\mathfrak{M}_1/\mathfrak{M}_2)$.

Note next that $\mathfrak{M}_1/\mathfrak{M}_3 = (\widetilde{Q}_3, \mathcal{V}, \mathcal{R}_1)$ for some thick class $\mathcal{V}$. Since the cores of $\mathfrak{M}_1/\mathfrak{M}_3 = (\widetilde{Q}_3, \mathcal{V}, \mathcal{R}_1)$ and $\mathfrak{M}_2 = (Q, \mathcal{W}_2, \mathcal{R}_2)$ must coincide we have

$$\mathrm{Ho}(\mathfrak{M}_1/\mathfrak{M}_3) \cong \mathrm{St}_\omega(\mathcal{A})\left(\widetilde{Q}_3 \cap \mathcal{R}_1\right) = \mathrm{St}_\omega(\mathcal{A})(Q \cap \mathcal{R}_2) \cong \mathrm{Ho}(\mathfrak{M}_2).$$

So the homotopy categories are equivalent, but we wish to show that the identity map $\mathfrak{M}_1/\mathfrak{M}_3 \to \mathfrak{M}_2$ is in fact a (left) Quillen equivalence. To see this, first note that it preserves cofibrant and trivially cofibrant objects, and this implies it is a left Quillen functor. To see it is a Quillen equivalence, we use Definition 7.15 to show that for all cofibrant $A \in \mathfrak{M}_1/\mathfrak{M}_3$ and fibrant $B \in \mathfrak{M}_2$, any map $f: A \to B$ is a weak equivalence in $\mathfrak{M}_2$ if and only if it is a weak equivalence in $\mathfrak{M}_1/\mathfrak{M}_3$. So let $f: A \to B$ be a map with $A \in \widetilde{Q}_3$ and $B \in \mathcal{R}_2$. Assuming it is a weak equivalence in $\mathfrak{M}_2$, we may factor it as $f = pi$ where $i: A \rightarrowtail C$ has $\mathrm{Cok}\, i \in \widetilde{Q}_2$ and $p: C \twoheadrightarrow B$ has $\mathrm{Ker}\, p \in \mathcal{W}_2 \cap \mathcal{R}_2 = \mathcal{W}_3 \cap \mathcal{R}_3$. Note then that $\mathrm{Ker}\, p \in \mathcal{R}_3 = \mathcal{V} \cap \mathcal{R}_1$ is automatically trivially fibrant in $\mathfrak{M}_1/\mathfrak{M}_3$. But since $\mathcal{W}_3$ is thick we also have $C \in \mathcal{W}_3$, and thus $\mathrm{Cok}\, i \in \mathcal{W}_3$. So $\mathrm{Cok}\, i \in \widetilde{Q}_2 \cap \mathcal{W}_3 = \widetilde{Q}_1 = \widetilde{Q}_3 \cap \mathcal{V}$. That is, $\mathrm{Cok}\, i$ is trivially cofibrant in $\mathfrak{M}_1/\mathfrak{M}_3$. Since we have shown f factors as a trivial cofibration followed by a trivial fibration in $\mathfrak{M}_1/\mathfrak{M}_3$ we conclude it is also a weak equivalence in $\mathfrak{M}_1/\mathfrak{M}_3$. Conversely, if f is a weak equivalence in $\mathfrak{M}_1/\mathfrak{M}_3$, the argument reverses and we see that f is a weak equivalence in $\mathfrak{M}_2$.

Now we wish to construct the recollement. Using that $\mathcal{R}_2 \subseteq \mathcal{R}_1$, we apply Proposition 8.29 to obtain a colocalization sequence which appears as the top horizontal row in the following diagram. However, we also have $\mathcal{R}_3 \subseteq \mathcal{R}_1$ and so we have the analogous colocalization sequence involving $\mathfrak{M}_3$. But we may rewrite this last colocalization "backwards" so that it appears as a localization sequence. This is the bottom row of the following diagram which, again, positions all left adjoints on top and all right adjoints on the bottom:

$$\begin{array}{ccccc}
\mathrm{Ho}(\mathfrak{M}_2) & \underset{R_2}{\overset{\bar{\gamma}_2}{\leftrightarrows}} & \mathrm{Ho}(\mathfrak{M}_1) & \underset{\bar{\gamma}_{1/2}}{\overset{\widetilde{Q}_2}{\leftrightarrows}} & \mathrm{Ho}(\mathfrak{M}_1/\mathfrak{M}_2) \\
\widetilde{Q}_3 \uparrow\downarrow R_2 & & \| & & \bar{\gamma}_3 \downarrow\uparrow \bar{\gamma}_{1/2} \\
\mathrm{Ho}(\mathfrak{M}_1/\mathfrak{M}_3) & \underset{\bar{\gamma}_{1/3}}{\overset{\widetilde{Q}_3}{\rightleftarrows}} & \mathrm{Ho}(\mathfrak{M}_1) & \underset{R_3}{\overset{\bar{\gamma}_3}{\rightleftarrows}} & \mathrm{Ho}(\mathfrak{M}_3).
\end{array} \qquad (***)$$

We turn to the vertical functors. The left-hand pair, $\left(\widetilde{Q}_3, R_2\right)$, represents one of the equivalences already argued earlier. It arises from Lemma 8.26. The pair $(\bar{\gamma}_3, \bar{\gamma}_{1/2})$ represents the other equivalence argued earlier, but here we use the notation of Lemma 8.27. Indeed this equivalence arises from the fact that $\mathfrak{M}_3$ and $\mathfrak{M}_1/\mathfrak{M}_2$ share the same weak equivalences. Hence the universal property in the Localization Theorem 5.21 induces such functors satisfying $\bar{\gamma}_3 \circ \bar{\gamma}_{1/2} \cong 1_{\mathrm{Ho}(\mathfrak{M}_3)}$ and $\bar{\gamma}_{1/2} \circ \bar{\gamma}_3 \cong 1_{\mathrm{Ho}(\mathfrak{M}_1/\mathfrak{M}_2)}$. Moreover, by the uniqueness properties of all the $\bar{\gamma}$ functors, we are assured that they are compatible under composition. Therefore, the right-hand square of the diagram may be collapsed and represented more succinctly by

$$\mathrm{Ho}(\mathfrak{M}_1) \begin{array}{c} \xleftarrow{\widetilde{Q}_2} \\ \xrightarrow{\bar{\gamma}_3} \\ \xleftarrow[R_3]{} \end{array} \mathrm{Ho}(\mathfrak{M}_3).$$

Next we consider the left square of Diagram $(***)$. The plan is to "glue" the functor $\mathrm{Ho}(\mathfrak{M}_2) \xrightarrow{R_2} \mathrm{Ho}(\mathfrak{M}_1)$ to the composite $\mathrm{Ho}(\mathfrak{M}_2) \xrightarrow{R_2} \mathrm{Ho}(\mathfrak{M}_1/\mathfrak{M}_3) \xrightarrow{\widetilde{Q}_3} \mathrm{Ho}(\mathfrak{M}_1)$ by showing the functors are naturally isomorphic. By Corollary 5.16 we only need to show that

$$R_2 \circ \gamma_2 \cong \left(\widetilde{Q}_3 \circ R_2\right) \circ \gamma_2.$$

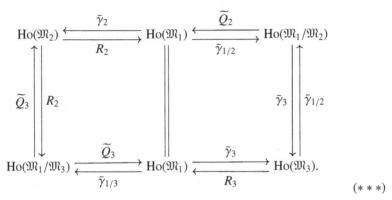

Using Lemma 8.26, the left-hand side can be rewritten $R_2 \circ \gamma_2 = \gamma_1 \circ R_2$. Similarly, for the right-hand side we compute:

$$\widetilde{Q}_3 \circ (R_2 \circ \gamma_2) = \widetilde{Q}_3 \circ (\gamma_{1/3} \circ R_2) = \left(\widetilde{Q}_3 \circ \gamma_{1/3}\right) \circ R_2 = \left(\gamma_1 \circ \widetilde{Q}_3\right) \circ R_2.$$

So given any morphism $f\colon A \to B$ in $\mathcal{A}$, the problem reduces to finding natural commutative squares where the isomorphisms are in $\mathrm{Ho}(\mathfrak{M}_1)$:

$$\begin{array}{ccc} \widetilde{Q}_3R_2A & \xrightarrow{\cong} & R_2A \\ {\scriptstyle \gamma_1(\widetilde{Q}_3(R_2(f)))}\downarrow & & \downarrow{\scriptstyle \gamma_1(R_2(f))} \\ \widetilde{Q}_3R_2B & \xrightarrow{\cong} & R_2B. \end{array}$$

The fibrant replacement R_2 acts by $f \mapsto R_2(f)$ where $R_2(f)$ is any map making the following diagram commute, where $R_2A, R_2B \in \mathcal{R}_2$ and $Q_A, Q_B \in \widetilde{Q}_2$:

$$\begin{array}{ccccc} A & \overset{j_A}{\rightarrowtail} & R_2A & \overset{q_A}{\twoheadrightarrow} & Q_A \\ {\scriptstyle f}\downarrow & & \downarrow{\scriptstyle R_2(f)} & & \\ B & \overset{j_B}{\rightarrowtail} & R_2B & \overset{q_B}{\twoheadrightarrow} & Q_B. \end{array}$$

Now applying $\mathrm{Ho}(\mathfrak{M}_1/\mathfrak{M}_3) \xrightarrow{\widetilde{Q}_3} \mathrm{Ho}(\mathfrak{M}_1)$ to $R_2(f)$ gives us $\widetilde{Q}_3(R_2(f))$ in the next diagram, where $R_A, R_B \in \mathcal{R}_3$ and $\widetilde{Q}_3R_2A, \widetilde{Q}_3R_2B \in \widetilde{Q}_3 = Q \cap \mathcal{W}_3$:

$$\begin{array}{ccccc} R_A & \overset{i_A}{\rightarrowtail} & \widetilde{Q}_3R_2A & \overset{p_A}{\twoheadrightarrow} & R_2A \\ & & {\scriptstyle \widetilde{Q}_3(R_2(f))}\downarrow & & \downarrow{\scriptstyle R_2(f)} \\ R_B & \overset{i_A}{\rightarrowtail} & \widetilde{Q}_3R_2B & \overset{p_B}{\twoheadrightarrow} & R_2B. \end{array}$$

But since $R_2A, R_2B \in \mathcal{R}_2 = \mathcal{W}_3 \cap \mathcal{R}_1$ and $\mathcal{W}_3$ is thick, we see that $R_A, R_B \in \mathcal{W}_3 \cap \mathcal{R}_3 = \mathcal{W}_1 \cap \mathcal{R}_1$. That is, R_A, R_B are trivial in $\mathfrak{M}_1$ and hence the maps $\{p_A\}$ are providing the desired natural isomorphism $\{\gamma_1(p_A)\}\colon \widetilde{Q}_3 \circ R_2 \cong R_2$.

Since the pair of functors $\left(\widetilde{Q}_3, R_2\right)$ on the left side of Diagram $(***)$ is an equivalence, they are both right and left adjoint to each other. So the isomorphism $\{\gamma_1(p_A)\}\colon \widetilde{Q}_3 \circ R_2 \cong R_2$ means that the functor $\mathrm{Ho}(\mathfrak{M}_2) \xrightarrow{R_2} \mathrm{Ho}(\mathfrak{M}_1)$ has right adjoint the composite $\mathrm{Ho}(\mathfrak{M}_1) \xrightarrow{\bar{\gamma}_{1/3}} \mathrm{Ho}(\mathfrak{M}_1/\mathfrak{M}_3) \xrightarrow{\widetilde{Q}_3} \mathrm{Ho}(\mathfrak{M}_2)$. Note that this composite commutes with the localization functors as claimed in the statement of the theorem: For composing it with γ_1 we have $\left(\widetilde{Q}_3 \circ \bar{\gamma}_{1/3}\right) \circ \gamma_1 = \widetilde{Q}_3 \circ (\bar{\gamma}_{1/3} \circ \gamma_1) = \widetilde{Q}_3 \circ \gamma_{1/3} = \gamma_2 \circ \widetilde{Q}_3$, where the last equality holds by Lemma 8.26. □

We have the following special case for injective model structures.

Corollary 8.31 (Injective Recollement Theorem) *Assume $(\mathcal{A}, \mathcal{E})$ possesses three injective model structures represented by the injective cotorsion pairs:*

$$\mathfrak{M}_1 = (\mathcal{W}_1, \mathcal{R}_1), \quad \mathfrak{M}_2 = (\mathcal{W}_2, \mathcal{R}_2), \quad \mathfrak{M}_3 = (\mathcal{W}_3, \mathcal{R}_3).$$

If $\mathcal{W}_3 \cap \mathcal{R}_1 = \mathcal{R}_2$ and $\mathcal{R}_3 \subseteq \mathcal{R}_1$ (or equivalently, $\mathcal{W}_2 \cap \mathcal{W}_3 = \mathcal{W}_1$ and $\mathcal{R}_2 \subseteq \mathcal{W}_3$), then $\mathrm{Ho}(\mathfrak{M}_1/\mathfrak{M}_2) \cong \mathrm{Ho}(\mathfrak{M}_3)$ and $\mathrm{Ho}(\mathfrak{M}_1/\mathfrak{M}_3) \cong \mathrm{Ho}(\mathfrak{M}_2)$. In fact, $\mathfrak{M}_1/\mathfrak{M}_3$ is Quillen equivalent to $\mathfrak{M}_2$, while $\mathfrak{M}_1/\mathfrak{M}_2$ is Quillen equivalent to $\mathfrak{M}_3$, and we even have a recollement:

$$\mathrm{Ho}(\mathfrak{M}_2) \begin{matrix} \xleftarrow{\;\bar{\gamma}_2\;} \\ \xrightarrow{\;R_2\;} \\ \xleftarrow[\;W_3\;]{} \end{matrix} \mathrm{Ho}(\mathfrak{M}_1) \begin{matrix} \xleftarrow{\;W_2\;} \\ \xrightarrow{\;\bar{\gamma}_3\;} \\ \xleftarrow[\;R_3\;]{} \end{matrix} \mathrm{Ho}(\mathfrak{M}_3),$$

where the functors behave as described in Theorem 8.30.

We also have the dual notion of left localization, denoted $\mathfrak{M}_2\backslash\mathfrak{M}_1$, and the dual statements of the aforementioned. In particular we have the following left recollement theorem and subsequent corollary for projective model structures.

Theorem 8.32 (Left Recollement Theorem) *Assume $(\mathcal{A}, \mathcal{E})$ possesses three hereditary model structures, sharing the same fibrant objects, and whose cores ω all coincide:*

$$\mathfrak{M}_1 = (\mathcal{Q}_1, \mathcal{W}_1, \mathcal{R}), \quad \mathfrak{M}_2 = (\mathcal{Q}_2, \mathcal{W}_2, \mathcal{R}), \quad \mathfrak{M}_3 = (\mathcal{Q}_3, \mathcal{W}_3, \mathcal{R}).$$

If $\mathcal{W}_3 \cap \mathcal{Q}_1 = \mathcal{Q}_2$ and $\mathcal{Q}_3 \subseteq \mathcal{Q}_1$ (or equivalently, $\widetilde{\mathcal{R}}_2 \cap \mathcal{W}_3 = \widetilde{\mathcal{R}}_1$ and $\mathcal{Q}_2 \subseteq \mathcal{W}_3$), then $\mathrm{Ho}(\mathfrak{M}_2\backslash\mathfrak{M}_1) \cong \mathrm{Ho}(\mathfrak{M}_3)$ and $\mathrm{Ho}(\mathfrak{M}_3\backslash\mathfrak{M}_1) \cong \mathrm{Ho}(\mathfrak{M}_2)$. In fact, $\mathfrak{M}_2\backslash\mathfrak{M}_1$ is Quillen equivalent to $\mathfrak{M}_3$, while $\mathfrak{M}_3\backslash\mathfrak{M}_1$ is Quillen equivalent to $\mathfrak{M}_2$, and we even have a recollement:

$$\mathrm{Ho}(\mathfrak{M}_2) \begin{matrix} \xleftarrow{\;\widetilde{R}_3\;} \\ \xrightarrow{\;Q_2\;} \\ \xleftarrow[\;\bar{\gamma}_2\;]{} \end{matrix} \mathrm{Ho}(\mathfrak{M}_1) \begin{matrix} \xleftarrow{\;Q_3\;} \\ \xrightarrow{\;\bar{\gamma}_3\;} \\ \xleftarrow[\;\widetilde{R}_2\;]{} \end{matrix} \mathrm{Ho}(\mathfrak{M}_3).$$

Here (for $i = 2, 3$) the quotient functors $\bar{\gamma}_i$ uniquely satisfy $\bar{\gamma}_i \circ \gamma_1 = \gamma_i$. Similarly, Q_i and $\widetilde{R}_i$ are induced by the approximation functors $Q_i, \widetilde{R}_i \colon \mathcal{A} \to \mathrm{St}_\omega(\mathcal{A})$ and are the unique functors commuting with the canonical localization functors.

Corollary 8.33 (Projective Recollement Theorem) *Assume $(\mathcal{A}, \mathcal{E})$ possesses three projective model structures represented by the projective cotorsion pairs:*

$$\mathfrak{M}_1 = (Q_1, \mathcal{W}_1), \quad \mathfrak{M}_2 = (Q_2, \mathcal{W}_2), \quad \mathfrak{M}_3 = (Q_3, \mathcal{W}_3).$$

If $\mathcal{W}_3 \cap Q_1 = Q_2$ and $Q_3 \subseteq Q_1$ (or equivalently, $\mathcal{W}_2 \cap \mathcal{W}_3 = \mathcal{W}_1$ and $Q_2 \subseteq \mathcal{W}_3$), then $\mathrm{Ho}(\mathfrak{M}_2 \backslash \mathfrak{M}_1) \cong \mathrm{Ho}(\mathfrak{M}_3)$ *and* $\mathrm{Ho}(\mathfrak{M}_3 \backslash \mathfrak{M}_1) \cong \mathrm{Ho}(\mathfrak{M}_2)$. *In fact, $\mathfrak{M}_3 \backslash \mathfrak{M}_1$ is Quillen equivalent to $\mathfrak{M}_2$, while $\mathfrak{M}_2 \backslash \mathfrak{M}_1$ is Quillen equivalent to $\mathfrak{M}_3$, and we even have a recollement:*

$$\mathrm{Ho}(\mathfrak{M}_2) \underset{\overleftarrow{\;\bar{\gamma}_2\;}}{\overset{\overleftarrow{\;W_3\;}}{\xrightarrow{\;Q_2\;}}} \mathrm{Ho}(\mathfrak{M}_1) \underset{\overleftarrow{\;W_2\;}}{\overset{\overleftarrow{\;Q_3\;}}{\xrightarrow{\;\bar{\gamma}_3\;}}} \mathrm{Ho}(\mathfrak{M}_3),$$

where the functors behave as described in Theorem 8.32.

Using known Hovey triples, an abundance of recollements involving homotopy categories of chain complexes can be constructed or recovered using the earlier recollement theorems. Let us point out just a few examples.

Example 8.34 Let R be a ring. It is known that there are three injective cotorsion pairs on $\mathrm{Ch}(R)$:

$$\mathfrak{M}_1 = \left(\mathcal{W}_{\mathrm{co}}, dw\widetilde{\mathcal{I}}\right), \quad \mathfrak{M}_2 = \left(\mathcal{W}_{\mathrm{st}}, ex\widetilde{\mathcal{I}}\right), \quad \mathfrak{M}_3 = \left(\widetilde{\mathcal{E}}, dg\widetilde{\mathcal{I}}\right).$$

Here, $dw\widetilde{\mathcal{I}}$ (resp. $ex\widetilde{\mathcal{I}}$) is the class of all chain complexes (resp. all exact chain complexes) of injective R-modules, and the complexes in $\mathcal{W}_{\mathrm{co}}$ are called *coacyclic (in the sense of Becker)*. $\mathfrak{M}_3 = \left(\widetilde{\mathcal{E}}, dg\widetilde{\mathcal{I}}\right)$ is the usual injective model structure for the derived category; see Example 4 of the Introduction and Main Examples. It is immediate that Corollary 8.31 applies and induces a recollement. This is a model category interpretation, first appearing in Becker [2014], of a well-known recollement $K_{\mathrm{ex}}(Inj) \to K(Inj) \to \mathcal{D}(R)$ from Krause [2005]. There are other variations of this based on abelian model structures having (co)fibrant objects based on complexes of projective modules, flat modules, and absolutely pure modules. We refer the reader to Gillespie [2016e], and the references therein which describe the connection to classical Grothendieck duality. The previous model structures and recollement were extended to complexes of quasi-coherent sheaves over a semiseparated quasi-compact scheme in Estrada and Gillespie [2024, corollary 4.2].

Example 8.35 Let $\mathrm{Ch}(R)_{dw}$ be the exact category of chain complexes of R-modules along with the degreewise split exact structure. There are complete cotorsion pairs $\mathfrak{M}_2 = \left(\text{K-}\mathcal{P}roj, \widetilde{\mathcal{E}}\right)$, and $\mathfrak{M}_3 = \left(\widetilde{\mathcal{E}}, \text{K-}\mathcal{I}nj\right)$, in $\text{Ch}(R)_{dw}$, where K-$\mathcal{P}roj$ is the class of all K-projective chain complexes and K-$\mathcal{I}nj$ is the class of all K-injective chain complexes. (In fact, in Section 11.1 we will show how these can be constructed as "mixed" model structures.) Moreover, these are each localizing cotorsion pairs in the sense of Exercise 8.1.3. In particular it means we have three injective model structures

$$\mathfrak{M}_1 = (\mathcal{A}ll, \mathcal{W}, \mathcal{A}ll)\,, \quad \mathfrak{M}_2^i = \left(\mathcal{A}ll, \text{K-}\mathcal{P}roj, \widetilde{\mathcal{E}}\right)\,, \quad \mathfrak{M}_3^i = \left(\mathcal{A}ll, \widetilde{\mathcal{E}}, \text{K-}\mathcal{I}nj\right)$$

in $\text{Ch}(R)_{dw}$ where $\mathfrak{M}_1$ is the standard Frobenius model structure for $K(R)$; see Example 6 of the Introduction and Main Examples. Once again Corollary 8.31 applies. The corresponding recollement is a standard one: It expresses the fact that the Verdier localization functor $K(R) \to \mathcal{D}(R)$ has a left adjoint given by taking K-projective resolutions and a right adjoint given by taking K-injective resolutions.

Note that we may also recover the recollement by way of the dual Projective Recollement Theorem. To see this, swap the role of $\mathfrak{M}_2$ and $\mathfrak{M}_3$ so that we have $\mathfrak{M}_3 = \left(\text{K-}\mathcal{P}roj, \widetilde{\mathcal{E}}\right)$, and $\mathfrak{M}_2 = \left(\widetilde{\mathcal{E}}, \text{K-}\mathcal{I}nj\right)$. Therefore we have three projective model structures on $\text{Ch}(R)_{dw}$

$$\mathfrak{M}_1 = (\mathcal{A}ll, \mathcal{W}, \mathcal{A}ll)\,, \quad \mathfrak{M}_2^p = \left(\widetilde{\mathcal{E}}, \text{K-}\mathcal{I}nj, \mathcal{A}ll\right)\,, \quad \mathfrak{M}_3^p = \left(\text{K-}\mathcal{P}roj, \widetilde{\mathcal{E}}, \mathcal{A}ll\right).$$

Now applying Corollary 8.33 yields what amounts to the same recollement.

Example 8.36 Again let R be a ring and now let $\text{Ch}(R)_{pur}$ denote the exact category of chain complexes of R-modules along with the degreewise pure exact structure. In Gillespie [2023a] and Gillespie [2023b] one will find similar recollements to the previous examples, but involving and the pure derived category, $\mathcal{D}_{pur}(R)$, and homotopy categories of (acyclic) complexes of pure–projectives and pure–injectives. We give an example of one such recollement here, but constructed in a different way. There are three abelian model structures on $\text{Ch}(R)_{pur}$ sharing the same fibrant objects:

$$\mathfrak{M}_1 = \left(\mathcal{A}ll, \mathcal{A}_{pur}, dw\widetilde{\mathcal{PI}}\right)\,, \qquad \mathfrak{M}_2 = \left(\widetilde{\mathcal{E}}, \text{K-}\mathcal{A}bs, dw\widetilde{\mathcal{PI}}\right)\,,$$

$$\mathfrak{M}_3 = \left(\text{K-}\mathcal{F}lat, \widetilde{\mathcal{E}}, dw\widetilde{\mathcal{PI}}\right).$$

The model $\mathfrak{M}_1$ appeared in Šťovíček [2015], and here $dw\widetilde{\mathcal{PI}}$ is the class of all chain complexes of pure-injective R-modules. K-$\mathcal{F}lat$ denotes the class of all K-flat complexes (in the sense of Spaltenstein), and $\mathfrak{M}_3$ is a model for the usual derived category, $\mathcal{D}(R)$; it is constructed in Gillespie [2023a, theorem 5.1]. The model $\mathfrak{M}_2$ was constructed in Estrada et al. [2024, proposition 6.8(2)], and here K-$\mathcal{A}bs$ denotes the class of all K-absolutely pure complexes from Emmanouil and Kaperonis [2024]. Applying the Left Recollement Theorem 8.32 yields a recollement $K(R)/\text{K-}Abs \to \mathcal{D}_{pur}(R) \to \mathcal{D}(R)$.

In Estrada et al. [2024, proposition 6.8(2)], the model structures in $\mathfrak{M}_1$ and $\mathfrak{M}_3$ are extended to complexes of quasi-coherent sheaves, but model $\mathfrak{M}_2 = \left(\widetilde{\mathcal{E}}, \text{K-}Abs, dw\widetilde{\mathcal{PI}}\right)$ was only constructed in the affine case. However, it can be extended to quasi-coherent sheaves by using the machinery of Theorem 8.16. Indeed it is shown in Estrada et al. [2024, theorems 5.3 and 6.6] that we have injective model structures $\left(\widetilde{\mathcal{E}}, \widetilde{\mathcal{E}}^{\perp}\right)$ and $\left(\mathcal{A}_{\otimes pur}, dw\widetilde{\mathcal{PI}}\right)$ with respect to the degreewise $\otimes$-pure exact structure. It follows from Theorem 8.16 that we have a unique thick class $\mathcal{V}$ such that $\left(\widetilde{\mathcal{E}}, \mathcal{V}, dw\widetilde{\mathcal{PI}}\right)$ is a Hovey triple. The uniqueness of $\mathcal{V}$ means that it coincides with K-Abs in the affine case.

Exercise 8.7.1 (Derived Identity Adjunctions) Generalize Lemma 8.26 as follows: Assume $(\mathcal{A}, \mathcal{E})$ has two abelian model structures, $\mathfrak{M}_1 = (\mathcal{Q}_1, \mathcal{W}_1, \mathcal{R}_1)$ on an exact structure $\mathcal{E}_1$, and $\mathfrak{M}_2 = (\mathcal{Q}_2, \mathcal{W}_2, \mathcal{R}_2)$ on an exact structure $\mathcal{E}_2 \subseteq \mathcal{E}_1$.

- The following conditions are equivalent.
  (1) Each $\mathfrak{M}_1$-cofibration is an $\mathfrak{M}_2$-cofibration and $\mathcal{R}_2 \subseteq \mathcal{R}_1$.
  (2) Each $\mathfrak{M}_1$-cofibration is an $\mathfrak{M}_2$-cofibration and $\widetilde{\mathcal{Q}}_1 \subseteq \widetilde{\mathcal{Q}}_2$.
  (3) $\mathcal{R}_2 \subseteq \mathcal{R}_1$ and $\widetilde{\mathcal{R}}_2 \subseteq \widetilde{\mathcal{R}}_1$.
  (4) The identity $\mathfrak{M}_1 \xrightarrow{\text{Id}} \mathfrak{M}_2$ is a left Quillen functor.
- The following conditions are equivalent.
  (1) Each $\mathfrak{M}_1$-fibration is an $\mathfrak{M}_2$-fibration and $\mathcal{Q}_2 \subseteq \mathcal{Q}_1$.
  (2) Each $\mathfrak{M}_1$-fibration is an $\mathfrak{M}_2$-fibration and $\widetilde{\mathcal{R}}_1 \subseteq \widetilde{\mathcal{R}}_2$.
  (3) $\mathcal{Q}_2 \subseteq \mathcal{Q}_1$ and $\widetilde{\mathcal{Q}}_2 \subseteq \widetilde{\mathcal{Q}}_1$.
  (4) The identity $\mathfrak{M}_2 \xrightarrow{\text{Id}} \mathfrak{M}_1$ is a left Quillen functor.

Show that in each case we obtain the analogous conclusion to the one in Lemma 8.26.

Exercise 8.7.2 Use Exercise 8.7.1 and the existence of the standard projective model structure for $\mathcal{D}(R)$, the derived category of a ring R, and the existence of the Frobenius model structure for the chain homotopy category $K(R)$, to deduce the existence of an adjunction

$$\mathcal{D}(R) \underset{\bar{\gamma}}{\overset{dgP}{\rightleftarrows}} K(R)$$

expressing the fact that the canonical Verdier quotient functor $K(R) \to \mathcal{D}(R)$ has a left adjoint which takes a complex to its DG-projective approximation.

On the other hand, show that the existence of the standard injective model structure for $\mathcal{D}(R)$ yields the existence of an adjunction

$$K(R) \underset{dgI}{\overset{\bar{\gamma}}{\rightleftarrows}} \mathcal{D}(R)$$

corresponding to the fact that DG-injective approximations provide a right adjoint to Verdier quotient functor $K(R) \to \mathcal{D}(R)$.

8.8 Compact Abelian Model Structures

We continue to consider an hereditary abelian model structure $\mathfrak{M} = (Q, \mathcal{W}, \mathcal{R})$ on a weakly idempotent complete exact category $(\mathcal{A}, \mathcal{E})$. We now also assume throughout this section that coproducts exist in $\mathcal{A}$, and that coproducts of short exact sequences are again short exact sequences. We will express this by saying that $(\mathcal{A}, \mathcal{E})$ has **exact coproducts**. The goal of this section is to define compactly generated categories and to give conditions on $\mathfrak{M}$ that will ensure that $\mathrm{Ho}(\mathfrak{M})$ is compactly generated. Some examples are given in the context of Grothendieck categories. The more general notion of a well-generated category is studied in Chapter 12.

Let $\{X_i\}_{i\in I}$ be a collection of objects and let $\left(\bigoplus_{i\in I} X_i, \{\eta_i\}_{i\in I}\right)$ denote the coproduct in $\mathcal{A}$; so each $X_i \xrightarrow{\eta_i} \bigoplus_{i\in I} X_i$ is the injection into the ith component. Applying the functor $\mathrm{Ext}^1_{\mathcal{E}}(C, -)$ for any given object C, we obtain abelian group homomorphisms $\widehat{\eta_i}\colon \mathrm{Ext}^1_{\mathcal{E}}(C, X_i) \to \mathrm{Ext}^1_{\mathcal{E}}\left(C, \bigoplus_{i\in I} X_i\right)$ which are defined via pushout along η_i (using the Yoneda description of Ext^1). The universal property of coproduct induces a canonical mapping

$$\xi\colon \bigoplus_{i\in I} \mathrm{Ext}^1_{\mathcal{E}}(C, X_i) \to \mathrm{Ext}^1_{\mathcal{E}}\left(C, \bigoplus_{i\in I} X_i\right) \qquad (\dagger)$$

uniquely satisfying $\xi \circ \alpha_i = \widehat{\eta_i}$, where now $\mathrm{Ext}^1_{\mathcal{E}}(C, X_i) \xrightarrow{\alpha_i} \bigoplus_{i\in I} \mathrm{Ext}^1_{\mathcal{E}}(C, X_i)$ is the canonical injection into the ith component.

Lemma 8.37 *Consider the abelian group homomorphism ξ in ($\dagger$).*

(1) *If the index set $I = \{1, 2, \ldots, n\}$ is finite, then ξ is an isomorphism and $\xi = [\,\widehat{\eta_1}\ \widehat{\eta_2}\ \cdots\ \widehat{\eta_n}\,]$.*

(2) *ξ is always a monomorphism, for any object C, and any index set I.*

(3) *ξ is an epimorphism (and hence an isomorphism) if and only if given any short exact sequence*

$$\mathbb{E} \equiv \bigoplus_{i\in I} X_i \rightarrowtail Y \twoheadrightarrow C,$$

there exists a finite subset $K \subseteq I$, and a short exact sequence

$$\mathbb{E}_K \equiv \bigoplus_{i \in K} X_i \rightarrowtail Y_K \twoheadrightarrow C$$

fitting into a commutative (pushout) diagram

$$\begin{array}{ccccc} \bigoplus_{i \in K} X_i & \rightarrowtail & Y_K & \twoheadrightarrow & C \\ {\scriptstyle \eta_K}\downarrow & & \downarrow & & \| \\ \bigoplus_{i \in I} X_i & \rightarrowtail & Y & \twoheadrightarrow & C, \end{array}$$

where η_K is the natural subcoproduct inclusion map.

Proof We first prove the finite case (1), and we will use it in the general proof of (2). So let $I = \{1, 2, \ldots, n\}$ be finite, and let π_i denote the projection of $\bigoplus_{i \in I} X_i \to X_i$ onto the ith coordinate. The notion of a biproduct as in Definition 1.2 extends from the case $n = 2$, to any finite $n > 2$, via the conditions: (i) $\pi_i \eta_i = 1_{X_i}$ for each i, (ii) $\pi_j \eta_i = 0$ whenever $i \neq j$, and (iii) $\eta_1 \pi_1 + \eta_2 \pi_2 + \cdots + \eta_n \pi_n = 1_{X_1 \oplus \cdots \oplus X_n}$. Since $\mathrm{Ext}^1_{\mathcal{E}}(C, -)$ is an additive functor, applying it to these equations yields a new set of biproduct equations: (i) $\widehat{\pi_i}\widehat{\eta_i} = 1_{\mathrm{Ext}^1_{\mathcal{E}}(C, X_i)}$ for each i, (ii) $\widehat{\pi_j}\widehat{\eta_i} = 0$ whenever $i \neq j$, and (iii) $\widehat{\eta_1}\widehat{\pi_1} + \widehat{\eta_2}\widehat{\pi_2} + \cdots + \widehat{\eta_n}\widehat{\pi_n} = 1_{\mathrm{Ext}^1_{\mathcal{E}}(C, \oplus_{i \in I} X_i)}$. This means $\mathrm{Ext}^1_{\mathcal{E}}\left(C, \bigoplus_{i \in I} X_i\right)$ is the biproduct of the finite set $\left\{\mathrm{Ext}^1_{\mathcal{E}}(C, X_i)\right\}_{i \in I}$. As in Lemma 1.3, the aforementioned equations immediately provide an isomorphism onto the external direct sum

$$\begin{bmatrix} \widehat{\pi_1} \\ \widehat{\pi_2} \\ \vdots \\ \widehat{\pi_n} \end{bmatrix} : \mathrm{Ext}^1_{\mathcal{E}}\left(C, \bigoplus_{i \in I} X_i\right) \to \bigoplus_{i \in I} \mathrm{Ext}^1_{\mathcal{E}}(C, X_i).$$

Its inverse is precisely the induced morphism ξ of (†):

$$\xi = [\,\widehat{\eta_1}\ \widehat{\eta_2}\ \cdots\ \widehat{\eta_n}\,] : \bigoplus_{i \in I} \mathrm{Ext}^1_{\mathcal{E}}(C, X_i) \to \mathrm{Ext}^1_{\mathcal{E}}\left(C, \bigoplus_{i \in I} X_i\right).$$

We turn to the general case (2). We want to show that ξ is a monomorphism for any index set I. Let us set some notation associated to any given subset $K \subseteq I$. First, we set $\eta_i^K : X_i \to \bigoplus_{i \in K} X_i$, for each $i \in K$. Note that there is an induced map uniquely satisfying the coproduct relations shown:

$$\eta_K : \bigoplus_{i \in K} X_i \to \bigoplus_{i \in I} X_i, \qquad \eta_K \circ \eta_i^K = \eta_i.$$

Similarly, we set $\alpha_i^K : \mathrm{Ext}^1_{\mathcal{E}}(C, X_i) \to \bigoplus_{i \in K} \mathrm{Ext}^1_{\mathcal{E}}(C, X_i)$, for $i \in K$, and note an induced inclusion of abelian groups:

$$\alpha_K : \bigoplus_{i \in K} \mathrm{Ext}^1_{\mathcal{E}}(C, X_i) \to \bigoplus_{i \in I} \mathrm{Ext}^1_{\mathcal{E}}(C, X_i), \qquad \alpha_K \circ \alpha_i^K = \alpha_i.$$

Lastly, for any given morphism $f\colon X \to Y$ in $\mathcal{A}$, we let $\widehat{f}$ denote the group homomorphism $\mathrm{Ext}^1_{\mathcal{E}}(C, f)\colon \mathrm{Ext}^1_{\mathcal{E}}(C, X) \to \mathrm{Ext}^1_{\mathcal{E}}(C, Y)$. Again, it is induced by pushout along f. The fact that it is a functor gives us $\widehat{gf} = \widehat{g}\widehat{f}$. (This is the fact that the composite of two consecutive pushouts is itself a pushout.)

Note that any homomorphism $\bigoplus_{i\in I} A_i \to B$ out of a coproduct of abelian groups is one-to-one if and only if it is one-to-one upon restricting to the summands $\bigoplus_{i\in K} A_i$, as K ranges over all *finite* subsets $K \subseteq I$. So in our previous notation set, our goal is to show that the composite $\xi \circ \alpha_K$ is one-to-one for any finite $K \subseteq I$. We claim that

$$\xi \circ \alpha_K = \widehat{\eta}_K \circ \xi^K, \tag{8.4}$$

where $\xi^K\colon \bigoplus_{i\in K} \mathrm{Ext}^1_{\mathcal{E}}(C, X_i) \to \mathrm{Ext}^1_{\mathcal{E}}\left(C, \bigoplus_{i\in K} X_i\right)$ is the homomorphism in (†) corresponding to the index set $K = \{k_1, k_2, \ldots, k_n\} \subseteq I$. To prove Equation (8.4) it is enough, by the universal property of a coproduct, to show that

$$\xi \circ \alpha_K \circ \alpha_i^K = \widehat{\eta}_K \circ \xi^K \circ \alpha_i^K$$

for all $i \in K$. Indeed using all of the properties listed in our earlier notation we have $\xi \circ (\alpha_K \circ \alpha_i^K) = \xi \circ \alpha_i = \widehat{\eta}_i$, and on the other hand, $\widehat{\eta}_K \circ (\xi^K \circ \alpha_i^K) = \widehat{\eta}_K \circ \widehat{\eta_i^K} = \widehat{\eta_K \circ \eta_i^K} = \widehat{\eta}_i$. This proves Equation (8.4).

So the problem of showing $\xi \circ \alpha_K$ to be one-to-one can be replaced with the problem of showing $\widehat{\eta}_K \circ \xi^K$ is one-to-one. But $K = \{k_1, k_2, \ldots, k_n\}$ is finite, so ξ^K must be an isomorphism. Thus we only need to show that $\widehat{\eta}_K$ is one-to-one. Finally, it is easy to see that $\widehat{\eta}_K$ is one-to-one. For suppose the bottom row of the pushout diagram

$$\begin{array}{ccccc}
\bigoplus_{i\in K} X_i & \rightarrowtail & Y & \twoheadrightarrow & C \\
{\scriptstyle \eta_K}\downarrow & & \downarrow & & \| \\
\bigoplus_{i\in I} X_i & \rightarrowtail & P & \twoheadrightarrow & C
\end{array}$$

splits. Then the top row also must split, since η_K itself is a split monomorphism: Note that η_K has the left inverse $\pi_K\colon \bigoplus_{i\in I} X_i \to \bigoplus_{i\in K} X_i$ induced by $\eta_i^K\colon X_i \to \bigoplus_{i\in K} X_i$ for each $i \in K$, along with the zero morphisms $0\colon X_i \to \bigoplus_{i\in K} X_i$ for all $i \notin K$. This completes the proof that $\widehat{\eta_K}$, and hence ξ, is a monomorphism.

Finally, let us prove statement (3) characterizing when ξ is onto. First, we note that any homomorphism $g\colon \bigoplus_{i\in I} A_i \to B$ out of a coproduct of abelian groups is onto if and only if for all elements $b \in B$, there exists a finite subset $K = \{k_1, k_2, \ldots, k_n\} \subseteq I$ such that $b = (g \circ \alpha_K)(a_{k_1}, a_{k_2}, \ldots, a_{k_n})$, where $(a_{k_1}, a_{k_2}, \ldots, a_{k_n})$ is some finite sequence of elements in the domain of the natural inclusion $\alpha_K\colon \bigoplus_{i\in K} A_i \to \bigoplus_{i\in I} A_i$. So in the current situation, ξ is onto if and only if for any short exact sequence $\mathbb{E} = \bigoplus_{i\in I} X_i \rightarrowtail Y \twoheadrightarrow C$, representing an element of $\operatorname{Ext}^1_{\mathcal{E}}\left(C, \bigoplus_{i\in I} X_i\right)$, there exists a finite subset $K = \{k_1, k_2, \ldots, k_n\} \subseteq I$ and an n-tuple of short exact sequences $(\mathbb{E}_{k_1}, \mathbb{E}_{k_2}, \ldots, \mathbb{E}_{k_n})$ in $\bigoplus_{i\in K} \operatorname{Ext}^1_{\mathcal{E}}(C, X_i)$ such that $\mathbb{E} = (\xi \circ \alpha_K)(\mathbb{E}_{k_1}, \mathbb{E}_{k_2}, \ldots, \mathbb{E}_{k_n})$, where

$$\alpha_K\colon \bigoplus_{i\in K} \operatorname{Ext}^1_{\mathcal{E}}(C, X_i) \to \operatorname{Ext}^1_{\mathcal{E}}\left(C, \bigoplus_{i\in I} X_i\right)$$

is the natural inclusion as already defined earlier. So if ξ is onto, using Equation (8.4) this data gives us $\mathbb{E} = \left(\widehat{\eta}_K \circ \xi^K\right)(\mathbb{E}_{k_1}, \mathbb{E}_{k_2}, \ldots, \mathbb{E}_{k_n})$, where recall $\xi^K\colon \bigoplus_{i\in K} \operatorname{Ext}^1_{\mathcal{E}}(C, X_i) \to \operatorname{Ext}^1_{\mathcal{E}}\left(C, \bigoplus_{i\in K} X_i\right)$ is an isomorphism. Setting $\mathbb{E}_K := \xi^K(\mathbb{E}_{k_1}, \mathbb{E}_{k_2}, \ldots, \mathbb{E}_{k_n})$, this produces a short exact sequence $\mathbb{E}_K$ for which $\mathbb{E} = \widehat{\eta}_K(\mathbb{E}_K)$. This says exactly that we have a commutative diagram as claimed. Conversely, if the commutative diagram property holds, then since ξ^K is an isomorphism we have an n-tuple of short exact sequence $(\mathbb{E}_{k_1}, \mathbb{E}_{k_2}, \ldots, \mathbb{E}_{k_n})$ for which $\xi^K(\mathbb{E}_{k_1}, \mathbb{E}_{k_2}, \ldots, \mathbb{E}_{k_n}) = \mathbb{E}_K$. So then given any $\mathbb{E}$, the pushout property gives us

$$\mathbb{E} = \widehat{\eta}_K(\mathbb{E}_K) = \widehat{\eta}_K\left(\xi^K(\mathbb{E}_{k_1}, \mathbb{E}_{k_2}, \ldots, \mathbb{E}_{k_n})\right) = (\xi \circ \alpha_K)(\mathbb{E}_{k_1}, \mathbb{E}_{k_2}, \ldots, \mathbb{E}_{k_n}),$$

where we again have used Equation (8.4). This proves ξ is onto. □

We will say that an object $C \in \mathcal{A}$ is **Ext compact**, if the map ξ displayed in (†) is always an epimorphism (and hence an isomorphism). Given a class $\mathcal{D}$ of objects in $\mathcal{A}$, we will say that C is **$\mathcal{D}$-Ext compact** if ξ is an epimorphism for any given collection $\{X_i\}_{i\in I} \subseteq \mathcal{D}$.

Definition 8.38 We will say that $\mathfrak{M} = (\mathcal{Q}, \mathcal{W}, \mathcal{R})$ is **compact** if each of the two associated cotorsion pairs are cogenerated by a set of $\mathcal{R}$-Ext compact objects.

Lemma 8.39 (Coproducts in Compact Model Structures) *Assume $(\mathcal{A}, \mathcal{E})$ has exact coproducts and $\mathfrak{M} = (\mathcal{Q}, \mathcal{W}, \mathcal{R})$ is compact. Then each of the following hold.*

(1) *Each of the classes $\mathcal{W}$ = trivial objects, $\mathcal{R}$ = fibrant objects, $\widetilde{\mathcal{R}}$ = trivially fibrant objects, $\mathcal{Q}$ = cofibrant objects, $\widetilde{\mathcal{Q}}$ = trivially cofibrant objects, $\mathcal{Q} \cap \mathcal{R}$ = bifibrant objects, and the core $\omega = \mathcal{Q} \cap \mathcal{W} \cap \mathcal{R}$, are all closed under coproducts.*
(2) *Given a coproduct object, $\bigoplus_{i\in I} A_i \in \mathcal{A}$, its cofibrant replacement may be taken to be $Q\left(\bigoplus_{i\in I} A_i\right) = \bigoplus_{i\in I} QA_i$. Similarly, its fibrant replacement may be taken as $R\left(\bigoplus_{i\in I} A_i\right) = \bigoplus_{i\in I} RA_i$, and $RQ\left(\bigoplus_{i\in I} A_i\right) = \bigoplus_{i\in I} RQA_i$ is a bifibrant replacement.*

(3) *Given a collection of objects* $\{A_i\}_{i\in I}$ *in* $\mathcal{A}$, *let* $A_i \xrightarrow{\eta_i} \bigoplus_{i\in I} A_i$ *denote the canonical injections into a coproduct of* $\{A_i\}_{i\in I}$ *in* $\mathcal{A}$. *Then*

$$\left(\bigoplus_{i\in I} A_i, \{\gamma(\eta_i)\}\right)$$

is a coproduct of $\{A_i\}_{i\in I}$, *in* $\mathrm{Ho}(\mathfrak{M})$. *Moreover,* $\gamma(\eta_i) = [\eta_{RQA_i}]$ *is represented by the homotopy class of* $\eta_{RQA_i} : RQA_i \to \bigoplus_{i\in I} RQA_i$, *the canonical injection or* RQA_i *into the coproduct* $\bigoplus_{i\in I} RQA_i$.

(4) $\mathrm{St}_\omega(Q \cap \mathcal{R})$, *the stable category of the Frobenius category* $Q \cap \mathcal{R}$, *has coproducts taken in the natural way. More precisely, let* $\{A_i\}_{i\in I}$ *be a collection of objects in* $Q \cap \mathcal{R}$ *and let* $\left(\bigoplus_{i\in I} A_i, \{\eta_i\}_{i\in I}\right)$ *denote their coproduct in* $\mathcal{A}$; *so each* $A_i \xrightarrow{\eta_i} \bigoplus_{i\in I} A_i$ *is the canonical injection into the ith component. Then* $\left(\bigoplus_{i\in I} A_i, \{[\eta_i]\}_{i\in I}\right)$ *is a coproduct of* $\{A_i\}_{i\in I}$ *in* $\mathrm{St}_\omega(Q \cap \mathcal{R})$.

Proof For (1), the classes Q and $Q_{\mathcal{W}}$ are always closed under coproducts, by Lemma 2.7. For the classes $\mathcal{R}$ and $\mathcal{R}_{\mathcal{W}}$, let $\mathcal{S}$ be a set of $\mathcal{R}$-Ext compact objects cogenerating $(Q_{\mathcal{W}}, \mathcal{R})$ (resp. $(Q, \mathcal{R}_{\mathcal{W}})$). Let $\{X_i\}_{i\in I}$ be a collection of objects from $\mathcal{R}$ (resp. $\mathcal{R}_{\mathcal{W}}$). The isomorphism

$$\bigoplus_{i\in I} \mathrm{Ext}^1_{\mathcal{E}}(S, X_i) \cong \mathrm{Ext}^1_{\mathcal{E}}\left(S, \bigoplus_{i\in I} X_i\right)$$

holds for each $S \in \mathcal{S}$, proving $\bigoplus_{i\in I} X_i \in \mathcal{R}$ (resp. $\bigoplus_{i\in I} X_i \in \mathcal{R}_{\mathcal{W}}$). So the class of fibrant objects, $\mathcal{R}$, and the class of trivially fibrant objects, $\mathcal{R}_{\mathcal{W}}$, are each closed under coproducts. It now follows easily, as shown in Proposition 5.30, that $\mathcal{W}$ is closed under coproducts. Finally, the class $Q \cap \mathcal{R}$ of bifibrant objects and the core ω are also clearly closed under coproducts.

For (2), it is shown in Proposition 5.30 that we may take $Q\left(\bigoplus_{i\in I} A_i\right) = \bigoplus_{i\in I} QA_i$, because $\mathcal{R}_{\mathcal{W}}$ is closed under coproducts. A similar argument shows that we also may take $R\left(\bigoplus_{i\in I} A_i\right) = \bigoplus_{i\in I} RA_i$, because here $\mathcal{R}$ is also closed under coproducts. Thus we may also take $RQ\left(\bigoplus_{i\in I} A_i\right) = \bigoplus_{i\in I} RQA_i$.

For (3), it was already shown in Proposition 5.30(2) that $\left(\bigoplus_{i\in I} A_i, \{\gamma(\eta_i)\}_{i\in I}\right)$ is a coproduct of $\{A_i\}_{i\in I}$, in $\mathrm{Ho}(\mathfrak{M})$. The proof uses the diagram of short exact sequences

$$\begin{array}{ccccc} R'_i & \rightarrowtail & QA_i & \xrightarrow{p_{A_i}} & A_i \\ \downarrow & & \downarrow{\scriptstyle \eta_{QA_i}} & & \downarrow{\scriptstyle \eta_i} \\ \bigoplus R'_i & \rightarrowtail & \bigoplus QA_i & \xrightarrow{\bigoplus p_{A_i}} & \bigoplus A_i, \end{array}$$

which shows that we can take $Q(\eta_i) = \eta_{QA_i}$. Hence $\gamma(\eta_i) = RQ[\eta_i] = R\left[\eta_{QA_i}\right]$. But now we also have $R\left(\bigoplus_{i\in I} QA_i\right) = \bigoplus_{i\in I} RQA_i$ is a bifibrant replacement of $\bigoplus A_i$. So continuing this argument we see that the commutative diagram

$$\begin{array}{ccccc} QA_i & \overset{j_{QA_i}}{\rightarrowtail} & RQA_i & \twoheadrightarrow & Q' \\ \Big\downarrow{\scriptstyle \eta_{QA_i}} & & \Big\downarrow{\scriptstyle \eta_{RQA_i}} & & \Big\downarrow \\ \bigoplus QA_i & \underset{\bigoplus j_{QA_i}}{\rightarrowtail} & \bigoplus RQA_i & \twoheadrightarrow & \bigoplus Q' \end{array}$$

shows that we can also take $R(\eta_{QA_i}) = \eta_{RQA_i}$. Hence $\gamma(\eta_i) = R[\eta_{QA_i}] = [\eta_{RQA_i}]$ as we have claimed.

Part (4) is clear since we now have $\gamma(\eta_i) = [\eta_i]$. □

Taking $B = \Omega B$ in Theorem 8.18, and using the isomorphism $\Sigma\Omega B \cong B$, in $\mathrm{Ho}(\mathfrak{M})$, we get the following.

Lemma 8.40 *Let $\mathfrak{M} = (Q, \mathcal{W}, \mathcal{R})$ be an hereditary abelian model structure on $(\mathcal{A}, \mathcal{E})$. Then for any cofibrant $C \in C$ and any $B \in \mathcal{A}$ we have an isomorphism*

$$\mathrm{Ho}(\mathfrak{M})(C, B) \cong \mathrm{Ext}^1_{\mathcal{E}}(C, R\Omega B).$$

Definition 8.41 If $\mathcal{T}$ is a triangulated category with coproducts, an object C is called **compact** if $\mathcal{T}(C, -)$ preserves coproducts. That is, the natural map $\bigoplus_{i\in I} \mathcal{T}(C, T_i) \to \mathcal{T}(C, \coprod_{i\in I} T_i)$ is an isomorphism for each collection of objects $\{T_i\}_{i\in I}$ in $\mathcal{T}$. We then say $\mathcal{T}$ is **compactly generated** if it possesses a set of compact weak generators; see Definition 8.20.

Theorem 8.42 *Assume $(\mathcal{A}, \mathcal{E})$ has exact coproducts and $\mathfrak{M} = (Q, \mathcal{W}, \mathcal{R})$ is a compact hereditary model structure. Then $\mathcal{R}$ and $\mathcal{W}$ are closed under coproducts and the homotopy category,* $\mathrm{Ho}(\mathfrak{M})$, *is compactly generated. Any set $\mathcal{S}$ of $\mathcal{R}$-Ext compact objects that cogenerates $(Q, \mathcal{R}_{\mathcal{W}})$ will serve as a set of compact weak generators for* $\mathrm{Ho}(\mathfrak{M})$.

Proof Let $\mathcal{S}$ be a set of $\mathcal{R}$-Ext compact objects that cogenerates $(Q, \mathcal{R}_{\mathcal{W}})$, and let $\{A_i\}_{i\in I}$ be a collection of objects in $\mathrm{Ho}(\mathfrak{M})$. By Lemma 8.39 we know that their usual coproduct $\bigoplus_{i\in I} A_i$ in $\mathcal{A}$ is also a coproduct $\mathrm{Ho}(\mathfrak{M})$. Taking a short exact sequence $\Omega A_i \rightarrowtail W_i \twoheadrightarrow A_i$ with $W_i \in \mathcal{W}$ for each $i \in I$, we obtain a short exact sequence $\bigoplus_{i\in I} \Omega A_i \rightarrowtail \bigoplus_{i\in I} W_i \twoheadrightarrow \bigoplus_{i\in I} A_i$. Since $\bigoplus W_i \in \mathcal{W}$ by Lemma 8.39, this directly shows $\Omega(\bigoplus_{i\in I} A_i) = \bigoplus_{i\in I} \Omega A_i$. Hence by Lemmas 8.40 and 8.39 we have isomorphisms:

$$\mathrm{Ho}(\mathfrak{M})\left(C, \bigoplus_{i\in I} A_i\right) \cong \mathrm{Ext}^1_{\mathcal{E}}\left(C, R\left(\bigoplus_{i\in I} \Omega A_i\right)\right) \cong \mathrm{Ext}^1_{\mathcal{E}}\left(C, \bigoplus_{i\in I} R\Omega A_i\right).$$

Now since each $C \in \mathcal{S}$ is $\mathcal{R}$-Ext compact, and again using Lemma 8.40, we continue the isomorphisms:

$$\cong \bigoplus \operatorname{Ext}^1_{\mathcal{E}}(C, R\Omega A_i) \cong \bigoplus_{i \in I} \operatorname{Ho}(\mathfrak{M})(C, A_i).$$

So each $C \in \mathcal{S}$ is compact. By Theorem 8.21, $\mathcal{S}$ is also a set of weak generators. This proves $\operatorname{Ho}(\mathfrak{M})$ is compactly generated by the set $\mathcal{S}$. □

In the spirit of Theorem 8.22 we have a sort of converse.

Proposition 8.43 *Assume $(\mathcal{A}, \mathcal{E})$ has exact coproducts and $\mathfrak{M} = (\mathcal{Q}, \mathcal{W}, \mathcal{R})$ is an hereditary model structure such that $(\mathcal{Q}_{\mathcal{W}}, \mathcal{R})$ is cogenerated by a set $\mathcal{U}$. If $\mathcal{R}$ and $\mathcal{W}$ are closed under coproducts and $\operatorname{Ho}(\mathfrak{M})$ is compactly generated by a suspension closed set $\mathcal{S}$ (i.e. $\Sigma^n\mathcal{S} \subseteq \mathcal{S}$ for all $n \in \mathbb{Z}$), then $\mathfrak{M}$ is compact.*

Proof It is easy to see that $\mathcal{R}$ is closed under coproducts if and only if each $U \in \mathcal{U}$ is $\mathcal{R}$-Ext compact. So to show that $\mathfrak{M}$ is compact, we only need to show that $(\mathcal{Q}, \mathcal{R}_{\mathcal{W}})$ is cogenerated by a set of $\mathcal{R}$-Ext compact objects. But by the assumption $\Sigma^n\mathcal{S} \subseteq \mathcal{S}$ for all $n \in \mathbb{Z}$, it follows from Theorem 8.22, that it is cogenerated by the set $\mathcal{U} \cup Q(\mathcal{S})$, where $Q(\mathcal{S}) = \{QS \mid S \in \mathcal{S}\}$. By Theorem 8.18 we have natural isomorphisms

$$\operatorname{Ext}^1_{\mathcal{E}}(QA, RB) \cong \operatorname{Ho}(\mathfrak{M})(A, \Sigma B).$$

So, for each $S \in \mathcal{S}$, and collection $\{B_i\}_{i \in I} \subseteq \mathcal{R}$ of fibrant objects, we have

$$\operatorname{Ext}^1_{\mathcal{E}}\left(QS, \bigoplus_{i \in I} B_i\right) \cong \operatorname{Ho}(\mathfrak{M})\left(S, \Sigma \bigoplus_{i \in I} B_i\right) \cong \operatorname{Ho}(\mathfrak{M})\left(S, \bigoplus_{i \in I} \Sigma B_i\right)$$

$$\cong \bigoplus_{i \in I} \operatorname{Ho}(\mathfrak{M})(S, \Sigma B_i) \cong \bigoplus_{i \in I} \operatorname{Ext}^1_{\mathcal{E}}(QS, RB_i) = \bigoplus_{i \in I} \operatorname{Ext}^1_{\mathcal{E}}(QS, B_i).$$

So $Q(\mathcal{S})$ and hence $\mathcal{U} \cup Q(\mathcal{S})$ is a set of $\mathcal{R}$-Ext compact objects, cogenerating $(\mathcal{Q}, \mathcal{R}_{\mathcal{W}})$. □

8.8.1 Examples in Locally Finitely Generated Grothendieck Categories

In applications, abelian model structures are often considered on Grothendieck categories that are *locally finitely generated*. Such a category is, by definition, a Grothendieck category possessing is a generating set consisting of finitely generated objects. We refer the reader to Stenström [1975, p. 121] for further background. The following lemma provides a source of Ext compact objects in such a setting.

Lemma 8.44 *Let $\mathcal{G}$ be a locally finitely generated Grothendieck category. Then any finitely presented object F is Ext compact. That is, the canonical homomorphism*

$$\xi : \bigoplus_{i \in I} \mathrm{Ext}^1_{\mathcal{G}}(F, X_i) \to \mathrm{Ext}^1_{\mathcal{G}}\left(F, \bigoplus_{i \in I} X_i\right)$$

of (†) is an isomorphism for any collection of objects $\{X_i\}_{i \in I}$.

Proof We only need to show that ξ satisfies property (3) of Lemma 8.37. So let $\mathbb{E} \equiv \bigoplus_{i \in I} X_i \rightarrowtail Y \twoheadrightarrow F$ be a given short exact sequence representing an element of $\mathrm{Ext}^1_{\mathcal{G}}\left(F, \bigoplus_{i \in I} X_i\right)$. Using the given set of finitely generated generators for $\mathcal{G}$, we can find a finitely generated G and a morphism $G \to Y$ such that the composition $G \to Y \to F$ is an epimorphism. Then taking the pullback of $\left(\bigoplus X_i \rightarrowtail Y \leftarrow G\right)$ we obtain a morphism of short exact sequences:

$$\begin{array}{ccccc} L & \rightarrowtail & G & \twoheadrightarrow & F \\ \downarrow & & \downarrow & & \| \\ \bigoplus_{i \in I} X_i & \rightarrowtail & Y & \twoheadrightarrow & F. \end{array}$$

Since F is finitely presented, L must be finitely generated. See Stenström [1975, p. 122] for details. Thus the image of $L \to \bigoplus_{i \in I} X_i$ is contained in some finite summand $\bigoplus_{i \in K} X_i \subseteq \bigoplus_{i \in I} X_i$ ($K \subseteq I$ some finite subset). This follows from Stenström [1975, p. 121, proposition 3.2], for $\bigoplus_{i \in I} X_i$ is the direct limit taken over all such finite summands (as $K \subseteq I$ ranges through all finite subsets of I). Taking a pushout we get another morphism of short exact sequences:

$$\begin{array}{ccccc} L & \rightarrowtail & G & \twoheadrightarrow & F \\ \downarrow & & \downarrow & & \| \\ \bigoplus_{i \in K} X_i & \rightarrowtail & Y_K & \twoheadrightarrow & F. \end{array}$$

Moreover, the maps $\eta_K : \bigoplus_{i \in K} X_i \to \bigoplus_{i \in I} X_i \to Y$, and, $G \to Y$ induce, via the universal property of the pushout Y_K, a commutative diagram with exact rows:

$$\begin{array}{ccccc} L & \rightarrowtail & G & \twoheadrightarrow & F \\ \downarrow & & \downarrow & & \| \\ \bigoplus_{i \in K} X_i & \rightarrowtail & Y_K & \twoheadrightarrow & F \\ {\scriptstyle \eta_K}\downarrow & & \downarrow & & \| \\ \bigoplus_{i \in I} X_i & \rightarrowtail & Y & \twoheadrightarrow & F, \end{array}$$

where the middle vertical composition agrees with $G \to Y$. The lower left square must also be a pushout by Proposition 1.12. This verifies property (3) of Lemma 8.37 and proves that ξ is an isomorphism. □

So the following corollary is immediate from Theorem 8.42 and Lemma 8.44.

Corollary 8.45 *Assume $\mathfrak{M} = (\mathcal{Q}, \mathcal{W}, \mathcal{R})$ is an hereditary abelian model structure on a locally finitely generated Grothendieck category $\mathcal{G}$. If each of the two associated cotorsion pairs are cogenerated by a set of finitely presented objects, then $\mathfrak{M}$ is compact in the sense of Definition 8.38. In particular, $\mathcal{R}$ and $\mathcal{W}$ must be closed under coproducts and the homotopy category,* $\mathrm{Ho}(\mathfrak{M})$*, is compactly generated. Any set $\mathcal{S}$ of finitely presented objects that cogenerates $(\mathcal{Q}, \mathcal{R}_{\mathcal{W}})$ serves as a set of compact weak generators for* $\mathrm{Ho}(\mathfrak{M})$.

Example 8.46 The category $\mathrm{Ch}(R)$, of chain complexes of modules over a ring R, is locally finitely generated. The set $\{D^n(R) \mid n \in \mathbb{Z}\}$ of disk complexes is a generating set, and each $D^n(R)$ is finitely presented. The standard projective model structure on $\mathrm{Ch}(R)$, see Example 3 of the Introduction and Main Examples, is compact in the sense of Definition 8.38. It is cogenerated by the set of sphere complexes, $\{S^n(R) \mid n \in \mathbb{Z}\}$, and each $S^n(R)$ is finitely presented. So this is a set of compact weak generators for the homotopy category, which is $\mathcal{D}(R)$, the derived category of R. See also Section 10.8 and Exercise 10.9.2.

Example 8.47 Recall from Example 5 of the Introduction and Main Examples, the Gorenstein projective model structure $\mathfrak{M}_{proj} = (\mathcal{GP}, \mathcal{W}, \mathcal{A}ll)$ on R-Mod, the category of (left) modules over an Iwanaga–Gorenstein ring R. Hovey shows in Hovey [2002] that $\mathrm{Ho}(\mathfrak{M}_{proj})$ is compactly generated. In fact the model structure is compact in the sense of Definition 8.38, as it is cogenerated by the set of syzygies

$$\mathcal{S} = \left\{\Omega^d(R/I) \mid I \subseteq R\right\},$$

where I ranges through all (left) ideals of R. Here, d is the injective dimension of R, and the Noetherian hypothesis implies that we may arrange for each $\Omega^d(R/I)$ to be finitely presented.

This model structure was generalized to *Ding–Chen rings*, a coherent generalization of Iwanaga–Gorenstein rings in Gillespie [2010], and similar arguments show it to be compact.

Exercise 8.8.1 Let $\mathrm{Ch}(R)$ be the category of chain complexes of modules over a ring R. If R is (left) Noetherian, then R/I is finitely presented for each (left) ideal $I \subseteq R$. Show that the set $\{D^n(R/I) \mid n \in \mathbb{Z}\}$, where I runs through all such ideals, cogenerates the injective model structure $\mathfrak{M}_1 = \left(\mathcal{W}_{\mathrm{co}}, dw\widetilde{\mathcal{I}}\right)$

from Example 8.34. Use the recollement in that example to show that $\mathfrak{M}_2 = \left(\mathcal{W}_{\text{st}}, ex\widetilde{\mathcal{I}}\right)$ has a compactly generated homotopy category.

Exercise 8.8.2 Let R be an Iwanaga–Gorenstein ring, and let $\mathfrak{M}_{inj} = (\mathcal{W}, \mathcal{GI})$ be the Gorenstein injective model structure on R-Mod; see Example 5 of the Introduction and Main Examples. Show that the sphere functor $S^0(-)\colon R\text{-Mod} \to \text{Ch}(R)$ is a (left) Quillen equivalence from $\mathfrak{M}_{inj}$ to the model structure $\mathfrak{M}_2 = \left(\mathcal{W}_{\text{st}}, ex\widetilde{\mathcal{I}}\right)$ from Examples 8.34 and Exercise 8.8.1.

9

Constructing Complete Cotorsion Pairs

The goal of this chapter is to develop a very general method for constructing (functorially) complete cotorsion pairs in exact categories. In essence we develop an algebraic version of Quillen's small object argument for cotorsion pairs. This also leads us to naturally consider cofibrantly generated cotorsion pairs and abelian model structures. The approach taken here is inspired by the notion of an *efficient exact category* as defined in Saorín and Šťovíček [2011]. These are exact categories satisfying weak analogs of Grothendieck's axioms for abelian categories. We generalize this idea a bit more by considering classes of objects that are *efficient* relative to the exact structure. The main idea is that with mild hypotheses on the exact category, any efficient set (not a proper class) of objects cogenerates a functorially complete cotorsion pair.

9.1 Transfinite Extensions and Eklof's Lemma

The concept of a transfinite extension, or filtration, is important to homological algebra. It allows us to describe objects that are built up from objects in some set, as special types of direct limits. In this section, we define transfinite extensions and prove the fundamental Eklof's Lemma, which is a theorem concerning the vanishing of $\mathrm{Ext}^1_{\mathcal{E}}$ for transfinite extensions. We formulate a version of Eklof's Lemma that holds in *any* exact category, and has the further advantage that we automatically obtain its (also useful) dual.

Let us first define the notion of transfinite extensions in an exact category $(\mathcal{A}, \mathcal{E})$. To do so, let λ be an ordinal. Thinking of it as a category, a functor $\lambda \to \mathcal{A}$ determines a *direct λ-system* of objects and morphisms $\{X_\alpha, i_{\alpha\beta}\}_{\alpha \leq \beta < \lambda}$ where $i_{\alpha\beta} \colon X_\alpha \to X_\beta$ is a morphism. Such a system satisfies (i) $i_{\alpha\alpha} = 1_{X_\alpha}$ for all $\alpha < \lambda$, and (ii) $i_{\alpha\gamma} = i_{\beta\gamma} \circ i_{\alpha\beta}$ for all $\alpha \leq \beta \leq \gamma < \lambda$. Since the source of such a functor $\lambda \to \mathcal{A}$ is a directed set, its colimit (if it exists) is written $\varinjlim_{\alpha<\lambda} X_\alpha$, and called the *direct limit* of $\{X_\alpha, i_{\alpha\beta}\}_{\alpha\leq\beta<\lambda}$. A direct λ-system $\{X_\alpha, i_{\alpha\beta}\}_{\alpha\leq\beta<\lambda}$ is called a *continuous direct λ-system* (or a *λ-sequence*) if the functor $\lambda \to \mathcal{A}$ preserves colimits. It means that for each limit ordinal $\kappa < \lambda$ we have that $(X_\kappa, i_{\alpha\kappa})_{\alpha<\kappa}$ is the direct limit of the direct κ-subsystem $\{X_\alpha, i_{\alpha\beta}\}_{\alpha\leq\beta<\kappa}$. That is, for each limit ordinal $\kappa < \lambda$, we have $X_\kappa = \varinjlim_{\alpha<\kappa} X_\alpha$ with the morphisms $i_{\alpha\kappa}: X_\alpha \to X_\kappa$ serving as the canonical morphisms into the colimit. Finally, assuming that it is exists, let $X := \varinjlim_{\alpha<\lambda} X_\alpha$ be the direct limit, and let $i_\alpha: X_\alpha \to X$ denote the canonical morphisms. Then the canonical map $i_0: X_0 \to X$ is called the **transfinite composition** of the direct λ-system $\{X_\alpha, i_{\alpha\beta}\}_{\alpha\leq\beta<\gamma}$.

So far there has been no mention of the given exact structure $\mathcal{E}$. Doing so brings us to the crucial notion of a *transfinite extension.*

Definition 9.1 Let $(\mathcal{A}, \mathcal{E})$ be any exact category and let $\{X_\alpha, i_{\alpha\beta}\}_{\alpha\leq\beta<\gamma}$ be a continuous direct λ-system with each transition morphism $i_{\alpha,\alpha+1}: X_\alpha \rightarrowtail X_{\alpha+1}$ an admissible monic as depicted:

$$X_0 \rightarrowtail X_1 \rightarrowtail X_2 \rightarrowtail \cdots \rightarrowtail X_\alpha \rightarrowtail X_{\alpha+1} \rightarrowtail \cdots . \tag{9.1}$$

Then we say that $\{X_\alpha, i_{\alpha\beta}\}_{\alpha\leq\beta<\lambda}$ is a **λ-extension sequence**. Assuming the direct limit $X := \varinjlim_{\alpha<\lambda} X_\alpha$ exists, then the *transfinite composition* $i_0: X_0 \to X$ of the λ-extension sequence is also called a **transfinite extension** of X_0. Moreover, assume $\mathcal{T}$ is a set (or even a class) of objects. If each cokernel $\operatorname{Cok} i_{\alpha,\alpha+1} = X_{\alpha+1}/X_\alpha \in \mathcal{T}$, then we say X is a *transfinite extension of* X_0 *by* $\mathcal{T}$. If in addition $X_0 \in \mathcal{T}$, then we say that X is a *transfinite extension of* $\mathcal{T}$.

We should note that a transfinite composition $i_0: X_0 \to X$ of a λ-extension sequence may not be an admissible monic. This is precisely the defining condition of the efficient classes introduced in Section 9.2.

Remark 9.2 A transfinite extension of $\mathcal{T}$ is commonly referred to as a **$\mathcal{T}$-filtered** object, and we will use both terminologies. The latter terminology typically uses the convention $X_0 = 0$. In this case, the λ-extension sequence depicted in (9.1) is referred to as a *$\mathcal{T}$-filtration* of X. The class of all $\mathcal{T}$-filtered objects (i.e. transfinite extensions of $\mathcal{T}$) will be denoted by $\operatorname{Filt}(\mathcal{T})$.

Exercise 9.1.1 Let $\mathcal{A}$ be an additive category and let us be given a continuous direct λ-system $\{X_\alpha, i_{\alpha\beta}\}_{\alpha\leq\beta<\lambda}$

$$X_0 \rightarrowtail X_1 \rightarrowtail X_2 \rightarrowtail \cdots \rightarrowtail X_\alpha \rightarrowtail X_{\alpha+1} \rightarrowtail \cdots$$

for which each $i_{\alpha,\alpha+1}: X_\alpha \rightarrowtail X_{\alpha+1}$ is a split monic with a cokernel C_α. Show that if the coproduct $\bigoplus_{\alpha<\lambda} C_\alpha$ exists, then the canonical split monomorphism

$\eta_{X_0} : X_0 \to X_0 \oplus \bigoplus_{\alpha<\lambda} C_\alpha$ is the transfinite composition of $\{X_\alpha, i_{\alpha\beta}\}_{\alpha\leq\beta<\lambda}$. Consequently, for any exact structure $(\mathcal{A}, \mathcal{E})$, any (existing) coproduct of objects in $\mathcal{T}$ is in $\mathrm{Filt}(\mathcal{T})$.

We now prove a general form of Eklof's Lemma. Note that it does not require $\mathcal{A}$ to always possess any particular type of direct limits, only those that already exist in some particularly given transfinite extension. It also does not require admissible monics to be closed under transfinite compositions.

Lemma 9.3 (Eklof's Lemma) *Let Y be any object in an exact category $(\mathcal{A}, \mathcal{E})$. Assume $\{X_\alpha, i_{\alpha\beta}\}_{\alpha\leq\beta<\lambda}$ is a λ-extension sequence for which the transfinite extension $X := \varinjlim_{\alpha<\lambda} X_\alpha$ exists. If $\mathrm{Ext}^1_{\mathcal{E}}(X_0, Y) = 0$ and $\mathrm{Ext}^1_{\mathcal{E}}(X_{\alpha+1}/X_\alpha, Y) = 0$ for all $\alpha < \lambda$, then $\mathrm{Ext}^1_{\mathcal{E}}(X, Y) = 0$ too.*

Equivalently, let $\mathcal{T}$ be any class of objects in $\mathcal{A}$. If $\mathrm{Ext}^1_{\mathcal{E}}(T, Y) = 0$ for all $T \in \mathcal{T}$, then $\mathrm{Ext}^1_{\mathcal{E}}(X, Y) = 0$ for all $X \in \mathrm{Filt}(\mathcal{T})$.

Proof Suppose we are given such a transfinite extension $X := \varinjlim_{\alpha<\lambda} X_\alpha$ corresponding to a λ-extension sequence $\{X_\alpha, i_{\alpha\beta}\}_{\alpha\leq\beta<\lambda}$

$$X_0 \rightarrowtail X_1 \rightarrowtail X_2 \rightarrowtail \cdots \rightarrowtail X_\alpha \rightarrowtail X_{\alpha+1} \rightarrowtail \cdots$$

with $\mathrm{Ext}^1_{\mathcal{E}}(X_0, Y) = 0$ and $\mathrm{Ext}^1_{\mathcal{E}}(X_{\alpha+1}/X_\alpha, Y) = 0$ for all $\alpha < \lambda$. We will use transfinite induction to show that $\mathrm{Ext}^1_{\mathcal{E}}(X_\beta, Y) = 0$ for all $\beta \leq \lambda$, where we set $X_\lambda := X = \varinjlim_{\alpha<\lambda} X_\alpha$. The base step of the induction is a given. For the successor ordinal step, we note that if $\mathrm{Ext}^1_{\mathcal{E}}(X_\alpha, Y) = 0$, then it follows that $\mathrm{Ext}^1_{\mathcal{E}}(X_{\alpha+1}, Y) = 0$ too, since $\mathrm{Ext}^1_{\mathcal{E}}(X_{\alpha+1}/X_\alpha, Y) = 0$ is also given.

For the limit ordinal step, suppose $\beta \leq \lambda$ is a limit ordinal and $\mathrm{Ext}^1_{\mathcal{E}}(X_\alpha, Y) = 0$ for all $\alpha < \beta$. An element of $\mathrm{Ext}^1_{\mathcal{E}}(X_\beta, Y)$ is represented by a short exact sequence:

$$\mathbb{E}: \qquad Y \overset{f}{\rightarrowtail} B \overset{p}{\twoheadrightarrow} X_\beta.$$

For each $\alpha < \beta$, taking the pullback of $i_{\alpha\beta}$ along p provides us with the following commutative diagram of short exact sequences:

$$\begin{array}{llcccc}
\mathbb{E}_\alpha: & Y & \overset{f_\alpha}{\rightarrowtail} & B_\alpha & \overset{p_\alpha}{\twoheadrightarrow} & X_\alpha \\
& \| & & \downarrow{\scriptstyle m_\alpha} & & \downarrow{\scriptstyle i_{\alpha\beta}} \\
\mathbb{E}: & Y & \underset{f}{\rightarrowtail} & B & \underset{p}{\twoheadrightarrow} & X_\beta.
\end{array}$$

In fact, we may construct a commutative diagram of short exact sequences as follows:

$$\begin{array}{llcccc}
\mathbb{E}_\alpha: & Y & \overset{f_\alpha}{\rightarrowtail} & B_\alpha & \overset{p_\alpha}{\twoheadrightarrow} & X_\alpha \\
 & \| & & \Big\downarrow{\scriptstyle m_{\alpha,\alpha+1}} & & \Big\downarrow{\scriptstyle i_{\alpha,\alpha+1}} \\
\mathbb{E}_{\alpha+1}: & Y & \underset{f_{\alpha+1}}{\rightarrowtail} & B_{\alpha+1} & \underset{p_{\alpha+1}}{\twoheadrightarrow} & X_{\alpha+1} \\
 & \| & & \Big\downarrow{\scriptstyle m_{\alpha+1}} & & \Big\downarrow{\scriptstyle i_{\alpha+1,\beta}} \\
\mathbb{E}: & Y & \underset{f}{\rightarrowtail} & B & \underset{p}{\twoheadrightarrow} & X_\beta.
\end{array}$$

Indeed the far right vertical composition is just $i_{\alpha\beta}$. So the universal property of the pullback $B_{\alpha+1}$ provides for a unique map $m_{\alpha,\alpha+1}$ making the upper right square commute and such that $m_\alpha = m_{\alpha+1} \circ m_{\alpha,\alpha+1}$. Moreover, by Proposition 1.12(2) the upper right square must be a pullback and so by Corollary 1.14(1) the morphism $m_{\alpha,\alpha+1}$ is an admissible monic (with $\operatorname{Cok} m_{\alpha,\alpha+1} = X_{\alpha+1}/X_\alpha$).

Claim There exist, for all $\alpha < \beta$, splittings $s_\alpha \colon X_\alpha \to B_\alpha$ of $p_\alpha \colon B_\alpha \to X_\alpha$ satisfying the relation $m_{\alpha,\alpha+1} s_\alpha = s_{\alpha+1} i_{\alpha,\alpha+1}$.

Keeping β fixed, we start a second transfinite induction to prove the *Claim*. For the base step of this new transfinite induction we start by simply taking a splitting s_0. For the successor ordinal step, suppose α is an ordinal and the required s_α has been constructed. Since $\alpha + 1 < \beta$, we have $\operatorname{Ext}^1_{\mathcal{E}}(Y, X_{\alpha+1}) = 0$, so we may certainly choose a splitting $t \colon X_{\alpha+1} \to B_{\alpha+1}$ for $p_{\alpha+1}$. We now modify t to a splitting $s_{\alpha+1}$ satisfying $m_{\alpha,\alpha+1} s_\alpha = s_{\alpha+1} i_{\alpha,\alpha+1}$. To do so, we note that $p_{\alpha+1}(m_{\alpha,\alpha+1} s_\alpha - t i_{\alpha,\alpha+1}) = 0$. Since $f_{\alpha+1} = \ker p_{\alpha+1}$, there is a map $h \colon X_\alpha \to Y$ such that $f_{\alpha+1} h = m_{\alpha,\alpha+1} s_\alpha - t i_{\alpha,\alpha+1}$. Now since

$$\operatorname{Hom}_{\mathcal{A}}(X_{\alpha+1}, Y) \xrightarrow{i^*_{\alpha,\alpha+1}} \operatorname{Hom}_{\mathcal{A}}(X_\alpha, Y) \to \operatorname{Ext}^1_{\mathcal{E}}(X_{\alpha+1}/X_\alpha, Y) = 0$$

is exact we have a map $g \colon X_{\alpha+1} \to Y$ such that $g i_{\alpha,\alpha+1} = h$. We set $s_{\alpha+1} = t + f_{\alpha+1} g$. Then $p_{\alpha+1} s_{\alpha+1} = p_{\alpha+1} t + p_{\alpha+1} f_{\alpha+1} g = p_{\alpha+1} t + 0 = 1_{X_{\alpha+1}}$. So $s_{\alpha+1}$ is a splitting of $p_{\alpha+1}$. Also $s_{\alpha+1}$ is compatible with the other s_α, for $s_{\alpha+1} i_{\alpha,\alpha+1} = (t + f_{\alpha+1} g) i_{\alpha,\alpha+1} = t i_{\alpha,\alpha+1} + f_{\alpha+1} h = t i_{\alpha,\alpha+1} + (m_{\alpha,\alpha+1} s_\alpha - t i_{\alpha,\alpha+1}) = m_{\alpha,\alpha+1} s_\alpha$. This shows the desired relation $m_{\alpha,\alpha+1} s_\alpha = s_{\alpha+1} i_{\alpha,\alpha+1}$, completing the successor ordinal step of the second transfinite induction. We turn to the limit ordinal step of the second transfinite induction: Given any limit ordinal $\gamma < \beta$, and compatible splittings s_α for all $\alpha < \gamma$, we take $s_\gamma \colon X_\gamma = \varinjlim_{\alpha<\gamma} X_\alpha, \to B_\gamma$ to be the map induced by the cocone $\{m_{\alpha\gamma} \circ s_\alpha\}_{\alpha<\gamma}$ above the subsystem determined by the γ-extension sequence $\{X_\alpha, i_{\alpha\alpha'}\}_{\alpha \leq \alpha' < \gamma}$. Here $m_{\alpha\gamma} \colon B_\alpha \to B_\gamma$ is the unique morphism satisfying $m_\alpha = m_\gamma \circ m_{\alpha\gamma}$ by the same pullback argument used previously to obtain the map $m_{\alpha,\alpha+1}$. The map s_γ is indeed a splitting for p_γ, for we have $(p_\gamma s_\gamma) i_{\alpha\gamma} = p_\gamma (s_\gamma i_{\alpha\gamma}) = p_\gamma (m_{\alpha\gamma} \circ s_\alpha) = (p_\gamma m_{\alpha\gamma}) s_\alpha = (i_{\alpha\gamma} \circ$

$p_\alpha)s_\alpha = i_{\alpha\gamma}1_{X_\alpha} = i_{\alpha\gamma}$. So the universal property of the colimit $(X_\gamma, \{i_{\alpha\gamma}\}_{\alpha<\gamma})$ gives us $p_\gamma s_\gamma = 1_{X_\gamma}$. This completes the second transfinite induction, and hence completes the proof of the aforementioned *Claim*.

So now we return to compete the limit ordinal step of the first transfinite induction. Let $\{s_\alpha\}_{\alpha<\beta}$ be a collection of splittings for p_α satisfying the relations $m_{\alpha,\alpha+1}s_\alpha = s_{\alpha+1}i_{\alpha,\alpha+1}$ for all $\alpha < \beta$. Composing on the left with $m_{\alpha+1}$ we get $m_\alpha s_\alpha = (m_{\alpha+1}s_{\alpha+1})i_{\alpha,\alpha+1}$, which means that the collection $\{m_\alpha s_\alpha\}_{\alpha<\beta}$ is a cocone above the subsystem determined by the β-extension sequence $\{X_\alpha, i_{\alpha\alpha'}\}_{\alpha\le\alpha'<\beta}$. Since $\left(X_\beta, \left\{i_{\alpha\beta}\right\}\right)$ is the colimit $X_\beta = \varinjlim_{\alpha<\beta} X_\alpha$, this gives rise to a unique map $s\colon X_\beta \to B$ satisfying $si_{\alpha\beta} = m_\alpha s_\alpha$ for all $\alpha < \beta$. Therefore we have $(ps)i_{\alpha\beta} = pm_\alpha s_\alpha = (i_{\alpha\beta} \circ p_\alpha)s_\alpha = i_{\alpha\beta}1_{X_\alpha} = i_{\alpha\beta}$. So the universal property of the colimit $\left(X_\beta, \left\{i_{\alpha\beta}\right\}\right)$ gives us $ps = 1_{X_\beta}$ and thus $\mathrm{Ext}^1_{\mathcal{E}}(X_\beta, Y) = 0$ as required. □

The reader will easily verify the following useful consequence of Eklof's Lemma.

Corollary 9.4 *Let $(\mathcal{A}, \mathcal{E})$ be an exact category and let $\mathcal{T}$ be any set, or even any class, of objects. Then* $\mathrm{Filt}(\mathcal{T}) \subseteq {}^\perp(\mathcal{T}^\perp)$.

Later we will show more. For a nice enough set of objects $\mathcal{T}$ we have that ${}^\perp(\mathcal{T}^\perp)$ consists precisely of direct summands of objects in $\mathrm{Filt}(\mathcal{T})$. See Theorem 9.34 and especially Corollary 9.40.

Next we discuss inverse transfinite extensions and the dual of Eklof's Lemma. Because of Eklof's Lemma and its dual, we obtain the following fundamental result about cotorsion pairs.

Corollary 9.5 *Let $(\mathcal{X}, \mathcal{Y})$ be any cotorsion pair in any exact category $(\mathcal{A}, \mathcal{E})$. Then $\mathcal{X}$ is closed under any transfinite extensions that may exist in $\mathcal{A}$, and $\mathcal{Y}$ is closed under any inverse transfinite extensions that may exist. In other words,* $\mathcal{X} = \mathrm{Filt}(\mathcal{X})$, *and* $\mathcal{Y} = \mathrm{CoFilt}(\mathcal{Y})$.

9.1.1 Inverse Transfinite Extensions and the Dual of Eklof's Lemma

The preceding presentation of Eklof's Lemma is readily dualizable, and it too can be useful. So we briefly describe inverse transfinite extensions and state the dual of Eklof's Lemma. Again, we let λ be an ordinal. A contravariant functor $\lambda \to \mathcal{A}$ determines an *inverse λ-system* of objects and morphisms $\left\{X_\alpha, p_{\alpha\beta}\right\}_{\alpha\le\beta<\lambda}$ where $p_{\alpha\beta}\colon X_\beta \to X_\alpha$ is a morphism. Such a system satisfies (i) $p_{\alpha\alpha} = 1_{X_\alpha}$ for all $\alpha < \lambda$, and (ii) $p_{\alpha\gamma} = p_{\alpha\beta} \circ p_{\beta\gamma}$ for all $\alpha \le \beta \le \gamma < \lambda$. An inverse λ-system $\left\{X_\alpha, p_{\alpha\beta}\right\}_{\alpha\le\beta<\lambda}$ is called a *continuous inverse λ-system* if the functor $\lambda \to \mathcal{A}$ takes colimits to limits. Here, we mean that for each limit ordinal $\kappa < \lambda$ we have that $(X_\kappa, p_{\alpha\kappa})_{\alpha<\kappa}$ is the inverse limit of the inverse κ-subsystem $\{X_\alpha, p_{\alpha\beta}\}_{\alpha\leq\beta<\kappa}$. That is, for each limit ordinal $\kappa < \lambda$, we have $X_\kappa = \varprojlim_{\alpha<\kappa} X_\alpha$ and the morphisms $p_{\alpha\kappa}\colon X_\kappa \to X_\alpha$ serve as the canonical morphisms out of the limit. Finally, assuming that it is exists, let $X := \varprojlim_{\alpha<\lambda} X_\alpha$ be the inverse limit, and denote the canonical morphisms out of the limit by $p_\alpha\colon X \to X_\alpha$. Then the canonical map $p_0\colon X \to X_0$ is called the *inverse transfinite composition* of the inverse λ-system $\{X_\alpha, p_{\alpha\beta}\}_{\alpha\leq\beta<\gamma}$.

Definition 9.6 Let $(\mathcal{A}, \mathcal{E})$ be an exact category and let $\{X_\alpha, p_{\alpha\beta}\}_{\alpha\leq\beta<\gamma}$ be a continuous inverse λ-system with each transition morphism $p_{\alpha,\alpha+1} : X_{\alpha+1} \twoheadrightarrow X_\alpha$ an admissible epic as depicted:

$$X_0 \twoheadleftarrow X_1 \twoheadleftarrow X_2 \twoheadleftarrow \cdots \twoheadleftarrow X_\alpha \twoheadleftarrow X_{\alpha+1} \twoheadleftarrow \cdots. \tag{9.2}$$

Then we say that $\{X_\alpha, p_{\alpha\beta}\}_{\alpha\leq\beta<\lambda}$ is an **inverse λ-extension sequence**. Assuming the limit $X := \varprojlim_{\alpha<\lambda} X_\alpha$ exists, then the *inverse transfinite composition* $p_0\colon X \to X_0$ is also called an **inverse transfinite extension** of X_0. Moreover, assume $\mathcal{T}$ is a set (or even any class) of objects. If each kernel $\operatorname{Ker} p_{\alpha,\alpha+1} = K_{\alpha+1} \in \mathcal{T}$, then we say X is an *inverse transfinite extension of X_0 by $\mathcal{T}$*. If in addition $X_0 \in \mathcal{T}$, then we say that X is an *inverse transfinite extension of $\mathcal{T}$*.

Remark 9.7 An inverse transfinite extension of $\mathcal{T}$ is also called a **$\mathcal{T}$-cofiltered** object and this typically uses the convention $X_0 = 0$. In this case, the inverse λ-extension sequence depicted in (9.2) is referred to as a *$\mathcal{T}$-cofiltration* of X. The class of all $\mathcal{T}$-cofiltered objects (i.e. inverse transfinite extensions of $\mathcal{T}$) will be denoted by $\operatorname{CoFilt}(\mathcal{T})$.

Lemma 9.8 (Dual of Eklof's Lemma) *Let X be any object in an exact category $(\mathcal{A}, \mathcal{E})$. Assume $\{Y_\alpha, p_{\alpha\beta}\}_{\alpha\leq\beta<\lambda}$ is an inverse λ-extension sequence for which the inverse transfinite extension $Y := \varprojlim_{\alpha<\lambda} Y_\alpha$ exists. If $\operatorname{Ext}^1_{\mathcal{E}}(X, Y_0) = 0$ and $\operatorname{Ext}^1_{\mathcal{E}}(X, \operatorname{Ker} p_{\alpha,\alpha+1}) = 0$ for all $\alpha < \lambda$, then $\operatorname{Ext}^1_{\mathcal{E}}(X, Y) = 0$ too.*

Equivalently, let $\mathcal{T}$ be any class of objects in $\mathcal{A}$. If $\operatorname{Ext}^1_{\mathcal{E}}(X, T) = 0$ for all $T \in \mathcal{T}$, then $\operatorname{Ext}^1_{\mathcal{E}}(X, Y) = 0$ for all $Y \in \operatorname{CoFilt}(\mathcal{T})$.

Corollary 9.9 *Let $(\mathcal{A}, \mathcal{E})$ be an exact category and let $\mathcal{T}$ be any set, or even any class, of objects. Then $\operatorname{CoFilt}(\mathcal{T}) \subseteq ({}^\perp\mathcal{T})^\perp$.*

Example 9.10 Any chain complex X, over any additive category, that is bounded below (resp. bounded above) can be expressed as a transfinite extension (resp. inverse transfinite extension) of "sphere" complexes. Furthermore, if X is a bounded above (resp. bounded below) exact complex, then it may be expressed as a transfinite extension (resp. inverse transfinite extension) of "disk" complexes. See Exercise 10.2.2.

9.2 Efficient Sets and Transfinite Compositions

Recall that we require the cofibrations in an abelian model structure to be admissible monics. This implies that if we want to "generate" a class of cofibrations by some given set of admissible monics, we need transfinite compositions of those monics to again be admissible monics. We deal with this by focusing on classes of objects that we call *efficient* relative to $\mathcal{E}$. Later in this chapter, efficient sets (with generators) will be shown to cogenerate complete cotorsion pairs relative to $\mathcal{E}$. For convenience, we *assume throughout this section that* $(\mathcal{A}, \mathcal{E})$ *is a weakly idempotent complete exact category.*

Recall that for any given class of objects $\mathcal{S}$, an $\mathcal{S}$*-cofibration* is an admissible monic possessing a cokernel in $\mathcal{S}$.

Definition 9.11 A class $\mathcal{S}$ of objects of $\mathcal{A}$ is said to be **efficient relative to** $\mathcal{E}$, or **efficient for** $\mathcal{E}$, or simply **efficient**, if transfinite compositions of $\mathcal{S}$-cofibrations exist and are admissible monics. Spelled out in more detail, for any λ-extension sequence $\{X_\alpha, i_{\alpha\beta}\}_{\alpha \leq \beta < \lambda}$

$$X_0 \rightarrowtail X_1 \rightarrowtail X_2 \rightarrowtail \cdots \rightarrowtail X_\alpha \rightarrowtail X_{\alpha+1} \rightarrowtail \cdots \tag{9.3}$$

for which each $i_{\alpha,\alpha+1} \colon X_\alpha \rightarrowtail X_{\alpha+1}$ is an $\mathcal{S}$-cofibration, we have the following.

- The colimit $X := \varinjlim_{\alpha<\lambda} X_\alpha$ exists.
- The canonical morphism $i_0 \colon X_0 \rightarrowtail X$ into the colimit, that is, the transfinite composition of $\{X_\alpha, i_{\alpha\beta}\}_{\alpha \leq \beta < \lambda}$, is an admissible monic.

Note in particular, that all transfinite extensions of an efficient class $\mathcal{S}$ exist. Again, we let Filt($\mathcal{S}$) denote the class of all such $\mathcal{S}$*-filtered* objects.

Example 9.12 Assume $\mathcal{A}$ has coproducts. Then any class $\mathcal{S}$ consisting of objects that are each projective relative to $\mathcal{E}$, is efficient for $\mathcal{E}$. In this case Filt($\mathcal{S}$) = Free($\mathcal{S}$) is the class of all coproducts of $\mathcal{S}$-objects. See Exercise 9.1.1.

In most applications, *every* class, including $\mathcal{S} = \mathcal{A}$, the class of all objects, is efficient relative to $\mathcal{E}$. Note that in this case, all transfinite compositions of admissible monics exist and are again admissible monics. We will say that such an exact category satisfies the **Transfinite Extensions Axiom (EF1)**. This will be the first axiom in the Definition of an *efficient exact category* given in Section 9.9. However, Definition 9.11 will allow us to define the general notion of a cofibrantly generated abelian model structure. For example, Section 10.8 gives a class of examples of cofibrantly generated abelian model categories on exact structures that need not satisfy Axiom (EF1).

Example 9.13 Let R-Mod be the category of (left) R-modules. Its standard exact structure, and its split exact structure, each satisfy the Transfinite Extensions Axiom (EF1); the latter statement is again Exercise 9.1.1. Also, the pure exact structure on R-Mod satisfies Axiom (EF1); see Example 9.43. However, for large cardinals λ, the λ-pure exact sequences of Adámek and Rosický [1994] need not satisfy Axiom (EF1).

Example 9.14 Assume $(\mathcal{A}, \mathcal{E})$ does not necessarily satisfy Axiom (EF1), but that $\mathcal{A}$ has another exact structure $\mathcal{E}' \subseteq \mathcal{E}$ for which $\mathcal{E}'$ does satisfy Axiom (EF1). Suppose $\mathcal{S}$ is a class of objects such that every $\mathcal{S}$-cofibration relative to $\mathcal{E}$ is also an $\mathcal{S}$-cofibration relative to $\mathcal{E}'$. Then $\mathcal{S}$ is not just efficient for $\mathcal{E}'$, it is also efficient for $\mathcal{E}$. For example, any class $\mathcal{S}$ consisting of flat R-modules is efficient relative to any exact structure containing the pure exact structure.

In the situation of Definition 9.11, we will typically denote the canonical morphisms into the colimit by $i_\alpha \colon X_\alpha \rightarrowtail X$. So note that we have the compatibilities $i_\alpha = i_\beta \circ i_{\alpha\beta}$ whenever $\alpha \le \beta < \lambda$. In fact, *each* canonical morphism $i_\alpha \colon X_\alpha \rightarrowtail X$ must be an admissible monic. The same is true of each transition morphism $i_{\alpha\beta} \colon X_\alpha \rightarrowtail X_\beta$. We will prove this in Lemma 9.15. The proof will call on the following exercise which one can either prove directly or cite cofinality.

Exercise 9.2.1 Let $\left\{X_\alpha, i_{\alpha\beta}\right\}_{\alpha \le \beta < \lambda}$ be any λ-extension sequence with existing colimit $X = \varinjlim_{\alpha<\lambda} X_\alpha$ and canonical maps $i_\alpha \colon X_\alpha \to X$. Show that for any particular $\alpha < \lambda$, we have $X = \varinjlim_{\alpha \le \gamma < \lambda} X_\gamma$. That is, X is the colimit of the direct subsequence $\{X_\gamma, i_{\gamma\gamma'}\}_{\alpha \le \gamma \le \gamma' < \lambda}$.

Lemma 9.15 *Let $\mathcal{S}$ be an efficient class of objects and $\left\{X_\alpha, i_{\alpha\beta}\right\}_{\alpha \le \beta < \lambda}$ be a λ-extension sequence of $\mathcal{S}$-cofibrations $i_{\alpha,\alpha+1} \colon X_\alpha \rightarrowtail X_{\alpha+1}$. Then the following hold.*

(1) *Each $i_{\alpha\beta} \colon X_\alpha \rightarrowtail X_\beta$ is also an admissible monic.*
(2) *Each $i_\alpha \colon X_\alpha \rightarrowtail X$ is an admissible monic.*

Proof To prove (2), truncate the entire λ-extension sequence $\left\{X_\alpha, i_{\alpha\beta}\right\}_{\alpha \le \beta < \lambda}$ at some particular α. We obtain the extension subsequence of $\mathcal{S}$-cofibrations:

$$\{X_\alpha \rightarrowtail X_{\alpha+1} \rightarrowtail X_{\alpha+2} \rightarrowtail \cdots \rightarrowtail X_\gamma \rightarrowtail X_{\gamma+1} \rightarrowtail \cdots\}_{\alpha \le \gamma < \lambda}.$$

We use the following standard set-theoretic fact: *Every well-ordered set is uniquely order isomorphic to a unique ordinal number.* So we may reindex by another ordinal λ'. In particular, there exists an ordinal λ' such that the previous extension sequence may be reindexed and realized as a λ'-extension sequence $\{X'_\sigma, j_{\sigma\tau}\}_{\sigma\leq\tau<\lambda'}$ as follows:

$$\{X'_0 = X_\alpha \rightarrowtail X'_1 = X_{\alpha+1} \rightarrowtail X'_2 = X_{\alpha+2} \rightarrowtail \cdots \rightarrowtail X'_\sigma \rightarrowtail X'_{\sigma+1} \rightarrowtail \cdots\}_{\sigma<\lambda'}.$$

By the assumption that transfinite compositions of $\mathcal{S}$-cofibrations are admissible monics, we have that $X' = \varinjlim_{\sigma<\lambda'} X_\sigma$ exists and the canonical morphism $j_0 \colon X'_0 \rightarrowtail X'$ is an admissible monic. But we also have that $X' = \varinjlim_{\sigma<\lambda'} X_\sigma$ coincides with $X = \varinjlim_{\alpha\leq\gamma<\lambda} X_\gamma$ (here we are using cofinality, or Exercise 9.2.1). In the process, the canonical morphisms $j_0 = i_\alpha$ identify. So i_α is an admissible monic.

Note that (1) follows easily from (2) by the Cancellation Law of Proposition 1.30. Indeed $i_\alpha = i_\beta \circ i_{\alpha\beta}$ being an admissible monic implies $i_{\alpha\beta}$ is an admissible monic. □

Remark 9.16 Our proof of part (1) of Lemma 9.15 uses Proposition 1.30 which relies on $\mathcal{A}$ being weakly idempotent complete. However, it can be proved without this assumption by using transfinite induction and the same set-theoretic fact mentioned in the previous proof.

Proposition 9.17 (Transfinite Compositions are Filt($\mathcal{S}$)-cofibrations) *Let $\mathcal{S}$ be an efficient class of objects of $(\mathcal{A}, \mathcal{E})$. Assume that $i \colon A \rightarrowtail X = \varinjlim_{\alpha<\lambda} X_\alpha$ is a transfinite composition of some λ-extension sequence*

$$A = X_0 \rightarrowtail X_1 \rightarrowtail X_2 \rightarrowtail \cdots \rightarrowtail X_\alpha \rightarrowtail X_{\alpha+1} \rightarrowtail \cdots \tag{9.4}$$

of $\mathcal{S}$-cofibrations $i_{\alpha,\alpha+1} \colon X_\alpha \rightarrowtail X_{\alpha+1}$. Then $\operatorname{Cok} i \cong \varinjlim_{\alpha<\lambda}(X_\alpha/A)$ *is the transfinite extension of the λ-extension sequence*

$$0 = X_0/A \rightarrowtail X_1/A \rightarrowtail X_2/A \rightarrowtail \cdots \rightarrowtail X_\alpha/A \rightarrowtail X_{\alpha+1}/A \rightarrowtail \cdots. \tag{9.5}$$

Consequently, $\varinjlim_{\alpha<\lambda}(X_\alpha/A) \in \operatorname{Filt}(\mathcal{S})$ *and so $i \colon A \rightarrowtail X$ is a* Filt($\mathcal{S}$)*-cofibration.*

Proof The object $X = \varinjlim_{\alpha<\lambda} X_\alpha$ sits as a colimit cone above the given λ-extension sequence. Again, $i_\alpha \colon X_\alpha \rightarrowtail X$ denotes the canonical morphism into the colimit X. It follows from Lemma 9.15 and the dual of Corollary 1.15 (Noether iso.) that we may take the quotient of the entire colimit diagram, by A, to obtain another λ-extension sequence $\left\{X_\alpha/A, \bar{i}_{\alpha\beta}\right\}_{\alpha\leq\beta<\lambda}$ as depicted in (9.5). We have $\operatorname{Cok} i = \operatorname{Cok} i_0 = X/A$ sitting above the diagram as a cocone, with compatible morphisms $\bar{i}_\alpha \colon X_\alpha/A \to X/A$. Note that $\bar{i}_\alpha \pi_\alpha = \pi i_\alpha$, where $\pi_\alpha \colon X_\alpha \xrightarrow{\operatorname{cok} i_{0\alpha}} X_\alpha/A$ and $\pi \colon X \xrightarrow{\operatorname{cok} i} X/A$ are the canonical maps. The goal is to show $X/A \cong \varinjlim_{\alpha<\lambda}(X_\alpha/A)$. To this end, suppose we are given any object T along with compatible morphisms $t_\alpha \colon X_\alpha/A \to T$ above Diagram (9.5). The universal property of the colimit X provides a unique morphism $\psi\colon X \to T$ such that $t_\alpha \pi_\alpha = \psi i_\alpha$. In particular (for $\alpha = 0$) we have $0 = t_0\pi_0 = \psi i_0 = \psi i$, so ψ factors uniquely through the cokernel $\pi\colon X \xrightarrow{\operatorname{cok} i} X/A$ as $\psi = \bar{\psi}\pi$. Therefore, $t_\alpha\pi_\alpha = \bar{\psi}\pi i_\alpha = \bar{\psi}\bar{i}_\alpha\pi_\alpha$. With π_α an epimorphism, it is right cancellable, yielding $t_\alpha = \bar{\psi}\bar{i}_\alpha$. As for the uniqueness of $\bar{\psi}$, suppose $\psi'\colon X/A \to T$ also satisfies $t_\alpha = \psi'\bar{i}_\alpha$. Then composing with π_α we obtain $t_\alpha\pi_\alpha = \psi'\bar{i}_\alpha\pi_\alpha \implies t_\alpha\pi_\alpha = (\psi'\pi)i_\alpha$. By the uniqueness property of ψ this implies $\psi'\pi = \psi$. Then by the uniqueness property of $\operatorname{cok} i = \pi$ we conclude $\psi' = \bar{\psi}$. This completes the proof that $\operatorname{Cok} i = \varinjlim_{\alpha<\lambda}(X_\alpha/A)$ is the colimit of the λ-extension sequence of Diagram (9.5). □

We leave it to the reader to verify that Proposition 9.17 generalizes as follows.

Corollary 9.18 *With the same hypotheses as Proposition 9.17 we have, more generally: For any particular $\alpha < \lambda$,*

$$\operatorname{Cok} i_\alpha \cong \varinjlim_{\alpha\le\beta<\lambda} \left(X_\beta/X_\alpha\right) \in \operatorname{Filt}(\mathcal{S}).$$

Consequently, $i_\alpha\colon X_\alpha \rightarrowtail X$ is a Filt($\mathcal{S}$)*-cofibration.*

Exercise 9.2.2 Prove Corollary 9.18 by reducing it to Proposition 9.17. (One can use the reindexing trick found in the proof of Lemma 9.15, along with Exercise 9.2.1.)

Proposition 9.17 shows that transfinite compositions of $\mathcal{S}$-cofibrations are Filt($\mathcal{S}$)-cofibrations. We now show the converse: If $i\colon A \rightarrowtail X$ is a Filt($\mathcal{S}$)-cofibration, then it may be identified with the transfinite composition of some λ-extension sequence of $\mathcal{S}$-cofibrations as depicted in (9.4) of Proposition 9.17.

Proposition 9.19 (Filt($\mathcal{S}$)-cofibrations are Transfinite Compositions) *Let $\mathcal{S}$ be an efficient class of objects of $(\mathcal{A}, \mathcal{E})$. Assume $j\colon A \rightarrowtail B$ is a* Filt($\mathcal{S}$)*-cofibration, so that we have a short exact sequence $\mathbb{E}\colon A \overset{j}{\rightarrowtail} B \overset{q}{\twoheadrightarrow} X$ where $X = \varinjlim_{\alpha<\lambda} X_\alpha$ has an $\mathcal{S}$-filtration $\left\{X_\alpha, i_{\alpha\beta}\right\}_{\alpha\le\beta<\lambda}$*

$$X_0 = 0 \rightarrowtail X_1 \rightarrowtail X_2 \rightarrowtail \cdots \rightarrowtail X_\alpha \rightarrowtail X_{\alpha+1} \rightarrowtail \cdots$$

with $X_0 = 0$. Then $j\colon A \rightarrowtail B$ identifies as the transfinite composition of a λ-extension sequence $\left\{B_\alpha, m_{\alpha\beta}\right\}_{\alpha\le\beta<\lambda}$

$$A = B_0 \rightarrowtail B_1 \rightarrowtail B_2 \rightarrowtail \cdots \rightarrowtail B_\alpha \rightarrowtail B_{\alpha+1} \rightarrowtail \cdots$$

with $B_0 = A$ and each $m_{\alpha,\alpha+1}\colon B_\alpha \rightarrowtail B_{\alpha+1}$ an $\mathcal{S}$-cofibration satisfying $\operatorname{Cok} m_{\alpha,\alpha+1} \cong \operatorname{Cok} i_{\alpha,\alpha+1}$ *for all $\alpha < \lambda$.*

Proof We let $i_\alpha \colon X_\alpha \rightarrowtail \varinjlim_{\alpha<\lambda} X_\alpha$ denote the canonical morphisms into the colimit, so that we have $i_\alpha = i_\beta \circ i_{\alpha\beta}$ for all $\alpha \leq \beta < \lambda$. Now for each $\alpha < \lambda$, taking the pullback of i_α along q provides us with a commutative diagram of short exact sequences:

$$\begin{array}{lccccc}
\mathbb{E}_\alpha: & A & \overset{j_\alpha}{\rightarrowtail} & B_\alpha & \overset{q_\alpha}{\twoheadrightarrow} & X_\alpha \\
 & \| & & \downarrow{\scriptstyle m_\alpha} & & \downarrow{\scriptstyle i_\alpha} \\
\mathbb{E}: & A & \underset{j}{\rightarrowtail} & B & \underset{q}{\twoheadrightarrow} & X,
\end{array}$$

and by Corollary 1.14(1) the morphism m_α is indeed an admissible monic. In fact, for each $\alpha \leq \beta < \lambda$ we can construct a commutative diagram of short exact sequences as follows:

$$\begin{array}{lccccc}
\mathbb{E}_\alpha: & A & \overset{j_\alpha}{\rightarrowtail} & B_\alpha & \overset{q_\alpha}{\twoheadrightarrow} & X_\alpha \\
 & \| & & \downarrow{\scriptstyle m_{\alpha\beta}} & & \downarrow{\scriptstyle i_{\alpha\beta}} \\
\mathbb{E}_\beta: & A & \underset{j_\beta}{\rightarrowtail} & B_\beta & \underset{q_\beta}{\twoheadrightarrow} & X_\beta \\
 & \| & & \downarrow{\scriptstyle m_\beta} & & \downarrow{\scriptstyle i_\beta} \\
\mathbb{E}: & A & \underset{j}{\rightarrowtail} & B & \underset{q}{\twoheadrightarrow} & X.
\end{array}$$

Indeed as already noted the far right vertical composition is just i_α. So the universal property of the pullback B_β provides for a unique map $m_{\alpha\beta}$ making the upper right square commute and such that $m_\alpha = m_\beta \circ m_{\alpha\beta}$. Again, all such $m_{\alpha\beta}$ are admissible monics, for by Proposition 1.12(2) we learn that each right square is bicartesian (a pushout as well as a pullback), so we conclude that

$$\operatorname{Cok} m_{\alpha\beta} \cong \operatorname{Cok} i_{\alpha\beta} \tag{9.6}$$

for all $\alpha \leq \beta < \lambda$. Note in particular that taking $\beta = \alpha + 1$ we have the isomorphism $\operatorname{Cok} m_{\alpha,\alpha+1} \cong \operatorname{Cok} i_{\alpha,\alpha+1}$ as the proposition claims.

Note that by the assumption $X_0 = 0$, we may take $B_0 = A$ and $j_0 = 1_A$ at the initial stage:

$$\begin{array}{lccccc}
\mathbb{E}_0: & A & \overset{j_0=1_A}{=\!=\!=} & B_0 & \twoheadrightarrow & 0 \\
 & \| & & \downarrow{\scriptstyle m_{01}} & & \downarrow \\
\mathbb{E}_1: & A & \underset{j_1}{\rightarrowtail} & B_1 & \underset{q_1}{\twoheadrightarrow} & X_1.
\end{array}$$

Consequently we have $m_{01} = j_1$, and $m_0 = j$. Now it is clear that we have what should be described as a λ-extension sequence

$$\mathbb{E}_0 \rightarrowtail \mathbb{E}_1 \rightarrowtail \mathbb{E}_2 \rightarrowtail \cdots \rightarrowtail \mathbb{E}_\alpha \rightarrowtail \mathbb{E}_{\alpha+1} \rightarrowtail \cdots$$

of (admissible) short exact sequences: $\left\{\mathbb{E}_\alpha, \left(1_A, m_{\alpha\beta}, i_{\alpha\beta}\right)\right\}_{\alpha\leq\beta<\lambda}$. The middle terms of the short exact sequences provide a λ-extension sequence

$$A = B_0 \rightarrowtail B_1 \rightarrowtail B_2 \rightarrowtail \cdots \rightarrowtail B_\alpha \rightarrowtail B_{\alpha+1} \rightarrowtail \cdots$$

of $\mathcal{S}$-cofibrations $m_{\alpha,\alpha+1}\colon B_\alpha \rightarrowtail B_{\alpha+1}$. So by the assumptions in Definition 9.11 we know that its colimit $\varinjlim_{\alpha<\lambda} B_\alpha$ exists. We will show that the colimit is B, as expected. To do so, let $\eta_\alpha\colon B_\alpha \rightarrowtail \varinjlim_{\alpha<\lambda} B_\alpha$ denote the canonical morphisms, each of which is an admissible monic by Lemma 9.15. The identities $m_\alpha = m_\beta \circ m_{\alpha\beta}$ induce a unique morphism $m\colon \varinjlim_{\alpha<\lambda} B_\alpha \to B$ satisfying $m\eta_\alpha = m_\alpha$ for all $\alpha < \lambda$. Moreover, Proposition 9.17 provides a short exact sequence:

$$A \overset{\eta_0}{\rightarrowtail} \varinjlim_{\alpha<\lambda} B_\alpha \overset{\pi}{\twoheadrightarrow} \varinjlim_{\alpha<\lambda} X_\alpha,$$

where π satisfies the relations $\pi\eta_\alpha = i_\alpha q_\alpha$ for all $\alpha < \lambda$ (which in the proof of Proposition 9.17 is the relations $\pi i_\alpha = \bar{i}_\alpha \pi_\alpha$). But $i_\alpha q_\alpha = q m_\alpha$, so we have $\pi\eta_\alpha = q m_\alpha$ for all $\alpha < \lambda$. Now we check that we have a commutative diagram of short exact sequences:

$$\begin{array}{lccccc} \varinjlim_{\alpha<\lambda} \mathbb{E}_\alpha: & A & \overset{\eta_0}{\rightarrowtail} & \varinjlim_{\alpha<\lambda} B_\alpha & \overset{\pi}{\twoheadrightarrow} & \varinjlim_{\alpha<\lambda} X_\alpha \\ & \| & & \downarrow{\scriptstyle m} & & \| \\ \mathbb{E}: & A & \overset{j}{\rightarrowtail} & B & \overset{q}{\twoheadrightarrow} & X. \end{array}$$

Indeed $m\eta_0 = m_0 = j$, so the left square commutes. For the right square we have $(qm)\eta_\alpha = q(m\eta_\alpha) = qm_\alpha = \pi\eta_\alpha$, for all $\alpha < \lambda$. So by the universal property of the colimit $\varinjlim_{\alpha<\lambda} B_\alpha$ we must have $qm = \pi$. By the Short Five Lemma 1.11 we conclude that the middle arrow m is an isomorphism. □

We obtain the following useful consequence of Proposition 9.19. It follows by iteration of the proposition and the fact that every well-ordered set is uniquely order isomorphic to a unique ordinal number.

Corollary 9.20 (Filt($\mathcal{S}$) is Closed Under Extensions) *Let $\mathcal{S}$ be an efficient class of objects of $(\mathcal{A}, \mathcal{E})$. Then transfinite compositions of* Filt($\mathcal{S}$)*-cofibrations exist and are again* Filt($\mathcal{S}$)*-cofibrations. In particular,* Filt($\mathcal{S}$) *is itself an efficient class and is closed under finite and transfinite extensions. In symbols,* Filt(Filt($\mathcal{S}$)) = Filt($\mathcal{S}$).

9.3 Exact Coproducts

In this section we develop some further properties of the cofibrations associated to an efficient class $\mathcal{S}$. In particular, the coproduct of any set of $\mathcal{S}$-cofibrations is still an admissible monic. As a consequence, Corollary 9.22 shows that coproducts exist for a very large class of exact categories, and their classes $\mathcal{E}$ of short exact sequences are themselves closed under coproducts. The reader familiar with Grothendieck's axioms for abelian categories will note that these are exact category versions of Grothendieck's Axiom (AB3), that coproducts exist, and (AB4), that short exact sequences are closed under coproducts.

We assume throughout this section that $(\mathcal{A}, \mathcal{E})$ is a weakly idempotent complete exact category.

The first part of the next result contains a converse to Exercise 9.1.1. Note that it doesn't use the existence of any coproducts, rather it proves existence of certain coproducts.

Proposition 9.21 *Let $\mathcal{S}$ be an efficient class of objects of $(\mathcal{A}, \mathcal{E})$. Then we have the following.*

- *Coproducts of objects in $\mathcal{S}$ exist and are in* Filt($\mathcal{S}$)*. Consequently, coproducts of objects in* Filt($\mathcal{S}$) *exist and are again in* Filt($\mathcal{S}$).
- *Assume all coproducts exist in $\mathcal{A}$. Then the coproduct of any set of $\mathcal{S}$-cofibrations is an admissible monic and so a* Filt($\mathcal{S}$)*-cofibration. In more detail, for any collection*

$$\{A_i \rightarrowtail B_i \twoheadrightarrow C_i\}_{i \in I}$$

of short exact sequences with each $C_i \in \mathcal{S}$, the coproduct

$$\bigoplus_{i \in I} A_i \rightarrowtail \bigoplus_{i \in I} B_i \twoheadrightarrow \bigoplus_{i \in I} C_i$$

is again a short exact sequence with $\bigoplus_{i \in I} C_\alpha \in$ Filt($\mathcal{S}$)*. Consequently, the class of all* Filt($\mathcal{S}$)*-cofibrations is closed under coproducts.*

Proof Our first goal is to show that coproducts of objects in $\mathcal{S}$ exist, and are in Filt($\mathcal{S}$). We may assume that we are given a set of objects $\{A_\alpha\}_{\alpha<\lambda}$ of $\mathcal{S}$ indexed by some ordinal λ. To construct $\bigoplus_{\alpha<\lambda} A_\alpha$, we inductively define a λ-extension sequence of $\mathcal{S}$-cofibrations as follows.

- (Base Step) $X_0 = 0$.
- (Successor Steps) Set $X_{\alpha+1} = X_\alpha \oplus A_\alpha$ and let $i_{\alpha,\alpha+1} \colon X_\alpha \to X_{\alpha+1}$ be the canonical split monic.
- (Limit Steps) Let $\alpha < \lambda$ be a limit ordinal. We may set $X_\alpha = \varinjlim_{\delta<\alpha} X_\delta$ for all $\delta < \alpha$; it exists since each $i_{\alpha,\alpha+1}$ is an $\mathcal{S}$-cofibration.

This produces a λ-extension sequence $\left\{X_\alpha, i_{\alpha\beta}\right\}_{\alpha\leq\beta<\lambda}$ of $\mathcal{S}$-cofibrations. Note that for all $\alpha < \lambda$ there is a canonical injection $j_\alpha\colon A_\alpha \to X_{\alpha+1}$. Again, $X_\lambda := \varinjlim_{\alpha<\lambda} X_\alpha$ exists, and we note that the composition $\eta_\alpha := i_{\alpha+1} j_\alpha$ determines morphisms $\eta_\alpha\colon A_\alpha \to X_\lambda$. We will show that $(X_\lambda, \{\eta_\alpha\})$ is a coproduct for $\{A_\alpha\}_{\alpha<\lambda}$.

Indeed suppose we are given a collection of morphisms $\left\{A_\alpha \xrightarrow{f_\alpha} Y\right\}_{\alpha<\lambda}$. Then it induces a corresponding cocone $\left\{X_\alpha \xrightarrow{g_\alpha} Y\right\}_{\alpha<\lambda}$ above the λ-extension sequence $\left\{X_\alpha, i_{\alpha\beta}\right\}_{\alpha\leq\beta<\lambda}$ as follows.

- (Base Step) $g_0 = 0$.
- (Successor Steps) Given g_α, set $g_{\alpha+1}$ to be $g_{\alpha+1}\colon X_\alpha \oplus A_\alpha \xrightarrow{[g_\alpha\ f_\alpha]} Y$. We note here that

$$g_{\alpha+1} i_{\alpha,\alpha+1} = g_\alpha \quad\text{and}\quad g_{\alpha+1} j_\alpha = f_\alpha. \tag{$*$}$$

- (Limit Steps) For limit ordinals $\alpha < \lambda$, we set $g_\alpha = \varinjlim_{\delta<\alpha} g_\delta$. That is, $g_\alpha\colon X_\alpha \to Y$ is the unique map satisfying, for all $\delta < \lambda$,

$$g_\alpha i_{\delta\alpha} = g_\delta. \tag{$**$}$$

By the universal property of the colimit X_λ, there also exists a unique map $\psi\colon X_\lambda \to Y$ such that $g_\alpha = \psi i_\alpha$ for all $\alpha < \lambda$. In particular, $g_{\alpha+1} = \psi i_{\alpha+1}$, and so $g_{\alpha+1} j_\alpha = \psi i_{\alpha+1} j_\alpha$ for all $\alpha < \lambda$. So by Equation $(*)$ and the definition of η, we have $f_\alpha = \psi\eta_\alpha$ for all $\alpha < \lambda$.

To prove ψ is unique with this property, suppose $\psi'\colon X_\lambda \to Y$ also satisfies $f_\alpha = \psi'\eta_\alpha$. We will prove uniqueness by showing $g_\alpha = \psi' i_\alpha$ for all $\alpha < \lambda$. Once again we use transfinite induction.

- (Base Step) $g_0 = 0$ and $i_0 = 0$.
- (Successor Steps) Suppose $g_\alpha = \psi' i_\alpha$. We wish to show $g_{\alpha+1} = \psi' i_{\alpha+1}$. Recall $g_{\alpha+1}$ is the unique map satisfying Equations $(*)$. But substituting $\psi' i_{\alpha+1}$ in place of $g_{\alpha+1}$ we see (i) $(\psi' i_{\alpha+1}) i_{\alpha,\alpha+1} = \psi' i_\alpha = g_\alpha$, and (ii) $(\psi' i_{\alpha+1}) j_\alpha = \psi'\eta_\alpha = f_\alpha$.
- (Limit Steps) Let α be a limit ordinal and assume $g_\delta = \psi' i_\delta$ for all $\delta < \alpha$. We need to show $g_\alpha = \psi' i_\alpha$. But g_α is the unique morphism satisfying Equation $(**)$. Substituting $\psi' i_\alpha$ in place of g_α we get $(\psi' i_\alpha) i_{\delta\alpha} = \psi' i_\delta = g_\delta$.

This completes the proof that $(X_\lambda, \{\eta_\alpha\})$ is a coproduct for $\{A_\alpha\}_{\alpha<\lambda}$. We conclude that coproducts of $\mathcal{S}$-objects exist. Moreover, the inductive construction of the coproduct $X_\lambda = \varinjlim_{\alpha<\lambda} X_\alpha$ shows it to be in Filt($\mathcal{S}$).

Now by Corollary 9.20, Filt($\mathcal{S}$) is itself an efficient class. So coproducts of Filt($\mathcal{S}$)-objects also exist and are in Filt(Filt($\mathcal{S}$)) = Filt($\mathcal{S}$).

Next, assume all coproducts exist in $\mathcal{A}$. We will show that coproducts of $\mathcal{S}$-cofibrations exist and are admissible monics, hence Filt($\mathcal{S}$)-cofibrations. So let $\{A_\alpha \rightarrowtail B_\alpha \twoheadrightarrow C_\alpha\}_{\alpha<\lambda}$ be a collection of short exact sequences with each $C_\alpha \in \mathcal{S}$, again indexed by some ordinal λ. There is a natural sequence

$$\bigoplus_{\alpha<\lambda} A_\alpha \xrightarrow{\oplus f_\alpha} \bigoplus_{\alpha<\lambda} B_\alpha \xrightarrow{\oplus g_\alpha} \bigoplus_{\alpha<\lambda} C_\alpha$$

and we will show $\oplus f_\alpha$ to be an admissible monic. This will prove the result, for $\oplus g_\alpha$ is its cokernel, and we have $\bigoplus_{\alpha<\lambda} C_\alpha \in \mathrm{Filt}(\mathcal{S})$ by what was just proved.

It is always true that $\mathcal{E}$ is closed under finite direct sums; see Exercise 1.5.1. Therefore, we may begin to construct a λ-extension sequence of $\mathcal{S}$-cofibrations

$$X_0 \rightarrowtail X_1 \rightarrowtail X_2 \rightarrowtail \cdots$$

that begins

$$A_0 \oplus \left(\bigoplus_{0<\delta<\lambda} A_\delta\right) \xrightarrow{f_0 \oplus 1} B_0 \oplus \left(\bigoplus_{0<\delta<\lambda} A_\delta\right) = B_0 \oplus A_1 \oplus \left(\bigoplus_{1<\delta<\lambda} A_\delta\right) \xrightarrow{1_{B_0} \oplus f_1 \oplus 1} \cdots .$$

We formalize this with transfinite induction as follows: Supposing we have obtained

$$X_\alpha = \left(\bigoplus_{\delta<\alpha} B_\delta\right) \oplus A_\alpha \oplus \left(\bigoplus_{\alpha<\delta<\lambda} A_\delta\right)$$

then we set

$$X_{\alpha+1} = \left(\bigoplus_{\delta<\alpha} B_\delta\right) \oplus B_\alpha \oplus \left(\bigoplus_{\alpha<\delta<\lambda} A_\delta\right)$$

connected along with the admissible monic $i_{\alpha,\alpha+1} \colon X_\alpha \xrightarrow{1 \oplus f_\alpha \oplus 1} X_{\alpha+1}$. Taking colimits at limit ordinals, one checks that this defines a λ-extension sequence $\{X_\alpha, i_{\alpha\beta}\}_{\alpha\leq\beta<\lambda}$ of $\mathcal{S}$-cofibrations, for which $X_\lambda := \varinjlim_{\alpha<\lambda} X_\alpha = \bigoplus_{\alpha<\lambda} B_\alpha$ and the canonical morphism $i_0 \colon X_0 \rightarrowtail X_\lambda$ coincides with $\bigoplus_{\alpha<\lambda} A_\alpha \xrightarrow{\oplus f_\alpha} \bigoplus_{\alpha<\lambda} B_\alpha$. So $\oplus f_\alpha$ is an admissible monic, in fact a $\mathrm{Filt}(\mathcal{S})$-cofibration by Proposition 9.17. One can show that $\mathrm{cok}\,(\oplus f_\alpha) = \oplus g_\alpha$. This also follows from Proposition 9.17 because we have the isomorphisms $X_{\alpha+1}/X_0 \cong \bigoplus_{\delta<\alpha} C_\delta$.

Now again by Corollary 9.20, $\mathrm{Filt}(\mathcal{S})$ is itself an efficient class. So it follows from what we just proved that the class of all $\mathrm{Filt}(\mathcal{S})$-cofibrations is closed under coproducts. □

For the special case that the class $\mathcal{S} = \mathcal{A}$ of all objects is efficient, we get the following.

Corollary 9.22 *Assume $(\mathcal{A}, \mathcal{E})$ satisfies the Transfinite Extensions Axiom (EF1). That is, assume that transfinite compositions of admissible monics always exist and are again admissible monics. Then $(\mathcal{A}, \mathcal{E})$ has exact coproducts. That is, we have the following.*

- *Coproducts exist in $\mathcal{A}$.*
- *The class $\mathcal{E}$ of all short exact sequences is closed under coproducts.*

We also get the following by combining Exercise 9.1.1 and Proposition 9.21.

Corollary 9.23 *Let $\mathcal{A}$ be an additive category, considered along with its canonical split exact structure. Then $\mathcal{A}$ has coproducts if and only if this exact structure satisfies the Transfinite Extensions Axiom (EF1).*

9.4 Smallness Relative to Transfinite Compositions

Assume throughout this section that $\mathcal{S}$ is an efficient class of objects in a weakly idempotent complete exact category $(\mathcal{A}, \mathcal{E})$. In the general theory of model categories the notion of *κ-small objects*, where κ is a cardinal number, is crucial to Quillen's small object argument. The argument is applied to provide factorizations in (cofibrantly generated) model categories. Translating this to the abelian case, we need the notion of κ-smallness relative to Filt($\mathcal{S}$)-cofibrations.

By Corollary 9.20, transfinite compositions of Filt($\mathcal{S}$)-cofibrations exist (and are again Filt($\mathcal{S}$)-cofibrations). So let us consider the colimit $X_\lambda = \varinjlim_{\alpha<\lambda} X_\alpha$ of any given λ-extension sequence $\{X_\alpha, i_{\alpha\beta}\}_{\alpha\le\beta<\lambda}$ of Filt($\mathcal{S}$)-cofibrations. It may be depicted as follows:

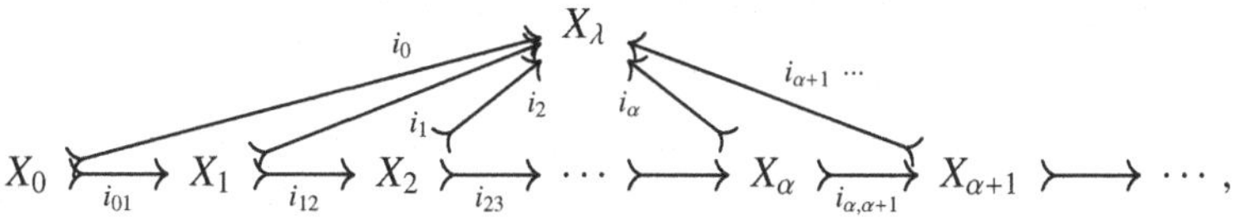

where $\{i_\alpha \colon X_\alpha \rightarrowtail X_\lambda\}_{\alpha<\lambda}$ are the canonical morphisms. Note that for another given object $A \in \mathcal{A}$, by applying $\mathrm{Hom}_{\mathcal{A}}(A, -)$ to the entire colimit diagram we obtain a *cocone*

$$\left\{\mathrm{Hom}_{\mathcal{A}}(A, X_\alpha) \xrightarrow{\mathrm{Hom}_{\mathcal{A}}(A, i_\alpha)} \mathrm{Hom}_{\mathcal{A}}(A, X_\lambda)\right\}_{\alpha<\lambda}$$

sitting atop the direct λ-system of abelian groups

$$\mathrm{Hom}_{\mathcal{A}}(A, X_0) \rightarrowtail \mathrm{Hom}_{\mathcal{A}}(A, X_1) \rightarrowtail \cdots \rightarrowtail \mathrm{Hom}_{\mathcal{A}}(A, X_\alpha) \rightarrowtail \cdots .$$

Thus for any given λ-extension we obtain a canonical map

$$\xi \colon \varinjlim_{\alpha<\lambda} \mathrm{Hom}_{\mathcal{A}}(A, X_\alpha) \to \mathrm{Hom}_{\mathcal{A}}(A, X_\lambda)$$

which acts on a morphism $f_\alpha \colon A \to X_\alpha$ representing an element of the colimit $\varinjlim_{\alpha<\lambda} \mathrm{Hom}_{\mathcal{A}}(A, X_\alpha)$ by $\xi(f_\alpha) = i_\alpha f_\alpha$. One should verify the following.

Exercise 9.4.1 In the previous situation, the canonical morphism ξ satisfies the following.

(1) ξ is always one-to-one.
(2) ξ is always onto, and therefore always an isomorphism, if and only if any given morphism $f\colon A \to X_\lambda$ factors as $f = i_\alpha f_\alpha$ for some $\alpha < \lambda$ and $f_\alpha\colon A \to X_\alpha$.

A *regular cardinal* is an infinite cardinal that is not the sum of a smaller number of smaller cardinals. For example, $\aleph_\omega = \sum_{n<\omega} \aleph_n$ is NOT a regular cardinal. However, every infinite successor cardinal is regular.

Definition 9.24 Let κ be a cardinal number. An object $A \in \mathcal{A}$ is called **κ-small relative to** Filt($\mathcal{S}$)**-cofibrations** if for every regular cardinal $\lambda > \kappa$, and any λ-extension sequence $\{X_\alpha, i_{\alpha\beta}\}_{\alpha\leq\beta<\lambda}$ of Filt($\mathcal{S}$)-cofibrations, the canonical map

$$\xi\colon \varinjlim_{\alpha<\lambda} \mathrm{Hom}_{\mathcal{A}}(A, X_\alpha) \to \mathrm{Hom}_{\mathcal{A}}(A, X_\lambda)$$

is an isomorphism. Equivalently, any given morphism $f\colon A \to X_\lambda$ factors as $f = i_\alpha f_\alpha$ for some $\alpha < \lambda$ and $f_\alpha\colon A \to X_\alpha$. An object A is called **small relative to** Filt($\mathcal{S}$)**-cofibrations** if there exists some cardinal κ for which it is κ-small in this sense.

Consider the special case that $(\mathcal{A}, \mathcal{E})$ satisfies the Transfinite Extensions Axiom (EF1). That is, $\mathcal{S} = \mathcal{A}$ is efficient, hence the class of admissible monics is closed under transfinite compositions. Then we use the terms **admissibly κ-small** and **admissibly small** for the concept in Definition 9.24. As observed in Exercise 9.4.1, we should think of an admissibly small object A as one with the following property: Any given morphism $f\colon A \to X_\lambda$ into a large enough λ-extension X_λ, must factor as $f = i_\alpha f_\alpha$ for some $\alpha < \lambda$ and morphism $f_\alpha\colon A \to X_\alpha$. That is, f must "land" in some X_α ($\alpha < \lambda$). The following exercises provide the motivating examples.

Exercise 9.4.2 (R-Modules) Let R be a ring and R-Mod the category of (left) R-modules. Consider it as an exact category with the usual short exact sequences. The class $\mathcal{S}$ of all R-modules is efficient. Let M be in R-Mod.

(1) Show that if M is finitely generated then it is admissibly n-small where n is any finite cardinal.
(2) More generally, show that if M has a generating set $\mathcal{T} \subseteq M$ of cardinality $\kappa = |\mathcal{T}|$, then M is admissibly κ-small.
(3) Conclude that every R-module is admissibly small.

Exercise 9.4.3 (Grothendieck Categories) Let $\mathcal{G}$ be a Grothendieck category and consider it as an exact category with the standard short exact sequences. Again, the class $\mathcal{S} = \mathcal{G}$ of all objects is efficient. Use the Gabriel–Popescu Theorem to explain why every object of $\mathcal{G}$ is admissibly small.

9.5 Generators for Exact Categories

Generating sets are essential to the utility of Grothendieck categories. Now that we have considered some mild exactness properties of short exact sequences, our next goal is to also extend the notion of generating sets to more general exact categories. The generating sets defined here will be called *admissible*, for they are generators relative to $\mathcal{E}$. They are in fact special generators for the underlying additive category $\mathcal{A}$.

Throughout this section we assume that $(\mathcal{A}, \mathcal{E})$ is a weakly idempotent complete exact category that has coproducts. In fact, any such $\mathcal{A}$ is then necessarily idempotent complete; see Proposition 1.47.

For a given object X, we will say that a morphism $g \colon A \to B$ is X**-epic** if $g_* \colon \mathrm{Hom}_{\mathcal{A}}(X, A) \xrightarrow{\mathrm{Hom}_{\mathcal{A}}(X,g)} \mathrm{Hom}_{\mathcal{A}}(X, B)$ is an epimorphism of abelian groups. It just means that any morphism $f \colon X \to B$ factors through g. More generally, for a given class of objects $\mathcal{X}$, we will say that a morphism $g \colon A \to B$ is $\mathcal{X}$**-epic** if $g_* \colon \mathrm{Hom}_{\mathcal{A}}(X, A) \xrightarrow{\mathrm{Hom}_{\mathcal{A}}(X,g)} \mathrm{Hom}_{\mathcal{A}}(X, B)$ is an epimorphism of abelian groups for all $X \in \mathcal{X}$.

Example 9.25 An *evaluation map* is a particularly useful example of an $\mathcal{X}$-epic morphism. For any given set $\mathcal{S} = \{S_i\}_{i \in I}$, we can define the evaluation map with codomain any other given object $A \in \mathcal{A}$. To do so, set $H_i = \mathrm{Hom}_{\mathcal{A}}(S_i, A)$, the group of all morphisms $f_i \colon S_i \to A$. By assumption, the coproduct $\bigoplus_{(i \in I, f_i \in H_i)} S_{f_i}$ exists, where here each $S_{f_i} = S_i$. Note that the universal property of the coproduct induces a canonical morphism

$$e \colon \bigoplus_{(i \in I, f_i \in H_i)} S_{f_i} \longrightarrow A$$

called the **evaluation map**. On any given component $f_i \in H_i$, e is simply the morphism $f_i \colon S_i \to A$. In particular we have $f_i = e \circ \eta_{f_i}$, where

$$\eta_{f_i} \colon S_{f_i} \rightarrowtail \bigoplus_{(i \in I, f_i \in H_i)} S_{f_i}$$

denotes the inclusion into the component $f_i \in H_i$. For example, in the case of modules over a ring, e acts on elements by the rule

$$\left(a_{f_i}\right)_{(i\in I, f_i\in H_i)} \mapsto \sum_{(i\in I, f_i\in H_i)} f_i\left(a_{f_i}\right).$$

A crucial point is that any such evaluation map e is an $\mathcal{S}$-epic morphism. This follows immediately from the relations $f_i = e \circ \eta_{f_i}$.

We are now ready to give the main definition.

Definition 9.26 An object U is an **admissible generator** for $(\mathcal{A}, \mathcal{E})$, or a **generator for** $\mathcal{E}$, if every U-epic morphism is an admissible epic. More generally, any class of objects $\mathcal{U}$ is a **class of admissible generators** if every $\mathcal{U}$-epic morphism is an admissible epic. In particular, by a **set of admissible generators** for $(\mathcal{A}, \mathcal{E})$ we mean an actual set $\mathcal{U} = \{U_i\}_{i\in I}$ (not a proper class) that is a generator for $\mathcal{E}$ in this sense.

For example, let R be a ring. Then the class of all finitely presented R-modules, which has a small skeleton, is a class of admissible generators for the proper class of all pure exact sequences of R-modules.

Exercise 9.5.1 Show that $\mathcal{U} = \{U_i\}_{i\in I}$ is a set of admissible generators for $(\mathcal{A}, \mathcal{E})$ if and only if the coproduct $U = \bigoplus_{i\in I} U_i$ is an admissible generator for $(\mathcal{A}, \mathcal{E})$.

Though we won't use it as often, note that each notion defined in Definition 9.26 has a dual. In particular, a set of objects $\mathcal{V} = \{V_i\}_{i\in I}$ is a **set of admissible cogenerators** for $(\mathcal{A}, \mathcal{E})$ if $f: A \to B$ being $\mathcal{V}$*-monic* implies f is an admissible monic. Explicitly, if f is an admissible monic whenever

$$f^*: \operatorname{Hom}_{\mathcal{A}}(B, V_i) \xrightarrow{\operatorname{Hom}_{\mathcal{A}}(f, V_i)} \operatorname{Hom}_{\mathcal{A}}(A, V_i)$$

is an epimorphism for all $V_i \in \mathcal{V}$.

To state the next proposition we introduce a notation that will be useful throughout the rest of this chapter: We denote by $\operatorname{Free}(\mathcal{U})$ the class of all set indexed coproducts (direct sums) $X = \bigoplus U_j$ in which each $U_j \in \mathcal{U}$.

Proposition 9.27 *Let $\mathcal{U} = \{U_i\}_{i\in I}$ be a set of objects. The following are equivalent.*

(1) *$\mathcal{U} = \{U_i\}_{i\in I}$ is a set of admissible generators for $(\mathcal{A}, \mathcal{E})$, equivalently, $U = \bigoplus_{i\in I} U_i$ is an admissible generator for $(\mathcal{A}, \mathcal{E})$.*
(2) *For each object $A \in \mathcal{A}$, the evaluation morphism*

$$e: \bigoplus_{(i\in I, f_i\in H_i)} U_{f_i} \longrightarrow A$$

(see Example 9.25) is an admissible epic. Equivalently, for the single generator $U = \bigoplus_{i \in I} U_i$ the evaluation morphism

$$e\colon \bigoplus_{f \in \mathrm{Hom}_{\mathcal{A}}(U,A)} U_f \longrightarrow A$$

is an admissible epic.

(3) *Every object $A \in \mathcal{A}$ is an admissible quotient of some set indexed coproduct of copies of objects from $\{U_i\}_{i \in I}$. That is, there exists an admissible epic $X \twoheadrightarrow A$ where $X \in$ Free($\mathcal{U}$).*

Proof (1) $\implies$ (2). Let $e\colon \bigoplus_{(i \in I, f_i \in H_i)} U_{f_i} \longrightarrow A$ be the evaluation map for any given object $A \in \mathcal{A}$. As is easy to see, and noted at the end of Example 9.5.1, this morphism is $\mathcal{U}$-epic. By hypothesis it is an admissible epic.

(3) follows trivially from (2), so it is left to show (3) $\implies$ (1). Suppose $g\colon A \to B$ is $\mathcal{U}$-epic. We need to prove g is an admissible epic. By hypothesis, we are given an admissible epic $\psi\colon X \twoheadrightarrow B$ with $X \in$ Free($\mathcal{U}$). By Proposition 1.30 it is enough to find a morphism $\alpha\colon X \to A$ such that $\psi = g\alpha$. Let us write $X = \bigoplus_{l \in L} U_l$ where each $U_l \in \mathcal{U}$. For each component $l \in L$, let us set $\psi_l := \psi \circ \eta_l$ where $\eta_l\colon U_l \to \bigoplus_{l \in L} U_l$ is the canonical injection. The supposition that g is $\mathcal{U}$-epic provides lifts α_l such that $\psi_l = g \circ \alpha_l$ as shown in the following diagram:

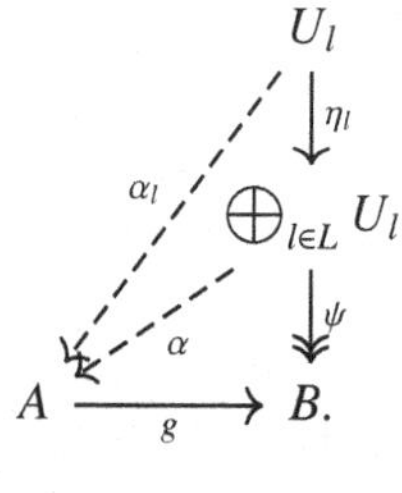

By the universal property of the coproduct $\bigoplus_{l \in L} U_l$, the α_l assemble to a morphism $\alpha\colon \bigoplus_{l \in L} U_l \to A$ satisfying $\alpha \circ \eta_l = \alpha_l$ for all $l \in L$. It follows that $g\alpha = \psi$. This proves, again because of Proposition 1.30, that g is an admissible epic. □

Note that while Definition 9.26 makes sense for any exact category, our proof of Proposition 9.27 uses Proposition 1.30, which requires that the underlying category $\mathcal{A}$ be weakly idempotent complete.

9.6 Generating Monics for Cotorsion Pairs

We now prove a result that is reminiscent of Baer's criterion for injectivity of modules. It will prove to be very useful for constructing functorially complete cotorsion pairs in exact categories.

Throughout this section, we continue to let $(\mathcal{A}, \mathcal{E})$ be a weakly idempotent complete exact category with coproducts. Moreover, we now also let $\mathcal{U} = \{U_i\}_{i \in I}$ denote a set of admissible generators and we will always let $U = \bigoplus_{i \in I} U_i$ denote the single generator obtained by taking their coproduct.

Definition 9.28 Let $\mathcal{Y}$ be a class of objects in $(\mathcal{A}, \mathcal{E})$. By a **set of generating monics for** $\mathcal{Y}$ we mean a set (not a proper class) I of admissible monics whose set $\mathcal{S} = \mathrm{Cok}\,(I)$ of cokernels satisfies $\mathcal{S}^{\perp} = \mathcal{Y}$, and such that $Y \in \mathcal{Y}$ if and only if given any $k \colon K \rightarrowtail X$ in I, there exists an extension of any given f over k as shown in the following diagram:

$$\begin{array}{ccc} K & \overset{k}{\rightarrowtail} & X \\ {\scriptstyle f}\downarrow & \swarrow & \\ Y. & & \end{array}$$

Said another way, every map $k \colon K \rightarrowtail X$ in I is Y-monic in the sense that $k^* \colon \mathrm{Hom}_{\mathcal{A}}(X, Y) \to \mathrm{Hom}_{\mathcal{A}}(K, Y)$ is surjective.

Baer's criterion states that a (left) R-module M over a ring R is injective precisely if for every (left) R-ideal D, any homomorphism $f \colon D \to M$ extends to a homomorphism $\bar{f} \colon R \to M$ along the inclusion $D \hookrightarrow R$. It follows that $\{R/D\}^{\perp}$ equals the class of injective R-modules and $\{D \hookrightarrow R\}$ is a set of generating monics for it.

Given any set $\mathcal{S}$, it is easy to construct a set of generating monics for the class $\mathcal{Y} = \mathcal{S}^{\perp}$ when the category has enough projectives. We leave this as an exercise.

Exercise 9.6.1 Suppose $(\mathcal{A}, \mathcal{E})$ has enough projectives. (This is the case for example if each $U_i \in \mathcal{U}$ is projective.) Show that for any set of objects $\mathcal{S}$ there is a set $I_{\mathcal{S}}$ of generating monics for $\mathcal{S}^{\perp}$.

Even if our category does not have enough projectives, we still have the following result. The point is that we need the set $\mathcal{U}$ of admissible generators. Recall that $\mathrm{Free}(\mathcal{U})$ denotes the class of all set indexed coproducts $X = \bigoplus U_j$ in which each $U_j \in \mathcal{U}$.

Proposition 9.29 (Generating Monics for Cotorsion Pairs) *Let $\mathcal{S}$ be any set (not a proper class) of objects in $(\mathcal{A}, \mathcal{E})$. Then there exists a set $I_{\mathcal{S}} = \{k\colon K \rightarrowtail X\}$ of generating monics for $\mathcal{S}^{\perp}$, with each codomain $X \in \mathrm{Free}(\mathcal{U})$, and with the property that each $\mathcal{S}$-cofibration arises as a pushout of some admissible monic $k \in I_{\mathcal{S}}$ as shown:*

$$\begin{array}{ccccc} K & \overset{k}{\rightarrowtail} & \bigoplus_{g\in J} U_g & \overset{\sigma}{\twoheadrightarrow} & S \\ \downarrow & & \downarrow & & \| \\ Y & \rightarrowtail & Z & \overset{q}{\twoheadrightarrow} & S. \end{array}$$

Proof We first set some notation: We have already set $U = \bigoplus_{i\in I} U_i$, and for convenience we set $H_A := \mathrm{Hom}_{\mathcal{A}}(U, A)$, for any given object A. Then given $S \in \mathcal{S}$, and any subset $J \subseteq H_S$, we will let $\sigma_J\colon \bigoplus_{g\in J} U_g \to S$ denote the *evaluation morphism*. The construction is analogous to Example 9.25: For each $g \in J$, we set $U_g = U$ and we have $\sigma_J \circ \eta_g = g$, where $\eta_g\colon U_g \rightarrowtail \bigoplus_{g\in J} U_g$ is the canonical injection. For some subsets J, for example if $J = H_S$, it will turn out that σ_J is an admissible epic. For all such subsets J, we let $k_J = \ker \sigma_J$, so that we have a short exact sequence $K_J \overset{k_J}{\rightarrowtail} \bigoplus_{g\in J} U_g \overset{\sigma_J}{\twoheadrightarrow} S$. Letting $J \subseteq H_S$ range through all the subsets for which σ_J is an admissible epic, we obtain a set $I_S := \left\{k_J\colon K_J \rightarrowtail \bigoplus_{g\in J} U_g\right\}$ (and it is not a proper class). Finally, we set $I_{\mathcal{S}} = \bigcup_{S\in\mathcal{S}} I_S$. Then $I_{\mathcal{S}}$ is a set of $\mathcal{S}$-cofibrations satisfying $\mathrm{Cok}\,(I_{\mathcal{S}}) = \mathcal{S}$ and with each codomain in $\mathrm{Free}(\mathcal{U})$. We will first prove that it satisfies the stated pushout property, and then use this to show that $I_{\mathcal{S}}$ is a set of generating monics for $\mathcal{S}^{\perp}$, as in Definition 9.28.

So let $Y \rightarrowtail Z \overset{q}{\twoheadrightarrow} S$ be any given short exact sequence with $S \in \mathcal{S}$. By Proposition 9.27, the evaluation morphism $e\colon \bigoplus_{f\in H_Z} U_f \longrightarrow Z$, where each $U_f = U$, is an admissible epimorphism. It satisfies

$$\text{(i)}\ \ e \circ \eta_f = f \text{ for all } f \in H_Z.$$

We consider the composite $qe\colon \bigoplus_{f\in H_Z} U_f \twoheadrightarrow S$, which is also an admissible epic. Let $J \subseteq H_S$ be the subset defined by

$$J = \{g \in H_S \mid g = qf \text{ for some } f \in H_Z\}.$$

Then in the notation we have already set, we have the evaluation map $\sigma_J\colon \bigoplus_{g\in J} U_g \to S$ satisfying

$$\text{(ii)}\ \ \sigma_J \circ \eta_g = g \text{ for all maps } g\colon U_g \to S \text{ in } J \subseteq H_S.$$

We will show there exists a morphism ψ making the following diagram commute:

$$\begin{array}{ccc} & & \bigoplus_{f \in H_Z} U_f \\ & \swarrow^{\psi} & \downarrow qe \\ \bigoplus_{g \in J} U_g & \xrightarrow[\sigma_J]{} & S. \end{array}$$

It will then follow from Proposition 1.30 that $\sigma_J \colon \bigoplus_{g \in J} U_g \twoheadrightarrow S$ is an admissible epic. Now, for all $f \in H_Z$ we have, upon composition with q, a morphism $g = qf \in J$. So by the universal property of $\bigoplus_{f \in H_Z} U_f$, the canonical injection morphisms $\eta_{qf} \colon U_f \to \bigoplus_{g \in J} U_g$ assemble to a unique morphism $\psi \colon \bigoplus_{f \in H_Z} U_f \to \bigoplus_{g \in J} U_g$ satisfying

$$\text{(iii)} \quad \psi \circ \eta_f = \eta_{qf} \text{ for all } f \in H_Z.$$

We claim: $\sigma_J \circ \psi = qe$. Indeed it is enough to show $\sigma_J \circ \psi \circ \eta_f = qe \circ \eta_f$, for all $f \in H_Z$. Using the relations (i), (ii), and (iii) already established earlier, we have $\sigma_J \circ \psi \circ \eta_f = \sigma_J \circ \eta_{qf} = qf$. And, on the other hand, $q(e \circ \eta_f) = qf$. This shows $\sigma_J \circ \psi = qe$ and so it follows from Proposition 1.30 that σ_J is an admissible epic. In particular, the top row of the following diagram is an admissible short exact sequence; whence $k_J \in I_S$:

$$\begin{array}{ccccc} K & \xrightarrow{k_J} & \bigoplus_{g \in J} U_g & \xrightarrow{\sigma_J} & S \\ \downarrow & & \downarrow h & & \| \\ Y & \longrightarrow & Z & \xrightarrow{q} & S. \end{array}$$

We continue to construct h, using the axiom of choice, to obtain the commutative diagram. To do so, for each $g \in J$, choose a morphism $f \colon U_g \to Z$ such that $g = qf$. The chosen morphisms $U_g \xrightarrow{f} Z$ assemble to a unique morphism $\bigoplus_{g \in J} U_g \xrightarrow{h} Z$ satisfying $h\eta_g = f$, for all $g \in J$:

$$\begin{array}{ccc} & & \bigoplus_{g \in J} U_g \\ & \nearrow^{\eta_g} & \downarrow h \\ U_g & \xrightarrow[f]{} & Z. \end{array}$$

So now, $qh = \sigma_J$ because we have $(qh)\eta_g = qf = g = \sigma_J \circ \eta_g$, for all $g \in J$. Finally, the vertical dashed arrow on the far left of the previous diagram exists by the universal property of the kernel $Y = \operatorname{Ker} q$. This completes the proof of the pushout property stated in the proposition.

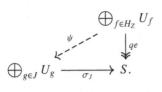

We now can easily prove the claim that I_S is a set of generating monics for $\mathcal{S}^\perp$. First, suppose I is *any* set of admissible monics with $[\text{Cok}\,(I)]^\perp = \mathcal{S}^\perp$. Then if $Y \in \mathcal{S}^\perp$ and $k\colon K \rightarrowtail X$ is in I, then we have $\text{Hom}_{\mathcal{A}}(X, Y) \xrightarrow{k^*} \text{Hom}_{\mathcal{A}}(K, Y) \to \text{Ext}^1_{\mathcal{E}}(\text{Cok}\,k, Y) = 0$. That is, $\text{Hom}_{\mathcal{A}}(X, Y) \xrightarrow{k^*} \text{Hom}_{\mathcal{A}}(K, Y)$ is surjective for every $k \in I$. So the point of Definition 9.28 is to show that the set of admissible monics I_S constructed previously provides a converse to this fact. So suppose Y has the property that for any given $k\colon K \rightarrowtail \bigoplus_{g\in J} U_g$ in I_S, there exists an extension of any f over k as shown:

$$\begin{array}{ccc} K & \overset{k}{\rightarrowtail} & \bigoplus_{g\in J} U_g \\ {\scriptstyle f}\downarrow & \swarrow & \\ Y. & & \end{array}$$

The goal is to show that any short exact sequence $Y \rightarrowtail Z \overset{q}{\twoheadrightarrow} S$ with $S \in \mathcal{S}$ must be split exact. But by what we just proved, this short exact sequence arises as a pushout of some morphism $k \in I_S$, and there is a commutative diagram as stated in the proposition. By hypothesis there exists a morphism making the upper left triangle of the diagram commute. But then the Homotopy Lemma 1.21 provides the existence of a map $S \to Z$ making the lower right triangle commute. This is precisely a splitting of the short exact sequence $Y \rightarrowtail Z \overset{q}{\twoheadrightarrow} S$, as desired. So I_S is a set of generating monics for $\mathcal{S}^\perp$. □

It is worth pointing out how Proposition 9.29 relates to some standard language from general model category theory. Let I be any set of morphisms in any cocomplete category C. A morphism $f\colon A \to B$ is called a *relative I-cell complex* if it equals a transfinite composition of some continuous direct λ-system $\left\{X_\alpha, i_{\alpha\beta}\right\}_{\alpha\leq\beta<\gamma}$ for which each $i_{\alpha,\alpha+1}$ is a pushout of some morphism in I. An object A is called an *I-cell* if $\phi \to A$ is a relative I-cell complex. The reader can easily do the following exercise by quoting previous propositions.

Exercise 9.6.2 Assume $\mathcal{S}$ is any efficient set of objects and let I_S be a set of generating monics for $\mathcal{S}^\perp$ as in Proposition 9.29. Then the class I_S*-cell*, of all relative I_S-cell complexes, coincides with the class of all Filt($\mathcal{S}$)-cofibrations. An object A is an I_S-cell if and only if $A \in \text{Filt}(\mathcal{S})$.

Exercise 9.6.3 Show that the Yoneda Ext groups $\text{Ext}^1_{\mathcal{E}}(X, Y)$ are sets (and not proper classes) whenever $(\mathcal{A}, \mathcal{E})$ is a weakly idempotent complete exact category with coproducts and a set $\mathcal{U}$ of admissible generators. (*Hint*: Think about the proof of Proposition 9.29.)

9.7 Cofibrantly Generated Cotorsion Pairs

We now define cofibrantly generated cotorsion pairs and show that they are complete cotorsion pairs. The proof is an algebraic version of Quillen's small object argument. However, for abelian model categories we need the small object argument to output *admissible* short exact sequences (conflations). We get these by working with a set of objects $\mathcal{S}$ that is efficient relative to the exact structure $\mathcal{E}$, and also contains admissible generators for $\mathcal{E}$. Because Quillen's small object argument produces functorial factorizations, the method here produces *functorially complete* cotorsion pairs; we return to make this point precise in Section 9.8.

Throughout this section, we again let $(\mathcal{A}, \mathcal{E})$ be a (weakly) idempotent complete exact category with coproducts. Moreover, we again assume that we have a set $\mathcal{U} = \{U_i\}_{i \in I}$ of admissible generators, and let $U = \bigoplus_{i \in I} U_i$ denote the single generator obtained by taking their coproduct.

We start by defining what we mean by a cofibrantly generated cotorsion pair.

Definition 9.30 Let $(\mathcal{X}, \mathcal{Y})$ be a cotorsion pair in $(\mathcal{A}, \mathcal{E})$. We say that $(\mathcal{X}, \mathcal{Y})$ is **cofibrantly generated** if there exists an efficient set (not a proper class) of objects, $\mathcal{S}$, such that each of the following hold.

(1) $\mathcal{S}$ cogenerates $(\mathcal{X}, \mathcal{Y})$. That is, $({}^{\perp}(\mathcal{S}^{\perp}), \mathcal{S}^{\perp}) = (\mathcal{X}, \mathcal{Y})$.
(2) There exists a set $I_{\mathcal{S}}$ of generating monics for $\mathcal{S}^{\perp}$ such that the domain of each map in $I_{\mathcal{S}}$ is small relative to Filt($\mathcal{S}$)-cofibrations.
(3) $\mathcal{X}$ contains a set $\mathcal{U} = \{U_i\}_{i \in I}$ of admissible generators for $(\mathcal{A}, \mathcal{E})$.

We say that $I_{\mathcal{S}} \cup \{0 \rightarrowtail U_i\}_{i \in I}$ is a set of **generating $\mathcal{X}$-cofibrations** for $(\mathcal{X}, \mathcal{Y})$.

Note that a complete cotorsion pair $(\mathcal{X}, \mathcal{Y})$ is equivalent to a trivial abelian model structure, $(\mathcal{X}, \mathcal{A}, \mathcal{Y})$. Then $(\mathcal{X}, \mathcal{Y})$ is cofibrantly generated if and only if $(\mathcal{X}, \mathcal{A}, \mathcal{Y})$ is cofibrantly generated in the sense of Definition 9.38 in Section 9.8.

Remark 9.31 In applications one would almost certainly be working in a category for which each object is "small" in some suitable sense. For example, this is the case in any efficient exact category in the sense of Section 9.9, or the locally presentable or accessible categories of Chapter 12. In all such cases, condition (2) of Definition 9.30 is *automatic*. Indeed Proposition 9.29 produces, from any admissible generating set $\mathcal{U}$, a set of generating monics for $\mathcal{S}^{\perp}$.

We first prove a couple of lemmas that will be crucial to our proof that such cotorsion pairs have enough injectives. For any set $\mathcal{S}$ of objects, Free($\mathcal{S}$) denotes the class of all set indexed coproducts (direct sums) of objects in $\mathcal{S}$.

Lemma 9.32 *Let $\mathcal{S}$ be an efficient set (not a proper class) of objects and suppose $I_{\mathcal{S}}$ is any set of generating monics for $\mathcal{S}^{\perp}$. Let $v\colon L \rightarrowtail V$ be the morphism obtained by taking the coproduct over all the maps in $I_{\mathcal{S}}$. Then v is a* Free($\mathcal{S}$)*-cofibration and the singleton $\{v\}$ is also a set of generating monics for $\{V/L\}^{\perp} = \mathcal{S}^{\perp}$.*

We call $v\colon L \rightarrowtail V$ the *single generating monic obtained from $I_{\mathcal{S}}$*.

Proof Write $I_{\mathcal{S}} = \{k\colon K \rightarrowtail X\}$. Since $\mathcal{S}$ is an efficient set, it follows from Proposition 9.21 that the coproduct $v\colon L \rightarrowtail V$ of $I_{\mathcal{S}}$ is a Free($\mathcal{S}$)-cofibration, because the cokernel $V/L = \bigoplus_{k\in I_{\mathcal{S}}} \operatorname{Cok} k$ is in the class Free($\mathcal{S}$). Now, since each $\operatorname{Cok} k$ is a direct summand, we have $\operatorname{Ext}^1_{\mathcal{E}}(V/L, Y) = 0$ implies $\operatorname{Ext}^1_{\mathcal{E}}(\operatorname{Cok} k, Y) = 0$ for each $k \in I_{\mathcal{S}}$, by Lemma 1.23. On the other hand, the converse holds by Lemma 1.24(1). Therefore, $\{V/L\}^{\perp} = [\operatorname{Cok}(I_{\mathcal{S}})]^{\perp} = \mathcal{S}^{\perp}$. Now one checks that $v\colon L \rightarrowtail V$ has the required property: Any given morphism $f\colon L \to Y$ is equivalent to a system of morphisms $\{f_k\colon K \to Y\}$. Each f_k extends over the corresponding $k\colon K \rightarrowtail X$, and these extensions assemble to an extension of f over $L \rightarrowtail V$. □

When we apply the next lemma, the morphism $v\colon L \rightarrowtail V$ will be the morphism that we just constructed in Lemma 9.32 (and we will take $\mathcal{S} =$ Filt($\mathcal{S}$)).

Lemma 9.33 *Assume $\mathcal{S}$ is any efficient class of objects in $(\mathcal{A}, \mathcal{E})$. Let A be any object of $\mathcal{A}$ and let $v\colon L \rightarrowtail V$ by any $\mathcal{S}$-cofibration (resp.* Filt($\mathcal{S}$)*-cofibration). Then there exists a* Free($\mathcal{S}$)*-cofibration (resp.* Filt($\mathcal{S}$)*-cofibration) $\bar{v}\colon A \rightarrowtail D$ with the property that any given morphism $f\colon L \to A$ extends to a morphism $\bar{f}\colon V \to D$ as in the following commutative diagram:*

$$\begin{array}{ccc} L & \overset{v}{\rightarrowtail} & V \\ {\scriptstyle f}\downarrow & & \downarrow{\scriptstyle \bar{f}} \\ A & \overset{\bar{v}}{\rightarrowtail} & D. \end{array}$$

Moreover, the cokernel D/A is isomorphic to a direct sum of copies of V/L.

Proof Consider the set $H = \operatorname{Hom}_{\mathcal{A}}(L, A)$ of all morphisms $f\colon L \to A$, and the corresponding induced evaluation morphism $e\colon \bigoplus_{f\in H} L_f \longrightarrow A$; see Example 9.25. So for each $f \in H$ we have $L_f = L$, and $f = e \circ \eta_f$, where $\eta_f\colon L \rightarrowtail \bigoplus_{f\in H} L_f$ denotes the canonical injection into component $f \in H$. Also, form the coproduct $\bigoplus_{f\in H} V_f$, where again each $V_f = V$. It follows from

Proposition 9.21 that the induced coproduct morphism $\bigoplus_{f\in H} L_f \rightarrowtail \bigoplus_{f\in H} V_f$ is a Free($\mathcal{S}$)-cofibration (resp. Filt($\mathcal{S}$)-cofibration), and that its cokernel is

$$\bigoplus_{f\in H} V_f \Big/ \bigoplus_{f\in H} L_f = \bigoplus_{f\in H} \big(V_f/L_f\big).$$

Note this is a direct sum of copies of V/L. By Lemma 1.10 we obtain the pushout diagram:

$$\begin{array}{ccccc}
\bigoplus_{f\in H} L_f & \rightarrowtail & \bigoplus_{f\in H} V_f & \twoheadrightarrow & \bigoplus_{f\in H}\big(V_f/L_f\big) \\
{\scriptstyle e}\downarrow & & {\scriptstyle \bar{e}}\downarrow & & \| \\
A & \rightarrowtail & D & \twoheadrightarrow & \bigoplus_{f\in H}\big(V_f/L_f\big).
\end{array}$$

We claim that $\bar{\nu}\colon A \rightarrowtail D$ has the desired property. Indeed given any morphism $f\colon L \to A$, we must construct an extension map $\bar{f}$ making the desired diagram commute. But the following diagram commutes:

$$\begin{array}{ccc}
L & \overset{\nu}{\rightarrowtail} & V \\
{\scriptstyle \eta_f}\downarrow & & \downarrow{\scriptstyle \eta_f} \\
\bigoplus_{f\in H} L_f & \overset{\oplus\nu}{\rightarrowtail} & \bigoplus_{f\in H} V_f \\
{\scriptstyle e}\downarrow & & \downarrow{\scriptstyle \bar{e}} \\
A & \overset{\bar{\nu}}{\rightarrowtail} & D,
\end{array}$$

and recall $f = e \circ \eta_f$. So by setting $\bar{f} := \bar{e} \circ \eta_f$, we have a map making the outer rectangle commute. □

We now prove the main theorem. Although the hypotheses of the statement are technical, it is a powerful tool for constructing complete cotorsion pairs in very general settings. A simplified (and very useful) version of it, for efficient exact categories, appears as Corollary 9.40 of Section 9.9. Another version, applicable in locally presentable or even accessible additive categories, is Corollary 12.4. The general form that follows will be used in Section 10.8.

Again, we denote by Free($\mathcal{U}$) the class of all set indexed direct sums $X = \bigoplus U_j$ in which each $U_j \in \mathcal{U}$.

Theorem 9.34 (Cofibrantly Generated Cotorsion Pairs are Complete) *Assume $(\mathcal{A}, \mathcal{E})$ is a weakly idempotent complete exact category with coproducts. Then any cofibrantly generated cotorsion pair $(\mathcal{X}, \mathcal{Y})$ is (functorially) complete. In more detail and generality, each of the following hold.*

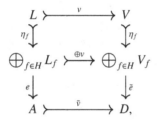

(1) *Let $\mathcal{S}$ and $I_{\mathcal{S}}$ be sets satisfying the first two conditions of Definition 9.30. Then the cotorsion pair $({}^{\perp}(\mathcal{S}^{\perp}), \mathcal{S}^{\perp})$ has enough injectives. In fact, for each $A \in \mathcal{A}$, there exists a short exact sequence*

$$A \rightarrowtail Y \twoheadrightarrow Y/A$$

with $Y \in \mathcal{S}^{\perp}$ and $Y/A \in \mathrm{Filt}(\mathcal{S}) \subseteq {}^{\perp}(\mathcal{S}^{\perp})$.

(2) *Assume moreover, that ${}^{\perp}(\mathcal{S}^{\perp})$ contains a set $\mathcal{U} = \{U_i\}$ of admissible generators. That is, $({}^{\perp}(\mathcal{S}^{\perp}), \mathcal{S}^{\perp})$ is a cofibrantly generated cotorsion pair. Then the cotorsion pair $({}^{\perp}(\mathcal{S}^{\perp}), \mathcal{S}^{\perp})$ also has enough projectives, and hence is a complete cotorsion pair. In fact , for each $A \in \mathcal{A}$, there exists a short exact sequence*

$$Y \rightarrowtail X \twoheadrightarrow A$$

with $Y \in \mathcal{S}^{\perp}$ and where X is an extension of an object in $\mathrm{Free}(\mathcal{U})$ *by an object in* $\mathrm{Filt}(\mathcal{S})$. *In particular, $X \in \mathrm{Filt}(\mathcal{S} \cup \mathcal{U}) \subseteq {}^{\perp}(\mathcal{S}^{\perp})$.*

Finally, each object $X \in \mathcal{X}$ of a cofibrantly generated cotorsion pair $(\mathcal{X}, \mathcal{Y}) = ({}^{\perp}(\mathcal{S}^{\perp}), \mathcal{S}^{\perp})$ is a direct summand of an extension of an object in $\mathrm{Free}(\mathcal{U})$ *by an object in* $\mathrm{Filt}(\mathcal{S})$. *In particular we have $\mathcal{X} = \overline{\mathrm{Filt}(\mathcal{S} \cup \mathcal{U})}$, the closure of* $\mathrm{Filt}(\mathcal{S} \cup \mathcal{U})$ *under taking direct summands.*

Remark 9.35 The functoriality is addressed in detail in Section 9.8. For now, we address the main points without mentioning the functoriality.

Proof Using Lemma 9.32, let $v\colon L \rightarrowtail V$ be the single generating monic obtained from $I_{\mathcal{S}}$. It is the coproduct of all the maps in $I_{\mathcal{S}}$, and $V/L \in \mathrm{Free}(\mathcal{S})$. Let $A \in \mathcal{A}$ be an arbitrary object. We first note that given *any* ordinal λ, we may repeatedly apply Lemma 9.33 to construct a λ-extension sequence

$$A = Y_0 \rightarrowtail Y_1 \rightarrowtail Y_2 \rightarrowtail \cdots \rightarrowtail Y_\alpha \rightarrowtail Y_{\alpha+1} \rightarrowtail \cdots \tag{9.7}$$

with the properties: (i) any morphism $f_\alpha\colon L \to Y_\alpha$ extends to a morphism $\bar{f}_\alpha\colon V \to Y_{\alpha+1}$ in the sense of having a commutative diagram

$$\begin{array}{ccc} L & \overset{v}{\rightarrowtail} & V \\ {\scriptstyle f_\alpha}\downarrow & & \downarrow{\scriptstyle \bar{f}_\alpha} \\ Y_\alpha & \overset{\bar{v}_{\alpha,\alpha+1}}{\rightarrowtail} & Y_{\alpha+1}, \end{array}$$

and (ii) $Y_{\alpha+1}/Y_\alpha$ is isomorphic to an object in $\mathrm{Free}(\mathcal{S}) \subseteq \mathrm{Filt}(\mathcal{S})$. For limit ordinal steps $\gamma \leq \lambda$ we set $Y_\gamma = \varinjlim_{\alpha<\gamma} Y_\alpha$. This constructs a λ-extension sequence $\{Y_\alpha, \bar{v}_{\alpha\beta}\}_{\alpha\leq\beta<\lambda}$ of $\mathrm{Free}(\mathcal{S})$-cofibrations $\bar{v}_{\alpha,\alpha+1}$, and so, by Corollary 9.20, the transfinite composition $\bar{v}_0\colon A \rightarrowtail Y_\lambda$ exists and is also a $\mathrm{Filt}(\mathcal{S})$-cofibration. Thus for any ordinal λ, we have constructed a short exact sequence

$$A \rightarrowtail Y_\lambda \twoheadrightarrow Y_\lambda/A$$

with $Y_\lambda/A \in \text{Filt}(\mathcal{S}) \subseteq {}^\perp(\mathcal{S}^\perp)$, where the containment follows from the version of Eklof's Lemma stated in Corollary 9.4.

To complete the proof of (1), it is left to show that there exists a (large enough) λ so that the construction produces a Y_λ with $Y_\lambda \in \mathcal{S}^\perp$. But by assumption, the domain of each morphism in $I_\mathcal{S}$ is small relative to Filt($\mathcal{S}$)-cofibrations. It follows that the domain L, being the coproduct over all the domains of $I_\mathcal{S}$, is κ-small relative to Filt($\mathcal{S}$)-cofibrations, for some cardinal κ. Indeed we may take κ to a cardinal satisfying $\kappa > |I_\mathcal{S}|$ and also satisfying that the domain of each map in $I_\mathcal{S}$ is κ-small relative to Filt($\mathcal{S}$)-cofibrations. For such a κ, we have that whenever $\lambda > \kappa$ is a regular cardinal, then any given map $f\colon L \to Y_\lambda$ must factor through some Y_α as a composition $L \xrightarrow{f_\alpha} Y_\alpha \xrightarrow{\bar{v}_\alpha} Y_\lambda$ (see Exercise 9.4.1.) We also have the compatibilities $\bar{v}_\alpha = \bar{v}_{\alpha+1} \circ \bar{v}_{\alpha,\alpha+1}$, so by the construction of Y_λ, we obtain a factorization:

$$\begin{array}{ccc} L & \overset{v}{\rightarrowtail} & V \\ {\scriptstyle f}\downarrow & \swarrow {\scriptstyle \bar{v}_{\alpha+1}\circ \bar{f}_\alpha} & \\ Y_\lambda & & \end{array}$$

and this proves that $Y_\lambda \in \mathcal{S}^\perp$, since $v : L \rightarrowtail V$ is the single generating monic for $\{V/L\}^\perp = \mathcal{S}^\perp$ as in Lemma 9.32. This completes the proof of (1).

We now turn to prove (2). So assume that the class ${}^\perp(\mathcal{S}^\perp)$ contains a set of admissible generators $\mathcal{U} = \{U_i\}$. To prove (2), we do an abstract form of a standard argument, originally due to Luigi Salce, and known as *Salce's Trick*. Let $A \in \mathcal{A}$ be given. Using the generating set $\mathcal{U}$ we may write a short exact sequence $K \rightarrowtail \bigoplus_{j\in J} U_j \twoheadrightarrow A$ with $\bigoplus_{j\in J} U_j \in \text{Free}(\mathcal{U})$. From Lemma 1.24 we know that ${}^\perp(\mathcal{S}^\perp)$ is closed under coproducts, so it follows that $\text{Free}(\mathcal{U}) \subseteq {}^\perp(\mathcal{S}^\perp)$. Using what we just proved in part (1), that the cotorsion pair has enough injectives, we obtain (by Lemma 1.10) a pushout diagram:

$$\begin{array}{ccccc} K & \rightarrowtail & \bigoplus_{j\in J} U_j & \twoheadrightarrow & A \\ \downarrow & & \downarrow & & \| \\ Y & \rightarrowtail & X & \twoheadrightarrow & A \\ \downarrow & & \downarrow & & \\ Y/K & = & Y/K, & & \end{array}$$

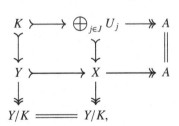

where $Y \in \mathcal{S}^{\perp}$ and $Y/K \in \text{Filt}(\mathcal{S})$. Since $\bigoplus_{j \in J} U_j \rightarrowtail X$ is a Filt($\mathcal{S}$)-cofibration, we have from Proposition 9.19 that it identifies as the transfinite composition of some λ-extension sequence of $\mathcal{S}$-cofibrations. But we also have $\bigoplus_{j \in J} U_j \in \text{Free}(\mathcal{U}) \subseteq \text{Filt}(\mathcal{U})$. Explicitly, letting $\kappa = |J|$, we can obtain $\bigoplus_{j \in J} U_j$ as a κ-iterated extension of objects in $\mathcal{U}$ (see Exercise 9.1.1 and Corollary 9.23). Then in the spirit of Corollary 9.20, X is a $(\kappa + \lambda)$-iterated extension of objects in $\mathcal{U} \cup \mathcal{S}$, and so $X \in \text{Filt}(\mathcal{S} \cup \mathcal{U})$. This proves the existence of a short exact sequence with the properties listed in (2).

Finally, suppose $A \in {}^{\perp}(\mathcal{S}^{\perp})$. By following the previous pushout diagram argument we are again led to a short exact sequence

$$Y \rightarrowtail X \twoheadrightarrow A$$

with $Y \in \mathcal{S}^{\perp}$ and with X an extension of an object in Free($\mathcal{U}$) by an object in Filt($\mathcal{S}$). But in this case the sequence splits, and A is a direct summand of $X \in \text{Filt}(\mathcal{S} \cup \mathcal{U})$. Since $\mathcal{S} \cup \mathcal{U} \subseteq \mathcal{X}$, it follows from Eklof's Lemma 9.3 that $\mathcal{X}$ equals the class of all direct summands of objects in the class Filt($\mathcal{S} \cup \mathcal{U}$). □

9.8 Functoriality and Cofibrantly Generated Models

We now wish to make clear that the construction of complete cotorsion pairs given in Theorem 9.34 is *functorial*. Through Hovey's Correspondence this translates to having functorial factorizations, and indeed cofibrantly generated abelian model structures. The resulting cofibrant and fibrant approximations constructed here will be used later in Chapter 12 to prove certain homotopy categories are well generated.

Let $(\mathcal{X}, \mathcal{Y})$ be a cotorsion pair in an exact category $(\mathcal{A}, \mathcal{E})$. We say that $(\mathcal{X}, \mathcal{Y})$ **functorially has enough injectives**, or **has functorial $\mathcal{Y}$-approximation sequences** if there are functors $Y : \mathcal{A} \to \mathcal{A}$ and $\Sigma : \mathcal{A} \to \mathcal{A}$, along with natural transformations $1_{\mathcal{A}} \xrightarrow{j} Y \xrightarrow{q} \Sigma$ such that each $A \overset{j_A}{\rightarrowtail} YA \overset{q_A}{\twoheadrightarrow} \Sigma A$ is a short exact sequence with $YA \in \mathcal{Y}$ and $\Sigma A \in \mathcal{X}$. The naturality means that for each morphism $f : A \to B$ we have a commutative diagram:

$$\begin{array}{ccccc} A & \overset{j_A}{\rightarrowtail} & YA & \overset{q_A}{\twoheadrightarrow} & \Sigma A \\ \downarrow f & & \downarrow Y(f) & & \downarrow \Sigma(f) \\ B & \overset{j_B}{\rightarrowtail} & YB & \overset{q_B}{\twoheadrightarrow} & \Sigma B. \end{array}$$

So (j, q) is to be thought of as a natural short exact sequence, and we call it a *functorial $\mathcal{Y}$-approximation sequence*. We also say that Y (although we really mean the natural admissible monic $j\colon 1_{\mathcal{A}} \rightarrowtail Y$), is a *functorial $\mathcal{Y}$-approximation.* We say that $(\mathcal{X}, \mathcal{Y})$ **functorially has enough projectives**, or **has functorial $\mathcal{X}$-approximation sequences**, if it satisfies the dual notion of having a functorial $\mathcal{X}$-approximation $X\colon \mathcal{A} \to \mathcal{A}$ along with a natural admissible epic $p\colon X \twoheadrightarrow 1_{\mathcal{A}}$ as indicated:

$$\begin{array}{ccccc} \Omega A & \overset{i_A}{\rightarrowtail} & XA & \overset{p_A}{\twoheadrightarrow} & A \\ \downarrow{\scriptstyle \Omega(f)} & & \downarrow{\scriptstyle X(f)} & & \downarrow{\scriptstyle f} \\ \Omega B & \overset{i_B}{\rightarrowtail} & XB & \overset{p_B}{\twoheadrightarrow} & B. \end{array}$$

The cotorsion pair is said to be **functorially complete** if it functorially has enough injectives and projectives

Throughout the rest of this section, we continue with the same assumptions as set at the beginning of the previous Section 9.7.

We start by considering the functoriality of Lemma 9.33 and its proof. The reader should verify the following version of that lemma, which is slightly more general and also functorial. To save space, throughout this section we will use the following notation: For any object $A \in \mathcal{A}$, we set $H^L_A := \mathrm{Hom}_{\mathcal{A}}(L, A)$.

Lemma 9.36 (Functoriality of Lemma 9.33) *Assume $\mathcal{S}$ is an efficient class of objects in $(\mathcal{A}, \mathcal{E})$. Let $v\colon L \rightarrowtail V$ by any $\mathcal{S}$-cofibration (resp.* Filt$(\mathcal{S})$*-cofibration) and let $F\colon \mathcal{A} \to \mathcal{A}$ be any endofunctor on $\mathcal{A}$. Then there is a functor $P^v\colon \mathcal{A} \to \mathcal{A}$ and a natural transformation $\bar{v}\colon F \rightarrowtail P^v$, with each component $\bar{v}_A$ a* Free$(\mathcal{S})$*-cofibration (resp.* Filt$(\mathcal{S})$*-cofibration), defined as follows.*

- *(Objects) Given any object $A \in \mathcal{A}$, we define P^vA and the component $\bar{v}_A\colon FA \rightarrowtail P^vA$ of the natural transformation $\bar{v}$ by way of the pushout:*

$$\begin{array}{ccc} \bigoplus_{t \in H^L_{FA}} L_t & \overset{\bigoplus v}{\rightarrowtail} & \bigoplus_{t \in H^L_{FA}} V_t \\ \downarrow{\scriptstyle e_{FA}} & & \downarrow{\scriptstyle \bar{e}_{FA}} \\ FA & \overset{\bar{v}_A}{\rightarrowtail} & P^vA. \end{array}$$

  Here $L_t = L$ and $V_t = V$, for each component $t \in H^L_{FA}$, and e_{FA} is the evaluation morphism uniquely defined by it satisfying the equation

$$e_{FA} \circ \eta^L_t = t \ \text{for all} \ t \in H^L_{FA}. \qquad (*)$$

- *(Morphisms) Given any morphism $f\colon A \to B$, there exist morphisms Ff^L_* and Ff^V_* making the top face and left face of the following cube commute:*

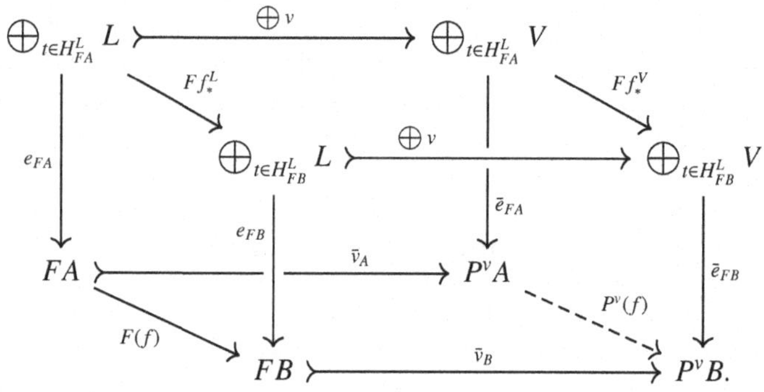

This in turn induces a unique morphism $P^v(f)\colon P^vA \to P^vB$ between the pushouts, providing the final two commutative squares. In all, we have a commutative cube, so the following equations hold:

$$e_{FB} \circ Ff^L_* = F(f) \circ e_{FA}, \tag{†}$$

$$\left(\oplus_{t \in H^L_{FB}} v\right) \circ Ff^L_* = Ff^V_* \circ \left(\oplus_{t \in H^L_{FA}} v\right), \tag{††}$$

$$P^v(f) \circ \bar{v}_A = \bar{v}_B \circ F(f), \quad P^v(f) \circ \bar{e}_{FA} = \bar{e}_{FB} \circ Ff^V_*. \tag{‡‡}$$

The natural transformation $\bar{v}\colon F \rightarrowtail P^v$ is such that each $\bar{v}_A$ is a Filt($\mathcal{S}$)-*cofibration and satisfies the property that any given morphism $t\colon L \to FA$ extends to a morphism $\bar{t}\colon V \to P^vA$ as in the commutative diagram:*

$$\begin{array}{ccc} L & \overset{v}{\rightarrowtail} & V \\ {\scriptstyle t}\downarrow & & \downarrow{\scriptstyle \bar{t}} \\ FA & \overset{\bar{v}_A}{\rightarrowtail} & P^vA. \end{array}$$

Each cokernel P^vA/FA is isomorphic to a direct sum of copies of V/L.

Exercise 9.8.1 Adapt the proof of Lemma 9.33 to prove Lemma 9.36 as follows: Define Ff^L_* to be the unique map out of the coproduct satisfying $Ff^L_* \circ \eta^L_t = \eta^L_{F(f)\circ t}$ for each $t \in H^L_{FA}$, and define Ff^V_* analogously. Verify that this makes Equations (†) and (††) hold.

Theorem 9.37 (Functoriality of Theorem 9.34) *Assume $(\mathcal{A}, \mathcal{E})$ is a (weakly) idempotent complete exact category with coproducts. Then any cofibrantly generated cotorsion pair $(\mathcal{X}, \mathcal{Y})$ is functorially complete. In fact, functorial versions of each of the more detailed statements of Theorem 9.34 hold.*

Proof We check that the arguments in the proof of Theorem 9.34 carry over to produce functorial versions. The key idea is to construct a (natural) transfinite extension of endofunctors on $\mathcal{A}$, rather than just objects, by replacing any reference to Lemma 9.33 with a reference to Lemma 9.36. In particular, taking $v\colon L \rightarrowtail V$ to be the single generating monic obtained from I_S (using Lemma 9.32), we may construct, for any given ordinal λ, a λ-extension sequence of endofunctors on $\mathcal{A}$

$$1_{\mathcal{A}} = P_0^v \rightarrowtail P_1^v \rightarrowtail P_\alpha^v \rightarrowtail \cdots \rightarrowtail P_\alpha^v \rightarrowtail P_{\alpha+1}^v \rightarrowtail \cdots$$

for which each component of each natural transformation $\bar{v}_{\alpha,\alpha+1}\colon P_\alpha^v \rightarrowtail P_{\alpha+1}^v$ is a Free($\mathcal{S}$)-cofibration. We do this by transfinite induction, taking $P_0^v = 1_{\mathcal{A}}$ for the base step. For the inductive step, if $P_\alpha^v\colon \mathcal{A} \to \mathcal{A}$ has been defined and $\alpha + 1 < \lambda$, then we apply Lemma 9.36, taking $F = P_\alpha^v$, to obtain a functor $P_{\alpha+1}^v\colon \mathcal{A} \to \mathcal{A}$ along with a natural $\bar{v}_{\alpha,\alpha+1}\colon P_\alpha^v \rightarrowtail P_{\alpha+1}^v$ with all of the stated properties. In the case of a limit ordinal $\gamma < \lambda$, we define $P_\gamma^v := \varinjlim_{\alpha<\gamma} P_\alpha^v$. That is, for each $A \in \mathcal{A}$, we set $P_\gamma^v A$ to be the transfinite extension $P_\gamma^v A := \varinjlim_{\alpha<\gamma} P_\alpha^v A$. This determines a functor $P_\gamma^v\colon \mathcal{A} \to \mathcal{A}$, thus we obtain the λ-extension sequence of functors $\{P_\alpha^v, \bar{v}_{\alpha\beta}\}_{\alpha\leq\beta<\lambda}$ as claimed. The rest of the argument in the proof of Theorem 9.34 carries over pointwise to construct a functor $P_\lambda^v := \varinjlim_{\alpha<\lambda} P_\alpha^v$ and a natural admissible monic $\bar{v}_0\colon 1_{\mathcal{A}} \rightarrowtail P_\lambda^v$ whose cokernel $P_\lambda^v/1_{\mathcal{A}}$ is a functor which takes values in Filt($\mathcal{S}$) $\subseteq {}^\perp(\mathcal{S}^\perp)$. Finally, by the smallness assumption, there exists a cardinal κ such that whenever $\lambda > \kappa$ is a regular cardinal, then the functor P_λ^v will take values in $\mathcal{S}^\perp$. So for any such regular cardinal λ, the natural transformation $\bar{v}_0\colon 1_{\mathcal{A}} \rightarrowtail P_\lambda^v$ provides functorial $\mathcal{S}^\perp$-approximation sequences. This proves $({}^\perp(\mathcal{S}^\perp), \mathcal{S}^\perp)$ functorially has enough injectives.

Salce's Trick is used at the end of the proof of Theorem 9.34 to show that we have enough projectives. This argument too can be made functorial. Indeed for each object $A \in \mathcal{A}$, we may use the evaluation morphism

$$e_A\colon \bigoplus_{t\in\mathrm{Hom}_{\mathcal{A}}(U,A)} U_t \twoheadrightarrow A$$

of Proposition 9.27(2). For any $f\colon A \to B$, define $f_*\colon \bigoplus_{t\in H_A^U} U \to \bigoplus_{t\in H_B^U} U$ by the rule $f_* \circ \eta_t = \eta_{ft}$. Then we have a commutative square $f \circ e_A = e_B \circ f_*$. Using this along with the functorial $\mathcal{S}^\perp$-approximation, $\bar{v}_0\colon 1_{\mathcal{A}} \rightarrowtail P_\lambda^v$, we obtain a functorial version of Salce's Trick. In particular, the functoriality of the pushout is similar to the statement of Lemma 6.26; see the last paragraph of the proof of that lemma. In this way, the construction produces a functorial $\mathcal{X}$-approximation sequence $\Omega \rightarrowtail X_\lambda \twoheadrightarrow 1_{\mathcal{A}}$. □

Exercise 9.8.2 (Complete Ext-Pairs) Verify that Theorems 9.34 and 9.37 may be stated for Ext-pairs in the sense of Definition 2.2. In particular, $(\mathrm{Filt}(\mathcal{S}), \mathcal{S}^{\perp})$ functorially has enough injectives and enough projectives whenever $\mathcal{U} \subseteq \mathrm{Filt}(\mathcal{S})$. We call $(\mathrm{Filt}(\mathcal{S}), \mathcal{S}^{\perp})$ the *Ext-pair cogenerated by* $\mathcal{S}$.

9.8.1 Cofibrantly Generated Abelian Model Categories

We now define *cofibrantly generated* abelian model structures. Our definition is essentially the same as the standard one for general model categories; see Exercises 9.6.2 and 9.8.4. The only difference is that we assume the existence of a set of trivially cofibrant generators for $\mathcal{E}$, so our definition is stronger. But this seems appropriate for the abelian case, and it reflects how cofibrantly generated abelian models arise in practice; see also Exercise 9.9.2.

Definition 9.38 Let $(\mathcal{A}, \mathcal{E})$ be a (weakly) idempotent complete exact category with coproducts and a set of admissible generators. An abelian model structure $\mathfrak{M} = (\mathcal{Q}, \mathcal{W}, \mathcal{R})$ is called **cofibrantly generated** if each of the associated cotorsion pairs, $(\mathcal{Q}, \mathcal{R}_{\mathcal{W}})$ and $(\mathcal{Q}_{\mathcal{W}}, \mathcal{R})$, are cofibrantly generated in the sense of Definition 9.30. We summarize all of this data as follows, and set some related notation and terminology.

- $\mathcal{Q}_{\mathcal{W}}$ contains a set of admissible generators for the exact structure $\mathcal{E}$. We let $\mathcal{U} = \{U_i\}$ denote such a set.
- We let $\mathcal{S}$ be an efficient set that cogenerates $(\mathcal{Q}, \mathcal{R}_{\mathcal{W}})$. It comes with some set $I_{\mathcal{S}}$ of generating monics for $\mathcal{S}^{\perp} = \mathcal{R}_{\mathcal{W}}$. We call $I_{\mathcal{S}} \cup \{0 \rightarrowtail U_i\}$ a set of **generating cofibrations** for $\mathfrak{M}$. We denote their coproduct by $v \colon L \rightarrowtail V$; it is a single generating cofibration for $\mathfrak{M}$, by Lemma 9.32.
- We let $\mathcal{T}$ be an efficient set that cogenerates $(\mathcal{Q}_{\mathcal{W}}, \mathcal{R})$. It comes with some set $J_{\mathcal{T}}$ of generating monics for $\mathcal{T}^{\perp} = \mathcal{R}$. We call $J_{\mathcal{T}} \cup \{0 \rightarrowtail U_i\}$ a set of **generating trivial cofibrations** for $\mathfrak{M}$. We denote their coproduct by $v' \colon L' \rightarrowtail V'$; it is a single generating trivial cofibration for $\mathfrak{M}$, by Lemma 9.32.

If $\mathcal{A}$ is a bicomplete category, then $(\mathcal{A}, \mathfrak{M})$ is called a **cofibrantly generated abelian model category**.

The following exercise indicates how a cofibrantly generated abelian model structure has *functorial factorizations*.

Exercise 9.8.3 (Functorial Factorizations) Verify that if $\mathfrak{C} = (\mathcal{X}, \mathcal{Y})$ is a functorially complete cotorsion pair (or even just an Ext-pair), then the factorizations constructed from them as in Proposition 2.13 are functorial as well. By this, we mean that the factorizations $f = qj$ described there may be assigned to each morphism $f = q_f j_f$ in the following way: For any commutative square $s_c f = f' s_d$, there is a morphism $x(s_d, s_c)$ providing a commutative diagram

$$\begin{array}{ccccc} A & \xrightarrow[j_f]{} & X & \xrightarrow[q_f]{} & B \\ {\scriptstyle s_d}\downarrow & & \downarrow{\scriptstyle x(s_d,\, s_c)} & & \downarrow{\scriptstyle s_c} \\ A' & \xrightarrow[j_{f'}]{} & X' & \xrightarrow[q_{f'}]{} & B' \end{array}$$

and in a way that is functorial in the sense that (i) $x(1_d, 1_c) = 1_X$ and (ii) $x(t_d s_d, t_c s_c) = x(t_d, t_c) \circ x(s_d, s_c)$ whenever we have another $t_c f' = f'' t_d$.

Exercise 9.8.4 Let $\mathfrak{M} = (Q, \mathcal{W}, \mathcal{R})$ be a cofibrantly generated abelian model structure, and let $I_S \cup \{0 \rightarrowtail U_i\}$ and $J_{\mathcal{T}} \cup \{0 \rightarrowtail U_i\}$ be the sets of generating cofibrations and trivial cofibrations. Show that a map $g\colon X \to Y$ is a fibration if and only if it satisfies the RLP (see Definition 2.12 and Proposition 2.11) with respect to $J_{\mathcal{T}}$. Similarly, show that g is a trivial fibration if and only if it satisfies the RLP with respect to I_S.

9.9 Efficient Exact Categories and Examples

The results of the previous sections are motivated by the *efficient exact categories* of [Saorín and Šťovíček, 2011]. This is a very convenient abstract setting where one can easily construct (functorially) complete cotorsion pairs.

Definition 9.39 An exact category $(\mathcal{A}, \mathcal{E})$ is called an **efficient exact category** if it satisfies the following axioms.
(EF0) The underlying additive category $\mathcal{A}$ is weakly idempotent complete.
(EF1) (*Transfinite Extensions Axiom*) For each ordinal λ and each λ-extension sequence $\{X_\alpha, i_{\alpha\beta}\}_{\alpha\leq\beta<\lambda}$

$$X_0 \rightarrowtail X_1 \rightarrowtail X_2 \rightarrowtail \cdots \rightarrowtail X_\alpha \rightarrowtail X_{\alpha+1} \rightarrowtail \cdots,$$

we have that the colimit $X := \varinjlim_{\alpha<\lambda} X_\alpha$ exists, and the canonical morphism $i_0\colon X_0 \rightarrowtail X$ into the colimit is an admissible monic.
(EF2) (*Smallness Axiom*) Each object $A \in \mathcal{A}$ is admissibly small, as defined after Definition 9.24. It means there exists a cardinal κ such that for every regular cardinal $\lambda > \kappa$, any morphism $f\colon A \to \varinjlim_{\alpha<\lambda} X_\alpha$ factors through some $f_\alpha\colon A \to X_\alpha$ ($\alpha < \lambda$), whenever we are given a λ-extension sequence $\{X_\alpha, i_{\alpha\beta}\}_{\alpha\leq\beta<\lambda}$.
(EF3) (*Generator Axiom*) There exists a set $\mathcal{U} = \{U_i\}_{i\in I}$ of admissible generators for $(\mathcal{A}, \mathcal{E})$; see Definition 9.26 and Proposition 9.27.

Note that Axiom (EF1) is equivalent to saying that *every* class $\mathcal{S}$ of objects is efficient for $\mathcal{E}$ in the sense of Definition 9.11. So we have already proved many results in this chapter for efficient exact categories. In particular, by Corollary 9.22, efficient exact categories have exact coproducts. That is, they satisfy the analog of Grothendieck's Axiom (AB3), that coproducts exist, and Axiom (AB4), that short exact sequences are closed under coproducts. The following useful corollary is immediate from Theorem 9.34 and Remark 9.31.

Corollary 9.40 (Efficient Exact Categories and Complete Cotorsion Pairs) *Let $\mathcal{S}$ be any set (not a proper class) of objects in an efficient exact category $(\mathcal{A}, \mathcal{E})$. Then the cotorsion pair $({}^{\perp}(\mathcal{S}^{\perp}), \mathcal{S}^{\perp})$ functorially has enough injectives. Moreover, if ${}^{\perp}(\mathcal{S}^{\perp})$ contains some set $\mathcal{U}$ of admissible generators, then this is a functorially complete cotorsion pair. In this case, we have ${}^{\perp}(\mathcal{S}^{\perp}) = \overline{\mathrm{Filt}(\mathcal{S} \cup \mathcal{U})}$, the closure of* Filt($\mathcal{S} \cup \mathcal{U}$) *under direct summands.*

Indeed a cotorsion pair $(\mathcal{X}, \mathcal{Y})$ in an efficient exact category $(\mathcal{A}, \mathcal{E})$ is cofibrantly generated (see Definition 9.30) if and only if it is cogenerated by a set $\mathcal{S}$, and $\mathcal{X}$ contains some set of admissible generators.

Example 9.41 For any ring R, the category R-Mod of all (left) R-modules is an efficient exact category. Transfinite extensions are just special types of direct limits, and Axiom (EF2) appears in Exercise 9.4.2. $\{R\}$ is a set of (admissible) generators. Note that Corollary 9.40 recovers Theorem 2 from the Introduction and Main Examples.

Example 9.42 More generally, every Grothendieck category is an efficient exact category. By definition, Grothendieck categories have exact direct limits, which implies (EF1), and they come with a set of (admissible) generators. Axiom (EF2) is asked in Exercise 9.4.3.

Example 9.43 Again let R be a ring and R-Mod the category of all (left) R-modules. But now let $\mathcal{E}$ be the proper class of all pure exact sequences. Recall that a short exact sequence is pure if it remains exact after applying $\mathrm{Hom}_R(F, -)$ for each finitely presented R-module F. Then $(R\text{-Mod}, \mathcal{E})$ is an efficient exact category. To check Axiom (EF1), let us be given a λ-extension sequence $\left\{X_\alpha, i_{\alpha\beta}\right\}_{\alpha \le \beta < \lambda}$

$$X_0 \rightarrowtail X_1 \rightarrowtail X_2 \rightarrowtail \cdots \rightarrowtail X_\alpha \rightarrowtail X_{\alpha+1} \rightarrowtail \cdots$$

with each $X_\alpha \rightarrowtail X_{\alpha+1}$ a pure monomorphism, and set $X_\lambda := \varinjlim_{\alpha<\lambda} X_\lambda$. We must show that the canonical morphism $i_0 \colon X_0 \to X_\lambda$ is also a pure monomorphism. Using transfinite induction we can see that each $X_0 \rightarrowtail X_\alpha$ $(\alpha \le \lambda)$ is a pure monomorphism. The base step and successor ordinal steps of the induction are clear. Assume $\gamma \le \lambda$ is any limit ordinal. Then the short exact sequence

$$X_0 \rightarrowtail X_\gamma \twoheadrightarrow X_\gamma/X_0$$

is the direct limit, taken over all $\alpha < \gamma$, of the short exact sequences

$$X_0 \rightarrowtail X_\alpha \twoheadrightarrow X_\alpha/X_0$$

and each of these are pure by inductive hypothesis. Since pure exact sequences are closed under direct limits, we get that $i_0 \colon X_0 \rightarrowtail X_\gamma$ is a pure monomorphism. Now Axiom (EF2) holds in the same way as in Example 9.41. For Axiom (EF3), it is easy to check that the set $\{F_i\}$ of "all" finitely presented R-modules is a set of admissible generators for (R-Mod, $\mathcal{E}$). (Really we obtain a set $\{F_i\}$ by taking one representative from each isomorphism class of finitely presented modules.)

Example 9.44 The purity Example 9.43 generalizes to any locally finitely presentable Grothendieck category. That is, to any Grothendieck category $\mathcal{G}$ possessing a generating set $\{F_i\}$ of finitely presented objects in the sense that each $\mathrm{Hom}_{\mathcal{G}}(F_i, -)$ preserves direct limits.

Example 9.45 Let $(\mathcal{A}, \mathcal{E})$ be an efficient exact category. Then the exact category $\mathrm{Ch}(\mathcal{A})_{\mathcal{E}}$ of all chain complexes, along with the degreewise $\mathcal{E}$-short exact sequences, is an efficient exact category. See Lemma 10.4 and Exercise 10.2.1.

Example 9.46 Let $(\mathcal{X}, \mathcal{Y})$ be an hereditary cotorsion pair in an efficient exact category $(\mathcal{A}, \mathcal{E})$. Consider $\mathcal{X}$ as a fully exact subcategory along with its inherited exact structure $\mathcal{E}_{\mathcal{X}}$, consisting of all short exact sequences in $\mathcal{E}$ having each term in $\mathcal{X}$. If $\mathcal{X}$ contains a set $\mathcal{U}$ of admissible generators for $\mathcal{E}$, then $(\mathcal{X}, \mathcal{E}_{\mathcal{X}})$ is also an efficient exact category. Axiom (EF1) follows from Eklof's Lemma, in particular Corollary 9.5; see also Proposition 9.17, and Lemma 9.15. Axiom (EF2) is inherited from $(\mathcal{A}, \mathcal{E})$; (EF3) holds by the hereditary hypothesis and the hypothesis that $\mathcal{X}$ contains $\mathcal{U}$.

For example, the category of all flat R-modules, and the category of all chain complexes of flat R-modules, are efficient exact categories.

Exercise 9.9.1 Let $(\mathcal{A}, \mathcal{E})$ be an efficient exact category with enough projectives. Any set $\mathcal{S}$ of objects cogenerates a cotorsion pair $(\mathcal{X}, \mathcal{Y}) = ({}^{\perp}(\mathcal{S}^{\perp}), \mathcal{S}^{\perp})$ that is cofibrantly generated in the sense of Definition 9.30. Give a description of the objects in ${}^{\perp}(\mathcal{S}^{\perp})$, and find a set of generating $\mathcal{X}$-cofibrations.

9.9.1 Abelian Model Structures on Efficient Exact Categories

The results of this chapter give us the following theorem for constructing cofibrantly generated abelian model structures on efficient exact categories.

Theorem 9.47 (Constructing Abelian Model Structures) *Let $(\mathcal{A}, \mathcal{E})$ be an efficient exact category. Assume we have a triple $\mathfrak{M} = (\mathcal{Q}, \mathcal{W}, \mathcal{R})$ of classes of objects and set $\mathcal{Q}_{\mathcal{W}} := \mathcal{Q} \cap \mathcal{W}$ and $\mathcal{R}_{\mathcal{W}} := \mathcal{W} \cap \mathcal{R}$. Then $\mathfrak{M}$ is a cofibrantly generated abelian model structure if and only if the following conditions hold.*

(1) *$\mathcal{W}$ satisfies the 2 out of 3 property.*
(2) *$\mathcal{Q}_{\mathcal{W}}$ contains a set $\mathcal{U} = \{U_i\}$ of admissible generators for $\mathcal{E}$.*
(3) *$(\mathcal{Q}, \mathcal{R}_{\mathcal{W}})$ is a cotorsion pair cogenerated by some set $\mathcal{S}$.*
(4) *$(\mathcal{Q}_{\mathcal{W}}, \mathcal{R})$ is a cotorsion pair cogenerated by some set $\mathcal{T}$.*

In this case, $I_{\mathcal{S}} \cup \{0 \rightarrowtail U_i\}$ is a set of generating cofibrations for $\mathfrak{M}$, where $I_{\mathcal{S}}$ is any set of generating monics for $\mathcal{S}^{\perp} = \mathcal{R}_{\mathcal{W}}$ (one exists by Proposition 9.29). Similarly, $J_{\mathcal{T}} \cup \{0 \rightarrowtail U_i\}$ is a set of generating trivial cofibrations for $\mathfrak{M}$, where $J_{\mathcal{T}}$ is any set of generating monics for $\mathcal{T}^{\perp} = \mathcal{R}$.

Proof Referring to Definition 9.38, this follows at once from Corollary 9.40 along with Theorem 4.25. □

Exercise 9.9.2 Let $\mathfrak{M} = (\mathcal{Q}, \mathcal{W}, \mathcal{R})$ be an abelian model structure on an efficient exact category. Show that there is a set of trivially cofibrant objects which is a set of admissible generators for $(\mathcal{A}, \mathcal{E})$. More generally, this is true whenever $(\mathcal{A}, \mathcal{E})$ has exact coproducts and possesses a set of admissible generators.

In the hereditary case the following method for constructing abelian model structures is implied by Theorem 8.16.

Corollary 9.48 (Constructing Hereditary Model Structures) *Let $(\mathcal{A}, \mathcal{E})$ be an efficient exact category. Assume the following conditions.*

(1) *There is a set of objects $\mathcal{S}$ that cogenerates an hereditary cotorsion pair $(\mathcal{Q}, \widetilde{\mathcal{R}})$.*
(2) *There is a set of objects $\mathcal{T}$ that cogenerates an hereditary cotorsion pair $(\widetilde{\mathcal{Q}}, \mathcal{R})$, and $\widetilde{\mathcal{Q}}$ contains a set $\mathcal{U} = \{U_i\}$ of admissible generators.*
(3) *$(\widetilde{\mathcal{Q}}, \widetilde{\mathcal{R}})$ is an Ext-pair. That is, $\mathrm{Ext}^1_{\mathcal{E}}(\widetilde{\mathcal{Q}}, \widetilde{\mathcal{R}}) = 0$.*
(4) *The two cotorsion pairs share the same core, $\widetilde{\mathcal{Q}} \cap \mathcal{R} = \mathcal{Q} \cap \widetilde{\mathcal{R}}$.*

Then there exists a cofibrantly generated and hereditary abelian model structure, $\mathfrak{M} = (\mathcal{Q}, \mathcal{W}, \mathcal{R})$, on $(\mathcal{A}, \mathcal{E})$. The class $\mathcal{W}$ of trivial objects is as characterized in Lemma 4.2. Sets of generating cofibrations and generating trivial cofibrations are as stated in Theorem 9.47.

Example 9.49 (Gorenstein Flat Model Structure) Let R be any ring and let $(\mathcal{F}, \mathcal{C})$ denote Enochs' flat cotorsion pair. So $\mathcal{F}$ is the class of all flat (left)

R-modules and $\mathcal{C} := \mathcal{F}^{\perp}$ is the class of all cotorsion R-modules. Let $\mathcal{GF}$ denote the class of all Gorenstein flat R-modules (see [Enochs and Jenda, 2000, definition 10.3.1]), and $\mathcal{GC} := \mathcal{GF}^{\perp}$ denote the class of all Gorenstein cotorsion R-modules. Šaroch and Šťovíček showed in [Šaroch and Šťovíček, 2020] that $(\mathcal{GF}, \mathcal{GC})$ is a hereditary cotorsion pair, cogenerated by a set, and that $\mathcal{GF} \cap \mathcal{GC} = \mathcal{F} \cap \mathcal{C}$. It was already known that $(\mathcal{F}, \mathcal{C})$ is cogenerated by a set too; see Exercise 9.9.4. So by Corollary 9.48 there exists a cofibrantly generated hereditary model structure $\mathfrak{M}_{\mathrm{G}}^{flat} = (\mathcal{GF}, \mathcal{V}, \mathcal{C})$ on the category of R-modules.

Exercise 9.9.3 (Proof of Eklof's Lemma for [EF1] Categories) Assume that $(\mathcal{A}, \mathcal{E})$ is an efficient exact category. Let $\mathcal{T}$ be any class of objects in $\mathcal{A}$. Assume $Y \in \mathcal{A}$ is any object for which $\operatorname{Ext}^1_{\mathcal{E}}(T, Y) = 0$ for all $T \in \mathcal{T}$. Prove the conclusion of Eklof's Lemma, that $\operatorname{Ext}^1_{\mathcal{E}}(X, Y) = 0$ for all $X \in \operatorname{Filt}(\mathcal{T})$, as follows.

(1) Let $\mathbb{E}\colon Y \overset{j}{\rightarrowtail} B \overset{q}{\twoheadrightarrow} X$ be any short exact sequence where $X \in \operatorname{Filt}(\mathcal{T})$. Use Proposition 9.19 to express $j\colon Y \rightarrowtail B$ as a λ-extension of $\{B_\alpha, m_{\alpha\beta}\}_{\alpha \leq \beta < \lambda}$ (with $B_0 = Y$) that is compatible with the chosen $\mathcal{T}$-filtration of X.

(2) Show that $\mathbb{E}$ must split as follows: Let $f_0\colon B_0 \xrightarrow{1_Y} Y$ be the base step of a transfinite induction argument that constructs compatible extensions $f_\alpha\colon B_\alpha \to Y$.

Exercise 9.9.4 Let R be any ring and $\kappa \geq |R|$ be a regular cardinal. It is known that for any R-module M and element $x \in M$, there exists a pure submodule $P \subseteq M$ with $x \in P$ and with cardinality $|P| \leq \kappa$. Use this fact and Corollary 9.40 to show that Enochs' flat cotorsion pair, $(\mathcal{F}, \mathcal{C})$, is a complete cotorsion pair cogenerated by the set of all (isomorphism representatives of) flat R-modules F of cardinality $|F| \leq \kappa$. (*Hint*: Use properties of purity and flatness such as those in Exercise 7.6.1 and Exercise 1.8.2.)

Exercise 9.9.5 Let $(\mathcal{A}, \otimes, \mathcal{H}om)$ be a symmetric monoidal structure on an efficient exact category $(\mathcal{A}, \mathcal{E})$ with an abelian model $\mathfrak{M} = (\mathcal{Q}, \mathcal{W}, \mathcal{R})$. Suppose $\mathcal{S}$ is a set that contains a set of admissible generators and that $\mathcal{S}$ cogenerates $(\mathcal{Q}, \mathcal{R}_{\mathcal{W}})$. Similarly, assume $\mathcal{T}$ is a set that contains a set of admissible generators and that $\mathcal{T}$ cogenerates $(\mathcal{Q}_{\mathcal{W}}, \mathcal{R})$. Assume condition **(⊗ 0)** from Section 7.6 holds. Show that condition **(⊗ 1)** can be replaced with the condition: If $P \in \mathcal{S}$ and $Q \in \mathcal{S}$, then $P \otimes Q \in \mathcal{Q}$. Similarly, condition **(⊗ 2)** can be replaced with: If $P \in \mathcal{S}$ and $Q \in \mathcal{T}$, then $P \otimes Q \in \mathcal{Q}_{\mathcal{W}}$.

Exercise 9.9.6 State a generalized version of Theorem 9.47 that constructs a (not necessarily closed) abelian model structure on an efficient exact category, from two Ext-pairs that are cogenerated by sets in the sense of Exercise 9.8.2. (A version of Theorem 9.40 holds for Ext-pairs. So simply refer to Definitions 4.1 and A.1 and verify that everything carries over by replacing Hovey triples with Hovey Ext-triples and using Theorem A.2 in place of Theorem 4.25.)

Exercise 9.9.7 Let R be a ring and $\mathrm{Ch}(R)$ denote the category of chain complexes of (left) R-modules. Let $\mathcal{S} = \{S^n(R) \mid n \in \mathbb{Z}\}$, where $S^n(R)$ denotes the chain complex consisting of R (viewed as a left module over itself) in degree n and 0 elsewhere. Use Exercise 9.9.6 to show that the Ext-pair cogenerated by $\mathcal{S}$, meaning $(\mathrm{Filt}(\mathcal{S}), \mathcal{S}^{\perp})$, determines a (not necessarily closed) abelian model structure on $\mathrm{Ch}(R)$. Note that each object in $\mathrm{Filt}(\mathcal{S})$ is a chain complex of free R-modules, and these are the cofibrant objects. The closure of $(\mathrm{Filt}(\mathcal{S}), \mathcal{S}^{\perp})$ is the cotorsion pair $({}^{\perp}(\mathcal{S}^{\perp}), \mathcal{S}^{\perp})$, which corresponds to $\mathrm{Ch}(R)_{proj}$, the standard *DG-projective model structure*. See Example 3 of the Introduction and Main Examples.

9.10 Direct Limits and (AB5) Exact Categories

As we have already noted, an efficient exact category satisfies the exact category analogs of Grothendieck's (AB3) and (AB4) axioms. In this section we consider an exact category analog of Grothendieck's (AB5) axiom, which is exactness of direct limits. Ultimately, our goal is to show that if $(\mathcal{A}, \mathcal{E})$ *has exact direct limits*, a condition that is stronger than (EF1), then the left-hand side of a cotorsion pair is automatically closed under direct limits whenever it is closed under taking cokernels of admissible monics. See Theorem 9.56. In particular, it implies that the class of trivial objects in any injective model structure is closed under direct limits.

We assume throughout this section that direct limits exist in our additive category $\mathcal{A}$. Before considering any exact structure on $\mathcal{A}$, we first need a technical result concerning the reduction of direct limits to well-ordered direct limits. The details behind this are crucial to proving our main theorems. Recall that direct limits are generally indexed by a directed set $(I, \leq)$. This is a nonempty set I together with a reflexive and transitive binary relation $\leq$ (i.e. a *preorder*) such that every finite subset has an upper bound. It is well known, and not so hard to see, that any given direct limit may be assumed to be indexed by a *directed poset* $(I, \leq)$. This is just a directed set $(I, \leq)$ with the extra condition that $\leq$ is antisymmetric.[1] So unless stated otherwise all direct limits we speak of are indexed by some directed poset $(I, \leq)$. Then a functor $(I, \leq) \to \mathcal{A}$ determines a *direct I-system* consisting of objects and morphisms $\{A_i, f_{ij}\}$, where $f_{ij}\colon A_i \to A_j$ is a morphism whenever $i \leq j$. Such a system satisfies (i) $f_{ii} = 1_{A_i}$ for all $i \in I$ and (ii) $f_{ik} = f_{jk} \circ f_{ij}$ whenever $i \leq j \leq k$. We are assuming that the colimit of any such direct I-system $\{A_i, f_{ij}\}$ exists. It is written $\varinjlim_I A_i$, and it is said to be its *direct limit*. Note that the direct λ-systems defined at the start of Section 9.1 are a special case of direct I-systems. We will need the following general result concerning the reduction of any direct limit to a direct limit of some (continuous) direct λ-system.

Lemma 9.50 (Corollary 1.7 of Adámek and Rosický [1994]) *Let $\mathcal{A}$ be an additive category in which direct limits exist. Let $\{A_i, f_{ij}\}$ be a direct I-system indexed by a directed poset $(I, \leq)$ of infinite cardinality $\lambda = |I|$. Then there exist directed subposets*

$$(I_\alpha, \leq) \subseteq (I, \leq),$$

one for each ordinal $\alpha < \lambda$, satisfying each of the following.

(1) *$|I_\alpha| < \lambda$, for each $\alpha < \lambda$.*

(2) *$(I_\alpha, \leq) \subseteq (I_\beta, \leq)$ whenever $\alpha < \beta$. Consequently, setting $A_\alpha := \varinjlim_{I_\alpha} A_i$, we have canonical maps*

$$f_{\alpha\beta}\colon A_\alpha \to A_\beta$$

defining the transition morphisms of a direct λ-system $\{A_\alpha, f_{\alpha\beta}\}$.

(3) *$I_\beta = \bigcup_{\alpha<\beta} I_\alpha$, for each limit ordinal $\beta \leq \lambda$, including $\beta = \lambda$ where we define $I_\lambda := I$.*

Consequently, the direct λ-system $\{A_\alpha, f_{\alpha\beta}\}$ is continuous (or smooth) in the sense that we have $A_\beta = \varinjlim_{\alpha<\beta} A_\alpha$ for all limit ordinals β. In particular, for the case $\beta = \lambda$ it means we have that

$$\varinjlim_{I} A_i = \varinjlim_{\alpha<\lambda} A_\alpha$$

is the direct limit of a continuous direct λ-system $\{A_\alpha, f_{\alpha\beta}\}$ where each $A_\alpha = \varinjlim_{I_\alpha} A_i$ is the direct limit of the direct I_α-subsystem $\{A_i, f_{ij}\}_{i\in I_\alpha}$, and the indexing set I_α is a directed subposet of I of cardinality $|I_\alpha| < \lambda$ (strictly smaller).

[1] The antisymmetric condition doesn't allow "loops", and the idea behind the reduction from preorders to directed posets is to identify all "loops" in the preorder.

Proof We may well order the index set as $I = \{ i_\alpha \mid \alpha < \lambda \}$. We use transfinite induction to define an increasing chain of directed subposets $I_0 \subseteq \cdots \subseteq I_\alpha \subseteq I_{\alpha+1} \subseteq \cdots$ with $\{ i_{\alpha'} \mid \alpha' < \alpha \} \subseteq I_\alpha$ for each α, and with properties (1)–(3).

- (Base Step) We start the induction by setting $I_0 = \phi$ and $I_1 = \{i_0\}$.
- (Successor Ordinal Step) Let $\alpha < \lambda$ be an ordinal and assume we have defined a chain $I_0 \subseteq I_1 \subseteq \cdots I_\alpha$ of directed subposets such that $\{ i_{\alpha'} \mid \alpha' < \alpha \} \subseteq I_\alpha$ and with properties (1)–(3) holding for $\alpha' < \alpha$. We need to define $I_{\alpha+1}$ containing i_α and possessing properties (1)–(3). If I_α is a finite set, we first take $I_\alpha \cup \{i_\alpha\}$. Then choose an upper bound of this set, say u, and define $I_{\alpha+1} := I_\alpha \cup \{i_\alpha\} \cup \{u\}$. Then $(I_{\alpha+1}, \leq) \subseteq (I, \leq)$ is a directed subposet and has all of the desired properties. But we need to deal with the general case of an infinite I_α. In this case, start again by considering the set $I_\alpha \cup \{i_\alpha\}$, which we denote by J_0. Now choose, for each pair of elements in J_0, an upper bound, and let J_1 be the set obtained by taking the union of J_0 with all of these upper bounds. Of course this subset may not determine a directed subposet, so we repeat this process to get an increasing chain of subsets $J_0 \subseteq J_1 \subseteq \cdots \subseteq J_n \subseteq \cdots$. Defining $I_{\alpha+1} := \bigcup_{n<\omega} J_n$ gives us a directed subposet $(I_{\alpha+1}, \leq) \subseteq (I, \leq)$ satisfying all the desired properties.
- (Limit Ordinal Step) Let $\beta < \lambda$ be a limit ordinal and assume we have defined a chain $I_0 \subseteq I_1 \subseteq \cdots I_\alpha \subseteq \cdots$ of directed subposets such that $\{ i_{\alpha'} \mid \alpha' < \alpha \} \subseteq I_\alpha$ for each $\alpha < \beta$ and with properties (1)–(3) holding for all $\alpha < \beta$. Then we simply set $I_\beta := \bigcup_{\alpha<\beta} I_\alpha$. We then have $\{ i_\alpha \mid \alpha < \beta \} \subseteq I_\beta$ and properties (1)–(3) hold for all $\alpha \leq \beta$.

Finally, we comment on the existence of the direct λ-system $\left\{ \varinjlim_{I_\alpha} A_i \, , f_{\alpha\beta} \right\}$ satisfying the stated properties. First, for each α, let $\eta_i^\alpha \colon A_i \to \varinjlim_{I_\alpha} A_i$ denote the canonical map into the colimit. Since $I_\alpha \subseteq I_\beta$ whenever $\alpha < \beta$, the collection $\{\eta_i^\beta\}_{I_\alpha}$ forms a cocone above the direct I_α-subsystem $\left\{A_i, f_{ij}\right\}_{I_\alpha}$. By the universal property, this means there is a unique morphism $f_{\alpha\beta} \colon \varinjlim_{I_\alpha} A_i \to \varinjlim_{I_\beta} A_i$ satisfying $f_{\alpha\beta} \circ \eta_i^\alpha = \eta_i^\beta$ for all $i \in I_\alpha$. It follows easily that $f_{\beta\gamma} \circ f_{\alpha\beta} = f_{\alpha\gamma}$ whenever $\alpha \leq \beta \leq \gamma$, and so $\left\{ \varinjlim_{I_\alpha} A_i \, , f_{\alpha\beta} \right\}$ is a direct λ-system. Property (3) implies $\varinjlim_{I_\beta} A_i = \varinjlim_{\alpha<\beta} \left(\varinjlim_{I_\alpha} A_i \right)$, that is, $A_\beta = \varinjlim_{\alpha<\beta} A_\alpha$ for all limit ordinals $\beta \leq \lambda$. This means $\left\{A_\alpha, f_{\alpha\beta}\right\}$ is a continuous direct λ-system. □

Still assuming that direct limits exist in $\mathcal{A}$, let us return to our usual setting of an exact category $(\mathcal{A}, \mathcal{E})$. Suppose we have a direct I-system of short exact sequences $\left\{\mathbb{E}_i, \left(\alpha_{ij}, \beta_{ij}, \gamma_{ij}\right)\right\}$ indexed by some directed poset $(I, \leq)$. In particular, for each $i \leq j$ in I, we have a morphism of short exact sequences in $\mathcal{E}$:

$$\begin{array}{llllll} \mathbb{E}_j: & A_j & \overset{f_j}{\rightarrowtail} & B_j & \overset{g_j}{\twoheadrightarrow} & C_j \\ & \uparrow{\scriptstyle \alpha_{ij}} & & \uparrow{\scriptstyle \beta_{ij}} & & \uparrow{\scriptstyle \gamma_{ij}} \\ \mathbb{E}_i: & A_i & \underset{f_i}{\rightarrowtail} & B_i & \underset{g_i}{\twoheadrightarrow} & C_i. \end{array}$$

By the universal properties of the involved colimits there is an induced sequence of morphisms:

$$\varinjlim_I \mathbb{E}_i: \qquad \varinjlim_I A_i \xrightarrow{\varinjlim_I f_i} \varinjlim_I B_i \xrightarrow{\varinjlim_I g_i} \varinjlim_I C_i.$$

Since direct limits commute with cokernels it is always true that $\varinjlim_I g_i$ is the cokernel of $\varinjlim_I f_i$. However, $\varinjlim_I f_i$ may not be a monomorphism. Grothendieck's (AB5) axiom is precisely the statement that $\varinjlim_I f_i$ is always a monomorphism, and consequently $\varinjlim_I \mathbb{E}_i$ is a short exact sequence in the (AB5) abelian case. A general exact category may also satisfy that $\varinjlim_I \mathbb{E}_i$ is always an (admissible) short exact sequence. We say that $\mathcal{E}$ is *closed under direct limits* if $\varinjlim_I \mathbb{E}_i \in \mathcal{E}$ for any such given direct system of short exact sequences $\{\mathbb{E}_i, (\alpha_{ij}, \beta_{ij}, \gamma_{ij})\}$.

Definition 9.51 We say that an exact category $(\mathcal{A}, \mathcal{E})$ **has exact direct limits** if (i) direct limits exist in $\mathcal{A}$ and (ii) the class $\mathcal{E}$ of short exact sequences is closed under direct limits.

For example, let R be a ring, R-Mod the category of all (left) R-modules, and $\mathcal{E}$ the proper class of all pure exact sequences. Then the exact category $(R\text{-Mod}, \mathcal{E})$ has exact direct limits. More generally, any locally finitely presentable Grothendieck category has such an exact structure; see Examples 9.43 and 9.44. In fact, the argument proving (EF1) in Example 9.43 generalizes to show that any exact category with exact direct limits is an (EF1) exact category:

Lemma 9.52 *Let $(\mathcal{A}, \mathcal{E})$ be an exact category having exact direct limits and let $\{A_i, f_{ij}\}$ be a direct I-system in $\mathcal{A}$. If each f_{ij} is an admissible monic, then the canonical injection $\eta_i: A_i \rightarrowtail \varinjlim_I A_i$ is an admissible monic for each $i \in I$.*

It follows that an exact category with exact direct limits satisfies the Transfinite Extensions Axiom (EF1).

Proof Let $i \in I$ be arbitrary, and define a subset $I_i \subseteq I$ by

$$I_i := \{\, j \in I \mid i \leq j \,\}.$$

Clearly I_i inherits from $\leq$ the structure of a directed subposet of I. In fact I_i is cofinal in I, for if $i' \in I$ is arbitrarily given, there exists an upper bound j for the pair $\{i, i'\}$. Therefore we have $\varinjlim_{I_i} A_j = \varinjlim_{I} A_i$. But also, the hypothesis that each transition morphism f_{ij} is an admissible monic leads to a direct I_i-system of short exact sequences $\left\{\mathbb{E}_j, \left(1_{A_i}, f_{jk}, \bar{f}_{jk}\right)\right\}$:

$$\begin{array}{llcccccc}
\mathbb{E}_k: & & A_i & \overset{f_{ik}}{\rightarrowtail} & A_k & \overset{c_k}{\twoheadrightarrow} & A_k/A_i \\
 & & \uparrow{\scriptstyle 1_{A_i}} & & \uparrow{\scriptstyle f_{jk}} & & \uparrow{\scriptstyle \bar{f}_{jk}} \\
\mathbb{E}_j: & & A_i & \underset{f_{ij}}{\rightarrowtail} & A_j & \underset{c_j}{\twoheadrightarrow} & A_j/A_i.
\end{array}$$

Becuase $(\mathcal{A}, \mathcal{E})$ has exact direct limits, we get a short exact sequence:

$$\varinjlim_{I_i} \mathbb{E}_j: \qquad \varinjlim_{I_i} A_i \overset{\varinjlim_{I_i} f_{ij}}{\rightarrowtail} \varinjlim_{I_i} A_j \overset{\varinjlim_{I_i} c_j}{\twoheadrightarrow} \varinjlim_{I_i} \left(A_j/A_i\right).$$

Since the colimit $\varinjlim_{I_i} A_i = A_i$ is trivial, the map $\varinjlim_{I_i} f_{ij}$ is, by definition, the unique morphism satisfying $\varinjlim_{I_i} f_{ij} = \eta_j \circ f_{ij}$ for all $j \in I_i$. Since $\eta_i = \eta_j \circ f_{ij}$ for all $j \in I_i$, we conclude $\varinjlim_{I_i} f_{ij} = \eta_i$ is an admissible monic.

In particular, for any given λ-extension sequence $\left\{X_\alpha, i_{\alpha\beta}\right\}_{\alpha \leq \beta < \lambda}$, an inductive argument (as in Example 9.43) shows that $i_{0,\alpha}: X_0 \rightarrowtail X_\alpha$ is an admissible monic for each $\alpha \leq \lambda$. In particular, $i_0: X_0 \rightarrowtail X_\lambda$ is an admissible monic. □

For the remainder of this section we will assume $(\mathcal{A}, \mathcal{E})$ has exact direct limits. The following continuation of Lemma 9.50 will be useful to us.

Lemma 9.53 *Suppose in the statement of Lemma 9.50 that $(\mathcal{A}, \mathcal{E})$ is an exact category with exact direct limits and that each transition morphism f_{ij} in the given direct I-system $\left\{A_i, f_{ij}\right\}$ is an admissible monic. Then the resulting direct λ-system $\left\{A_\alpha, f_{\alpha\beta}\right\}_{\alpha \leq \beta < \lambda}$ is a λ-extension sequence. That is, each transition morphism $f_{\alpha,\alpha+1}$ is an admissible monic, and in particular*

$$A_\lambda := \varinjlim_{I} A_i = \varinjlim_{\alpha < \lambda} A_\alpha$$

is the transfinite extension of the cokernels $\operatorname{Cok} f_{\alpha,\alpha+1} = A_{\alpha+1}/A_\alpha$.

Proof We return to and continue the proof of Lemma 9.50, where we have constructed the continuous direct λ-system $\left\{A_\alpha, f_{\alpha\beta}\right\}_{\alpha \leq \beta < \lambda}$. Let α, β be given ordinals such that $\alpha \leq \beta < \lambda$. Using the notation set in the last paragraph of the proof, we now have that the canonical injections $\eta_i^\beta: A_i \rightarrowtail A_\beta$ are all admissible monics, by Lemma 9.52. Letting $i \leq j$ range through the set I_α, we obtain a direct I_α-system of short exact sequences $\{\mathbb{E}_i, (f_{ij}, 1_{A_\beta}, -)\}$:

$$\begin{array}{llllll} \mathbb{E}_j: & A_j & \overset{\eta_j^\beta}{\rightarrowtail} & A_\beta & \twoheadrightarrow & A_\beta/A_i \\ & {\scriptstyle f_{ij}}\uparrow & & \uparrow{\scriptstyle 1_{A_\beta}} & & \uparrow \\ \mathbb{E}_i: & A_i & \underset{\eta_i^\beta}{\rightarrowtail} & A_\beta & \twoheadrightarrow & A_\beta/A_i. \end{array}$$

Becuase $(\mathcal{A}, \mathcal{E})$ has exact direct limits we get a short exact sequence

$$\varinjlim_{I_\alpha} \mathbb{E}_i: \qquad \varinjlim_{I_\alpha} A_i \overset{\varinjlim_{I_\alpha} \eta_i^\beta}{\rightarrowtail} \varinjlim_{I_\alpha} A_\beta \twoheadrightarrow \varinjlim_{I_\alpha} (A_\beta/A_i).$$

By definition, we have $\varinjlim_{I_\alpha} A_i = A_\alpha$, and the colimit $\varinjlim_{I_\alpha} A_\beta = A_\beta$ is trivial. So by definition, $\varinjlim_{I_\alpha} \eta_i^\beta : A_\alpha \rightarrowtail A_\beta$ is the unique morphism satisfying $\varinjlim_{I_\alpha} \eta_i^\beta \circ \eta_i^\alpha = \eta_i^\beta$ for all $i \in I_\alpha$. Since by construction we have $f_{\alpha\beta} \circ \eta_i^\alpha = \eta_i^\beta$ for all $i \in I_\alpha$, we conclude $\varinjlim_{I_\alpha} \eta_i^\beta = f_{\alpha\beta}$ is an admissible monic. In particular, each transition morphism $f_{\alpha,\alpha+1}$ is an admissible monic, so $\{A_\alpha, f_{\alpha\beta}\}_{\alpha\leq\beta<\lambda}$ is a λ-extension sequence and $A_\lambda := \varinjlim_I A_i = \varinjlim_{\alpha<\lambda} A_\alpha$ is the transfinite extension of the cokernels $\operatorname{Cok} f_{\alpha,\alpha+1} = A_{\alpha+1}/A_\alpha$. □

As a first step to proving Theorem 9.56 we are now able to prove a weaker version where the direct system consists of admissible monics.

Proposition 9.54 *Let $(\mathcal{X}, \mathcal{Y})$ be a cotorsion pair in an exact category $(\mathcal{A}, \mathcal{E})$ with exact direct limits. If $\mathcal{X}$ is closed under cokernels of admissible monics, then $\mathcal{X}$ is closed under direct limits of admissible monics. That is, for any direct I-system $\{A_i, f_{ij}\}$ of admissible monics, $\varinjlim_I A_i \in \mathcal{X}$ whenever each $A_i \in \mathcal{X}$.*

Proof Let $\{A_i, f_{ij}\}$ be an direct I-system of admissible monics $f_{ij} : A_i \rightarrowtail A_j$ with each $A_i \in \mathcal{X}$. We proceed by transfinite induction on the cardinality $\lambda = |I|$ of the directed poset $(I, \leq)$, to show that the direct limit $\varinjlim_I A_i$ must be in the class $\mathcal{X}$.

First, we note that if λ is finite, then the direct limit $\varinjlim_I A_i$ must coincide with the particular A_i corresponding to the unique maximal element of I. So clearly $\varinjlim_I A_i = A_i \in \mathcal{X}$ in this case.

For the induction step, let us be given such a direct limit $\varinjlim_I A_i$ with $\lambda = |I|$ an infinite cardinal. The induction hypothesis is this: Let $(J, \leq)$ be any directed poset of cardinality $|J| < \lambda$ (strictly less). Then for any direct J-system $\{X_i, g_{ij}\}$ of admissible monics $g_{ij} : X_i \rightarrowtail X_j$ we have $\varinjlim_J X_i \in \mathcal{X}$ whenever each $X_i \in \mathcal{X}$. By Lemma 9.53 we may express $\varinjlim_I A_i$ as a transfinite extension $A_\lambda :=$

$\varinjlim_{\alpha<\lambda} A_\alpha$ of a λ-extension sequence $\{A_\alpha, f_{\alpha\beta}\}_{\alpha\le\beta<\lambda}$ where each $A_\alpha = \varinjlim_{I_\alpha} A_i$ is itself the direct limit of an I_α-directed subsystem $\{A_i, f_{ij}\}_{i\in I_\alpha}$, and the indexing set I_α is a directed subposet of I of strictly smaller cardinality $|I_\alpha| < \lambda$. By the induction hypothesis, each $A_\alpha \in \mathcal{X}$. Moreover, each inclusion $f_{\alpha,\alpha+1}: A_\alpha \rightarrowtail A_{\alpha+1}$ is an admissible monic, so by hypothesis each $A_{\alpha+1}/A_\alpha \in \mathcal{X}$ too. It means $A_\lambda := \varinjlim_I A_i = \varinjlim_{\alpha<\lambda} A_\alpha$ is a transfinite λ-extension of objects in $\mathcal{X}$. So it follows from Eklof's Lemma 9.3, that $\varinjlim_I A_i \in \mathcal{X}$. □

A special case of a direct I-system of admissible monics arises when we have a **direct I-system of admissible subobjects** of some fixed object $A \in \mathcal{A}$. By this we mean a direct I-system $\{A_i, f_{ij}\}$ of admissible monics f_{ij} where each $a_i: A_i \rightarrowtail A$ is also an admissible monic (i.e. admissible subobject of A) and we have the compatibilities $a_i = a_j f_{ij}$. Note that in the typical case that $\mathcal{A}$ is weakly idempotent complete, it is enough to specify that each $a_i: A_i \rightarrowtail A$ is an admissible monic. For then it is automatic that each f_{ij} is an admissible monic, by Proposition 1.30. In any case, we obtain in an obvious way a direct I-system of short exact sequences $\{\mathbb{E}_i\}_{i\in I}$ and our assumption that direct limits are exact implies that we have an admissible monic $\varinjlim_I a_i: \varinjlim_I A_i \rightarrowtail A$. We may opt to use the notation $\bigcup_{i\in I} A_i := \varinjlim_I A_i$, and one could even denote this admissible monic by $\bigcup_{i\in I} A_i \subseteq_{\mathcal{E}} A$. We may also opt to refer to the direct limit $\bigcup_{i\in I} A_i$ as the *direct union* of the admissible subobjects $\{A_i, f_{ij}\}$.

Theorem 9.55 *Assume $(\mathcal{A}, \mathcal{E})$ is an exact category that has exact direct limits. Let $\{A_i, f_{ij}\}$ be any given direct I-system in $\mathcal{A}$. Then there exists an $\mathcal{E}$-short exact sequence*

$$\bigcup_{J\in\mathcal{S}} A_J \xrightarrow{\varinjlim \phi_J} \bigoplus_{i\in I} A_i \xrightarrow{\pi} \varinjlim_I A_i,$$

where $(\mathcal{S}, \subseteq)$ is a directed poset consisting of finite subposets of $(I, \le)$, and

$$\bigcup_{J\in\mathcal{S}} A_J := \varinjlim_{J\in\mathcal{S}} A_J$$

denotes the direct limit (i.e. direct union) of some direct $\mathcal{S}$-system $\{A_J, D_{JJ'}\}$ of admissible subobjects $\phi_J: A_J \rightarrowtail \bigoplus_{i\in I} A_i$. In fact, each $D_{JJ'}: A_J \rightarrowtail A_{J'}$ and each $\phi_J: A_J \rightarrowtail \bigoplus_{i\in I} A_i$ are admissible split monics.

Proof Consider the set $\mathcal{S} = \{J \subseteq I \mid 1 < |J| < \omega\}$ of all finite directed subposets $J \subseteq I$ with $|J| > 1$. Being finite, each such J has a unique maximal element which we will denote by j. For each $J \in \mathcal{S}$ we consider the mapping

$$\phi_J: \bigoplus_{i\in J\setminus\{j\}} A_i \to \bigoplus_{i\in I} A_i$$

induced by $e_i - e_j f_{ij} \colon A_i \to \bigoplus_{i\in I} A_i$ on each summand of $\bigoplus_{i\in J\setminus\{j\}} A_i$, where $e_i \to \bigoplus_{i\in I} A_i$ denotes the canonical injection into the coproduct. For precision in what will come later, let us also denote by $e_i^J \colon A_i \to \bigoplus_{i\in J\setminus\{j\}} A_i$ the canonical injection into the (finite) coproduct $\bigoplus_{i\in J\setminus\{j\}} A_i$, for each $J \in \mathcal{S}$. So then we have the identity

$$\phi_J \circ e_i^J = e_i - e_j f_{ij}, \quad \text{for each } i \in J\setminus\{j\}. \tag{9.8}$$

Observe that $(\mathcal{S}, \subseteq)$ is a directed poset and we define a direct $\mathcal{S}$-system

$$\{A_J, D_{JJ'}\}$$

by $A_J := \bigoplus_{i\in J\setminus\{j\}} A_i$ and for an inclusion $J \subseteq J'$ we define the map $D_{JJ'}$ on the summand A_i via $e_i^{J'}$ if $j = j'$ (that is, just a natural inclusion if J and J' have the same maximal element), but via $e_i^{J'} - e_j^{J'} f_{ij}$ if $j < j'$. Thus we have the relations

$$D_{JJ'} \circ e_i^J = \begin{cases} e_i^{J'}, & \text{if } j = j', \\ e_i^{J'} - e_j^{J'} f_{ij}, & \text{if } j < j', \end{cases} \tag{9.9}$$

and the reader verifies (exercise!) that this determines a direct $\mathcal{S}$-system.

Next, we observe that each ϕ_J is a admissible split monic, sitting in a biproduct diagram:

$$\mathbb{E}_J\colon \qquad \bigoplus_{i\in J\setminus\{j\}} A_i \underset{\phi_J}{\overset{\pi_J}{\rightleftarrows}} \bigoplus_{i\in I} A_i \underset{\pi_{\overline{J}}}{\overset{\eta_{\overline{J}}}{\rightleftarrows}} \bigoplus_{i\in \overline{J\setminus\{j\}}} A_i. \tag{9.10}$$

The retraction π_J is defined by

$$\pi_J \circ e_i = \begin{cases} e_i^J, & \text{if } i \in J\setminus\{j\}, \\ 0, & \text{if } i \notin J\setminus\{j\}, \end{cases}$$

while $\pi_{\overline{J}}$ is defined by

$$\pi_{\overline{J}} \circ e_i = \begin{cases} e_j^{\overline{J}} \circ f_{ij}, & \text{if } i \in J\setminus\{j\}, \\ e_i^{\overline{J}}, & \text{if } i \notin J\setminus\{j\}, \end{cases}$$

and $\eta_{\overline{J}}$ is a straight inclusion, defined by

$$\eta_{\overline{J}} \circ e_i^{\overline{J}} = e_i \quad \text{for all } i \notin J\setminus\{j\}.$$

We leave for the reader to verify (again, exercise!) that these maps determine a biproduct diagram $\mathbb{E}_J$ for each $J \in \mathcal{S}$. In fact for each $J \subseteq J'$, the morphism $D_{JJ'}$ is also an admissible split monic with similar retraction and we have the compatibilities $\phi_J = \phi_{J'} \circ D_{JJ'}$. In particular, $\{A_J, D_{JJ'}\}$ is a direct $\mathcal{S}$-system of admissible subobjects $\phi_J \colon A_J \rightarrowtail \bigoplus_{i\in I} A_i$ (which are even direct summands with complements as claimed). This provides us with a direct $\mathcal{S}$-system of split exact sequences $\left\{\mathbb{E}_J, \left(D_{JJ'}, 1, \pi_{\overline{JJ'}}\right)\right\}$ as indicated:

$$\begin{array}{llllll} \mathbb{E}_{J'}: & \bigoplus_{i\in J'\setminus\{j'\}} A_i & \xrightarrow{\phi_{J'}} & \bigoplus_{i\in I} A_i & \xrightarrow{\pi_{\overline{J'}}} & \bigoplus_{i\in \overline{J'\setminus\{j'\}}} A_i \\ & \uparrow D_{JJ'} & & \| & & \uparrow \exists!\, \pi_{\overline{JJ'}} \\ \mathbb{E}_J: & \bigoplus_{i\in J\setminus\{j\}} A_i & \xrightarrow{\phi_J} & \bigoplus_{i\in I} A_i & \xrightarrow{\pi_{\overline{J}}} & \bigoplus_{i\in \overline{J\setminus\{j\}}} A_i. \end{array}$$

By our assumption that $(\mathcal{A}, \mathcal{E})$ has exact direct limits we get an admissible monic

$$\varinjlim \phi_J \colon \bigcup_{J\in\mathcal{S}} A_J \rightarrowtail \bigoplus_{i\in I} A_i,$$

where we are letting $\bigcup_{J\in\mathcal{S}} A_J := \varinjlim_{\mathcal{S}} A_J$ denote the existing direct limit of the direct $\mathcal{S}$-system $\{A_J, D_{JJ'}\}$ of admissible subobjects $A_J := \bigoplus_{i\in J\setminus\{j\}} A_i$. There are insertion morphisms $\eta_{A_J} : A_J \to \bigcup_{J\in\mathcal{S}} A_J$ and $\varinjlim \phi_J$ is the unique morphism satisfying $\left(\varinjlim \phi_J\right) \circ \eta_{A_J} = \phi_J$ for all $J \in \mathcal{S}$. We claim that the existing cokernel, $\pi := \operatorname{cok}\left(\varinjlim \phi_J\right)$, serves as the direct limit of the originally given direct I-system $\left\{A_i, f_{ij}\right\}$ in $\mathcal{A}$, proving that we have an admissible short exact sequence:

$$\bigcup_{J\in\mathcal{S}} A_J \overset{\varinjlim \phi_J}{\rightarrowtail} \bigoplus_{i\in I} A_i \overset{\pi}{\twoheadrightarrow} \varinjlim_I A_i.$$

More precisely, let $\pi\colon \bigoplus_{i\in I} A_i \to C$ be a cokernel of $\varinjlim \phi_J$. We will show that $(C, \{\pi e_i\})$ is a universal cocone above the direct I-system $\left\{A_i, f_{ij}\right\}$. First, we note that any morphism $p\colon \bigoplus_{i\in I} A_i \to X$ satisfies $p \circ (\varinjlim \phi_J) = 0$ if and only if $(X, \{pe_i\})$ is a cocone above the direct I-system $\left\{A_i, f_{ij}\right\}$. Indeed by the universal property, any two morphisms existing a colimit are equal if and only if they are equal upon composition with the canonical injections. In particular, $p \circ \left(\varinjlim \phi_J\right) = 0$ if and only if $0 = p \circ \left(\varinjlim \phi_J\right) \circ \eta_{A_J} = p \circ \phi_J$ for all $J \in \mathcal{S}$. But for the same reason, $p \circ \phi_J = 0$ if and only if $0 = p \circ \phi_J \circ e_i^J = p \circ \left(e_i - e_j f_{ij}\right) = pe_i - pe_j f_{ij}$, for each $i \in J\setminus\{j\}$. So, $p \circ \left(\varinjlim \phi_J\right) = 0$ if and only if $pe_i = pe_j f_{ij}$, for each $i \in J\setminus\{j\}$. This shows $p \circ \left(\varinjlim \phi_J\right) = 0$ if and only if $(X, \{pe_i\})$ is a cocone above the direct I-system $\left\{A_i, f_{ij}\right\}$. From this it is immediate that $\pi\colon \bigoplus_{i\in I} A_i \to C$ is a cokernel of $\varinjlim \phi_J$ if and only if $(C, \{\pi e_i\})$ is a universal cocone above the direct I-system $\left\{A_i, f_{ij}\right\}$. □

Exercise 9.10.1 Verify the two details flagged (exercise!) within the proof of Theorem 9.55.

Finally we are able to state and prove the main result of this section.

Theorem 9.56 *Let $(\mathcal{X}, \mathcal{Y})$ be a cotorsion pair in an exact category $(\mathcal{A}, \mathcal{E})$ with exact direct limits. If $\mathcal{X}$ is closed under cokernels of admissible monics, then $\mathcal{X}$ is closed under all direct limits. In particular, if $\mathcal{X}$ is thick then it must be closed under direct limits.*

Proof Let $(I, \leq)$ be a directed poset and let $\{A_i, f_{ij}\}$ be any given direct I-system in $\mathcal{A}$. By Theorem 9.55 there exists an $\mathcal{E}$-short exact sequence

$$\bigcup_{J \in \mathcal{S}} A_J \overset{\varinjlim \phi_J}{\rightarrowtail} \bigoplus_{i \in I} A_i \overset{\pi}{\twoheadrightarrow} \varinjlim_{I} A_i,$$

where $(\mathcal{S}, \subseteq)$ is a directed poset and

$$\bigcup_{J \in \mathcal{S}} A_J := \varinjlim_{J \in \mathcal{S}} A_J$$

is the direct limit (i.e. direct union) of some direct $\mathcal{S}$-system $\{A_J, D_{JJ'}\}$ of admissible subobjects $\phi_J \colon A_J \to \bigoplus_{i \in I} A_i$, where each $D_{JJ'} \colon A_J \rightarrowtail A_{J'}$ and each $\phi_J \colon A_J \rightarrowtail \bigoplus_{i \in I} A_i$ are even admissible split monics (so direct summands). Since $\mathcal{X}$ is closed under direct sums and direct summands, we have $\bigoplus_{i \in I} A_i \in \mathcal{X}$ and each $A_J \in \mathcal{X}$. So by Proposition 9.54 we have $\bigcup_{J \in \mathcal{S}} A_J \in \mathcal{X}$. Finally, the assumption that $\mathcal{X}$ is closed under cokernels of admissible monics implies that $\varinjlim_I A_i \in \mathcal{X}$. □

In relation to abelian model structures we have the following useful consequence.

Corollary 9.57 *Let $\mathfrak{M} = (\mathcal{W}, \mathcal{R})$ be an injective cotorsion pair, equivalently an injective model structure, on an exact category $(\mathcal{A}, \mathcal{E})$ with exact direct limits. Then the class $\mathcal{W}$ of trivial objects is closed under direct limits.*

The following is an equivalent form of Theorem 9.56 that may be applied to cotorsion pairs on a full exact subcategory $\mathcal{F} \subseteq \mathcal{A}$ that is closed under direct limits.

Theorem 9.58 *Let $(\mathcal{A}, \mathcal{E})$ be an exact category with exact direct limits. Assume $\mathcal{F}$ is a class of objects that is closed under $\mathcal{E}$-extensions and direct limits in $\mathcal{A}$. Let $\mathcal{E}_{\mathcal{F}}$ denote the inherited Quillen exact structure on $\mathcal{F}$. If $(\mathcal{X}, \mathcal{Y})$ is a cotorsion pair on the exact category $(\mathcal{F}, \mathcal{E}_{\mathcal{F}})$, and $\mathcal{X}$ is closed under cokernels of $\mathcal{E}_{\mathcal{F}}$-admissible monics, then $\mathcal{X}$ is closed under direct limits.*

Proof Since $\mathcal{F}$ is closed under $\mathcal{E}$-extensions it naturally inherits an exact structure $(\mathcal{F}, \mathcal{E}_{\mathcal{F}})$. Since $\mathcal{F}$ is closed under direct limits, this exact structure also has exact direct limits. So the result follows from Theorem 9.56. Conversely, we recover Theorem 9.56 by taking $\mathcal{F} = \mathcal{A}$ to be the class of all objects. □

Example 9.59 Let R be a ring and let R-Mod denote the category of all (left) R-modules. One can use Theorem 9.58 to show that each cycle module of any acyclic complex of cotorsion modules must also be a cotorsion module. See Exercises 10.9.4 and 10.9.5. In particular, each cycle module of any acyclic complex of injective modules must be cotorsion. Gorenstein injective modules are examples of such modules.

In a similar way, Theorem 9.56 may be used to show that every pure–acyclic chain complex with pure–injective components must be contractibe. See Exercises 10.9.7 and 10.9.8.

Exercise 9.10.2 Let $\mathfrak{M} = (\mathcal{W}, \mathcal{R})$ be an injective cotorsion pair (i.e. injective model structure) on R-Mod. Show that the class $\mathcal{W}$ of all trivial objects contains all modules of finite flat dimension. The same is true for an injective model structure on $\mathrm{Ch}(R)$.

Exercise 9.10.3 (Kaplansky Classes) Let $(\mathcal{A}, \mathcal{E})$ be a weakly idempotent complete exact category with exact direct limits. Assume each object of $\mathcal{A}$ has (up to isomorphism) only a set of $\mathcal{E}$-admissible subobjects. Let $\mathcal{X}$ be a class of objects satisfying the following conditions.

(1) $\mathcal{X}$ is closed under $\mathcal{E}$-subobjects and $\mathcal{E}$-quotients. That is, whenever $S \rightarrowtail X$ is an admissible monic with $X \in \mathcal{X}$, then $S, X/S \in \mathcal{X}$.
(2) There exists a set (not a proper class) $\mathcal{S} \subseteq \mathcal{X}$ such that for each nonzero $X \in \mathcal{X}$, there exists a nonzero $S \in \mathcal{S}$ and an admissible monic $S \rightarrowtail X$ (i.e. an admissible subobject $S \subseteq_{\mathcal{E}} X$).

Use the properties in Exercises 1.8.1 and 1.8.2 to show that $\mathcal{X} = \mathrm{Filt}_{\mathcal{E}}(\mathcal{S})$.

10

Abelian Model Structures on Chain Complexes

This chapter is dedicated to the central example of chain complexes over additive and exact categories. After giving a detailed construction of the standard model structure for $K(\mathcal{A})$, the chain homotopy category of an additive category $\mathcal{A}$, we look at general properties of abelian model structures on chain complexes. We will identify formal (Quillen) homotopy categories of complexes with traditional triangulated subcategories of $K(\mathcal{A})$. Later in the chapter we construct abelian models for derived categories. Theorem 10.40 constructs an abelian model structure for the derived category of any exact category with coproducts, kernels, and a set of (small) projective generators. In the last section we see how an hereditary cotorsion pair on a Grothendieck category $\mathcal{G}$ may be lifted to a hereditary abelian model structure on $\mathrm{Ch}(\mathcal{G})$.

10.1 Chain Complexes on Additive Categories

The material in this section will be familiar to most readers, though perhaps not in this generality. Unless other hypotheses are specified, $\mathcal{A}$ will denote a general additive category throughout this section. We discuss basic notions of chain complexes over additive categories, including the chain homotopy relation and the homotopy category, $K(\mathcal{A})$.

We start by noting that the notion of a chain complex makes sense for any given additive category $\mathcal{A}$. A **chain complex** X is a $\mathbb{Z}$-indexed sequence of objects and morphisms called *differentials*

$$\cdots \to X_{n+1} \xrightarrow{d_{n+1}} X_n \xrightarrow{d_n} X_{n-1} \to \cdots \tag{10.1}$$

such that $d_n \circ d_{n+1} = 0$ for all n. A morphism of chain complexes $f\colon X \to Y$, called a *chain map*, is a $\mathbb{Z}$-indexed collection of morphisms $\{f_n\colon X_n \to Y_n\}$ making all the squares commute: $d_n^Y f_n = f_{n-1} d_n^X$. Thus we obtain the category

$\mathrm{Ch}(\mathcal{A})$ of all chain complexes over $\mathcal{A}$. It is straightforward to verify that $\mathrm{Ch}(\mathcal{A})$ is also an additive catgory, and we let $\mathrm{Hom}_{\mathrm{Ch}(\mathcal{A})}(X, Y)$ denote the abelian group of all chain maps from X to Y.

For any given object $A \in \mathcal{A}$, we denote the **n-disk on** A by $D^n(A)$. This is the chain complex consisting only of $A \xrightarrow{1_A} A$ concentrated in degrees n and $n-1$, and 0 for all other n. We denote the **n-sphere on** A by $S^n(A)$. It is the complex consisting only of A in degree n, and 0 elsewhere.

If the differential d_n of a chain complex X admits a kernel, we will denote it (really its domain) by Z_nX, and if d_{n+1} admits a cokernel we will denote it by C_nX. If the underlying category happens to be an *abelian* category, then every d_{n+1} has an image, denoted B_nX, and we have $C_nX = X_n/B_nX$. We may go farther in the abelian case and define the *nth-homology group* of X to be $H_nX = Z_nX/B_nX$. A chain complex X over an abelian category is said to be *exact* (or *acyclic*) if $H_nX = 0$ for all $n \in \mathbb{Z}$. In particular, these ideas apply to chain complexes of abelian groups. So in any *additive* category $\mathcal{A}$, given any object $A \in \mathcal{A}$ and chain complex $X \in \mathrm{Ch}(\mathcal{A})$, applying the functor $\mathrm{Hom}_{\mathcal{A}}(A, -)$ yields a chain complex $\mathrm{Hom}_{\mathcal{A}}(A, X)$ of abelian groups

$$\cdots \to \mathrm{Hom}_{\mathcal{A}}(A, X_{n+1}) \xrightarrow{(d_{n+1})_*} \mathrm{Hom}_{\mathcal{A}}(A, X_n) \xrightarrow{(d_n)_*} \mathrm{Hom}_{\mathcal{A}}(A, X_{n-1}) \xrightarrow{(d_{n-1})_*} \cdots .$$

Being an abelian category, each $(d_n)_*$ certainly has a kernel and a cokernel and in the above notation we have

$$C_n[\mathrm{Hom}_{\mathcal{A}}(A, X)] := \mathrm{Cok}\,(d_{n+1})_* = \mathrm{Hom}_{\mathcal{A}}(A, X_n)/B_n[\mathrm{Hom}_{\mathcal{A}}(A, X)]$$

and

$$Z_n[\mathrm{Hom}_{\mathcal{A}}(A, X)] := \mathrm{Ker}\,(d_n)_*.$$

Many authors instead work with complexes in which the differential raises degree instead of lowering degree. One usually calls this a *cochain complex.* We use superscript notation instead of subscript notation: A cochain complex looks like

$$\cdots \to X^{n-1} \xrightarrow{d^{n-1}} X^n \xrightarrow{d^n} X^{n+1} \to \cdots$$

and satisfies $d^n \circ d^{n-1} = 0$ for all n. We will almost exclusively work with chain complexes, but no matter which convention is used, the other idea arises naturally. For example, given an object $A \in \mathcal{A}$ and chain complex $X \in \mathrm{Ch}(\mathcal{A})$ as in (10.1), applying the contravariant $\mathrm{Hom}_{\mathcal{A}}(-, A)$ to X yields what is best thought of as a cochain complex $\mathrm{Hom}_{\mathcal{A}}(X, A)$ of abelian groups

$$\cdots \leftarrow \mathrm{Hom}_{\mathcal{A}}(X_{n+1}, A) \xleftarrow{d^*_{n+1}} \mathrm{Hom}_{\mathcal{A}}(X_n, A) \xleftarrow{d^*_n} \mathrm{Hom}_{\mathcal{A}}(X_{n-1}, A) \xleftarrow{d^*_{n-1}} \cdots .$$

One might then use the cochain (superscript) notations Z^n, B^n, and H^n to target the kernel, image, and *cohomology* in a particular degree n. For example, we have $Z^n[\mathrm{Hom}_{\mathcal{A}}(X, A)] := \mathrm{Ker}\, d^*_{n+1}$ for the previous cochain complex.

The reader should verify each statement in the following lemma.

Lemma 10.1 (Hom Isomorphisms for Disks and Spheres) *Let $A \in \mathcal{A}$ be an object of an additive category and let $X \in \mathrm{Ch}(\mathcal{A})$ be a chain complex.*

(1) *There is an isomorphism of abelian groups*

$$\mathrm{Hom}_{\mathrm{Ch}(\mathcal{A})}(D^n(A), X) \cong \mathrm{Hom}_{\mathcal{A}}(A, X_n).$$

(2) *There is an isomorphism of abelian groups*

$$\mathrm{Hom}_{\mathrm{Ch}(\mathcal{A})}\left(X, D^{n+1}(A)\right) \cong \mathrm{Hom}_{\mathcal{A}}(X_n, A).$$

(3) *There is an isomorphism of abelian groups*

$$\mathrm{Hom}_{\mathrm{Ch}(\mathcal{A})}(S^n(A), X) \cong Z_n[\mathrm{Hom}_{\mathcal{A}}(A, X)].$$

If $Z_n X := \mathrm{Ker}\, d_n$ exists, this can be expressed as

$$\mathrm{Hom}_{\mathrm{Ch}(\mathcal{A})}(S^n(A), X) \cong \mathrm{Hom}_{\mathcal{A}}(A, Z_n X).$$

(4) *There is an isomorphism of abelian groups*

$$\mathrm{Hom}_{\mathrm{Ch}(\mathcal{A})}(X, S^n(A)) \cong Z^n[\mathrm{Hom}_{\mathcal{A}}(X, A)].$$

If $C_n X := \mathrm{Cok}\, d_{n+1}$ exists, this can be expressed as

$$\mathrm{Hom}_{\mathrm{Ch}(\mathcal{A})}(X, S^n(A)) \cong \mathrm{Hom}_{\mathcal{A}}(C_n X, A).$$

Exercise 10.1.1 Prove Lemma 10.1.

Two chain maps $f, g \colon X \to Y$ are called **chain homotopic** if there exists a collection of morphisms $\{s_n : X_n \to Y_{n+1}\}$ such that

$$d^Y_{n+1} s_n + s_{n-1} d^X_n = g_n - f_n.$$

If f and g are chain homotopic we write $f \sim g$, and a chain map f is called **null homotopic** if $f \sim 0$. We leave it as an exercise for the reader to check that $\sim$ is an equivalence relation on any morphism set $\mathrm{Hom}_{\mathrm{Ch}(\mathcal{A})}(X, Y)$; it is called the **chain homotopy relation**. The chain homotopy equivalence class of a morphism f is denoted by $[f]$. Moreover, composition of chain homotopy classes is well defined, and all told we may define $K(\mathcal{A})$, the **chain homotopy category** of $\mathcal{A}$. It is the category whose objects are the same as $\mathrm{Ch}(\mathcal{A})$ but whose morphisms are homotopy classes of chain maps. That is,

$$\mathrm{Hom}_{K(\mathcal{A})}(X, Y) := \mathrm{Hom}_{\mathrm{Ch}(\mathcal{A})}(X, Y)/\sim.$$

A chain map $f: X \to Y$ is called a **homotopy equivalence** if $[f]$ is an isomorphism in $K(\mathcal{A})$. It means there exists a chain map $g: Y \to X$ such that $gf \sim 1_X$ and $fg \sim 1_Y$.

Exercise 10.1.2 Verify that the chain homotopy relation $\sim$ is an equivalence relation, that it respects composition, and that $K(\mathcal{A})$ is an additive category.

Given a chain complex X, the **suspension** of X, denoted ΣX, is the complex given in degree $k \in \mathbb{Z}$ by the shift $(\Sigma X)_k = X_{k-1}$ and $d_k^{\Sigma X} = -d_{k-1}$. More generally, we define $\Sigma^n X$ for any integer $n \in \mathbb{Z}$ by $(\Sigma^n X)_k = X_{k-n}$ and $d_k^{\Sigma^n X} = (-1)^n d_{k-n}$. Clearly, Σ determines an automorphism functor on both $\mathrm{Ch}(\mathcal{A})$ and $K(\mathcal{A})$, with inverse Σ^{-1}.

Definition 10.2 Given two chain complexes X and Y we define $Hom(X, Y)$ to be the complex of abelian groups

$$\cdots \to \prod_{k \in \mathbb{Z}} \mathrm{Hom}_{\mathcal{A}}(X_k, Y_{k+n}) \xrightarrow{\delta_n} \prod_{k \in \mathbb{Z}} \mathrm{Hom}_{\mathcal{A}}(X_k, Y_{k+n-1}) \to \cdots,$$

where in degree n, the differential δ_n is defined by the rule

$$\delta_n(\{f_k : X_k \to Y_{k+n}\}_{k \in \mathbb{Z}}) = \left\{d_{k+n} f_k - (-1)^n f_{k-1} d_k\right\}_{k \in \mathbb{Z}}.$$

More succinctly, $(\delta_n f)_k = d_{k+n} f_k - (-1)^n f_{k-1} d_k$.

One verifies that $Hom(X, Y)$ is indeed a chain complex of abelian groups and as a result we obtain a bifunctor

$$Hom(-,-): \mathrm{Ch}(\mathcal{A})^{\mathrm{op}} \times \mathrm{Ch}(\mathcal{A}) \to \mathrm{Ch}(\mathbb{Z}).$$

Being a chain complex of abelian groups, we can take the homology groups of $Hom(X, Y)$,

$$H_n[Hom(X, Y)] := Z_n[Hom(X, Y)]/B_n[Hom(X, Y)],$$

and we have the following useful computation.

Lemma 10.3 *The nth homology group of the chain complex $Hom(X, Y)$ is precisely*

$$H_n[Hom(X, Y)] = K(\mathcal{A})(X, \Sigma^{-n} Y).$$

Therefore, the complex $Hom(X, Y)$ is exact if and only if for any $n \in \mathbb{Z}$, all chain maps $f: X \to \Sigma^n Y$ are null homotopic (equivalently, all chain maps $f: \Sigma^n X \to Y$ are null homotopic).

Proof From the definition, it follows that a degree n element

$$\{f_k : X_k \to Y_{k+n}\}_{k\in\mathbb{Z}} \in Hom(X,Y)_n$$

is in $\operatorname{Ker}(\delta_n)$ if and only if $(-1)^{-n}d_{k-(-n)}f_k = f_{k-1}d_k$. So, if and only if it is a chain map $f: X \to \Sigma^{-n}Y$. That is, the group of n-cycles is precisely $Z_nHom(X,Y) = \operatorname{Hom}_{\operatorname{Ch}(\mathcal{A})}(X, \Sigma^{-n}Y)$. On the other hand, such a chain map

$$f = \{f_k : X_k \to Y_{k+n}\}_{k\in\mathbb{Z}} \in \operatorname{Hom}_{\operatorname{Ch}(\mathcal{A})}(X, \Sigma^{-n}Y)$$

is in $\operatorname{Im}(\delta_{n+1})$ if and only if there exists some

$$\{h_k : X_k \to Y_{k+n+1}\}_{k\in\mathbb{Z}} \in Hom(X,Y)_{n+1}$$

such that

$$f_k = d_{k+n+1} \circ h_k - (-1)^{n+1} h_{k-1} \circ d_k.$$

Taking $s_k := (-1)^n h_k$ we get

$$f_k = (-1)^n d_{k+n+1} \circ (-1)^n h_k + (-1)^n h_{k-1} \circ d_k = (-1)^n d_{k+n+1} \circ s_k + s_{k-1} \circ d_k,$$

which shows f to be null homotopic. Therefore, the nth homology group is precisely

$$H_nHom(X,Y) = \operatorname{Hom}_{\operatorname{Ch}(\mathcal{A})}(X, \Sigma^{-n}Y)/\sim\ = \operatorname{Hom}_{K(\mathcal{A})}(X, \Sigma^{-n}Y).$$

□

Exercise 10.1.3 Show that limits and colimits are taken degreewise in $\operatorname{Ch}(\mathcal{A})$. Therefore $\operatorname{Ch}(\mathcal{A})$ possesses any type of limit or colimit that $\mathcal{A}$ has.

10.2 The Exact Category of Chain Complexes

Later in this chapter we will be considering abelian model structures on chain complexes. We note here how common properties of exact structures lift to the category of chain complexes, and prove a few isomorphisms of Ext^1 groups. The reader should verify the statements listed in the following lemma.

Lemma 10.4 *Let $\mathcal{A}$ be any additive category. Then* $\operatorname{Ch}(\mathcal{A})$ *is also additive and the following statements hold.*

(1) $\operatorname{Ch}(\mathcal{A})$ *is weakly idempotent complete (resp. idempotent complete) if and only if $\mathcal{A}$ is such.*
(2) *If $\mathcal{A}$ is an abelian (resp. Grothendieck) category, then* $\operatorname{Ch}(\mathcal{A})$ *is an abelian (resp. Grothendieck) category.*

(3) *If* $(\mathcal{A}, \mathcal{E})$ *is an exact category, then* $\mathrm{Ch}(\mathcal{A})_{\mathcal{E}} := (\mathrm{Ch}(\mathcal{A}), \mathrm{Ch}(\mathcal{E}))$ *is also an exact category. Here* $\mathrm{Ch}(\mathcal{E})$ *denotes the class of all sequences of chain maps that are* $\mathcal{E}$*-short exact sequences in each degree* $n \in \mathbb{Z}$*.*

(4) *If* $(\mathcal{A}, \mathcal{E})$ *is an efficient exact category, then so is the exact category* $\mathrm{Ch}(\mathcal{A})_{\mathcal{E}}$*.*

Exercise 10.2.1 Verify each statement of Lemma 10.4.

Recall that if $\mathcal{A}$ is any additive category, then it possesses an exact structure consisting of all the split exact sequences. (See Exercise 1.3.1.) So if $(\mathcal{A}, \mathcal{E})$ is any exact category, then Lemma 10.4 automatically provides us with the exact structures, $\mathrm{Ch}(\mathcal{A})_{\mathcal{E}}$ and $\mathrm{Ch}(\mathcal{A})_{dw}$. The latter denotes the exact category of complexes along with the sequences that are split exact in each degree $n \in \mathbb{Z}$. We study $\mathrm{Ch}(\mathcal{A})_{dw}$ in Section 10.3, and it will play a role throughout this chapter. We denote the Yoneda Ext groups associated to $\mathrm{Ch}(\mathcal{A})_{\mathcal{E}}$ by $\mathrm{Ext}^1_{\mathrm{Ch}(\mathcal{E})}(X, Y)$. Its elements are equivalence classes of degreewise $\mathcal{E}$-short exact sequences. Since $\mathrm{Ch}(\mathcal{E})$ clearly contains all degreewise split short exact sequences, the Yoneda Ext group associated to $\mathrm{Ch}(\mathcal{A})_{dw}$, which we will denote by $\mathrm{Ext}^1_{dw}(X, Y)$, is the subgroup of $\mathrm{Ext}^1_{\mathrm{Ch}(\mathcal{E})}(X, Y)$ consisting of the degreewise $\mathcal{E}$-short exact sequences which are split exact in each degree. Similarly, $\mathrm{Ext}^n_{dw}(X, Y) \subseteq \mathrm{Ext}^n_{\mathrm{Ch}(\mathcal{E})}(X, Y)$.

For the remainder of the section we let $(\mathcal{A}, \mathcal{E})$ be any exact category and we let $\mathrm{Ch}(\mathcal{A})_{\mathcal{E}}$ be the exact category of chain complexes along with the standard exact structure as in Lemma 10.4. We will prove some useful isomorphisms relating the $\mathrm{Ext}^1_{\mathrm{Ch}(\mathcal{E})}$ functor on disk and sphere complexes to the $\mathrm{Ext}^1_{\mathcal{E}}$ functor on the ground category.

Lemma 10.5 (Ext Isomorphisms for Disk Complexes) *Let* $A \in \mathcal{A}$ *be an object and let* $X \in \mathrm{Ch}(\mathcal{A})$ *be a chain complex.*

(1) *There is an isomorphism of abelian groups*

$$\mathrm{Ext}^1_{\mathrm{Ch}(\mathcal{E})}(D^n(A), X) \cong \mathrm{Ext}^1_{\mathcal{E}}(A, X_n).$$

(2) *There is an isomorphism of abelian groups*

$$\mathrm{Ext}^1_{\mathrm{Ch}(\mathcal{E})}\left(X, D^{n+1}(A)\right) \cong \mathrm{Ext}^1_{\mathcal{E}}(X_n, A).$$

Proof Statements (1) and (2) are dual. We will prove (1). To do so, define a function

$$\mathrm{Ext}^1_{\mathrm{Ch}(\mathcal{E})}(D^n(A), X) \to \mathrm{Ext}^1_{\mathcal{E}}(A, X_n)$$

by taking a short exact sequence $X \rightarrowtail Y \twoheadrightarrow D^n(A)$ to the short exact sequence $X_n \rightarrowtail Y_n \twoheadrightarrow A$. It is clearly well defined. It has an inverse which works by taking any given $\mathcal{E}$-short exact sequence $X_n \rightarrowtail Y_n \twoheadrightarrow A$ and constructing the pushout along the differential $d_n\colon X_n \to X_{n-1}$:

$$\begin{array}{ccccc} X_n & \rightarrowtail & Y_n & \twoheadrightarrow & A \\ {\scriptstyle d_n}\downarrow & & \downarrow & & \| \\ X_{n-1} & \rightarrowtail & Y_{n-1} & \twoheadrightarrow & A. \end{array}$$

This diagram may be extended in an obvious way to a short exact sequence

$$X \rightarrowtail Y \twoheadrightarrow D^n(A)$$

in the exact category $\mathrm{Ch}(\mathcal{A})_{\mathcal{E}}$. On the other hand, by Proposition 1.12, any such short exact sequence is necessarily a pushout diagram in degrees n and $n-1$. Therefore, the described mappings are mutually inverse. □

Definition 10.6 Given any exact category $(\mathcal{A}, \mathcal{E})$ and a chain complex $X \in \mathrm{Ch}(\mathcal{A})$, we say that X is **acyclic**, or **exact**, (or for clarity *$\mathcal{E}$-acyclic* or *$\mathcal{E}$-exact*) if its differentials d_n each factor as $X_n \twoheadrightarrow Z_{n-1}X \rightarrowtail X_{n-1}$ in such a way that each

$$Z_nX \rightarrowtail X_n \twoheadrightarrow Z_{n-1}X$$

is a short exact sequence in $\mathcal{E}$.

Note that the differentials of an acyclic complex X each have a kernel, a cokernel, and an image. In particular, $Z_nX \rightarrowtail X_n$ equals $\ker d_n$, while $X_n \twoheadrightarrow Z_{n-1}X$ equals $\operatorname{cok} d_{n+1}$. Moreover, the image, $\operatorname{im} d_{n+1} := \ker(\operatorname{cok} d_{n+1})$, is precisely $Z_nX \rightarrowtail X_n$. So, $\ker d_n = \operatorname{im} d_{n+1}$ for each n. However, the converse need not be true. For example, an acyclic complex of abelian groups need not be pure acyclic. This is why we might sometimes emphasize the exact structure by saying a complex is $\mathcal{E}$-acyclic.

Lemma 10.7 (Ext Isomorphisms for Sphere Complexes) *Let $A \in \mathcal{A}$ be an object and let $X \in \mathrm{Ch}(\mathcal{A})$ be a chain complex.*

(1) *If X is $\mathcal{E}$-acyclic, then there is an isomorphism of abelian groups*

$$\mathrm{Ext}^1_{\mathrm{Ch}(\mathcal{E})}(S^nA, X) \cong \mathrm{Ext}^1_{\mathcal{E}}(A, Z_nX),$$

where $Z_nX \rightarrowtail X_n$ is the kernel of $X_n \xrightarrow{d_n} X_{n-1}$.

(2) *If X is $\mathcal{E}$-acyclic, then there is an isomorphism of abelian groups*

$$\mathrm{Ext}^1_{\mathrm{Ch}(\mathcal{E})}(X, S^nA) \cong \mathrm{Ext}^1_{\mathcal{E}}(Z_{n-1}X, A),$$

where $X_n \twoheadrightarrow Z_{n-1}X$ is the cokernel of $X_{n+1} \xrightarrow{d_{n+1}} X_n$.

Proof Statements (1) and (2) are dual. We again will only prove (1). Suppose

$$\mathcal{S}\colon\; Z_nX \overset{f}{\rightarrowtail} D \overset{g}{\twoheadrightarrow} A$$

is a short exact sequence representing an element of $\operatorname{Ext}^1_{\mathcal{E}}(A, Z_nX)$. Let $d_n = k_{n-1} \circ d'_n$ represent the factorizations guaranteed by Definition 10.6. Taking the pushout of f along $k_n \colon Z_nX \rightarrowtail X_n$ we get a commutative diagram:

$$\begin{array}{ccccc}
Z_nX & \overset{f}{\rightarrowtail} & D & \overset{g}{\twoheadrightarrow} & A \\
{\scriptstyle k_n}\big\downarrow & & \big\downarrow{\scriptstyle k'} & & \big\| \\
X_n & \overset{f'}{\rightarrowtail} & P & \overset{g'}{\twoheadrightarrow} & A \\
{\scriptstyle d'_n}\big\downarrow & & \big\downarrow{\scriptstyle h'} & & \\
Z_{n-1}X & = & Z_{n-1}X. & &
\end{array}$$

This allows us to construct the short exact sequence of complexes indicated in the following:

$$\begin{array}{ccccc}
\vdots & & \vdots & & \vdots \\
\downarrow & & \downarrow & & \downarrow \\
X_{n+1} & = & X_{n+1} & \twoheadrightarrow & 0 \\
{\scriptstyle d_{n+1}}\big\downarrow & & \big\downarrow{\scriptstyle f'd_{n+1}} & & \big\downarrow \\
X_n & \overset{f'}{\rightarrowtail} & P & \underset{g'}{\twoheadrightarrow} & A \\
{\scriptstyle d_n}\big\downarrow & & {\scriptstyle k_{n-1}h'}\big\downarrow & & \big\downarrow \\
X_{n-1} & = & X_{n-1} & \twoheadrightarrow & 0. \\
\downarrow & & \downarrow & & \downarrow \\
\vdots & & \vdots & & \vdots
\end{array}$$

This short exact sequence is an element of $\operatorname{Ext}^1_{\operatorname{Ch}(\mathcal{E})}(S^nA, X)$ which we denote by $\phi(\mathcal{S})$, and it provides a mapping $\phi\colon \operatorname{Ext}^1_{\mathcal{E}}(A, Z_nX) \to \operatorname{Ext}^1_{\operatorname{Ch}(\mathcal{E})}(S^nA, X)$. The reader may check that ϕ is a well-defined homomorphism of abelian groups. Using that X is $\mathcal{E}$-exact, we now construct an inverse ψ for ϕ. Indeed let $\mathcal{T}$ be a given short exact sequence $X \rightarrowtail Y \twoheadrightarrow S^nA$ in $\operatorname{Ext}^1_{\operatorname{Ch}(\mathcal{E})}(S^nA, X)$. Then we may assume it has the following form:

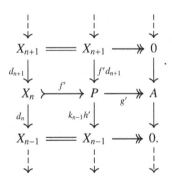

$$
\begin{array}{ccccc}
\vdots & & \vdots & & \vdots \\
\downarrow & & \downarrow & & \downarrow \\
X_{n+1} & = & Y_{n+1} & \twoheadrightarrow & 0 \\
{\scriptstyle d_{n+1}}\downarrow & & {\scriptstyle \alpha d_{n+1}}\downarrow & & \downarrow \\
X_n & \stackrel{\alpha}{\rightarrowtail} & Y_n & \stackrel{\beta}{\twoheadrightarrow} & A \\
{\scriptstyle d_n}\downarrow & & {\scriptstyle h}\downarrow & & \downarrow \\
X_{n-1} & = & Y_{n-1} & \twoheadrightarrow & 0. \\
\vdots\,{\scriptstyle d_{n-1}} & & \vdots\,{\scriptstyle d_{n-1}} & & \vdots \\
\downarrow & & \downarrow & & \downarrow
\end{array}
$$

Since Y is a complex we have $d_{n-1}h = 0$. Therefore, h factors uniquely through $Z_{n-1}X = \operatorname{Ker} d_{n-1}$, so that $h = k_{n-1}h'$ for some $h' \colon Y_n \to Z_{n-1}X$. Applying the Five Lemma (see Exercise 1.4.2) we may conclude that h' must be an admissible epimorphism. Note that $\ker h' = \ker h$ exists, and we set $Z_nY := \operatorname{Ker} h$. Thus, by the (3×3)-Lemma 1.16, $\mathcal{T}$ induces a commutative diagram with exact rows and columns:

$$
\begin{array}{ccccc}
Z_nX & \rightarrowtail & Z_nY & \twoheadrightarrow & A \\
{\scriptstyle k_n}\downarrow & & \downarrow & & \| \\
X_n & \stackrel{\alpha}{\rightarrowtail} & Y_n & \stackrel{\beta}{\twoheadrightarrow} & A \\
{\scriptstyle d'_n}\downarrow & & \downarrow{\scriptstyle h'} & & \\
Z_{n-1}X & = & Z_{n-1}X. & &
\end{array}
\qquad (10.2)
$$

So we may define ψ by the rule

$$\psi(\mathcal{T}) := \quad Z_nX \rightarrowtail Z_nY \twoheadrightarrow A$$

and clearly $\psi(\phi(\mathcal{S})) = \mathcal{S}$ for the originally given short exact sequence

$$\mathcal{S}\colon \; Z_nX \stackrel{f}{\rightarrowtail} D \stackrel{g}{\twoheadrightarrow} A$$

coming from $\operatorname{Ext}^1_{\mathcal{E}}(A, Z_nX)$. On the other hand, by Proposition 1.12(1), the upper left square of Diagram (10.2) must always be a pushout. Therefore we have $\phi(\psi(\mathcal{T})) = \mathcal{T}$. So ϕ and ψ are inverse isomorphisms. □

Exercise 10.2.2 Let $(\mathcal{A}, \mathcal{E})$ be any exact category and let $(\mathcal{X}, \mathcal{Y})$ be a cotorsion pair of chain complexes in $\operatorname{Ch}(\mathcal{A})_{\mathcal{E}}$. Suppose C is some given class of objects in $\mathcal{A}$.

(1) If the spheres $S^n(C)$ are in $\mathcal{X}$ whenever C is in C, then any bounded below complex with entries in C is also in $\mathcal{X}$.

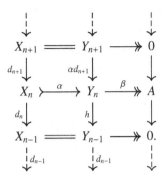

(2) If the disks $D^n(C)$ are in $\mathcal{X}$ whenever C is in C, then any bounded above $\mathcal{E}$-acyclic complex with cycles in C is also in $\mathcal{X}$.
(3) If the spheres $S^n(C)$ are in $\mathcal{Y}$ whenever C is in C, then any bounded above complex with entries in C is also in $\mathcal{Y}$.
(4) If the disks $D^n(C)$ are in $\mathcal{Y}$ whenever C is in C, then any bounded below $\mathcal{E}$-acyclic complex with cycles in C is also in $\mathcal{Y}$.

Hint: Use the Eklof Lemma 9.3 and/or its dual Lemma 9.8.

10.3 Spilt Exact and Contractible Complexes

Throughout this section, $\mathcal{A}$ will denote any given additive category. Our main goal is to characterize the chain homotopy relation in terms of factorizations through contractible complexes. We also clarify the difference between contractible and split exact complexes. They are the same thing if and only if $\mathcal{A}$ is idempotent complete.

Again, $\mathcal{A}$ may be viewed as an exact category when we consider it along with the class of all split exact sequences, and this lifts to $\text{Ch}(\mathcal{A})_{dw}$, the exact category of chain complexes along with the degreewise split exact sequences. Following Definition 10.6, we will say that a chain complex X is **split exact** if its differentials d_n each factor as $X_n \twoheadrightarrow Z_{n-1}X \rightarrowtail X_{n-1}$ in such a way that each

$$Z_nX \rightarrowtail X_n \twoheadrightarrow Z_{n-1}X$$

is a split exact sequence.

For any given collection of objects $\{A_n\}_{n\in\mathbb{Z}}$, note that

$$\bigoplus_{n\in\mathbb{Z}} D^n(A_n) = \prod_{n\in\mathbb{Z}} D^n(A_n)$$

always exists because it is really just the biproduct $A_{n+1} \bigoplus A_n = A_{n+1} \prod A_n$ in each degree n. The reader should verify that the split exact complexes are characterized as direct sums of such n-disks as follows.

Lemma 10.8 *A chain complex X is split exact if and only if X is isomorphic to a complex of the form $\bigoplus_{n\in\mathbb{Z}} D^n(A_n)$ for some collection of $\mathcal{A}$-objects $\{A_n\}_{n\in\mathbb{Z}}$. In particular, if X is split exact then it is isomorphic to the direct sum of n-disks on its cycles $Z_nX :=$ Ker d_n. Explicitly,*

$$X \cong \bigoplus_{n\in\mathbb{Z}} D^n(Z_{n-1}X) = \prod_{n\in\mathbb{Z}} D^n(Z_{n-1}X).$$

Exercise 10.3.1 Prove Lemma 10.8. (*Hint*: The differentials of a split exact complex X factor as $d_n = k_{n-1}d'_n$ and we have biproduct diagrams

$$Z_nX \underset{k_n}{\overset{r_n}{\rightleftarrows}} X_n \underset{d'_n}{\overset{s_n}{\rightleftarrows}} Z_{n-1}X.$$

Use Lemma 1.3 to construct a chain isomorphism from X to $\bigoplus_{n\in\mathbb{Z}} D^n(Z_{n-1}X)$.)

The following lemma is key to characterizing the chain homotopy relation in terms of split exact complexes. The reader can verify it by direct computation.

Lemma 10.9 *Let X and Y be arbitrary chain complexes, and let C be a split exact complex. So we may assume $C = \bigoplus_{n\in\mathbb{Z}} D^n(A_n)$ for some collection of $\mathcal{A}$-objects $\{A_n\}_{n\in\mathbb{Z}}$.*

(1) *Any collection of maps $\left\{ u_n : X_n \to A_{n+1} \right\}$ determines a chain map*

$$\beta = \left\{ \begin{bmatrix} u_n \\ u_{n-1}d_n^X \end{bmatrix} \right\} : X \to C.$$

Conversely, any chain map $\beta : X \to C$ must take the form $\beta_n = \begin{bmatrix} u_n \\ u_{n-1}d_n^X \end{bmatrix}$ for some collection of maps $\left\{ u_n : X_n \to A_{n+1} \right\}$.

(2) *Any collection of maps $\left\{ q_n : A_n \to Y_n \right\}$ determines a chain map*

$$p = \left\{ \begin{bmatrix} d_{n+1}^Y q_{n+1} & q_n \end{bmatrix} \right\} : C \to Y.$$

Conversely, any chain map $p : C \to Y$ must take the form $p_n = \left\{ \begin{bmatrix} d_{n+1}^Y q_{n+1} & q_n \end{bmatrix} \right\}$ for some collection of maps $\left\{ q_n : A_n \to Y_n \right\}$.

Exercise 10.3.2 Prove Lemma 10.9.

Proposition 10.10 *Let $f, g : X \to Y$ be chain maps. Then f and g are chain homotopic, that is $f \sim g$, if and only if their difference $g - f$ factors through a split exact complex.*

Proof It is enough to show f is null homotopic if and only if f factors through a split exact complex. So assume $f \sim 0$. Then there exists a collection of maps $\{s_n : X_n \to Y_{n+1}\}$ such that $d_{n+1}^Y s_n + s_{n-1}d_n^X = f_n$ for each n. By part (1) of Lemma 10.9, the collection $\{s_n : X_n \to Y_{n+1}\}$ determines a chain map

$$\beta = \left\{ \begin{bmatrix} s_n \\ s_{n-1}d_n^X \end{bmatrix} \right\} : X \to \bigoplus_{n\in\mathbb{Z}} D^n(Y_n).$$

Likewise, by part (2) of Lemma 10.9, the identity maps $\left\{ 1_{Y_n} : Y_n \to Y_n \right\}$ determine a chain map

$$p = \left\{ \begin{bmatrix} d_{n+1}^Y & 1_{Y_n} \end{bmatrix} \right\} : \bigoplus_{n\in\mathbb{Z}} D^n(Y_n) \to Y.$$

This shows that f factors through the split exact complex $\bigoplus_{n\in\mathbb{Z}} D^n(Y_n)$ since

$$p_n\beta_n = \begin{bmatrix} d^Y_{n+1} & 1_{Y_n} \end{bmatrix}\begin{bmatrix} s_n \\ s_{n-1}d^X_n \end{bmatrix} = f_n.$$

On the other hand, suppose f factors through some split exact complex $\bigoplus_{n\in\mathbb{Z}} D^n(A_n)$. So $f = p\beta$ where

$$\beta = \left\{\begin{bmatrix} u_n \\ u_{n-1}d^X_n \end{bmatrix}\right\}: X \to \bigoplus_{n\in\mathbb{Z}} D^n(A_n)$$

for some collection of morphisms $\{ u_n : X_n \to A_{n+1} \}$, and

$$p = \left\{\begin{bmatrix} d^Y_{n+1}q_{n+1} & q_n \end{bmatrix}\right\}: \bigoplus_{n\in\mathbb{Z}} D^n(A_n) \to Y$$

for some collection of morphisms $\{ q_n : A_n \to Y_n \}$. Composing we get

$$p_n\beta_n = \begin{bmatrix} d^Y_{n+1}q_{n+1} & q_n \end{bmatrix}\begin{bmatrix} u_n \\ u_{n-1}d^X_n \end{bmatrix} = d^Y_{n+1}q_{n+1}u_n + q_nu_{n-1}d^X_n.$$

Now setting $s_n = q_{n+1}u_n$ we get a collection of maps $\{ s_n : X_n \to Y_{n+1} \}$ satisfying $d^Y_{n+1}s_n + s_{n-1}d^X_n = f_n$ for each n. By definition, $f \sim 0$. □

It is immediate from Proposition 10.10 that if X is a split exact complex, then $1_X \sim 0$. More generally, the proposition implies that a chain complex C satisfies $1_C \sim 0$ if and only if 1_C factors through some split exact complex X. In other words, $1_C \sim 0$ if and only if C is a retract, in $\mathrm{Ch}(\mathcal{A})$, of a split exact complex X. Such chain complexes C are called contractible.

Definition 10.11 A chain complex C is called **contractible** if $1_C \sim 0$, that is, its identity map is null homotopic.

The next lemma is useful whenever one wants to reduce retracts in $K(\mathcal{A})$ to retracts in $\mathrm{Ch}(\mathcal{A})$.

Lemma 10.12 (Retracts in $K(\mathcal{A})$) *Suppose a complex X is a retract, in $K(\mathcal{A})$, of a complex Y. Then X is a retract, in* $\mathrm{Ch}(\mathcal{A})$*, of the complex*

$$Y \bigoplus \left(\bigoplus_{n\in\mathbb{Z}} D^{n+1}(X_n) \right),$$

which we note is homotopy equivalent to Y.

A specific retraction is constructed, from a given one, in the proof.

Proof Say that $f : X \to Y$ is a split monomorphism, in $K(\mathcal{A})$. By definition, it means there exists a homotopy retraction $r : Y \to X$. That is, $rf \sim 1_X$. Let $\{s_n : X_n \to X_{n+1}\}$ be the homotopy, so that $r_nf_n - 1_{X_n} = d^X_{n+1}s_n + s_{n-1}d^X_n$ for each n. There is an inclusion $X \to Y \bigoplus \left(\bigoplus_{n\in\mathbb{Z}} D^{n+1}(X_n)\right)$ defined in degree n by the matrix

$$\begin{bmatrix} f_n \\ 1_{X_n} \\ d_n^X \end{bmatrix} : X_n \to Y_n \bigoplus X_n \bigoplus X_{n-1}.$$

(Actually, this morphism is just the one induced by $f\colon X \to Y$ along with the natural inclusion $\begin{bmatrix} 1 \\ d \end{bmatrix} : X \to \bigoplus_{n\in\mathbb{Z}} D^{n+1}(X_n)$ appearing ahead in Diagram (10.3) of Lemma 10.16.) One can check directly that this is a chain map, and that it has a left inverse in $\mathrm{Ch}(\mathcal{A})$ defined by

$$\begin{bmatrix} r_n & -d_{n+1}^X s_n & -s_{n-1} \end{bmatrix} : Y_n \bigoplus X_n \bigoplus X_{n-1} \to X_n.$$

We leave these straightforward computations as an exercise. □

Exercise 10.3.3 Complete the proof of Lemma 10.12 by verifying that the constructed maps are indeed chain maps and that their composite equals 1_X.

Corollary 10.13 (Characterizations of Contractible Complexes) *The following are equivalent for a chain complex C.*

(1) *C is contractible. That is, if $1_C \sim 0$.*
(2) *C is a 0 object in $K(\mathcal{A})$.*
(3) *C is a retract, in* $\mathrm{Ch}(\mathcal{A})$*, of some split exact complex X.*
(4) *C is a retract, in $K(\mathcal{A})$, of some split exact complex X.*

Proof In any additive category, an object is a 0 object if and only if its identity morphism coincides with its 0 morphism. In particular, (1) and (2) are equivalent. We already noted before Definition 10.11 that (1) and (3) are equivalent by Proposition 10.10. Clearly (3) implies (4). Conversely, suppose C is a retract, in $K(\mathcal{A})$, of a split exact complex $\bigoplus_{n\in\mathbb{Z}} D^{n+1}(A_n)$. Then by Lemma 10.12, C is a retract, in $\mathrm{Ch}(\mathcal{A})$, of the split exact complex

$$\left(\bigoplus_{n\in\mathbb{Z}} D^{n+1}(A_n)\right) \bigoplus \left(\bigoplus_{n\in\mathbb{Z}} D^{n+1}(C_n)\right) \cong \left(\bigoplus_{n\in\mathbb{Z}} D^{n+1}\left(A_n \bigoplus C_n\right)\right).$$

□

The following categorical characterization of the chain homotopy relation is quite useful.

Theorem 10.14 (Characterizations of Chain Homotopy) *Let $f, g\colon X \to Y$ be chain maps. The following are equivalent.*

(1) *f and g are chain homotopic; that is, $f \sim g$.*
(2) *Their difference $g - f$ factors through some split exact complex.*
(3) *Their difference $g - f$ factors through some contractible complex.*

Proof Assume $g - f$ factors as $g - f = \beta\alpha$ through a contractible complex C. We can also factor $1_C : C \xrightarrow{i} X \xrightarrow{r} C$ for some split exact X. Therefore $g - f = \beta r i \alpha$ factors through X. So the result follows from Proposition 10.10. □

10.3.1 Contractible Complexes and Idempotent Completeness

In applications, the notion of a contractible complex typically coincides with the notion of a split exact complex. For example, this is well known for chain complexes of modules over a ring. It is interesting to note, however, that for a general additive category $\mathcal{A}$, this nice behavior is equivalent to whether or not $\mathcal{A}$ is *idempotent complete* in the sense of Section 1.11.

Proposition 10.15 (Idempotent Completeness and Contractible Complexes) *The following are equivalent for an additive category* $\mathcal{A}$.

(1) *$\mathcal{A}$ is idempotent complete.*

(2) *Every contractible complex is split exact, and therefore the class of contractible complexes coincides with the class of split exact complexes.*

(3) *The class of split exact complexes is closed under retracts.*

Proof (1) $\implies$ (2). Assume $\mathcal{A}$ is idempotent complete and let $A \in \mathrm{Ch}(\mathcal{A})$ be a contractible complex. The goal is to show A is split exact. We may write A as a retraction $1_A : A \xrightarrow{f} X \xrightarrow{r} A$ of some split exact complex X. Then $e := fr : X \to X$ is a (split) idempotent, so, by Corollary 1.43 and Exercise 10.1.3, X has a direct sum decomposition

$$\left[\begin{smallmatrix} r \\ g \end{smallmatrix}\right] : X \xrightarrow{\cong} A \bigoplus C,$$

where g appears in the splitting of its complement $1_X - fr = sg$. Moreover, note that for each n there is an induced map $e'_n : Z_nX \to Z_nX$ on the cycles of X, satisfying $k_n e'_n = e_n k_n$ for the split inclusion $k_n : Z_nX \rightarrowtail X_n$. Of course this map too must be an idempotent: For $k_n e'_n = e_n k_n = e_n^2 k_n = e_n k_n e'_n = k_n(e'_n)^2$, and k_n is left cancellable. By Corollary 1.43, all of the e'_n also admit direct sum decompositions and one checks that they are compatible with the decompositions for each e_n, as indicated by the following diagram:

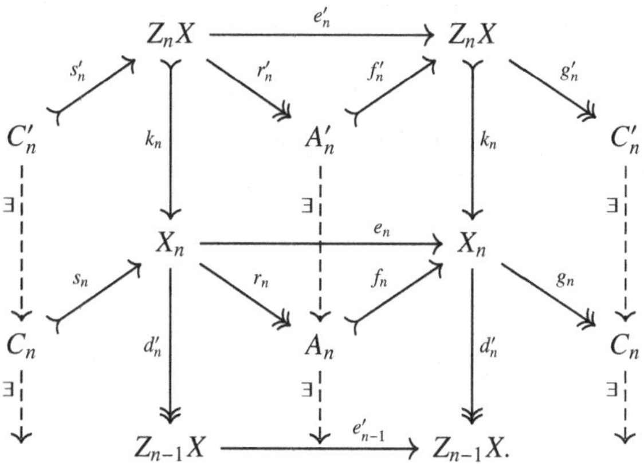

So each differential $d_n = k_{n-1}d'_n$ has a direct sum decomposition $d_n = d_n^A \oplus d_n^C$ with $\operatorname{Ker} d_n = Z_nX = A'_n \bigoplus C'_n$ and $\operatorname{Im} d_n = Z_{n-1}X = A'_{n-1} \bigoplus C'_{n-1}$. But short exact sequences are closed under direct summands; see Exercise 1.5.1. In particular, A is a split exact complex with $Z_nA = A'_n$.

It is clear from Corollary 10.13 that contractible complexes are always closed under retracts. Therefore, (2) $\implies$ (3). Let us show (3) $\implies$ (1). To do so, let $e \colon B \to B$ be any idempotent in $\mathcal{A}$. By Proposition 1.46, it is enough to show that e has a kernel. We see that the chain complex

$$C \equiv \cdots \xrightarrow{1-e} B \xrightarrow{e} B \xrightarrow{1-e} B \xrightarrow{e} \cdots$$

is a retract of the split exact complex $\bigoplus_{n\in\mathbb{Z}} D^{n+1}(B)$. Indeed there is a degree-wise split short exact sequence $C \rightarrowtail \bigoplus_{n\in\mathbb{Z}} D^{n+1}(B) \twoheadrightarrow \Sigma C$ which in successive degrees is as shown:

$$
\begin{array}{ccccc}
B & \xrightarrow{\left[\begin{smallmatrix}1\\e\end{smallmatrix}\right]} & B \bigoplus B & \xrightarrow{[-e\ 1]} & B \\
\scriptstyle e \downarrow & \scriptstyle \left[\begin{smallmatrix}1\\1-e\end{smallmatrix}\right] & \downarrow \left[\begin{smallmatrix}0\ 1\\0\ 0\end{smallmatrix}\right] & & \downarrow \scriptstyle -(1-e) \\
B & \rightarrowtail & B \bigoplus B & \twoheadrightarrow & B.
\end{array}
$$

$$\quad\quad\quad\quad\quad\quad\quad\quad\quad [e-1\ 1]$$

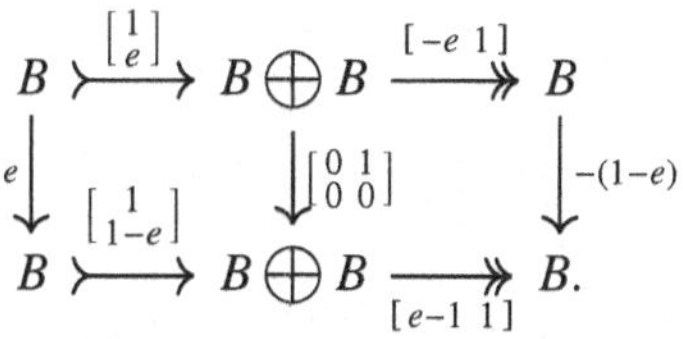

(This is another instance of the short exact sequence in Diagram (10.3) of Lemma 10.16.) However, $C \rightarrowtail \bigoplus_{n\in\mathbb{Z}} D^{n+1}(B)$ has a left inverse. Indeed $[\ 1-e\ 1\ ]$ is a retraction for the monic in the top row of the diagram and $[\ e\ 1\ ]$ is a retraction for the monic in the bottom row of the diagram. These do, in fact, assemble to a chain map $\bigoplus_{n\in\mathbb{Z}} D^{n+1}(B) \twoheadrightarrow C$ providing the claimed left inverse (retraction). By hypothesis, this means C must be a split exact complex. In particular, each differential must admit a kernel. So $\mathcal{A}$ is idempotent complete. $\square$

Recall that in the case $\mathcal{A}$ is *weakly* idempotent complete, then retracts are the same thing as direct summands. So in this case, contractible complexes are precisely the direct summands of split exact complexes. But if $\mathcal{A}$ is not idempotent complete then there must exist an idempotent e without a kernel. The complex C constructed in the proof of Proposition 10.15 provides an example of a contractible complex that is not split exact.

10.4 The Frobenius Structure on Chain Complexes

In Section 8.1 we discussed Frobenius categories. One standard example is $\mathrm{Ch}(R)_{dw}$, the category of chain complexes of R-modules along with the degreewise split exact sequences; see Exercise 8.1.1. In this section we show that the same is true for chain complexes over any additive category $\mathcal{A}$, regardless of whether or not it is idempotent complete. We describe the standard triangulated structure on the homotopy category $K(\mathcal{A})$ from the model category point of view.

So we continue to let $\mathcal{A}$ denote an additive category throughout this section. First we note that suspensions fit into short exact sequences in $\mathrm{Ch}(\mathcal{A})_{dw}$.

Lemma 10.16 *For any chain complex X, there is a natural degreewise split short exact sequence*

$$\mathbb{W}_X: \qquad X \xrightarrow{\left[\begin{smallmatrix} 1 \\ d \end{smallmatrix}\right]} \bigoplus_{n\in\mathbb{Z}} D^{n+1}(X_n) \xrightarrow{[-d \ 1]} \Sigma X, \tag{10.3}$$

which in degree n is given by the matrices $\left[\begin{smallmatrix} 1_{X_n} \\ d_n \end{smallmatrix}\right] : X_n \rightarrowtail X_n \bigoplus X_{n-1}$, and $[\, -d_n \ 1_{X_{n-1}} \,] : X_n \bigoplus X_{n-1} \twoheadrightarrow X_{n-1}$.

On the other hand, there is a natural degreewise split short exact sequence

$$\mathbb{W}^X: \qquad \Sigma^{-1}X \xrightarrow{\left[\begin{smallmatrix} 1 \\ -d \end{smallmatrix}\right]} \bigoplus_{n\in\mathbb{Z}} D^{n}(X_n) \xrightarrow{[d \ 1]} X. \tag{10.4}$$

Proof Recall that the differential of $\bigoplus_{n\in\mathbb{Z}} D^{n+1}(X_n)$ is given in degree n by the matrix

$$\begin{bmatrix} 0 & 1_{X_{n-1}} \\ 0 & 0 \end{bmatrix} : X_n \bigoplus X_{n-1} \to X_{n-1} \bigoplus X_{n-2}.$$

So one easily checks we have chain maps. Observe that $[\, -d_n \ 1_{X_{n-1}} \,]$ is split epic with right inverse $\left[\begin{smallmatrix} 0 \\ 1_{X_{n-1}} \end{smallmatrix}\right]$. On the other hand, $\left[\begin{smallmatrix} 1_{X_n} \\ d_n \end{smallmatrix}\right]$ is split monic with left inverse $[\, 1_{X_n} \ 0 \,]$. □

Proposition 10.17 *For any additive category $\mathcal{A}$, the exact category* $\mathrm{Ch}(\mathcal{A})_{dw}$ *is Frobenius. The projective–injective objects are precisely the contractible complexes.*

Proof One can easily check directly that any chain map $D^n(A) \to Z$ must lift over a degreewise split epimorphism $Y \twoheadrightarrow Z$; see also Lemma 10.1. So by Proposition 1.25, any n-disk $D^n(A)$ is projective in $\mathrm{Ch}(\mathcal{A})_{dw}$. Similarly, any $D^n(A)$ is injective in $\mathrm{Ch}(\mathcal{A})_{dw}$. Now for any given collection of objects $\{A_n\}_{n\in\mathbb{Z}}$, as noted in the previous section, the direct sum $\bigoplus_{n\in\mathbb{Z}} D^n(A_n)$ exists in the category $\mathrm{Ch}(\mathcal{A})$. Since each summand is projective, so is $\bigoplus_{n\in\mathbb{Z}} D^n(A_n)$, by Lemma 1.26. But for that matter we have that the product $\prod_{n\in\mathbb{Z}} D^n(A_n) = \bigoplus_{n\in\mathbb{Z}} D^n(A_n)$ is injective as well. So by Lemma 10.8, any split exact complex $\bigoplus_{n\in\mathbb{Z}} D^n(A_n)$ is projective–injective. Therefore, by Corollary 10.13 and Lemma 1.27, any contractible complex must also be projective–injective.

For any given chain complex X, the short exact sequence in Diagram (10.3) of Lemma 10.16 shows that $\mathrm{Ch}(\mathcal{A})_{dw}$ has enough injectives. The dual in (10.4) gives us enough projectives. If X itself is injective in $\mathrm{Ch}(\mathcal{A})_{dw}$, then the admissible monomorphism $X \rightarrowtail \bigoplus_{n\in\mathbb{Z}} D^{n+1}(X_n)$ splits. In particular, any injective complex X is contractible. Dually all projective complexes must be contractible. Hence the exact category $\mathrm{Ch}(\mathcal{A})_{dw}$ is Frobenius. □

Definition 10.18 Let $\mathcal{A}$ be any additive category and $f\colon X \to Y$ any chain map in $\mathrm{Ch}(\mathcal{A})$. We define the **mapping cone** of f, denoted $\mathrm{Cone}(f)$, to be the chain complex whose degree n component is the direct sum $Y_n \bigoplus X_{n-1}$, and whose differential in degree n is given by $\begin{bmatrix} d_n^Y & f_{n-1} \\ 0 & -d_{n-1}^X \end{bmatrix}$. In short, $\begin{bmatrix} d_Y & f \\ 0 & -d_X \end{bmatrix}$.

Sign conventions for the differential of $\mathrm{Cone}(f)$ vary in the literature.[1] Regardless of the sign convention used, the important thing is that $\mathrm{Cone}(f)$ fits into the pushout construction of the next lemma.

Lemma 10.19 *Let $f\colon X \to Y$ be any chain map in* $\mathrm{Ch}(\mathcal{A})$*. Then* $\mathrm{Cone}(f)$ *fits into a pushout diagram*

$$\begin{array}{lccccc}
\mathbb{W}_X\colon & X & \overset{\left[\begin{smallmatrix}1\\ d\end{smallmatrix}\right]}{\rightarrowtail} & \bigoplus_{n\in\mathbb{Z}} D^{n+1}(X_n) & \overset{[-d\ 1]}{\twoheadrightarrow} & \Sigma X \\
& {\scriptstyle f}\downarrow & & \downarrow & & \| \\
\mathbb{P}_f\colon & Y & \overset{\left[\begin{smallmatrix}1\\ 0\end{smallmatrix}\right]}{\rightarrowtail} & \mathrm{Cone}(f) & \overset{[0\ 1]}{\twoheadrightarrow} & \Sigma X,
\end{array} \tag{10.5}$$

where the top row is the natural degreewise split exact sequence from Lemma 10.16 and the bottom row is given in degree n by

[1] For example, Weibel's text uses the sign convention $\begin{bmatrix} d_Y & -f \\ 0 & -d_X \end{bmatrix}$. See [Weibel, 1994].

$$Y_n \overset{\left[\begin{smallmatrix}1_{Y_n}\\ 0\end{smallmatrix}\right]}{\rightarrowtail} Y_n \bigoplus X_{n-1} \overset{[0\ 1_{X_{n-1}}]}{\twoheadrightarrow} X_{n-1}.$$

Proof We define a chain map $\left[\begin{smallmatrix} f & 0 \\ -d & 1\end{smallmatrix}\right]\colon \bigoplus_{n\in\mathbb{Z}} D^{n+1}(X_n) \to \mathrm{Cone}(f)$ which gives us a commutative diagram in each degree n:

$$\begin{array}{ccccc}
X_n & \overset{\left[\begin{smallmatrix}1_{X_n}\\ d_n\end{smallmatrix}\right]}{\rightarrowtail} & X_n \bigoplus X_{n-1} & \overset{[-d_n\ 1]}{\twoheadrightarrow} & X_{n-1} \\
{\scriptstyle f_n}\downarrow & & \downarrow{\scriptstyle \left[\begin{smallmatrix} f_n & 0\\ -d_n & 1\end{smallmatrix}\right]} & & \Vert \\
Y_n & \underset{}{\overset{\left[\begin{smallmatrix}1_{Y_n}\\ 0\end{smallmatrix}\right]}{\rightarrowtail}} & Y_n \bigoplus X_{n-1} & \underset{[0\ 1]}{\twoheadrightarrow} & X_{n-1}.
\end{array}$$

We easily check $\left[\begin{smallmatrix} f & 0 \\ -d & 1\end{smallmatrix}\right]\colon \bigoplus_{n\in\mathbb{Z}} D^{n+1}(X_n) \to \mathrm{Cone}(f)$ is in fact a chain map by computing:

$$\begin{bmatrix} d_n & f_{n-1} \\ 0 & -d_{n-1}\end{bmatrix}\begin{bmatrix} f_n & 0 \\ -d_n & 1_{X_{n-1}}\end{bmatrix} = \begin{bmatrix} d_nf_n - f_{n-1}d_n & f_{n-1} \\ +d_{n-1}d_n & -d_{n-1}\end{bmatrix} = \begin{bmatrix} 0 & f_{n-1} \\ 0 & -d_{n-1}\end{bmatrix}$$

and

$$\begin{bmatrix} f_{n-1} & 0 \\ -d_{n-1} & 1_{X_{n-2}}\end{bmatrix}\begin{bmatrix} 0 & 1_{X_{n-1}} \\ 0 & 0\end{bmatrix} = \begin{bmatrix} 0 & f_{n-1} \\ 0 & -d_{n-1}\end{bmatrix}.$$

The rules defining the maps in the bottom row of the diagram are clearly chain maps providing a degreewise split short exact sequence. By Proposition 1.12, we have a pushout diagram in each degree n, and since pushouts of complexes are taken degreewise we are done. □

From the abelian model category point of view we have the following result. It recovers the standard triangulated structure on $K(\mathcal{A})$ from Theorem 6.34.

Theorem 10.20 (Model Structure for $K(\mathcal{A})$) *Let $\mathcal{A}$ be a (weakly idempotent complete) additive category and $\mathcal{W}$ denote the class of all contractible complexes. Then $\mathfrak{M}_K = (\mathrm{Ch}(\mathcal{A}), \mathcal{W}, \mathrm{Ch}(\mathcal{A}))$ is an abelian model structure on $\mathrm{Ch}(\mathcal{A})_{dw}$, the exact category of chain complexes along with the degreewise split exact sequences, and the following basic properties are satisfied.*

1. *A morphism is a weak equivalence if and only if it is a homotopy equivalence. The associated homotopy category is $K(\mathcal{A})$ and the canonical localization functor is precisely the functor $\pi\colon \mathrm{Ch}(\mathcal{A}) \to K(\mathcal{A})$ taking a morphism f to its chain homotopy equivalence class $[f]$.*
2. *The usual suspension functor Σ of shifting indices serves as a formal suspension functor on the homotopy category.*
3. *The usual mapping cone $\mathrm{Cone}(f)$ serves as a formal mapping cone $C(f)$.*

(4) *The formal triangulated structure coincides with the usual triangulation on $K(\mathcal{A})$: A distinguished triangle is a diagram in $K(\mathcal{A})$ that is isomorphic to a strict triangle arising from a cofiber sequence*

$$X \xrightarrow{f} Y \xrightarrow{\left[\begin{smallmatrix}1\\0\end{smallmatrix}\right]} \mathrm{Cone}(f) \xrightarrow{[0\ 1]} \Sigma X$$

as constructed in Lemma 10.19.

Proof It follows from Theorems 8.2 and 10.14 and Proposition 10.17 that the triple determines a Frobenius model structure with homotopy category $K(\mathcal{A})$, satisfying the properties in (1). For (2), it is clear from Lemma 10.16 that the usual suspension functor Σ may be chosen to serve as the formal suspension functor in the sense of Section 6.4. Consequently, by Lemma 10.19 the usual mapping cone, $\mathrm{Cone}(f)$, serves as a natural choice for $C(f)$, the formal mapping cone of f. So (3) and (4) hold. □

Remark 10.21 For concreteness, unless stated otherwise, our convention will be that the usual suspension functor Σ (shifting indices) will always serve as our choice for the formal suspension functor on $K(\mathcal{A})$. Similarly, we take Σ^{-1} to be the formal loop functor Ω.

The following useful result is an immediate consequence of Theorems 8.18 and 10.20.

Corollary 10.22 *For all chain complexes X and Y and $n \geq 1$, we have natural isomorphisms:*

$$K(\mathcal{A})(\Sigma^{-n}X, Y) \cong \mathrm{Ext}^n_{dw}(X, Y) \cong K(\mathcal{A})(X, \Sigma^n Y).$$

In particular, it follows that for all integers $n \in \mathbb{Z}$ we have

$$\mathrm{Ext}^1_{dw}\left(X, \Sigma^{(-n-1)}Y\right) \cong H_n Hom(X, Y) = K(\mathcal{A})(X, \Sigma^{-n}Y).$$

Proof The first group of isomorphisms is an immediate application of Theorem 8.18, applied to the Frobenius model $\mathfrak{M}_K$ of Theorem 10.20. Taking $n = 1$ we have $\mathrm{Ext}^1_{dw}(X, Y) \cong K(\mathcal{A})(X, \Sigma Y)$. But then substituting $\Sigma^{(-n-1)}Y$ for Y (for arbitrary $n \in Z$), we get

$$\mathrm{Ext}^1_{dw}\left(X, \Sigma^{(-n-1)}Y\right) \cong K(\mathcal{A})(X, \Sigma^{-n}Y) = H_n[Hom(X, Y)]$$

where the last equality holds by Lemma 10.3. □

It is fair to think of $K(\mathcal{A})$ as the derived category of $\mathcal{A}$ with respect to the split exact structure. Indeed as we noted at the start of Section 10.3, the acyclic complexes in $\mathrm{Ch}(\mathcal{A})_{dw}$ are precisely the split exact complexes. In this way $\mathfrak{M}_K = (\mathrm{Ch}(\mathcal{A}), \mathcal{W}, \mathrm{Ch}(\mathcal{A}))$ serves as a model for the derived category of $\mathcal{A}$

along with its split exact structure. The class $\mathcal{W}$ consists precisely of the (direct summands of) acyclic complexes.

Remark 10.23 Technically we need $(\mathcal{A}, \mathcal{E})$ to be weakly idempotent complete to obtain the abelian model structure in Theorem 10.20. However, even if $\mathcal{A}$ is not weakly idempotent complete we still get a classical abelian model structure in the sense of Appendix A. See Proposition A.5.

Exercise 10.4.1 Let A be an object of some exact category $(\mathcal{A}, \mathcal{E})$, and let $X \in \mathrm{Ch}(\mathcal{A})$ be a chain complex such that $\mathrm{Ext}^1_{\mathcal{E}}(A, X_n) = 0$. Use Corollary 10.22 to show that for each integer n we have isomorphisms of abelian groups

$$\mathrm{Ext}^1_{\mathrm{Ch}(\mathcal{E})}(S^n(A), X) \cong K(\mathcal{A})(S^{n-1}(A), X) \cong H_{n-1}[\mathrm{Hom}_{\mathcal{A}}(A, X)].$$

In particular, the complex $\mathrm{Hom}_{\mathcal{A}}(A, X)$ is exact whenever $\mathrm{Ext}^1_{\mathrm{Ch}(\mathcal{E})}(S^n(A), X) = 0$ for each integer n.

10.5 Homotopy Categories as Triangulated Subcategories of $K(\mathcal{A})$

In this section we will see that, under some very natural conditions, the homotopy category of an abelian model structure on $\mathrm{Ch}(\mathcal{A})$ is equivalent to a triangulated subcategory of the chain homotopy category $K(\mathcal{A})$. Throughout the section we assume $(\mathcal{A}, \mathcal{E})$ is an exact category and that the underlying additive category $\mathcal{A}$ is weakly idempotent complete. $\mathrm{Ch}(\mathcal{A})_{\mathcal{E}}$ denotes the category of chain complexes along with the degreewise exact structure, as described in Lemma 10.4.

Proposition 10.24 *Let $\mathfrak{M} = (Q, \mathcal{W}, \mathcal{R})$ be an abelian model structure on the exact category $\mathrm{Ch}(\mathcal{A})_{\mathcal{E}}$. Assume that $\mathcal{W}$ contains the class of all contractible chain complexes. Then we have the following.*

1. *If $f\colon X \to Y$ is an isomorphism in $K(\mathcal{A})$, that is, f is a homotopy equivalence, then $X \in \mathcal{W}$ if and only if $Y \in \mathcal{W}$.*
2. *We may choose the usual suspension functor Σ (i.e. shifting indices) to serve as the formal suspension functor on the homotopy category $\mathrm{Ho}(\mathfrak{M})$. Then the usual mapping cone, $\mathrm{Cone}(f)$, serves as the formal mapping cone $C(f)$.*
3. *With this Σ as our choice of suspension functor, the triangulated structure (Δ, Σ) on $\mathrm{Ho}(\mathfrak{M})$ may be described as follows: A triangle in $\mathrm{Ho}(\mathfrak{M})$ is a*

distinguished triangle if and only if it is isomorphic in $\mathrm{Ho}(\mathfrak{M})$ *to the* γ-*image of a cofiber sequence (strict triangle)*

$$X \xrightarrow{f} Y \xrightarrow{\left[\begin{smallmatrix}1\\0\end{smallmatrix}\right]} \mathrm{Cone}(f) \xrightarrow{[0\ 1]} \Sigma X$$

on some chain map $f\colon X \to Y$.

(4) *A chain map f is a weak equivalence if and only if* $\mathrm{Cone}(f) \in \mathcal{W}$.

Because of this proposition, we can and will continue to follow the convention of Remark 10.21. That is, as long as $\mathcal{W}$ contains all contractible complexes, we will take the formal suspension functor to be the usual Σ and the formal cone on f to be the usual mapping cone, $\mathrm{Cone}(f)$.

Proof First let us show (1), that if $f\colon X \to Y$ is a homotopy equivalence, then $X \in \mathcal{W}$ if and only if $Y \in \mathcal{W}$. Indeed by Theorem 10.20, such an f is a weak equivalence in the Frobenius model structure on $\mathrm{Ch}(\mathcal{A})_{dw}$, and so it factors as $f = pi$ where i is a degreewise split monomorphism with contractible cokernel and p is a degreewise split epimorphism with contractible kernel. Since $\mathcal{W}$ is a thick class in $\mathrm{Ch}(\mathcal{A})_{\mathcal{E}}$ and contains all contractible complexes, it follows that $X \in \mathcal{W}$ if and only if $Y \in \mathcal{W}$.

Statements (2) and (3) follow from the results on general suspension functors discussed at the end of Section 6.6. Indeed the usual chain complex suspension functor, Σ, arises from the assignment $\mathbb{W}\colon X \mapsto \mathbb{W}_X$ of general suspension sequences given in Lemma 10.16. So Lemma 6.39 and Corollary 6.42 apply. We see from Lemma 10.19 that $C_{\mathbb{W}}(f)$, the formal cone on f (relative to this choice $\mathbb{W}$) coincides with $\mathrm{Cone}(f)$, the usual chain complex mapping cone. Consequently, a diagram of the form

$$X \xrightarrow{f} Y \xrightarrow{\left[\begin{smallmatrix}1\\0\end{smallmatrix}\right]} \mathrm{Cone}(f) \xrightarrow{[0\ 1]} \Sigma X$$

is precisely a cofiber sequence (relative to $\mathbb{W}$). Therefore, Corollary 6.42 shows that the class of distinguished triangles is precisely the class of all Σ-triangles in $\mathrm{Ho}(\mathfrak{M})$ that are isomorphic to the γ-image of such a cofiber sequence. So we have shown (3).

Now let us show (4), that f is a weak equivalence if and only if $\mathrm{Cone}(f) \in \mathcal{W}$. Recall from Proposition 4.7 that an admissible monic $i\colon X \rightarrowtail Z$ is a $\mathfrak{M}$-weak equivalence if and only if $\mathrm{Cok}\, i \in \mathcal{W}$. Referring to the pushout construction of $\mathrm{Cone}(i)$ shown in Lemma 10.19, one can see that in this case we are led to a short exact sequence

$$\bigoplus_{n\in\mathbb{Z}} D^{n+1}(X_n) \rightarrowtail \mathrm{Cone}(i) \twoheadrightarrow \mathrm{Cok}\, i.$$

Since $\bigoplus_{n\in\mathbb{Z}} D^{n+1}(X_n) \in \mathcal{W}$ we conclude that $\mathrm{Cone}(i) \in \mathcal{W}$ if and only if $\mathrm{Cok}\, i \in \mathcal{W}$. So the statement is true for any admissible monic $i\colon X \rightarrowtail Z$. Similarly, for any admissible epimorphism $p\colon Z \twoheadrightarrow Y$, by using Corollary 1.14(2) we obtain a short exact sequence

$$\mathrm{Ker}\, p \rightarrowtail \bigoplus_{n\in\mathbb{Z}} D^{n+1}(Z_n) \twoheadrightarrow \mathrm{Cone}(p).$$

So in this case, $\mathrm{Cone}(p) \in \mathcal{W}$ if and only if $\mathrm{Ker}\, p \in \mathcal{W}$ and therefore the desired statement is true of any admissible epic $p\colon Z \twoheadrightarrow Y$. Now by the factorization axiom, any morphism f factors as $f = pi$ where i is an admissible monic and p is an admissible epic with kernel $\mathrm{Ker}\, p \in \mathcal{W}$, equivalently, $\mathrm{Cone}(p) \in \mathcal{W}$. By the 2 out of 3 Axiom, we have that f is a weak equivalence if and only if $\mathrm{Cok}\, i \in \mathcal{W}$ too. By the octahedral axiom, there is a distinguished triangle (exact triangle) $\mathrm{Cone}(i) \to \mathrm{Cone}(f) \to \mathrm{Cone}(p) \to \Sigma\mathrm{Cone}(i)$ in $K(\mathcal{A})$. By Theorem 10.20, it must be isomorphic in $K(\mathcal{A})$ to a strict triangle arising from a cofiber sequence

$$A \xrightarrow{h} B \xrightarrow{\left[\begin{smallmatrix}1\\0\end{smallmatrix}\right]} \mathrm{Cone}(h) \xrightarrow{[0\ 1]} \Sigma A$$

on some chain map $h\colon A \to B$. Now recalling that f is a weak equivalence if and only if $\mathrm{Cok}\, i \in \mathcal{W}$, we note now that this occurs if and only if $\mathrm{Cone}(i) \cong_{K(\mathcal{A})} A$ is in $\mathcal{W}$. Since $\mathcal{W}$ contains all contractible complexes, this is the case if and only if $\Sigma A \in \mathcal{W}$. Finally, because of the short exact sequence

$$B \xrightarrow{\left[\begin{smallmatrix}1\\0\end{smallmatrix}\right]} \mathrm{Cone}(h) \xrightarrow{[0\ 1]} \Sigma A$$

having $\mathrm{Cone}(h) \cong_{K(\mathcal{A})} \mathrm{Cone}(p)$ in $\mathcal{W}$, we conclude that f is a weak equivalence if and only $B \cong_{K(\mathcal{A})} \mathrm{Cone}(f)$ is in $\mathcal{W}$. □

The main theorem that follows identifies certain homotopy categories of complexes with triangulated subcategories of $K(\mathcal{A})$. Let us pause to make more precise what is meant by a triangulated subcategory.

Definition 10.25 Let $\mathcal{T} = (\mathcal{T}, \Sigma, \Delta)$ be a triangulated category. A full additive subcategory $\mathcal{S} \subseteq \mathcal{T}$ is a called a **triangulated subcategory** if the following hold.

(1) $\mathcal{S}$ is replete. That is, any object of $\mathcal{T}$ that is isomorphic to one in $\mathcal{S}$ must itself be in $\mathcal{S}$. This is also often expressed by saying that $\mathcal{S}$ is a *strictly full* subcategory of $\mathcal{T}$.
(2) $\mathcal{S}$ is closed under applying the autoequivalence Σ and its inverse Ω.
(3) $\mathcal{S}$ is closed under mapping cones. That is, for any distinguished triangle $A \xrightarrow{f} B \to C \to \Sigma A$ with $f \in \mathcal{S}$, we have $C \in \mathcal{S}$.

Note that along with the Rotation Axiom, **(Tr2)**, the three conditions imply that whenever two out of three terms in a distinguished triangle are in $\mathcal{S}$, then so is the third. Note too that condition (2) implies that the autoequivalence (Σ, Ω) restricts to an autoequivalence of $\mathcal{S}$. Therefore, $(\mathcal{S}, \Sigma, \Delta_{\mathcal{S}})$ is a triangulated category, where $\Delta_{\mathcal{S}}$ is the class of all distinguished triangles in Δ whose all three terms are in $\mathcal{S}$. Clearly, the inclusion functor from $\mathcal{S}$ into $\mathcal{T}$ is a triangulated functor.

Now, given some class $\mathcal{X}$ of chain complexes in $\mathrm{Ch}(\mathcal{A})$, we let $K(\mathcal{X})$ denote the full subcategory of all complexes in $K(\mathcal{A})$ isomorphic (i.e. homotopy equivalent) to one in $\mathcal{X}$. We call $K(\mathcal{X})$ the **strictly full subcategory generated by** $\mathcal{X}$.

Lemma 10.26 *Let $\mathcal{X}$ be any class of chain complexes containing 0, closed under degreewise split extensions, and closed under the usual suspension Σ and its inverse Σ^{-1}. Then $K(\mathcal{X})$, the strictly full subcategory of $K(\mathcal{A})$ generated by $\mathcal{X}$, is a triangulated subcategory of $K(\mathcal{A})$. In particular, a distinguished triangle in $K(\mathcal{X})$ is just a distinguished triangle $X \xrightarrow{[f]} Y \xrightarrow{[g]} Z \xrightarrow{[h]} \Sigma X$ in $K(\mathcal{A})$ in which all three terms are homotopy equivalent to one in $\mathcal{X}$.*

See Theorem 10.20 to recall the standard triangulation $(K(\mathcal{A}), \Delta, \Sigma)$ on the homotopy category of complexes.

Proof Observe that since $0 \in \mathcal{X}$, all contractible complexes are in $K(\mathcal{X})$; see Corollary 10.13. Since $\mathcal{X}$ is closed under Σ and under degreewise split extensions, $\mathrm{Cone}(f) \in \mathcal{X}$ whenever $f\colon X \to Y$ is a chain map between complexes in $\mathcal{X}$; see Lemma 10.19. It follows that each condition of Definition 10.25 holds for $K(\mathcal{X}) \subseteq K(\mathcal{A})$. □

Theorem 10.27 *Let $\mathfrak{M} = (\mathcal{Q}, \mathcal{W}, \mathcal{R})$ be an abelian model structure on the exact category $\mathrm{Ch}(\mathcal{A})_{\mathcal{E}}$. Assume that $\mathcal{W}$ contains all contractible complexes and that each of the three classes $\mathcal{Q}, \mathcal{W}$, and $\mathcal{R}$ are closed under the usual suspension Σ and its inverse Σ^{-1}. Then, in addition to the four properties of Proposition 10.24, each of the following hold.*

(5) *Bifibrant replacement induces a triangulated equivalence*

$$\mathrm{Ho}(R) \circ \mathrm{Ho}(Q)\colon \mathrm{Ho}(\mathfrak{M}) \to K(\mathcal{Q} \cap \mathcal{R})$$

onto the strictly full subcategory of $K(\mathcal{A})$ generated by $\mathcal{Q} \cap \mathcal{R}$. Moreover, we have

$$\mathrm{Ho}(\mathfrak{M})(X, Y) \cong K(\mathcal{A})(QX, RY).$$

(6) *Each complex in the core $\omega := \mathcal{Q} \cap \mathcal{W} \cap \mathcal{R}$ is contractible and $K(\mathcal{Q} \cap \mathcal{R}) \cap \mathcal{W}$ is precisely the class of contractible complexes.*

Proof Let us first show (6). The hypotheses imply that ω is closed under applying Σ, so we have $\mathrm{Ext}^1_{\mathrm{Ch}(\mathcal{E})}(\Sigma W, W) = 0$ for any $W \in \omega$. In particular, using Corollary 10.22, we have

$$0 = \mathrm{Ext}^1_{dw}(\Sigma W, W) = K(\mathcal{A})(\Sigma W, \Sigma W) = K(\mathcal{A})(W, W).$$

This means all chain maps $W \to W$ are null homotopic, so W is contractible. It is clear that the class of all contractible complexes is in $K(\mathcal{Q} \cap \mathcal{R}) \cap \mathcal{W}$. On the other hand, suppose $X \in K(\mathcal{Q} \cap \mathcal{R}) \cap \mathcal{W}$. Then there is a homotopy equivalence $X \xrightarrow{f} Y$ for some $Y \in \mathcal{Q} \cap \mathcal{R}$. By Proposition 10.24(1) we know $Y \in \mathcal{W}$ too. So $Y \in \omega$. We just saw that this implies Y is contractible. So X too is contractible.

Next, we show that there is an isomorphism $\mathrm{Ho}(\mathfrak{M})(X, Y) \cong K(\mathcal{A})(QX, RY)$. By the fundamental Corollary 5.17, we know

$$\mathrm{Ho}(\mathfrak{M})(X, Y) \cong \underline{\mathrm{Hom}}(QX, RY) := \mathrm{Hom}_{\mathrm{Ch}(\mathcal{A})}(QX, RY)/ \sim^{\omega},$$

where $\sim^{\omega}$ is the formal homotopy relation. That is, $f \sim^{\omega} g$ if and only if $g - f$ factors through some chain complex $W \in \omega$. Since any such W must be contractible, Theorem 10.14 implies that f is chain homotopic to g whenever we have the formal relation $f \sim^{\omega} g$. On the other hand, Theorem 10.14 guarantees $g - f$ factors through some contractible complex C whenever f and g are chain homotopic. Since $\mathcal{W}$ contains all contractible complexes, $g - f$ factors through an object of $\mathcal{W}$. Moreover, the domain of f and g is cofibrant while their codomain is fibrant, so we have $f \sim^{\omega} g$, by Lemma 4.11. We conclude $\mathrm{Ho}(\mathfrak{M})(X, Y) \cong K(\mathcal{A})(QX, RY)$.

We turn to the statement concerning the bifibrant replacement functor

$$\mathrm{Ho}(\mathcal{R}) \circ \mathrm{Ho}(\mathcal{Q})\colon \mathrm{Ho}(\mathfrak{M}) \to \mathrm{St}_{\omega}(\mathcal{Q} \cap \mathcal{R}).$$

We already know from Proposition 5.25 that this functor provides a category equivalence to the full subcategory of bifibrant objects $\mathrm{St}_{\omega}(\mathcal{Q} \cap \mathcal{R}) \subseteq \mathrm{Ho}(\mathfrak{M})$. Moreover, with $\sim$ denoting the usual chain homotopy relation, we have just seen that we may make the identification

$$\mathrm{St}_{\omega}(\mathcal{Q} \cap \mathcal{R}) = (\mathcal{Q} \cap \mathcal{R})/ \sim \subseteq K(\mathcal{Q} \cap \mathcal{R}) \subseteq K(\mathcal{A}).$$

So $\mathrm{Ho}(\mathcal{R}) \circ \mathrm{Ho}(\mathcal{Q})$ determines a functor to $K(\mathcal{A})$ with essential image (i.e. isomorphic closure of its image) precisely $K(\mathcal{Q} \cap \mathcal{R}) \subseteq K(\mathcal{A})$. By Lemma 10.26, $K(\mathcal{Q} \cap \mathcal{R})$ is a triangulated subcategory of $K(\mathcal{A})$, because $\mathcal{Q} \cap \mathcal{R}$ is closed under Σ, Σ^{-1}, and degreewise split extensions.

To show that $\mathrm{Ho}(\mathcal{R}) \circ \mathrm{Ho}(\mathcal{Q})$ is a triangulated functor we first show that our convention of choosing Σ to be the usual suspension functor provides, not just a natural equivalence, but an actual equality

$$\Sigma_{\mathrm{Ho}(\mathfrak{M})} \circ [\mathrm{Ho}(\mathcal{R}) \circ \mathrm{Ho}(Q)] = [\mathrm{Ho}(\mathcal{R}) \circ \mathrm{Ho}(Q)] \circ \Sigma_{\mathrm{Ho}(\mathfrak{M})}. \tag{10.6}$$

By Corollary 5.16, it is enough to establish an equality

$$\Sigma_{\mathrm{Ho}(\mathfrak{M})} \circ [\mathrm{Ho}(\mathcal{R}) \circ \mathrm{Ho}(Q)] \circ \gamma = [\mathrm{Ho}(\mathcal{R}) \circ \mathrm{Ho}(Q)] \circ \Sigma_{\mathrm{Ho}(\mathfrak{M})}] \circ \gamma$$

upon composing with the canonical functor $\gamma\colon \mathrm{Ch}(\mathcal{A}) \to \mathrm{Ho}(\mathfrak{M})$. Let $f\colon X \to Y$ be a chain map. Working through the definitions involved (see Definition 5.11 for γ, the paragraphs before Proposition 5.25 for $\mathrm{Ho}(\mathcal{R}) \circ \mathrm{Ho}(Q)$, and the proof of Proposition 10.24 implicitly sets the relation $\Sigma_{\mathrm{Ho}(\mathfrak{M})} \circ \gamma = \gamma \circ \Sigma_{\mathrm{Ch}(\mathcal{A})}$, where $\Sigma_{\mathrm{Ch}(\mathcal{A})} = \Sigma$ is the usual suspension of shifting indices), the composition on the left-hand side acts by sending f to the chain homotopy class

$$[\Sigma(R(Q(f)))]\colon \Sigma RQX \to \Sigma RQY.$$

On the other hand, the composition on the right-hand side acts by sending f to the chain homotopy class

$$[R(Q(\Sigma(f)))]\colon RQ\Sigma X \to RQ\Sigma Y.$$

But since $Q, \mathcal{W}$, and $\mathcal{R}$ are each closed under Σ, it is easy to see that we may choose $\Sigma(R(Q(f)))\colon \Sigma RQX \to \Sigma RQY$ to play the role of $R(Q(\Sigma(f)))\colon RQ\Sigma X \to RQ\Sigma Y$. This proves the desired Equation (10.6).

So now let us show that bifibrant replacement $\mathrm{Ho}(\mathcal{R}) \circ \mathrm{Ho}(Q)$ preserves triangles. By Proposition 10.24, it is enough to show that $\mathrm{Ho}(\mathcal{R}) \circ \mathrm{Ho}(Q) \circ \gamma$ takes any cofiber sequence

$$X \xrightarrow{f} Y \xrightarrow{\left[\begin{smallmatrix}1\\0\end{smallmatrix}\right]} \mathrm{Cone}(f) \xrightarrow{[0\ 1]} \Sigma X$$

on any chain map $f\colon X \to Y$, to a distinguished triangle in $K(Q \cap \mathcal{R})$. In fact the functor $\mathrm{Ho}(\mathcal{R}) \circ \mathrm{Ho}(Q) \circ \gamma$ takes such a cofiber sequence to the triangle

$$RQX \xrightarrow{[R(Q(f))]} RQY \xrightarrow{[R(Q(\left[\begin{smallmatrix}1\\0\end{smallmatrix}\right]))]} RQ\mathrm{Cone}(f) \xrightarrow{[R(Q([0\ 1]))]} RQ\Sigma X = \Sigma RQX.$$

This is certainly a distinguished triangle in $\mathrm{Ho}(\mathfrak{M})$ due to the natural isomorphism $\mathrm{Ho}(\mathcal{R}) \circ \mathrm{Ho}(Q) \cong 1_{\mathrm{Ho}(\mathfrak{M})}$. But the point is that it is also a distinguished triangle in $K(Q \cap \mathcal{R})$, for it is homotopy equivalent to the strict triangle on $\alpha = R(Q(f))\colon RQX \to RQY$. Indeed, apply the standard construction of $\mathrm{Cone}(\alpha)$,

$$\begin{array}{ccccc} RQX & \xrightarrow{\left[\begin{smallmatrix}1\\ d\end{smallmatrix}\right]} & \bigoplus_{n\in\mathbb{Z}} D^{n+1}((RQX)_n) & \xrightarrow{[-d\ 1]} & \Sigma RQX \\ {\scriptstyle\alpha}\downarrow & & \downarrow & & \| \\ RQY & \xrightarrow{\left[\begin{smallmatrix}1\\ 0\end{smallmatrix}\right]} & \mathrm{Cone}(\alpha) & \xrightarrow{[0\ 1]} & \Sigma RQX, \end{array}$$

coming from Lemma 10.19. Since $\mathcal{Q} \cap \mathcal{R}$ is closed under Σ, and under extensions in $\mathrm{Ch}(\mathcal{A})_{\mathcal{E}}$, we see that each complex in the diagram is actually an object of $\mathcal{Q} \cap \mathcal{R}$. Since γ is nothing more than taking chain homotopy classes when restricted to $\mathcal{Q} \cap \mathcal{R}$, we see that its image under γ is, by definition, a distinguished triangle in $K(\mathcal{Q} \cap \mathcal{R})$; in fact, one with bifibrant components. On the other hand, its image under γ is also a distinguished triangle in $\mathrm{Ho}(\mathfrak{M})$, which by the Five Lemma must be isomorphic in $\mathrm{Ho}(\mathfrak{M})$ to the (γ-image of the) previous cofiber sequence on $R(Q(f))$ having cone $RQ\mathrm{Cone}(f)$. Since all the complexes involved are bifibrant, the isomorphism $\mathrm{Cone}(\alpha) \to RQ\mathrm{Cone}(f)$ must be represented by an ω-homotopy equivalence. But again, such a map is just a chain homotopy equivalence. In particular note that $\mathrm{Cone}(\alpha)$ is chain homotopy equivalent to $RQ\mathrm{Cone}(f)$. □

10.6 Models for Verdier Quotients of Complexes

Continuing under the assumptions of the previous section, we proceed to show that the homotopy category of $\mathfrak{M} = (\mathcal{Q}, \mathcal{W}, \mathcal{R})$ coincides with the Verdier quotient of $K(\mathcal{A})$ modulo the class $\mathcal{W}$ of trivial objects.

By definition, a thick subcategory of $K(R)$, the chain homotopy category of a ring R, is a strictly full subcategory that is closed under suspensions, mapping cones and direct summands. We make the same definition for $K(\mathcal{A})$.

Definition 10.28 A **thick subcategory** of $K(\mathcal{A})$ is a strictly full subcategory that is closed under suspensions, mapping cones and direct summands. In other words, a triangulated subcategory that is closed under direct summands.

Theorem 10.29 (Models for Verdier Quotients) *Let $\mathfrak{M} = (\mathcal{Q}, \mathcal{W}, \mathcal{R})$ be an abelian model structure on the exact category $\mathrm{Ch}(\mathcal{A})_{\mathcal{E}}$. Assume that $\mathcal{W}$ contains the class of all contractible chain complexes. Then $\mathcal{W}$ is a thick subcategory of $K(\mathcal{A})$, and the canonical functor γ factors as*

$$\gamma\colon \mathrm{Ch}(\mathcal{A}) \xrightarrow{\pi} K(\mathcal{A}) \xrightarrow{\bar{\gamma}} \mathrm{Ho}(\mathfrak{M}),$$

where $\pi(f) = [f]$ and the functor $\bar{\gamma}\colon K(\mathcal{A}) \to \mathrm{Ho}(\mathfrak{M})$ is the Verdier quotient $K(\mathcal{A})/\mathcal{W}$. That is, $\bar{\gamma}$ satisfies the universal property summarized as follows.

(1) $\bar{\gamma}\colon K(\mathcal{A}) \to \mathrm{Ho}(\mathfrak{M})$ *is a triangulated functor annihilating* $\mathcal{W}$.

(2) *Given any triangulated functor* $F\colon K(\mathcal{A}) \to \mathcal{T}$ *with* $\mathcal{W} \subseteq \operatorname{Ker} F$, *there exists a unique functor* $\bar{F}\colon \mathrm{Ho}(\mathfrak{M}) \to \mathcal{T}$ *satisfying* $F = \bar{F} \circ \bar{\gamma}$.

Moreover, if each of the three classes $\mathcal{Q}$, $\mathcal{W}$, *and* $\mathcal{R}$ *are closed under the usual suspension* Σ *and its inverse* Σ^{-1}, *then bifibrant replacement induces a triangulated equivalence*

$$\mathrm{Ho}(R) \circ \mathrm{Ho}(Q)\colon \mathrm{Ho}(\mathfrak{M}) \to K(\mathcal{Q} \cap \mathcal{R})$$

onto the strictly full subcategory of $K(\mathcal{A})$ *generated by* $\mathcal{Q} \cap \mathcal{R}$, *and we have isomorphisms of hom-sets*

$$\mathrm{Ho}(\mathfrak{M})(X, Y) \cong K(\mathcal{A})(QX, RY).$$

Proof Let us check that the thickness of $\mathcal{W}$ in $\mathrm{Ch}(\mathcal{A})_{\mathcal{E}}$ implies its thickness, in the sense of Definition 10.28, in $K(\mathcal{A})$. First, $\mathcal{W}$ determines a strictly full subcategory of $K(\mathcal{A})$ by Proposition 10.24(1). It is closed under suspensions and mapping cones by Lemmas 10.16 and 10.19. Finally, we see from Lemma 10.12 that $\mathcal{W}$ is closed under direct summands in $K(\mathcal{A})$.

Note that the canonical functor $\gamma\colon \mathrm{Ch}(\mathcal{A}) \to \mathrm{Ho}(\mathfrak{M})$ factors through $K(\mathcal{A})$. Indeed since $\mathcal{W}$ contains all contractible complexes, any homotopy equivalence must be a weak equivalence. So applying the Localization Theorem 5.21 to the Frobenius model structure on $\mathrm{Ch}(\mathcal{A})_{dw}$, the universal property provides a unique factorization $\gamma = \bar{\gamma} \circ \pi$:

$$\mathrm{Ch}(\mathcal{A}) \xrightarrow{\pi} K(\mathcal{A}) \xrightarrow{\bar{\gamma}} \mathrm{Ho}(\mathfrak{M}).$$

Clearly $\bar{\gamma}$ simply acts by $\bar{\gamma}([f]) = \gamma(f)$. Let us now show that $\bar{\gamma}$ is a triangulated functor. First, it is required that there be a natural equivalence

$$\bar{\gamma} \circ \Sigma_{K(\mathcal{A})} \cong \Sigma_{\mathrm{Ho}(\mathfrak{M})} \circ \bar{\gamma}.$$

By Corollary 5.16, it is enough to have a natural equivalence

$$\bar{\gamma} \circ \Sigma_{K(\mathcal{A})} \circ \pi \cong \Sigma_{\mathrm{Ho}(\mathfrak{M})} \circ \bar{\gamma} \circ \pi = \Sigma_{\mathrm{Ho}(\mathfrak{M})} \circ \gamma.$$

But by the formal identity $\Sigma_{\mathrm{Ho}(\mathfrak{M})} \circ \gamma = \gamma \circ \Sigma_{\mathrm{Ch}(\mathcal{A})}$ that emerges from the proof of Proposition 10.24, it means we need a natural equivalence

$$\bar{\gamma} \circ (\pi \circ \Sigma_{\mathrm{Ch}(\mathcal{A})}) \cong \gamma \circ \Sigma_{\mathrm{Ch}(\mathcal{A})}.$$

In fact, we have an equality $\bar{\gamma} \circ (\pi \circ \Sigma_{\mathrm{Ch}(\mathcal{A})}) = (\bar{\gamma} \circ \pi) \circ \Sigma_{\mathrm{Ch}(\mathcal{A})} = \gamma \circ \Sigma_{\mathrm{Ch}(\mathcal{A})}$. So now it is immediate from Theorem 10.20 and Proposition 10.24 that $\bar{\gamma}$ is a triangulated functor.

Finally, we show that $\bar{\gamma}\colon K(\mathcal{A}) \to \mathrm{Ho}(\mathfrak{M})$ satisfies the universal property described, making it the Verdier quotient $K(\mathcal{A})/\mathcal{W}$. Indeed suppose $F\colon K(\mathcal{A}) \to \mathcal{T}$ is a triangulated functor with $\operatorname{Ker} F \supseteq \mathcal{W}$. Then given any $\mathfrak{M}$-weak equivalence f, we have $\mathrm{Cone}(f) \in \mathcal{W}$, by Proposition 10.24(4). Moreover, by Theorem 10.20(4) there is a strict triangle in $K(\mathcal{A})$

$$X \xrightarrow{\pi(f)} Y \xrightarrow{\pi\left[\begin{smallmatrix}1\\0\end{smallmatrix}\right]} \mathrm{Cone}(f) \xrightarrow{\pi[0\ 1]} \Sigma X.$$

Since F is a triangulated, and $F(\mathrm{Cone}(f)) = 0$, we get that $F\pi(f)$ is an isomorphism. So by the Localization Theorem 5.21, $F \circ \pi$ induces a unique functor $\bar{F}\colon \mathrm{Ho}(\mathfrak{M}) \to \mathcal{T}$ satisfying $\bar{F} \circ \gamma = F \circ \pi$. Now we have $F \circ \pi = \bar{F} \circ \bar{\gamma} \circ \pi$, and π is right cancellable. Therefore $F = \bar{F} \circ \bar{\gamma}$, proving $K(\mathcal{A}) \xrightarrow{\bar{\gamma}} \mathrm{Ho}(\mathfrak{M})$ satisfies the universal propery unique to the Verdier quotient $K(\mathcal{A})/\mathcal{W}$.

The last statement of the theorem is a reiteration of Theorem 10.27(5). □

Example 10.30 The results of Sections 10.5 and 10.6 apply to an abundance of abelian model structures on (unbounded) chain complexes. As just one example, consider the Hovey triple $\left(All, \mathcal{W}_{\mathrm{co}}, dw\widetilde{\mathcal{I}}\right)$ from Example 8.34. Here $dw\widetilde{\mathcal{I}}$ is the class of all chain complexes of injective R-modules, and $\mathcal{W}_{\mathrm{co}}$ is the class of all coacyclic complexes. By Theorem 10.29 its homotopy category identifies with the Verdier quotient $K(R)/\mathcal{W}_{\mathrm{co}}$. By Theorem 10.27, it is triangle equivalent to $K(Inj)$, the chain homotopy category of all complexes (homotopy equivalent to a complex) of injective R-modules.

10.7 Lifting Cotorsion Pairs to Chain Complexes

In this section we define the classes of chain complexes that typically appear as the (trivially) fibrant and cofibrant objects in abelian model structures for derived categories. In particular, we will show that if $(\mathcal{Q}, \mathcal{R})$ is a cotorsion pair in $(\mathcal{A}, \mathcal{E})$, then it lifts to two cotorsion pairs, denoted $\left(dg\widetilde{\mathcal{Q}}, \widetilde{\mathcal{R}}\right)$ and $\left(\widetilde{\mathcal{Q}}, dg\widetilde{\mathcal{R}}\right)$, in $\mathrm{Ch}(\mathcal{A})_{\mathcal{E}}$. Throughout this section, $(\mathcal{A}, \mathcal{E})$ may be any exact category unless explicitly specified otherwise.

Given any class of objects $\mathcal{C} \subseteq \mathcal{A}$, we say that a chain complex X is $\mathrm{Hom}_{\mathcal{A}}(\mathcal{C}, -)$**-acyclic** if $\mathrm{Hom}_{\mathcal{A}}(C, X)$ is an acyclic complex of abelian groups for all $C \in \mathcal{C}$. Similarly, we say X is $\mathrm{Hom}_{\mathcal{A}}(-, \mathcal{C})$**-acyclic** if the cochain complex $\mathrm{Hom}_{\mathcal{A}}(X, C)$ is acyclic for all $C \in \mathcal{C}$.

Definition 10.31 Suppose $(\mathcal{Q}, \mathcal{R})$ is a cotorsion pair in $(\mathcal{A}, \mathcal{E})$. We set the following notation.

- $\widetilde{\mathcal{R}}$ denotes the class of all $\mathrm{Hom}_{\mathcal{A}}(\mathcal{Q}, -)$-acyclic complexes X with components $X_n \in \mathcal{R}$. (Typically, $\widetilde{\mathcal{R}}$ coincides with the class all $\mathcal{E}$-acyclic complexes X with cycles $Z_nX \in \mathcal{R}$; see Lemma 10.35.)
- We then set $dg\widetilde{\mathcal{Q}} := {}^{\perp}\widetilde{\mathcal{R}}$, its left Ext-orthogonal in the exact category $\mathrm{Ch}(\mathcal{A})_{\mathcal{E}}$.

On the other hand, we have the following.

- $\widetilde{\mathcal{Q}}$ denotes the class of all $\mathrm{Hom}_{\mathcal{A}}(-, \mathcal{R})$-acyclic complexes X with components $X_n \in \mathcal{Q}$. (Typically, $\widetilde{\mathcal{Q}}$ coincides with the class all $\mathcal{E}$-acyclic complexes X with cycles $Z_nX \in \mathcal{Q}$; by the dual of Lemma 10.35.)
- We then set $dg\widetilde{\mathcal{R}} := \widetilde{\mathcal{Q}}^{\perp}$, its right Ext-orthogonal in the exact category $\mathrm{Ch}(\mathcal{A})_{\mathcal{E}}$.

Lemma 10.32 *Let $\mathcal{Q}$ be a class of objects in $\mathcal{A}$, and let $X \in \mathrm{Ch}(\mathcal{A})$ be a chain complex. The following are equivalent.*

(1) *X is $\mathrm{Hom}_{\mathcal{A}}(\mathcal{Q}, -)$-acyclic. That is, $\mathrm{Hom}_{\mathcal{A}}(Q, X)$ is acyclic for all $Q \in \mathcal{Q}$.*
(2) *Every chain map $f \colon S^n(Q) \to X$ with $Q \in \mathcal{Q}$ extends over the inclusion $S^n(Q) \rightarrowtail D^{n+1}(Q)$ to a chain map $D^{n+1}(Q) \to X$.*

Moreover, the following condition implies conditions (1) and (2) and is equivalent to them whenever $\mathrm{Ext}^1_{\mathcal{E}}(Q, X_n) = 0$ for all n and $Q \in \mathcal{Q}$.

(3) *$\mathrm{Ext}^1_{\mathrm{Ch}(\mathcal{E})}(S^n(Q), X) = 0$ for all $Q \in \mathcal{Q}$.*

Proof By Lemma 10.1, there is an abelian group homomorphism

$$\mathrm{Hom}_{\mathrm{Ch}(\mathcal{A})}(S^n(Q), X) \cong Z_n[\mathrm{Hom}_{\mathcal{A}}(Q, X)].$$

It associates a chain map $S^n(Q) \to X$ to a morphism $t \colon Q \to X_n$ belonging to $\mathrm{Ker}\,(d_n)_* = Z_n[\mathrm{Hom}_{\mathcal{A}}(Q, X)]$. Exactness of $\mathrm{Hom}_{\mathcal{A}}(Q, X)$ in degree n means every such t is an element of $\mathrm{Im}\,(d_{n+1})_*$. This translates to the existence of a morphism $s \colon Q \to X_{n+1}$ such that $d_{n+1}s = t$. By the other isomorphism of Lemma 10.1,

$$\mathrm{Hom}_{\mathrm{Ch}(\mathcal{A})}\left(D^{n+1}(Q), X\right) \cong \mathrm{Hom}_{\mathcal{A}}(Q, X_{n+1}),$$

such a morphism s is equivalent to a chain map $D^{n+1}(Q) \to X$ extending over the inclusion $S^n(Q) \rightarrowtail D^{n+1}(Q)$. In this way, (1) and (2) are equivalent.

For condition (3), consider the obvious short exact sequence

$$S^n(Q) \rightarrowtail D^{n+1}(Q) \twoheadrightarrow S^{n+1}(Q).$$

Being degreewise split it is a short exact sequence in the exact category $\mathrm{Ch}(\mathcal{A})_{\mathcal{E}}$. Applying $\mathrm{Hom}_{\mathrm{Ch}(\mathcal{A})}(-, X)$ induces an exact sequence of abelian groups

$$\mathrm{Hom}_{\mathrm{Ch}(\mathcal{A})}\left(D^{n+1}(Q), X\right) \to \mathrm{Hom}_{\mathrm{Ch}(\mathcal{A})}(S^n(Q), X) \to \mathrm{Ext}^1_{\mathrm{Ch}(\mathcal{E})}\left(S^{n+1}(Q), X\right)$$

$$\to \mathrm{Ext}^1_{\mathrm{Ch}(\mathcal{E})}\left(D^{n+1}(Q), X\right) \cong \mathrm{Ext}^1_{\mathcal{E}}(Q, X_{n+1}),$$

where the last isomorphism comes from Lemma 10.5. Now if (3) is true, then $\mathrm{Hom}_{\mathrm{Ch}(\mathcal{A})}\left(D^{n+1}(Q), X\right) \to \mathrm{Hom}_{\mathrm{Ch}(\mathcal{A})}(S^n(Q), X)$ is onto, proving (3) implies (2). On the other hand, we see (2) implies (3) under the added assumption that

$$\mathrm{Ext}^1_{\mathcal{E}}(Q, X_n) = 0$$

for all n and $Q \in \mathcal{Q}$. Indeed exactness of the long exact sequence forces

$$\mathrm{Ext}^1_{\mathrm{Ch}(\mathcal{E})}(S^{n+1}(Q), X) = 0.$$

□

Lemma 10.33 *Suppose $(\mathcal{Q}, \mathcal{R})$ is a cotorsion pair in an exact category $(\mathcal{A}, \mathcal{E})$.*

(1) *$\widetilde{\mathcal{R}}$ contains all disks $D^n(R)$ on objects $R \in \mathcal{R}$; more generally, it contains all contractible complexes with components in $\mathcal{R}$.*
(2) *$dg\widetilde{\mathcal{Q}}$ contains all spheres $S^n(Q)$ and disks $D^n(Q)$ on objects $Q \in \mathcal{Q}$, and all contactible complexes with components in $\mathcal{Q}$.*

Proof We prove (1). For any object $A \in \mathcal{A}$ and any contractible complex C, we have

$$H_n[\mathrm{Hom}_{\mathcal{A}}(A, C)] = H_n\left[\mathit{Hom}\left(S^0(A), C\right)\right] = K(\mathcal{A})\left(S^0(A), \Sigma^{-n}C\right) = 0$$

by Lemma 10.3 and Corollary 10.13. So in particular we will have $\mathrm{Hom}_{\mathcal{A}}(Q, C)$ is acyclic for any $Q \in \mathcal{Q}$ and contractible C with components in $\mathcal{R}$.

Now we prove (2). We recall that $dg\widetilde{\mathcal{Q}} := {}^{\perp}\widetilde{\mathcal{R}}$, the left Ext-orthogonal in the exact category $\mathrm{Ch}(\mathcal{A})_{\mathcal{E}}$. Given $S^n(Q)$ with $Q \in \mathcal{Q}$, we need to show

$$\mathrm{Ext}^1_{\mathrm{Ch}(\mathcal{E})}(S^n(Q), X) = 0$$

for any $X \in \widetilde{\mathcal{R}}$. But for such an X we have

$$\mathrm{Ext}^1_{\mathrm{Ch}(\mathcal{E})}(S^n(Q), X) = \mathrm{Ext}^1_{dw}(S^n(Q), X) = \mathrm{Ext}^1_{dw}\left(S^0(Q), \Sigma^{-n}X\right)$$

$$\cong H_{n-1}\left[\mathit{Hom}\left(S^0(Q), X\right)\right] = H_{n-1}[\mathrm{Hom}_{\mathcal{A}}(Q, X)] = 0,$$

where the isomorphism comes from Corollary 10.22. Now if C is a contractible complex with components in $\mathcal{Q}$, then similarly for any $X \in \widetilde{\mathcal{R}}$ we have

$$\mathrm{Ext}^1_{\mathrm{Ch}(\mathcal{E})}(C, X) = \mathrm{Ext}^1_{dw}(C, X) \cong K(\mathcal{A})\left(C, \Sigma^1 X\right) = 0$$

by Corollaries 10.13 and 10.22. □

Note that Lemmas 10.32 and 10.33 each have duals which we leave for the reader to formulate. We get the following result.

Proposition 10.34 (Lifting Cotorsion Pairs to Chain Complexes) *Let $(\mathcal{Q}, \mathcal{R})$ be a cotorsion pair in an exact category $(\mathcal{A}, \mathcal{E})$. Then $\left(dg\widetilde{\mathcal{Q}}, \widetilde{\mathcal{R}}\right)$ and $\left(\widetilde{\mathcal{Q}}, dg\widetilde{\mathcal{R}}\right)$ are each cotorsion pairs in the exact category* $\mathrm{Ch}(\mathcal{A})_{\mathcal{E}}$. *Their cores, $dg\widetilde{\mathcal{Q}} \cap \widetilde{\mathcal{R}} = \widetilde{\mathcal{Q}} \cap dg\widetilde{\mathcal{R}}$, each coincide with the class of all contractible complexes with components in $\mathcal{Q} \cap \mathcal{R}$.*

Proof We prove that $\left(dg\widetilde{\mathcal{Q}}, \widetilde{\mathcal{R}}\right)$ is a cotorsion pair. By definition, we already have $dg\widetilde{\mathcal{Q}} = {}^{\perp}\widetilde{\mathcal{R}}$. So we wish to show $\left(dg\widetilde{\mathcal{Q}}\right)^{\perp} = \widetilde{\mathcal{R}}$. Note that $\widetilde{\mathcal{R}} \subseteq \left(dg\widetilde{\mathcal{Q}}\right)^{\perp}$ is automatic, so the crucial containment is to show $\left(dg\widetilde{\mathcal{Q}}\right)^{\perp} \subseteq \widetilde{\mathcal{R}}$. So assume X satisfies $\mathrm{Ext}^1_{\mathcal{E}}(C, X) = 0$ for all complexes $C \in dg\widetilde{\mathcal{Q}}$. By Lemma 10.33 we may take $C = D^n(Q)$ for arbitrary $Q \in \mathcal{Q}$, and by applying Lemma 10.5 we have

$$0 = \mathrm{Ext}^1_{\mathrm{Ch}(\mathcal{E})}(D^n(Q), X) \cong \mathrm{Ext}^1_{\mathcal{E}}(Q, X_n).$$

Therefore each $X_n \in \mathcal{R}$. Next, by Lemma 10.33 we may take $C = S^n(Q)$ for arbitrary $Q \in \mathcal{Q}$, and by Lemma 10.32 we may conclude $\mathrm{Hom}_{\mathcal{A}}(Q, X)$ is acyclic for all $Q \in \mathcal{Q}$. This proves $X \in \widetilde{\mathcal{R}}$, and completes the proof that $\left(dg\widetilde{\mathcal{Q}}, \widetilde{\mathcal{R}}\right)$ is a cotorsion pair.

Next we show that the core, $dg\widetilde{\mathcal{Q}} \cap \widetilde{\mathcal{R}}$, is precisely the class of all contractible complexes with components in $\mathcal{Q} \cap \mathcal{R}$. First, it follows from Lemma 10.33 that any contractible complex with components in $\mathcal{Q} \cap \mathcal{R}$ must be in $dg\widetilde{\mathcal{Q}} \cap \widetilde{\mathcal{R}}$. On the other hand, suppose $X \in dg\widetilde{\mathcal{Q}} \cap \widetilde{\mathcal{R}}$. Then $\Sigma^{-1}X \in dg\widetilde{\mathcal{Q}} \cap \widetilde{\mathcal{R}}$ too. Since $\left(dg\widetilde{\mathcal{Q}}, \widetilde{\mathcal{R}}\right)$ is a cotorsion pair we have

$$\mathrm{Ext}^1_{\mathrm{Ch}(\mathcal{E})}\left(X, \Sigma^{-1}X\right) = 0,$$

and so Corollary 10.22 gives us

$$0 = \mathrm{Ext}^1_{dw}\left(X, \Sigma^{-1}X\right) \cong H_0 Hom(X, X) = K(\mathcal{A})(X, X).$$

Therefore X is a contractible complex and certainly each X_n must be in $\mathcal{Q} \cap \mathcal{R}$.

On the other hand, Lemmas 10.32 and 10.33 have duals as previously noted. Then dualizing the preceding proof we see that $\left(\widetilde{\mathcal{Q}}, dg\widetilde{\mathcal{R}}\right)$ is also a cotorsion pair with the same core as $\left(dg\widetilde{\mathcal{Q}}, \widetilde{\mathcal{R}}\right)$. □

Exercise 10.7.1 Show that $dg\widetilde{\mathcal{R}}$ coincides with the class of all complexes Y with $Y_n \in \mathcal{R}$ such that any chain map $X \to Y$ with $X \in \widetilde{\mathcal{Q}}$ is null homotopic. The dual holds for $dg\widetilde{\mathcal{Q}}$.

Generating Sets and Hom-Acyclicity In abelian categories, we often use a set of generators to detect acyclicity of a chain complex. In particular, if $\mathcal{U}$ is a set of generators, then $\mathrm{Hom}_{\mathcal{A}}(\mathcal{U}, -)$-acyclicity of a complex X implies the usual acyclicity of X. We would like to be able to do the same, detect acyclicity in the sense of Definition 10.6, by testing for $\mathrm{Hom}_{\mathcal{A}}(\mathcal{U}, -)$-acyclicity where $\mathcal{U}$ is some set of admissible generators for $(\mathcal{A}, \mathcal{E})$. We can do this fairly easily if we are willing to assume that each morphism of our underlying category $\mathcal{A}$ admits a kernel. Under this assumption, given a complex X, let us use the notation $Z_nX := \mathrm{Ker}\, d_n$ and denote the (not necessarily admissible) monomorphism by $k_n : Z_nX \to X_n$. Being a complex, we have $d_{n-1}d_n = 0$, and so for each n we have a unique morphism

$$d'_n : X_n \to Z_{n-1}X$$

satisfying $k_{n-1}d'_n = d_n$. Dual statements hold if each morphism of $\mathcal{A}$ admits a cokernel. In this case, given a complex X, we use the notation from Section 10.1, $C_nX := \mathrm{Cok}\, d_{n+1}$. Denoting the (not necessarily admissible) epimorphism by $c_{n+1} : X_n \to C_nX$, we obtain unique morphisms $l_n : C_nX \to X_{n-1}$ satisfying $d_n = l_n c_{n+1}$ for each n.

Lemma 10.35 *Assume each morphism of the underlying additive category $\mathcal{A}$ admits a kernel and that $\mathcal{Q}$ contains a class of admissible generators $\mathcal{U} = \{U_i\}_{i \in I}$. Then $\widetilde{\mathcal{R}}$ coincides with the class all $\mathcal{E}$-acyclic complexes X with $Z_nX \in \mathcal{R}$ (equivalently, $C_nX \in \mathcal{R}$).*

Dually, assume each morphism of $\mathcal{A}$ admits a cokernel and $\mathcal{R}$ contains a class of admissible cogenerators $\mathcal{V} = \{V_i\}_{i \in I}$. Then $\widetilde{\mathcal{Q}}$ coincides with the class of all $\mathcal{E}$-acyclic complexes X with $C_nX \in \mathcal{Q}$ (equivalently, $Z_nX \in \mathcal{Q}$).

Proof We will prove the first statement. Using the earlier notation set we will deduce that each d'_n is an admissible epimorphism with kernel k_n. This will prove k_n is an admissible monic and that X is acyclic in the sense of Definition 10.6.

Now since k_{n-1} is a kernel, it must be a monomorphism (left cancellable) and so it is easy to see that $k_n : Z_nX \to X_n$ is not just the kernel of $d_n : X_n \to X_{n-1}$ but also the kernel of $d'_n : X_n \to Z_{n-1}X$. Given any morphism $f' : U_i \to Z_{n-1}X$, define $f : U_i \to X_{n-1}$ to be $f = k_{n-1}f'$. Then f satisfies $d_{n-1}f = 0$, and since $\mathrm{Hom}_{\mathcal{A}}(U_i, X)$ is an acyclic complex of abelian groups, there exists a lift $\hat{f} : U_i \to X_n$ such that $d_n\hat{f} = f$. Therefore, $k_{n-1}d'_n\hat{f} = k_{n-1}f'$. Since k_{n-1} is left cancellable we get $d'_n\hat{f} = f'$. This shows that $d'_n : X_n \to Z_{n-1}X$ is $\mathcal{U}$-epic in the sense of Definition 9.26. Since $\mathcal{U}$ is a class of admissible generators it means that $d'_n : X_n \twoheadrightarrow Z_{n-1}X$ is an admissible epic. Since its kernel is k_n, we conclude it must be an admissible monomorphism. All together we see that d_n is the composite

$$X_n \xrightarrow{d'_n} Z_{n-1}X \xrightarrow{k_{n-1}} X_{n-1}$$

of an admissible monic with an admissible epic. This proves that X is acyclic with respect to the exact structure.

Now it is easy to check that $\widetilde{\mathcal{R}}$ is precisely the class of all acyclic complexes X with cycles $Z_nX \in \mathcal{R}$: Consider the long exact sequence obtained by applying $\mathrm{Hom}_{\mathcal{A}}(Q,-)$ to the short exact sequence $Z_nX \overset{k_n}{\rightarrowtail} X_n \overset{d'_n}{\twoheadrightarrow} Z_{n-1}X$. □

Recall that an additive category $\mathcal{A}$ is called *pre-abelian* if every morphism has a kernel and a cokernel. This class of categories includes not just abelian categories but also the quasi-abelian and semi-abelian categories appearing in topology and functional analysis.

Corollary 10.36 (Lifted Cotorsion Pairs on Pre-abelian Categories) *Let $(\mathcal{Q},\mathcal{R})$ be a cotorsion pair on an exact category $(\mathcal{A},\mathcal{E})$ whose underlying category $\mathcal{A}$ is pre-abelian. Assume $\mathcal{Q}$ contains a class of admissible generators for $\mathcal{E}$ and $\mathcal{R}$ contains a class of admissible cogenerators for $\mathcal{E}$. Then we have the following.*

(1) *$\left(dg\widetilde{\mathcal{Q}},\widetilde{\mathcal{R}}\right)$ is a cotorsion pair in the exact category $\mathrm{Ch}(\mathcal{A})_{\mathcal{E}}$ and $\widetilde{\mathcal{R}}$ coincides with the class all $\mathcal{E}$-acyclic complexes X with $Z_nX \in \mathcal{R}$.*
(2) *$\left(\widetilde{\mathcal{Q}},dg\widetilde{\mathcal{R}}\right)$ is a cotorsion pair in the exact category $\mathrm{Ch}(\mathcal{A})_{\mathcal{E}}$ and $\widetilde{\mathcal{Q}}$ coincides with the class of all $\mathcal{E}$-acyclic complexes X with $Z_nX \in \mathcal{Q}$.*

Of course we would like to know whether or not the lifted cotorsion pairs are compete whenever $(\mathcal{Q},\mathcal{R})$ is complete. Although this is not an easy question for $\left(\widetilde{\mathcal{Q}},dg\widetilde{\mathcal{R}}\right)$, it is rather easy for $\left(dg\widetilde{\mathcal{Q}},\widetilde{\mathcal{R}}\right)$ if we are working with efficient exact categories.

Proposition 10.37 *Let $(\mathcal{A},\mathcal{E})$ be an efficient exact category. Assume $(\mathcal{Q},\mathcal{R})$ is the complete cotorsion pair cogenerated by a set $\mathcal{S}$ containing a set $\mathcal{U} = \{U_i\}_{i\in I}$ of admissible generators. Then $\left(dg\widetilde{\mathcal{Q}},\widetilde{\mathcal{R}}\right)$ is a complete cotorsion pair in the exact category $\mathrm{Ch}(\mathcal{A})_{\mathcal{E}}$, and it is cogenerated by the set $\{S^n(Q)\,|\,n \in \mathbb{Z},\ Q \in \mathcal{S}\}$.*

If each morphism of the underlying additive category $\mathcal{A}$ admits a kernel, then $\widetilde{\mathcal{R}}$ coincides with the class of all $\mathcal{E}$-acyclic complexes X with cycles $Z_nX \in \mathcal{R}$.

Proof Note that by Lemma 10.33, the class $dg\widetilde{\mathcal{Q}}$ contains $\{D^n(U_i)\}$, which is a set of admissible generators for the exact category $\mathrm{Ch}(\mathcal{A})_{\mathcal{E}}$. So by Corollary 9.40 we only need to show

$$\widetilde{\mathcal{R}} = \{S^n(Q)\,|\,n \in \mathbb{Z},\ Q \in \mathcal{S}\}^{\perp}$$

to conclude that $\left(dg\widetilde{\mathcal{Q}},\widetilde{\mathcal{R}}\right)$ is a complete cotorsion pair in $\mathrm{Ch}(\mathcal{A})_{\mathcal{E}}$. We can argue the key containment, $\{S^n(Q)\,|\,n \in \mathbb{Z},\ Q \in \mathcal{S}\}^{\perp} \subseteq \widetilde{\mathcal{R}}$, analogously to the proof of Proposition 10.34. Indeed for each n and $Q \in \mathcal{Q}$, we have the obvious degreewise split short exact sequence

$$S^{n-1}(Q) \rightarrowtail D^n(Q) \twoheadrightarrow S^n(Q)$$

in the exact category $\mathrm{Ch}(\mathcal{A})_{\mathcal{E}}$. Applying $\mathrm{Ext}^1_{\mathrm{Ch}(\mathcal{E})}(-,X)$ for any

$$X \in \{ S^n(Q) \mid n \in \mathbb{Z},\ Q \in \mathcal{S} \}^{\perp}$$

yields an exact sequence of abelian groups

$$\mathrm{Ext}^1_{\mathrm{Ch}(\mathcal{E})}(S^n(Q), X) \to \mathrm{Ext}^1_{\mathrm{Ch}(\mathcal{E})}(D^n(Q), X) \to \mathrm{Ext}^1_{\mathrm{Ch}(\mathcal{E})}\left(S^{n-1}(Q), X\right).$$

Therefore, $\mathrm{Ext}^1_{\mathrm{Ch}(\mathcal{E})}(D^n(Q), X) = 0$, which in turn implies $\mathrm{Ext}^1_{\mathcal{E}}(Q, X_n) = 0$, by Lemma 10.5. It follows that each $X_n \in \mathcal{R}$, and moreover the implication (3) $\Longrightarrow$ (1) of Lemma 10.32 applies. Therefore, $\{ S^n(Q) \mid n \in \mathbb{Z},\ Q \in \mathcal{S} \}^{\perp} \subseteq \widetilde{\mathcal{R}}$. □

Exercise 10.7.2 Show that any bounded below complex with components in $\mathcal{Q}$ is in the class $dg\widetilde{\mathcal{Q}}$. On the other hand, any bounded above complex with components in $\mathcal{R}$ is in the class $dg\widetilde{\mathcal{R}}$.

Exercise 10.7.3 Let $(\mathcal{A}, \mathcal{E})$ be an efficient exact category. Assume $(\mathcal{Q}, \mathcal{R})$ is the complete cotorsion pair cogenerated by a set $\mathcal{S}$ containing a set $\mathcal{U} = \{U_i\}_{i \in I}$ of admissible generators. Show that $\left({}^{\perp}dw\widetilde{\mathcal{R}}, dw\widetilde{\mathcal{R}}\right)$ is cogenerated by a set, and hence a complete cotorsion pair in $\mathrm{Ch}(\mathcal{A})_{\mathcal{E}}$. Here, $dw\widetilde{\mathcal{R}}$ is the class of all complexes X with each $X_n \in \mathcal{R}$.

10.8 Projective Generators and the Derived Category

The goal of this section is to construct a projective model structure on chain complexes when the underlying exact category possesses a nice set of projective generators. This model structure is a generalization of the standard projective model structure on unbounded chain complexes of modules over a ring. Our approach is quite general, so the result applies to some of the exact structures arising in Functional Analysis; recall Example 1.35. In general, following an idea of Jack Kelly, if kernels exist in our underlying additive category $\mathcal{A}$, then we obtain a model for the derived category of the exact structure $(\mathcal{A}, \mathcal{E})$.

Throughout this section, we let $(\mathcal{A}, \mathcal{E})$ denote a weakly idempotent complete exact category.

Proposition 10.38 *Let* $\mathrm{Ch}(\mathcal{A})_{\mathcal{E}}$ *be the category of chain complexes with the usual degreewise exact structure.*

(1) *The projective objects in the exact category* $\mathrm{Ch}(\mathcal{A})_{\mathcal{E}}$ *are precisely the contractible complexes having projective components. If* $(\mathcal{A}, \mathcal{E})$ *has enough projectives then so does* $\mathrm{Ch}(\mathcal{A})_{\mathcal{E}}$.
(2) *The injective objects in the exact category* $\mathrm{Ch}(\mathcal{A})_{\mathcal{E}}$ *are precisely the contractible complexes having injective components. If* $(\mathcal{A}, \mathcal{E})$ *has enough injectives then so does* $\mathrm{Ch}(\mathcal{A})_{\mathcal{E}}$.

Proof We prove (1). P is projective in $\mathrm{Ch}(\mathcal{A})_{\mathcal{E}}$ if and only if $\mathrm{Ext}^1_{\mathrm{Ch}(\mathcal{E})}(P, X) = 0$ for all complexes X. So if P is projective, then for all objects $A \in \mathcal{A}$ we have, by Lemma 10.5,

$$0 = \mathrm{Ext}^1_{\mathrm{Ch}(\mathcal{E})}\left(P, D^{n+1}(A)\right) \cong \mathrm{Ext}^1_{\mathcal{E}}(P_n, A)$$

for all n. Therefore, each component P_n is projective. Moreover, since

$$\mathrm{Ext}^1_{\mathrm{Ch}(\mathcal{E})}\left(P, \Sigma^{-1} P\right) = 0,$$

Corollary 10.22 gives us

$$0 = \mathrm{Ext}^1_{dw}\left(P, \Sigma^{-1} P\right) \cong H_0 Hom(P, P) = K(\mathcal{A})(P, P).$$

Therefore P is contractible. On the other hand, if P is a contractible complex with each P_n projective, then for any complex X we have

$$\mathrm{Ext}^1_{\mathrm{Ch}(\mathcal{E})}(P, X) = \mathrm{Ext}^1_{dw}(P, X) \cong H_0 Hom(P, \Sigma X) = K(\mathcal{A})(P, \Sigma X) = 0.$$

Now we show that $\mathrm{Ch}(\mathcal{A})_{\mathcal{E}}$ has enough projectives whenever $(\mathcal{A}, \mathcal{E})$ does. Given any chain complex X, by Lemma 10.16 we have a degreewise split epimorphism $\bigoplus_{n \in \mathbb{Z}} D^n(X_n) \twoheadrightarrow X$. Of course, this is an admissible epimorphism in $\mathrm{Ch}(\mathcal{A})_{\mathcal{E}}$. Since $(\mathcal{A}, \mathcal{E})$ has enough projectives we can also find admissible epimorphisms $P_n \twoheadrightarrow X_n$, with each P_n projective. Then $\bigoplus_{n \in \mathbb{Z}} D^n(P_n) \twoheadrightarrow \bigoplus_{n \in \mathbb{Z}} D^n(X_n)$ is an admissible epimorphism in $\mathrm{Ch}(\mathcal{A})_{\mathcal{E}}$. The composition of the two admissible epimorphisms proves the result. □

A set $\{P_i\}_{i \in I}$ of admissible generators with each object P_i projective will be called a set of **projective generators** for $(\mathcal{A}, \mathcal{E})$. Since P is projective if and only if $\mathrm{Ext}^1_{\mathcal{E}}(P, A) = 0$ for all objects $A \in \mathcal{A}$, the set $\{P_i\}_{i \in I}$ cogenerates a cotorsion pair $(\mathcal{P}, \mathcal{A})$ where $\mathcal{P}$ is the class of all projectives. Assuming $\mathcal{A}$ has coproducts, Proposition 9.27 together with Lemma 1.26 shows that this is a complete cotorsion pair.

Now consider the cotorsion pair, $\left(dg\widetilde{\mathcal{P}}, \widetilde{\mathcal{A}}\right)$, on $\mathrm{Ch}(\mathcal{A})_{\mathcal{E}}$, obtained by lifting the projective cotorsion pair $(\mathcal{P}, \mathcal{A})$. Then $\widetilde{\mathcal{A}}$ is the class of all complexes Y such that each $\mathrm{Hom}_{\mathcal{A}}(P_i, Y)$ is an exact complex of abelian groups. This certainly includes all $\mathcal{E}$-acyclic complexes Y, but unless $\mathcal{A}$ admits kernels, there may be more; see Exercise 10.8.2. The next theorem gives conditions guaranteeing that $(dg\widetilde{\mathcal{P}}, \widetilde{\mathcal{A}})$ determines a (cofibrantly generated) projective model structure. It is based on the following notion of smallness in additive categories with coproducts.

Definition 10.39 Let $\mathcal{A}$ be an additive category with coproducts and let κ be a regular cardinal. We say that an object $S \in \mathcal{A}$ is **κ-coproduct small**, or simply **κ-small**, if given any morphism $f : S \to \bigoplus_{i\in I} A_i$ into the coproduct of any family of objects $\{A_i\}_{i\in I}$, there exists a subset $J \subseteq I$ with $|J| < \kappa$ such that f factors as $S \xrightarrow{f'} \bigoplus_{i\in J} A_i \xrightarrow{\eta_J} \bigoplus_{i\in I} A_i$ where η_J is the canonical subcoproduct insertion. (Note that η_J is the unique morphism satisfying $\eta_J \circ \eta_i^J = \eta_i$ for each $i \in J$, where $\eta_i^J : A_i \to \bigoplus_{i\in J} A_i$ and $\eta_i : A_i \to \bigoplus_{i\in I} A_i$ are the canonical insertions.)

Theorem 10.40 *Let $\{P_i\}_{i\in I}$ be a set of projective generators for a (weakly) idempotent complete exact category $(\mathcal{A}, \mathcal{E})$ with coproducts. Assume each P_i is κ_i-small for some κ_i. Then $\left(dg\widetilde{\mathcal{P}}, \widetilde{\mathcal{A}}\right)$ is a projective cotorsion pair relative to the exact category* $\mathrm{Ch}(\mathcal{A})_{\mathcal{E}}$. *It is a cofibrantly generated cotorsion pair in the sense of Definition 9.30 as follows.*

(1) *It is cogenerated by the (efficient) set $\mathcal{S} = \{\, S^n(P_i) \mid n \in \mathbb{Z},\ i \in I \,\}$.*
(2) *A set of generating cofibrations is given by*

$$\left\{ S^n(P_i) \rightarrowtail D^{n+1}(P_i) \mid n \in \mathbb{Z},\ i \in I \right\} \cup \{\, 0 \rightarrowtail D^n(P_i) \mid n \in \mathbb{Z},\ i \in I \,\}.$$

Each complex in $dg\widetilde{\mathcal{P}}$ is a direct summand of a complex in $\mathrm{Filt}(\mathcal{S})$, *the class of all transfinite extensions of complexes in $\mathcal{S}$.*

Moreover, if each morphism of the underlying additive category $\mathcal{A}$ admits a kernel, then the class of trivial objects, $\widetilde{\mathcal{A}}$, coincides with the class of all $\mathcal{E}$-acyclic complexes. So in this case, the homotopy category associated to $\left(dg\widetilde{\mathcal{P}}, \widetilde{\mathcal{A}}\right)$ is the derived category of the exact category $(\mathcal{A}, \mathcal{E})$.

Remark 10.41 The associated abelian model structure $\mathfrak{M} = \left(dg\widetilde{\mathcal{P}}, \widetilde{\mathcal{A}}, All\right)$ satisfies the hypotheses of Theorems 10.27 and 10.29. A chain complex in $dg\widetilde{\mathcal{P}}$ is said to be a *DG-projective complex.*

Proof $\mathrm{Ch}(\mathcal{A})_{\mathcal{E}}$ is certainly weakly idempotent complete and has coproducts since $(\mathcal{A}, \mathcal{E})$ is such. So Theorem 9.34 can be applied; it says that any cofibrantly generated cotorsion pair is (functorially) complete. Now it follows from Lemma 10.32 that the set $\mathcal{S} = \{S^n(P_i)\}_{n\in\mathbb{Z}}$ satisfies $\mathcal{S}^\perp = \widetilde{\mathcal{A}}$, and moreover that $\left\{ S^n(P_i) \rightarrowtail D^{n+1}(P_i) \mid n \in \mathbb{Z},\ i \in I \right\}$ is a set of generating monics (Definition 9.28) for $\mathcal{S}^\perp = \widetilde{\mathcal{A}}$. We claim that $\mathcal{S}$ is efficient relative to $\mathrm{Ch}(\mathcal{A})_{\mathcal{E}}$, in the sense of Definition 9.11. To see this, note that any $\mathcal{S}$-cofibration is necessarily an admissible monic in $\mathrm{Ch}(\mathcal{A})_{dw}$ (i.e. a degreewise split monic), because each P_i is projective. Since coproducts exist, any transfinite composition of degreewise split monics also exists and is again a degreewise split monic; see Exercise 9.1.1. So $\mathcal{S}$ is indeed an efficient set of objects. Moreover, as pointed out in Example 9.12, in the ground category $(\mathcal{A}, \mathcal{E})$, we have $\mathrm{Filt}(\{P_i\}) = \mathrm{Free}(\{P_i\})$, the class of all coproducts of the projective generators. Since all such objects are again projective, we have that even all $\mathrm{Filt}(\mathcal{S})$-cofibrations are necessarily degreewise split monics with a projective cokernel in each degree. As a result, suppose that $A \rightarrowtail X_\lambda$ is the the transfinite composition of some λ-extension sequence

$$A = X_0 \rightarrowtail X_1 \rightarrowtail X_2 \rightarrowtail \cdots \rightarrowtail X_\alpha \rightarrowtail X_{\alpha+1} \rightarrowtail \cdots$$

of $\mathrm{Filt}(\mathcal{S})$-cofibrations. Then in each degree n, the complex X_λ is merely a coproduct whose summands may be indexed by the ordinal type $1 + \lambda = \lambda = \{\alpha \mid \alpha < \lambda\}$; again see Exercise 9.1.1. It follows that $S^n(P_i)$ is small relative to $\mathrm{Filt}(\mathcal{S})$-cofibrations: Indeed suppose λ is any regular cardinal such that $\lambda \geq \kappa_i$ and consider a morphism $S^n(P_i) \to X_\lambda$ into the direct limit of some λ-extension sequence. Then the degree-n component, $P_i \to (X_\lambda)_n$, must factor through a subcoproduct of $(X_\lambda)_n$, indexed by some subset $J \subsetneq \lambda = \{\alpha \mid \alpha < \lambda\}$ of cardinality $|J| < \kappa_i \leq \lambda$. By the regularity of λ, the set J has an upper bound $\alpha < \lambda$. Therefore, $S^n(P_i) \to X_\lambda$ factors as $S^n(P_i) \to X_\alpha \to X_\lambda$, through some stage of the λ-extension sequence.

At this point we have verified the first two conditions of Definition 9.30. As for the last condition, it is clear that the set $\mathcal{U} = \{D^n(P_i)\}_{i\in I, n\in\mathbb{Z}}$ of projective generators is in ${}^\perp(\mathcal{S}^\perp)$, so we have shown that $\left(dg\widetilde{\mathcal{P}}, \widetilde{\mathcal{A}}\right) = ({}^\perp(\mathcal{S}^\perp), \mathcal{S}^\perp)$ is a cofibrantly generated cotorsion pair. So this is a functorially complete cotorsion pair by Theorem 9.34, and every complex in $dg\widetilde{\mathcal{P}}$ is a direct summand of a complex in $\mathrm{Filt}(\{\, S^n(P_i), D^m(P_i) \mid n, m \in \mathbb{Z}\,,\ i \in I\,\})$. But note that

$$\mathrm{Filt}(\{\, S^n(P_i), D^m(P_i) \mid n, m \in \mathbb{Z}\,,\ i \in I\,\}) = \mathrm{Filt}(\mathcal{S})$$

because each $D^n(P_i)$ is easily seen to be an extension of spheres in $\mathcal{S}$. Therefore $D^n(P_i) \in \mathrm{Filt}(\mathcal{S})$. So it follows from the statement $\mathrm{Filt}(\mathrm{Filt}(\mathcal{S})) = \mathrm{Filt}(\mathcal{S})$, from Corollary 9.20. So each complex in $dg\widetilde{\mathcal{P}}$ is a direct summand of one in $\mathrm{Filt}(\mathcal{S})$.

Let us show that $\widetilde{\mathcal{A}}$ is thick. For this, consider a short exact sequence $X \rightarrowtail Y \twoheadrightarrow Z$ in the exact category $\mathrm{Ch}(\mathcal{A})_{\mathcal{E}}$. For projectives P, the functor $\mathrm{Hom}_{\mathcal{A}}(P, -)$ sends short exact sequences in $\mathcal{E}$ to short exact sequences of abelian groups. Therefore, we have a short exact sequence

$$0 \to \mathrm{Hom}_{\mathcal{A}}(P, X) \to \mathrm{Hom}_{\mathcal{A}}(P, Y) \to \mathrm{Hom}_{\mathcal{A}}(P, Z) \to 0$$

of complexes of abelian groups. By the corresponding long exact sequence of homology groups, if any two out of three of these complexes are exact, then so must be the third. It is also easy to see that the class $\widetilde{\mathcal{A}}$, of all $\mathrm{Hom}_{\mathcal{A}}(\mathcal{P}, -)$-acyclic complexes, is closed under direct summands. Therefore, the class $\widetilde{\mathcal{A}}$ is thick.

It follows from Propositions 10.34 and 10.38 that $dg\widetilde{\mathcal{P}} \cap \widetilde{\mathcal{A}}$ coincides with the class of projectives in the exact category $\mathrm{Ch}(\mathcal{A})_{\mathcal{E}}$, and moreover that $\mathrm{Ch}(\mathcal{A})_{\mathcal{E}}$ has enough projectives. Therefore we have shown that $\left(dg\widetilde{\mathcal{P}}, \widetilde{\mathcal{A}}\right)$ is a projective cotorsion pair relative to $\mathrm{Ch}(\mathcal{A})_{\mathcal{E}}$. □

We obtain the following statement for pre-abelian categories. These include the locally presentable additive categories discussed in Chapter 12.

Corollary 10.42 *Let $\{P_i\}_{i\in I}$ be a set of projective generators for a pre-abelian exact category $(\mathcal{A}, \mathcal{E})$ with coproducts. Assume each P_i is κ_i-small for some κ_i. Then $\left(dg\widetilde{\mathcal{P}}, \widetilde{\mathcal{A}}\right)$ is a projective cotorsion pair on* $\mathrm{Ch}(\mathcal{A})_{\mathcal{E}}$ *and determines a cofibrantly generated abelian model structure. Its homotopy category is $\mathcal{D}(\mathcal{A}, \mathcal{E})$, the derived category of $(\mathcal{A}, \mathcal{E})$.*

We note that the projective model structure of Corollary 10.42 exists on any Grothendieck category possessing a set of projective generators. (In fact, even on locally presentable categories with projective generators; Exercise 12.1.1.) In particular, it recovers the standard projective model structure for the derived category of a ring R; see Example 3 from the Introduction and Main Examples. More generally, we have the following.

Example 10.43 (The Derived Category with Respect to a Generator) Let $\mathcal{G}$ be a Grothendieck category and let $\mathcal{U} = \{U_i\}_{i\in I}$ denote of any set of generators. Then $U = \bigoplus_{i\in I} U_i$ is a generator. Let $R = \mathrm{Hom}_{\mathcal{G}}(U, U)$ be the endomorphism ring. For each $A \in \mathcal{G}$, set $GA = \mathrm{Hom}_{\mathcal{G}}(U, A)$. Then GA is a right R-module, and for each morphism $f \colon A \to B$ we obtain a homomorphism $f_* \colon \mathrm{Hom}_{\mathcal{G}}(U, A) \to \mathrm{Hom}_{\mathcal{G}}(U, B)$ of right R-modules. Part of the Gabriel–Popescu Theorem is that the functor $G \colon \mathcal{G} \to \text{Mod-}R$ has a left adjoint. Using this one can deduce each of the following.

(1) Let $\mathcal{E}$ denote the proper class of all short exact sequences $\mathbb{E}$ for which $\mathrm{Hom}_{\mathcal{G}}(U, \mathbb{E})$ remains a short exact sequence. Then $(\mathcal{G}, \mathcal{E})$ is an exact category; see Exercise 1.5.3.
(2) An object is projective in the exact category $(\mathcal{G}, \mathcal{E})$ if and only if it is a direct summand of some coproduct of elements of $\mathcal{U} = \{U_i\}_{i\in I}$. In particular, $\mathcal{U} = \{U_i\}_{i\in I}$ is a set of projective generators for $(\mathcal{G}, \mathcal{E})$.
(3) The hypotheses of Corollary 10.42 hold. The corresponding derived category is the triangulated localization with respect to the class of all acyclic

chain complexes Y for which $\mathrm{Hom}_{\mathcal{G}}(U_i, Y)$ remains exact for all $U_i \in \mathcal{U}$. We think of it as the derived category with respect to the generating set $\mathcal{U}$.

Exercise 10.8.1 Show that the homotopy category of the projective model structure on $\mathrm{Ch}(\mathcal{A})_{\mathcal{E}}$, from Theorem 10.40, is compactly generated whenever the set $\{P_i\}_{i\in I}$ of projective generators has each P_i finitely generated. Use Theorem 8.42 and Exercise 10.4.1.

The point of the next exercise is to give an easy example of an efficient exact category with a projective generator for which $\widetilde{\mathcal{A}}$ strictly contains the class of $\mathcal{E}$-acyclic complexes.

Exercise 10.8.2 Let R be a ring and consider R-Mod, the category of all (say left) R-modules. Let Flat(R) be the full subcategory of R-Mod consisting of all flat R-modules. Consider Flat(R) along with its inherited exact structure: The short exact sequences are the usual ones but with all three terms in the subcategory Flat(R). Note that a morphism in Flat(R) may not necessarily admit a (flat) kernel. Show, however, that $\{R\}$ is an (admissible) projective generator for the exact structure and describe the complexes in $dg\widetilde{\mathcal{P}}$ and $\widetilde{\mathcal{A}}$ arising in the cotorsion pair from Theorem 10.40.

Example 10.44 (Functional Analysis) There are examples of nonabelian but still quasi-abelian categories arising in Functional Analysis that satisfy the hypotheses of Corollary 10.42. See Jack Kelly's work in Kelly [2024].

10.9 Models for the Derived Category of a Grothendieck Category

Throughout this section we let $\mathcal{G}$ denote a Grothendieck category, and let $\mathrm{Ch}(\mathcal{G})$ denote its associated category of chain complexes. We will consider the problem of lifting a cotorsion pair $(\mathcal{Q}, \mathcal{R})$ in $\mathcal{G}$ to an abelian model structure $\left(dg\widetilde{\mathcal{Q}}, \widetilde{\mathcal{E}}, dg\widetilde{\mathcal{R}}\right)$ on $\mathrm{Ch}(\mathcal{G})$. Since $\widetilde{\mathcal{E}}$ denotes the class of all exact (i.e. acyclic) chain complexes, this is a model for $\mathcal{D}(\mathcal{G})$, the derived category of $\mathcal{G}$. This section assumes a little bit more of the reader in terms of their familiarity with the basics of Grothendieck categories. However, with the exception of taking Theorem 10.45 for granted, the requirements are still quite minimal.

In particular, we use the following well-known fact without proof. It is equivalent to the statement that every chain complex $X \in \mathrm{Ch}(\mathcal{G})$ is quasi-isomorphic to a K-injective complex with injective components (i.e. a DG-injective complex).

Theorem 10.45 (Existence of Exact Precovers) *Every chain complex $X \in \mathrm{Ch}(\mathcal{G})$ has a special exact precover. That is, there exists a short exact sequence $J \rightarrowtail E \twoheadrightarrow X$ with $E \in \widetilde{\mathcal{E}}$ and $J \in \widetilde{\mathcal{E}}^{\perp}$.*

The class $\widetilde{\mathcal{E}}^{\perp} = dg\widetilde{\mathcal{I}}$ is precisely the class of DG-injective complexes. So the statement is asserting that the cotorsion pair $\left(\widetilde{\mathcal{E}}, dg\widetilde{\mathcal{I}}\right)$, which is the result of lifting to complexes the canonical injective cotorsion pair $(All, \mathcal{I})$ on $\mathcal{G}$, has enough projectives. The reader is urged to work out the following exercise which outlines a proof of Theorem 10.45 for the case that $\mathcal{G}$ is the category of modules over a ring R. The argument may be imitated for other concrete Gothendieck categories. A general proof of Theorem 10.45 can be found, for example, in Gillespie [2007, corollary 7.1], but we won't cloud the simplicity behind the main idea here.

Exercise 10.9.1 Prove Theorem 10.45 for the case that $\mathcal{G}$ is the category of modules over a ring R, as follows: Let κ be a regular cardinal number satisfying $\kappa \geq \max\{|R|, \omega\}$. Show that for each nonzero exact chain complex $E \in \widetilde{\mathcal{E}}$, there exists a nonzero exact subcomplex $E' \subseteq E$ with each $|E'_n| \leq \kappa$. Deduce that there exists a set $\widetilde{\mathcal{E}}_{\kappa}$ for which each $E \in \widetilde{\mathcal{E}}$ is a transfinite extension of complexes in $\widetilde{\mathcal{E}}_{\kappa}$. Conclude that $\left(\widetilde{\mathcal{E}}, \widetilde{\mathcal{E}}^{\perp}\right)$ is necessarily a complete cotorsion pair in $\mathrm{Ch}(R)$.

Now suppose we are given a complete cotorsion pair, $(\mathcal{Q}, \mathcal{R})$, in $\mathcal{G}$. Corollary 10.36 implies that it lifts to two cotorsion pairs, $\left(dg\widetilde{\mathcal{Q}}, \widetilde{\mathcal{R}}\right)$ and $\left(\widetilde{\mathcal{Q}}, dg\widetilde{\mathcal{R}}\right)$, on $\mathrm{Ch}(\mathcal{G})$. Indeed any Grothendieck category possesses an injective cogenerator, and we can use that $(\mathcal{Q}, \mathcal{R})$ has enough projectives to show that $\mathcal{Q}$ possesses a generating set for $\mathcal{G}$.

Lemma 10.46 *Every chain map $f : X \to Y$ with $X \in \widetilde{\mathcal{Q}}$ and $Y \in \widetilde{\mathcal{R}}$ is homotopic to 0. In particular, $\mathrm{Ext}^1_{\mathrm{Ch}(\mathcal{G})}\left(\widetilde{\mathcal{Q}}, \widetilde{\mathcal{R}}\right) = 0$.*

Proof Let $X \in \widetilde{\mathcal{Q}}$ and $Y \in \widetilde{\mathcal{R}}$, and let $f : X \to Y$ be a chain map. The proof is in two stages. First we show that we can replace f with a homotopic map g which satisfies $d_n g_n = 0$ and $g_n d_{n+1} = 0$. Then we show that any map $g : X \to Y$ with this property is homotopic to 0.

The map $f_n : X_n \to Y_n$ restricts to $\hat{f}_n : Z_n X \to Z_n Y$, and

$$Z_{n+1}Y \rightarrowtail Y_{n+1} \twoheadrightarrow Z_n Y$$

is an exact sequence of objects in $\mathcal{R}$. So

$$0 \to \mathrm{Hom}_{\mathcal{G}}(Z_n X, Z_{n+1} Y) \to \mathrm{Hom}_{\mathcal{G}}(Z_n X, Y_{n+1}) \to \mathrm{Hom}_{\mathcal{G}}(Z_n X, Z_n Y) \to 0$$

is a short exact sequence in **Ab**. Therefore we have $\alpha_n \colon Z_nX \to Y_{n+1}$ such that $d_{n+1}\alpha_n = \hat{f}_n$.

Now also $Z_nX \rightarrowtail X_n \twoheadrightarrow Z_{n-1}X$ is an exact sequence of objects in $\mathcal{Q}$. So $0 \to \mathrm{Hom}_{\mathcal{G}}(Z_{n-1}X, Y_{n+1}) \to \mathrm{Hom}_{\mathcal{G}}(X_n, Y_{n+1}) \to \mathrm{Hom}_{\mathcal{G}}(Z_nX, Y_{n+1}) \to 0$ is exact and there exists $\beta_n \colon X_n \to Y_{n+1}$ which equals α_n when restricted to Z_nX.

Now set $g_n = f_n - (d_{n+1}\beta_n + \beta_{n-1}d_n)$. It is easy to see that $g = \{g_n\}_{n\in\mathbb{Z}}$ is a chain map. It is homotopic to f since $f_n - g_n = d_{n+1}\beta_n + \beta_{n-1}d_n$. Furthermore, a straightforward computation shows it satisfies $d_ng_n = 0$ and $g_nd_{n+1} = 0$.

The remainder of the proof shows that whenever we have a chain map g such that $dg = 0 = gd$, then g is homotopic to 0. Indeed we know that $\mathrm{Im}\, g_n \subset \mathrm{Ker}\, d_n$ and $\mathrm{Im}\, d_{n+1} = \mathrm{Ker}\, d_n \subset \mathrm{Ker}\, g_n$. This allows us to define a map $\bar{g}_n \colon X_n/Z_nX \to Z_nY$ which makes the following diagram commute:

$$\begin{array}{ccccc} X_n & = & X_n & = & X_n \\ {\scriptstyle d_n}\downarrow & & {\scriptstyle \pi}\downarrow & & \downarrow{\scriptstyle g_n} \\ Z_{n-1}X & \xleftarrow[\bar{d}_n]{\cong} & X_n/Z_nX & \xrightarrow[\bar{g}_n]{} & Z_nY. \end{array}$$

If we set $\hat{g}_n := \bar{g}_n\bar{d}_n^{-1}$, then $\hat{g}_n \colon Z_{n-1}X \to Z_nY$ and $\hat{g}_nd_n = g_n$. Now (using the argument as previously to obtain the maps α_n) there exists a map $\delta_n \colon Z_{n-1}X \to Y_{n+1}$ such that $d_{n+1}\delta_n = \hat{g}_n$. One can easily check that the maps $\delta_nd_n \colon X_n \to Y_{n+1}$ are a homotopy from g to 0. □

Lemma 10.47 *If $(\mathcal{Q}, \mathcal{R})$ is a complete hereditary cotorsion pair on $\mathcal{G}$, then $\left(\widetilde{\mathcal{Q}}, dg\widetilde{\mathcal{R}}\right)$ is also a complete hereditary cotorsion pair on* $\mathrm{Ch}(\mathcal{G})$. *Its core, $\widetilde{\mathcal{Q}} \cap dg\widetilde{\mathcal{R}}$, is precisely the class of all contractible complexes with components in $\mathcal{Q} \cap \mathcal{R}$.*

Proof We already know $\left(\widetilde{\mathcal{Q}}, dg\widetilde{\mathcal{R}}\right)$ is a cotorsion pair. Let us first show that it has enough projectives. Using the result in Theorem 10.45, there exists a short exact sequence of chain complexes $J \rightarrowtail E \twoheadrightarrow X$, with $E \in \widetilde{\mathcal{E}}$ and $J \in \widetilde{\mathcal{E}}^{\perp}$. For each integer n, using enough projectives of the cotorsion pair $(\mathcal{Q}, \mathcal{R})$, we choose a short exact sequence $Z_nR \rightarrowtail Z_nQ \twoheadrightarrow Z_nE$ with $Z_nR \in \mathcal{R}$ and $Z_nQ \in \mathcal{Q}$. (We use the Z_n notation because we will see shortly that these will become the cycles of acyclic complexes.) Since each $Z_nE \rightarrowtail E_n \twoheadrightarrow Z_{n-1}E$ is exact and $(\mathcal{Q}, \mathcal{R})$ is hereditary, the Generalized Horseshoe Lemma 2.18 allows us to extend each of these to a commutative diagram

$$\begin{array}{ccccc} Z_nR & \rightarrowtail & Z_nQ & \twoheadrightarrow & Z_nE \\ \downarrow & & \downarrow & & \downarrow \\ R_n & \rightarrowtail & Q_n & \twoheadrightarrow & E_n \\ \downarrow & & \downarrow & & \downarrow \\ Z_{n-1}R & \rightarrowtail & Z_{n-1}Q & \twoheadrightarrow & Z_{n-1}E \end{array}$$

whose rows and columns are all short exact sequences. We can paste all these diagrams together in an obvious way to obtain a short exact sequence of chain complexes $R \rightarrowtail Q \twoheadrightarrow E$ with $R \in \widetilde{\mathcal{R}}$ and $Q \in \widetilde{\mathcal{Q}}$. We claim that the composite $Q \twoheadrightarrow E \twoheadrightarrow X$ is a special $\widetilde{\mathcal{Q}}$-precover of X. That is, we claim its kernel is in $\widetilde{\mathcal{Q}}^{\perp} = dg\widetilde{\mathcal{R}}$. Indeed by Corollary 1.15 the kernel of the composite, denote it by K, sits in a short exact sequence $R \rightarrowtail K \twoheadrightarrow J$. We have $R \in dg\widetilde{\mathcal{R}}$ since $\widetilde{\mathcal{R}} \subseteq dg\widetilde{\mathcal{R}}$, by Lemma 10.46. Since $\widetilde{\mathcal{Q}} \subseteq \widetilde{\mathcal{E}}$, we also have $J \in dg\widetilde{\mathcal{R}}$. Since $dg\widetilde{\mathcal{R}}$ is closed under extensions we conclude $K \in dg\widetilde{\mathcal{R}}$, proving that the cotorsion pair $(\widetilde{\mathcal{Q}}, dg\widetilde{\mathcal{R}})$ has enough projectives.

We now can use Salce's Trick to show that $(\widetilde{\mathcal{Q}}, dg\widetilde{\mathcal{R}})$ also has enough injectives. To do so, we start by using Lemma 10.16 to write the short exact sequence

$$X \rightarrowtail \bigoplus_{n\in\mathbb{Z}} D^{n+1}(X_n) \twoheadrightarrow \Sigma X$$

for an arbitrary complex X. By embedding each $X_n \rightarrowtail R_n$ in an object of $\mathcal{R}$, we obtain a monomorphism $\bigoplus_{n\in\mathbb{Z}} D^{n+1}(X_n) \rightarrowtail \bigoplus_{n\in\mathbb{Z}} D^{n+1}(R_n)$, and so the composite $X \rightarrowtail \bigoplus_{n\in\mathbb{Z}} D^{n+1}(X_n) \rightarrowtail \bigoplus_{n\in\mathbb{Z}} D^{n+1}(R_n)$ is a monomorphism. So we have a short exact sequence $X \rightarrowtail \bigoplus_{n\in\mathbb{Z}} D^{n+1}(R_n) \twoheadrightarrow C$, where C is the cokernel of the composite. Using what was shown earlier we can write a short exact sequence $R \rightarrowtail Q \twoheadrightarrow C$ where $Q \in \widetilde{\mathcal{Q}}$ and $R \in dg\widetilde{\mathcal{R}}$. Constructing the pullback diagram

$$\begin{array}{ccccc} & & R & = & R \\ & & \downarrow & & \downarrow \\ X & \rightarrowtail & P & \twoheadrightarrow & Q \\ \| & & \downarrow & & \downarrow \\ X & \rightarrowtail & \bigoplus_{n\in\mathbb{Z}} D^{n+1}(R_n) & \twoheadrightarrow & C, \end{array}$$

we have $R \in dg\widetilde{\mathcal{R}}$ and $\bigoplus_{n\in\mathbb{Z}} D^{n+1}(R_n) \in dg\widetilde{\mathcal{R}}$ too, by Lemma 10.33. So the extension P is in $dg\widetilde{\mathcal{R}}$. Therefore, the middle row shows that $(\widetilde{\mathcal{Q}}, dg\widetilde{\mathcal{R}})$ also has enough injectives.

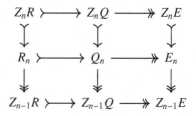

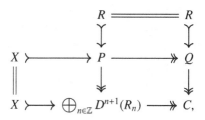

Finally, by Proposition 10.34 we have that the core, $\widetilde{Q} \cap dg\widetilde{\mathcal{R}}$, is precisely the class of all contractible complexes with components in $Q \cap \mathcal{R}$. □

Lemma 10.48 *If $(Q, \mathcal{R})$ is an hereditary cotorsion pair on a Grothendieck category, then $dg\widetilde{Q} \cap \widetilde{\mathcal{E}} = \widetilde{Q}$.*

Proof We have $\widetilde{Q} \subseteq dg\widetilde{Q} \cap \widetilde{\mathcal{E}}$, by Lemma 10.46. Conversely, let $X \in dg\widetilde{Q} \cap \widetilde{\mathcal{E}}$. To show $X \in \widetilde{Q}$, we will show $\mathrm{Ext}^1_{\mathrm{Ch}(\mathcal{G})}(Z_n X, R) = 0$ for all $R \in \mathcal{R}$. Let $I_\circ$ be an augmented injective coresolution of R,

$$I_\circ = 0 \to R \to I_{-1} \to I_{-2} \to I_{-3} \to \cdots$$

considered as a chain complex with R concentrated in degree 0. Since $(Q, \mathcal{R})$ is hereditary, $I_\circ \in \widetilde{\mathcal{R}}$. Now $Z_{n+1}X \rightarrowtail X_{n+1} \twoheadrightarrow Z_n X$ is a short exact sequence and $\mathrm{Ext}^1_{\mathcal{G}}(X_n, R) = 0$ for all $R \in \mathcal{R}$, so we will be done if we can show that any morphism $f\colon Z_{n+1}X \to R$ extends over X_{n+1}. But any $f\colon Z_{n+1}X \to R$ induces a chain map $X \to \Sigma^{n+2} I_\circ$. (Check! This follows just like the proof of the usual "comparison theorem" for injective coresolutions. Use the fact that X is exact, and the injectivity of each I_i, to inductively build extensions downward.) Since $X \in dg\widetilde{Q}$ and $\Sigma^{n+2} I_\circ \in \widetilde{\mathcal{R}}$, this chain map must be homotopic to 0. The existence of a chain homotopy $\{s_n\}$ implies that $f\colon Z_{n+1}X \to R$ does extend over $Z_{n+1}X \rightarrowtail X_{n+1}$. □

Theorem 10.49 *Let $\mathcal{G}$ be a Grothendieck category. Assume $(Q, \mathcal{R})$ is an hereditary cotorsion pair, cogenerated by a set S containing a set $\mathcal{U} = \{U_i\}_{i \in I}$ of generators for $\mathcal{G}$. Then $(Q, \mathcal{R})$ lifts to an hereditary abelian model structure on* $\mathrm{Ch}(\mathcal{G})$*:*

$$\mathfrak{M} = \left(dg\widetilde{Q}, \widetilde{\mathcal{E}}, dg\widetilde{\mathcal{R}}\right).$$

Its homotopy category is $\mathcal{D}(\mathcal{G})$, the usual derived category of $\mathcal{G}$, and

$$\{ S^n(Q) \mid n \in \mathbb{Z},\ Q \in S \}$$

is a set of weak generators for $\mathcal{D}(\mathcal{G})$. Moreover, we have isomorphisms

$$\mathcal{D}(\mathcal{G})(X, Y) \cong K(\mathcal{G})(QX, RY),$$

and bifibrant replacement induces a triangulated equivalence

$$\mathcal{D}(\mathcal{G}) \to K\left(dg\widetilde{Q} \cap dg\widetilde{\mathcal{R}}\right)$$

onto the strictly full subcategory of $K(\mathcal{G})$ generated by $dg\widetilde{Q} \cap dg\widetilde{\mathcal{R}}$.

Proof By Proposition 10.37, $\left(dg\widetilde{Q}, \widetilde{\mathcal{R}}\right)$ is a complete cotorsion pair on $\mathrm{Ch}(\mathcal{G})$, cogenerated by the set $\{ S^n(Q) \mid n \in \mathbb{Z},\ Q \in S \}$. Since $(Q, \mathcal{R})$ is hereditary it follows that $\left(dg\widetilde{Q}, \widetilde{\mathcal{R}}\right)$ is also hereditary. Once we see that the model structure exists, it follows at once from Theorem 8.21 that $\{S^n(Q) \mid n \in \mathbb{Z},\, Q \in \mathcal{S}\}$ is a set of weak generators for its homotopy category.

By Lemma 10.47 we also have that $\left(\widetilde{Q}, dg\widetilde{\mathcal{R}}\right)$ is a complete hereditary cotorsion pair on $\mathrm{Ch}(\mathcal{G})$. Moreover, $\mathrm{Ext}^1_{\mathrm{Ch}(\mathcal{G})}(\widetilde{Q}, \widetilde{\mathcal{R}}) = 0$ by Lemma 10.46. So the existence of a model structure $\mathfrak{M} = \left(dg\widetilde{Q}, \mathcal{W}, dg\widetilde{\mathcal{R}}\right)$ is immediate from Theorem 8.16, where here the class $\mathcal{W}$ of trivial objects is characterized as

$$\mathcal{W} = \left\{X \in \mathrm{Ch}(\mathcal{G}) \mid \exists \text{ short exact sequence } X \rightarrowtail \tilde{R} \twoheadrightarrow \tilde{Q} \text{ with } \tilde{R} \in \widetilde{\mathcal{R}}, \tilde{Q} \in \widetilde{Q}\right\}$$
$$= \left\{X \in \mathrm{Ch}(\mathcal{G}) \mid \exists \text{ short exact sequence } \tilde{R} \rightarrowtail \tilde{Q} \twoheadrightarrow X \text{ with } \tilde{R} \in \widetilde{\mathcal{R}}, \tilde{Q} \in \widetilde{Q}\right\}.$$

We need to show that $\mathcal{W} = \widetilde{\mathcal{E}}$, the class of all acyclic chain complexes. First, if two out of three complexes in a short exact sequence are exact, then so is the third. So clearly $\mathcal{W} \subseteq \widetilde{\mathcal{E}}$. Conversely, given an exact complex E, use enough injectives of the cotorsion pair $\left(dg\widetilde{Q}, \widetilde{\mathcal{R}}\right)$ to write a short exact sequence $E \rightarrowtail \tilde{R} \twoheadrightarrow Q$ with $\tilde{R} \in \widetilde{\mathcal{R}}$ and $Q \in dg\widetilde{Q}$. By the 2 out of 3 property, we have $Q \in \widetilde{\mathcal{E}}$. So by Lemma 10.48 we have $Q \in dg\widetilde{Q} \cap \widetilde{\mathcal{E}} = \widetilde{Q}$, proving $E \in \mathcal{W}$.

We have shown that $\mathfrak{M} = \left(dg\widetilde{Q}, \widetilde{\mathcal{E}}, dg\widetilde{\mathcal{R}}\right)$ is an abelian model structure on $\mathrm{Ch}(\mathcal{G})$. By Proposition 10.24(4), a chain map f is a weak equivalence in $\mathfrak{M}$ if and only if $\mathrm{Cone}(f)$ is acyclic. Therefore, by Theorem 10.29, the homotopy category $\mathrm{Ho}(\mathfrak{M})$ coincides with the usual derived category, $\mathcal{D}(\mathcal{G})$, and promises the final statement of the theorem too. □

Example 10.50 (Modules over a Ring) Let R be a ring. Then the abelian model structure $\mathfrak{M} = \left(dg\widetilde{Q}, \widetilde{\mathcal{E}}, dg\widetilde{\mathcal{R}}\right)$ exists for any complete hereditary cotorsion pair $(Q, \mathcal{R})$ that is cogenerated by a set. There is an abundance of such cotorsion pairs because *any* set $\mathcal{S}$ of R-modules generates one as follows: For each $S \in \mathcal{S}$, choose a projective resolution of $P_\circ \twoheadrightarrow S$ and then take each cycle module of the complex to obtain a set of all n-syzygies

$$\Omega^n(\mathcal{S}) := \{\Omega^n(S) \mid n \geq 0,\, S \in \mathcal{S}\}.$$

Then $\Omega^n(\mathcal{S})$ cogenerates a complete hereditary cotorsion pair $(Q, \mathcal{R})$ in R-Mod. A detailed proof of this can be found in Göbel and Trlifaj [2006, corollary 2.2.11].

Exercise 10.9.2 Show that Example 8.46 generalizes to $\mathrm{Ch}(\mathcal{G})$ whenever $\mathcal{G}$ is a Grothendieck category possessing a set of finitely generated projective generators.

Exercise 10.9.3 Let $\mathfrak{M} = \left(dg\widetilde{Q}, \widetilde{\mathcal{E}}, dg\widetilde{\mathcal{R}}\right)$ be an abelian model structure on $\mathrm{Ch}(R)$ as in Example 10.50. Use the existence of the model structure to show that

$$\mathrm{Ext}^n_R(M, N) \cong K(R)(Q_\circ, \Sigma^n R_\circ),$$

where $Q_\circ \twoheadrightarrow M$ is any resolution of M by objects in Q and $N \rightarrowtail R_\circ$ is a coresolution of N by objects in $\mathcal{R}$.

Exercise 10.9.4 Let $(\mathcal{F}, C)$ be Enochs' flat cotorsion pair in R-Mod; see Exercise 9.9.4. It is known that any chain complex in $\widetilde{\mathcal{F}}$, the class of all acyclic complexes with flat cycles, is a direct limit of contractible complexes with projective components. Use this and Theorem 9.58 to show that $\left(\widetilde{\mathcal{F}}, dw\widetilde{C}\right)$ is a complete cotorsion pair in Ch(R), where here $dw\widetilde{C}$ denotes the class of all chain complexes that are degreewise cotorsion R-modules. That is, show $dg\widetilde{C} = dw\widetilde{C}$, and so $\mathfrak{M}_{flat} = \left(dg\widetilde{\mathcal{F}}, \widetilde{\mathcal{E}}, dw\widetilde{C}\right)$ is an hereditary abelian model structure for the derived category. (*Hint*: Take $\mathcal{F}$ in the statement of Theorem 9.58 to be the class Ch($\mathcal{F}$), of all chain complexes of flat modules. Show that we have a complete cotorsion pair $\left({}^{\perp}dw\widetilde{C}, dw\widetilde{C}\right)$ in Ch(R), and that it restricts to a complete cotorsion pair $\left({}^{\perp}dw\widetilde{C}, dw\widetilde{C} \cap dw\widetilde{\mathcal{F}}\right)$ on Ch($\mathcal{F}$).)

Exercise 10.9.5 Continuing Exercise 10.9.4, show that any acyclic complex of cotorsion R-modules necessarily has cotorsion cycle modules.

Exercise 10.9.6 Assume R is a commutative ring. Show that the flat model structure $\mathfrak{M}_{flat} = \left(dg\widetilde{\mathcal{F}}, \widetilde{\mathcal{E}}, dw\widetilde{C}\right)$ on Ch(R) is a monoidal model structure.

Exercise 10.9.7 Let R be a ring, R-Mod$_{pur}$ the exact category of R-modules along with the pure exact sequences, and let Ch$(R)_{pur}$ be the associated category of chain complexes along with the degreewise pure exact sequences. Let $\mathcal{PI}$ denote the class of all pure–injective R-modules. It is known that $(All, \mathcal{PI})$ is a complete hereditary cotorsion pair in R-Mod$_{pur}$, cogenerated by a set $\mathcal{S}$. It is also known that any chain complex in $\mathcal{A}_{pur}$, the class of all pure acyclic chain complexes, is a direct limit of contractible complexes; see [Emmanouil, 2016, proposition 2.2]. Use these facts and Theorem 9.56 to show that $\left(\mathcal{A}_{pur}, dw\widetilde{\mathcal{PI}}\right)$ is a (complete) cotorsion pair in Ch$(R)_{pur}$. Here, $dw\widetilde{\mathcal{PI}}$ is the class of all chain complexes that are degreewise pure–injective R-modules.

Exercise 10.9.8 Continuing Exercise 10.9.7, show that any pure acyclic complex of pure–injective R-modules must be contractible.

Exercise 10.9.9 Assume R is a commutative ring. Continuing Exercise 10.9.7, show that $\mathfrak{M}_{pur} = \left(All, \mathcal{A}_{pur}, dw\widetilde{\mathcal{PI}}\right)$ is an abelian monoidal model structure on Ch$(R)_{pur}$.

Example 10.51 (Quasi-coherent Sheaves) Let $\mathbb{X}$ be a scheme, and let Qco($\mathbb{X}$) denote the category of quasi-coherent sheaves on $\mathbb{X}$. It is known that Qco($\mathbb{X}$) is a Grothendieck category. However, it rarely has a set of projective generators. So the standard projective model structure on chain complexes of modules over a ring does not generalize to Ch($\mathbb{X}$), the category of chain complexes of quasi-coherent sheaves. But it is known that if $\mathbb{X}$ is semiseparated and quasi-compact then the class $\mathcal{F}$ of flat quasi-coherent sheaves contains a set of generators for Qco($\mathbb{X}$). Setting $C := \mathcal{F}^{\perp}$, it is shown in Enochs and Estrada [2005] that $(\mathcal{F}, C)$ is a complete cotorsion pair, cogenerated by a set. Therefore, Theorem 10.49 implies the existence of a flat model structure $\mathfrak{M}_{flat} = \left(dg\widetilde{\mathcal{F}}, \widetilde{\mathcal{E}}, dg\widetilde{C}\right)$ on Ch($\mathbb{X}$) for the derived category $\mathcal{D}$(Qco($\mathbb{X}$)).

In fact, each of the results in Exercises 10.9.4–10.9.9 generalize to this setting; see Christensen et al. [2021], Estrada et al. [2024], and Positselski and Šťovíček [2023]. In particular, $dg\widetilde{C} = dw\widetilde{C}$ is simply the class of all complexes of cotorsion sheaves, and $\mathfrak{M}_{flat} = \left(dg\widetilde{\mathcal{F}}, \widetilde{\mathcal{E}}, dw\widetilde{C}\right)$.

11

Mixed Model Structures and Examples

Michael Cole described how two suitably related model structures induce a third "mixed" model structure. We describe here an abelian version of this which was worked out by Timothée Moreau. The theorem provides a method to construct interesting examples of abelian model structures. Assume throughout that $\mathcal{A}$ is a weakly idempotent complete additive category.

11.1 Mixing Abelian Model Structures

Suppose that $\mathfrak{M}_h = (Q_h, \mathcal{W}_h, \mathcal{R}_h)$ is an abelian model structure relative to $(\mathcal{A}, \mathcal{E}_h)$. Moreover, assume $\mathcal{E}_d$ is another exact structure on $\mathcal{A}$, and that we have another abelian model structure $\mathfrak{M}_d = (Q_d, \mathcal{W}_d, \mathcal{R}_d)$ on $(\mathcal{A}, \mathcal{E}_d)$. Cole's theorem, for abelian model categories, is the statement that if $\mathcal{E}_h \subseteq \mathcal{E}_d$, $\mathcal{W}_h \subseteq \mathcal{W}_d$, and $\mathcal{R}_h \subseteq \mathcal{R}_d$, then there exists an abelian model structure $\mathfrak{M}_m = (Q_m, \mathcal{W}_d, \mathcal{R}_h)$ on $(\mathcal{A}, \mathcal{E}_h)$, called their **mixed model structure**. Note that it has the same trivial objects as $\mathfrak{M}_d$ but the same fibrant objects as $\mathfrak{M}_h$. A similar statement holds by duality. The goal of this section is to prove this; see Theorem 11.7.

But first we point out the canonical example which motivates our use of the subscripts h and d. The h references the chain *homotopy* category, $K(R)$, of chain complexes of R-modules. So we have in mind the Frobenius model structure $\mathfrak{M}_h = (All, \mathcal{W}, All)$ on $\text{Ch}(R)_{dw}$, described in Example 6 of the Introduction and Main Examples, and generalized in Theorem 10.20. Here, $\mathcal{W}$ is the class of all contractible chain complexes of R-modules, and the exact structure $\mathcal{E}_h$ is the class of all degreewise split short exact sequences. For the second model structure, the d references the *derived* category, $\mathcal{D}(R)$, and we take $\mathfrak{M}_d$ to be the standard projective model structure $\mathfrak{M}_d = \left(dg\widetilde{\mathcal{P}}, \widetilde{\mathcal{E}}, All\right)$ of Example 3 of the Introduction and Main Examples. Here, $\widetilde{\mathcal{E}}$ denotes the class of all exact chain complexes, and $dg\widetilde{\mathcal{P}}$ denotes the class of all DG-projective complexes.

The exact structure $\mathcal{E}_d$ is the class of all (usual) short exact sequences of chain complexes. In this case, Cole's theorem produces the "mixed" model structure, $\mathfrak{M}_m = \left(\text{K-}Proj, \widetilde{\mathcal{E}}, All\right)$, on the degreewise split exact structure $\mathcal{E}_h$, where the cofibrant objects are Spaltenstein's *K-projective* chain complexes. Every K-projective complex is homotopy equivalent to some DG-projective chain complex. It turns out that, in general, each cofibrant object in $\mathfrak{M}_m$ is h-weakly equivalent to one in $\mathfrak{M}_d$. See Proposition 11.9.

The two statements listed in Cole's Theorem 11.7 are dual to one another. We now adopt the set up of statement (1) of that theorem to prove some needed results along the way. So for the remainder of this section, assume $\mathfrak{M}_h = (\mathcal{Q}_h, \mathcal{W}_h, \mathcal{R}_h)$ is an abelian model structure on $(\mathcal{A}, \mathcal{E}_h)$, and $\mathfrak{M}_d = (\mathcal{Q}_d, \mathcal{W}_d, \mathcal{R}_d)$ is another one on $(\mathcal{A}, \mathcal{E}_d)$, all satisfying $\mathcal{E}_h \subseteq \mathcal{E}_d$, $\mathcal{W}_h \subseteq \mathcal{W}_d$, and $\mathcal{R}_h \subseteq \mathcal{R}_d$. To distinguish between the two model structures we adopt the following language and notation throughout: A short exact sequence in $\mathcal{E}_h$ will be called an *h-exact sequence* while one in $\mathcal{E}_d$ will be called a *d-exact sequence*. The Yoneda Ext groups $\text{Ext}^1_{\mathcal{E}_h}(A, B)$ and $\text{Ext}^1_{\mathcal{E}_d}(A, B)$ will be denoted simply by $\text{Ext}^1_h(A, B)$ and $\text{Ext}^1_d(A, B)$. Note that $\text{Ext}^1_h(A, B)$ is a subgroup of $\text{Ext}^1_d(A, B)$. We also use language such as *h-monics*, *d-epics*, *h-fibrant* objects, *d-cofibrations*, *h-trivial monics*, *d-weak equivalences*, etc. Finally, we will need to distinguish between left and right Ext orthogonal classes with respect to $\mathcal{E}_d$ and $\mathcal{E}_h$. We do so by using notation such as ${}^{\perp_h}C$, or $C^{\perp_d}$, etc.

Now considering what we wish to prove, the following definition is essentially forced upon us.

Definition 11.1 We define $\mathcal{Q}_m := {}^{\perp_h}(\mathcal{W}_d \cap \mathcal{R}_h)$. This is the class of **mixed cofibrant** objects, or more briefly, **m-cofibrant** objects.

It follows from the definition that $\mathcal{Q}_d \subseteq \mathcal{Q}_m \subseteq \mathcal{Q}_h$. We prove this in the next lemma.

Lemma 11.2 *The following class containments hold.*

(1) *For any class of objects C, we have ${}^{\perp_d}C \subseteq {}^{\perp_h}C$ and $C^{\perp_d} \subseteq C^{\perp_h}$.*
(2) $\mathcal{Q}_d \subseteq \mathcal{Q}_m \subseteq \mathcal{Q}_h$.

Proof Each $\text{Ext}^1_h(X, C)$ is a subgroup of $\text{Ext}^1_d(X, C)$. So if $\text{Ext}^1_d(X, C) = 0$ for all $C \in C$, then certainly $\text{Ext}^1_h(X, C) = 0$ too for all those C. So ${}^{\perp_d}C \subseteq {}^{\perp_h}C$. Similarly, $C^{\perp_d} \subseteq C^{\perp_h}$.

For (2), we are given that $\mathcal{R}_h \subseteq \mathcal{R}_d$, so $\mathcal{W}_d \cap \mathcal{R}_h \subseteq \mathcal{W}_d \cap \mathcal{R}_d$, and hence ${}^{\perp_h}(\mathcal{W}_d \cap \mathcal{R}_h) \supseteq {}^{\perp_h}(\mathcal{W}_d \cap \mathcal{R}_d)$. So now using (1), we get

$$\mathcal{Q}_d = {}^{\perp_d}(\mathcal{W}_d \cap \mathcal{R}_d) \subseteq {}^{\perp_h}(\mathcal{W}_d \cap \mathcal{R}_d) \subseteq {}^{\perp_h}(\mathcal{W}_d \cap \mathcal{R}_h) = \mathcal{Q}_m.$$

To see $\mathcal{Q}_m \subseteq \mathcal{Q}_h$, we just note $\mathcal{W}_h \cap \mathcal{R}_h \subseteq \mathcal{W}_d \cap \mathcal{R}_h$, and so we get

$$\mathcal{Q}_m = {}^{\perp_h}(\mathcal{W}_d \cap \mathcal{R}_h) \subseteq {}^{\perp_h}(\mathcal{W}_h \cap \mathcal{R}_h) = \mathcal{Q}_h.$$

□

We will prove Moreau's version of Cole's Theorem 11.7 by continuing to prove a sequence of enlightening lemmas. The next one tells us that the trivial cofibrations in the alleged model structure $\mathfrak{M}_m$ coincide with the trivial cofibrations of $\mathfrak{M}_h$.

Lemma 11.3 $\mathcal{Q}_m \cap \mathcal{W}_d = \mathcal{Q}_h \cap \mathcal{W}_h$.

Proof ($\subseteq$) Let $Q \in \mathcal{Q}_m \cap \mathcal{W}_d$. Since $(\mathcal{Q}_h \cap \mathcal{W}_h, \mathcal{R}_h)$ has enough projectives we may write an h-exact sequence $A \rightarrowtail B \twoheadrightarrow Q$ where $A \in \mathcal{R}_h$ and $B \in \mathcal{Q}_h \cap \mathcal{W}_h$. Since $\mathcal{W}_h \subseteq \mathcal{W}_d$ and $\mathcal{E}_h \subseteq \mathcal{E}_d$, and $\mathcal{W}_d$ satisfies the 2 out of 3 property on d-exact sequences we conclude $A \in \mathcal{W}_d \cap \mathcal{R}_h$. Thus the sequence must be split exact because $Q \in \mathcal{Q}_m := {}^{\perp_h}(\mathcal{W}_d \cap \mathcal{R}_h)$. So Q is a direct summand of B, proving $Q \in \mathcal{Q}_h \cap \mathcal{W}_h$.

($\supseteq$) We have $\mathcal{W}_d \cap \mathcal{R}_h \subseteq \mathcal{R}_h$, so ${}^{\perp_h}\mathcal{R}_h \subseteq {}^{\perp_h}(\mathcal{W}_d \cap \mathcal{R}_h)$. That is, $\mathcal{Q}_h \cap \mathcal{W}_h \subseteq \mathcal{Q}_m$. But also, $\mathcal{Q}_h \cap \mathcal{W}_h \subseteq \mathcal{W}_h \subseteq \mathcal{W}_d$. Hence $\mathcal{Q}_h \cap \mathcal{W}_h \subseteq \mathcal{Q}_m \cap \mathcal{W}_d$. □

The following lemma will be used to show that one of the alleged cotorsion pairs, in the new model structure $\mathfrak{M}_m$, is complete. It will also help us to characterize the m-cofibrant objects later in Proposition 11.9.

Lemma 11.4 *For any object $A \in \mathcal{A}$, there is a bicartesian (both a pullback and a pushout) diagram in $\mathcal{E}_d$,*

$$\begin{array}{ccccc} \tilde{R}_d & \rightarrowtail & Q_d & \overset{p}{\twoheadrightarrow} & A \\ \downarrow & & \downarrow{\scriptstyle j} & & \| \\ R_m & \rightarrowtail & Q_m & \overset{q}{\twoheadrightarrow} & A \\ \downarrow & & \downarrow & & \\ \tilde{Q}_h & = & \tilde{Q}_h, & & \end{array}$$

with middle row and middle column in $\mathcal{E}_h$, and with $Q_d \in \mathcal{Q}_d$, $\tilde{R}_d \in \mathcal{W}_d \cap \mathcal{R}_d$, $Q_m \in \mathcal{Q}_m$, $R_m \in \mathcal{W}_d \cap \mathcal{R}_h$, and $\tilde{Q}_h \in \mathcal{Q}_m \cap \mathcal{W}_d = \mathcal{Q}_h \cap \mathcal{W}_h$.

Proof Since the cotorsion pair $(\mathcal{Q}_d, \mathcal{W}_d \cap \mathcal{R}_d)$ has enough projectives we can start by writing a d-exact sequence as in the top row of the diagram. Then using the factorizations available in $\mathfrak{M}_h$, we may write $p = qj$, where j is an h-trivial cofibration, so with cokernel $\tilde{Q}_h \in \mathcal{Q}_h \cap \mathcal{W}_h = \mathcal{Q}_m \cap \mathcal{W}_d$, and q is an h-fibration, so with kernel $R_m \in \mathcal{R}_h$. This constructs the right commutative square of the diagram and the top two rows and middle column. Note that the middle column expresses Q_m as an h-extension of $Q_d \in \mathcal{Q}_d \subseteq \mathcal{Q}_m$ and $\tilde{Q}_h \in \mathcal{Q}_m$. It follows that $Q_m \in \mathcal{Q}_m := {}^{\perp_h}(\mathcal{W}_d \cap \mathcal{R}_h)$, since any Ext^1_h-orthogonal class must be closed under h-extensions. Next, there exists a unique arrow $\tilde{R}_d \to R_m$ making the upper left square commute, by the universal property of $\ker q$. Setting $R_m \to \tilde{Q}_h$ to be the composition $R_m \rightarrowtail Q_m \twoheadrightarrow \tilde{Q}_h$, this forms the left column and it must be d-exact by the (3×3)-Lemma 1.16. The upper left square is bicartesian by Proposition 1.12. As alredy noted $R_m \in \mathcal{R}_h$, but now note that R_m is a d-extension of two objects in $\mathcal{W}_d$, so we do have $R_m \in \mathcal{W}_d \cap \mathcal{R}_h$. So we have constructed the desired diagram. □

Lemma 11.5 *Any d-exact sequence $A \rightarrowtail B \twoheadrightarrow C$ with $C \in \mathcal{Q}_d$ is necessarily an h-exact sequence. Consequently,* $\mathrm{Ext}^1_d(C, X) = \mathrm{Ext}^1_h(C, X)$ *whenever* $C \in \mathcal{Q}_d$.

Proof Using enough injectives of $(\mathcal{Q}_h, \mathcal{W}_h \cap \mathcal{R}_h)$ we write an h-exact sequence $A \rightarrowtail R \twoheadrightarrow Q$ with $Q \in \mathcal{Q}_h$ and $R \in \mathcal{W}_h \cap \mathcal{R}_h$. Form the pushout diagram as shown:

$$\begin{array}{ccccc} A & \rightarrowtail & B & \twoheadrightarrow & C \\ \downarrow & & \downarrow & & \| \\ R & \rightarrowtail & P & \twoheadrightarrow & C \\ \downarrow & & \downarrow & & \\ Q & = & Q. & & \end{array}$$

Note that all the columns are h-exact, the middle column in particular because of the pushout axiom for exact categories. Furthermore, the middle row is an element of $\mathrm{Ext}^1_d(C, R)$. But since $R \in \mathcal{W}_h \cap \mathcal{R}_h \subseteq \mathcal{W}_d \cap \mathcal{R}_d$ we get $\mathrm{Ext}^1_d(C, R) = 0$, so this middle row is split exact. So now all three columns and the bottom two rows of the diagram are h-exact. So the remaining top row must also be h-exact by the (3×3)-Lemma 1.16. □

Remark 11.6 Since $\mathcal{Q}_d \subseteq \mathcal{Q}_m$, it follows from Lemma 11.5 that any (trivial) cofibration in $\mathfrak{M}_d$ is also a (trivial) cofibration in the alleged mixed model structure $\mathfrak{M}_m$.

Theorem 11.7 (Cole's Theorem for Abelian Model Categories) *Let $\mathcal{A}$ be an additive category with two exact structures, $(\mathcal{A}, \mathcal{E}_h)$ and $(\mathcal{A}, \mathcal{E}_d)$, satisfying $\mathcal{E}_h \subseteq \mathcal{E}_d$. Assume we have an abelian model structure $\mathfrak{M}_h = (\mathcal{Q}_h, \mathcal{W}_h, \mathcal{R}_h)$ on $(\mathcal{A}, \mathcal{E}_h)$, an abelian model structure $\mathfrak{M}_d = (\mathcal{Q}_d, \mathcal{W}_d, \mathcal{R}_d)$ on $(\mathcal{A}, \mathcal{E}_d)$, and their trivial objects satisfy the containment $\mathcal{W}_h \subseteq \mathcal{W}_d$.*

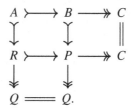

(1) *If* $\mathcal{R}_h \subseteq \mathcal{R}_d$, *then there exists an abelian model structure* $\mathfrak{M}_m = (\mathcal{Q}_m, \mathcal{W}_d, \mathcal{R}_h)$ *on* $(\mathcal{A}, \mathcal{E}_h)$, *having the same trivial objects as* $\mathfrak{M}_d$ *but the same fibrant objects as* $\mathfrak{M}_h$.

(2) *If* $\mathcal{Q}_h \subseteq \mathcal{Q}_d$, *then there exists an abelian model structure* $\mathfrak{M}_m = (\mathcal{Q}_h, \mathcal{W}_d, \mathcal{R}_m)$ *on* $(\mathcal{A}, \mathcal{E}_h)$, *with the same trivial objects as* $\mathfrak{M}_d$ *but the same cofibrant objects as* $\mathfrak{M}_h$.

In either case, we call $\mathfrak{M}_m$ *the* ***mixed model structure*** *of* $\mathfrak{M}_h$ *and* $\mathfrak{M}_d$.

Proof We use the aforementioned results to prove (1). Statement (2) follows from duality. So assuming $\mathcal{R}_h \subseteq \mathcal{R}_d$ we need to show that $\mathfrak{M}_m = (\mathcal{Q}_m, \mathcal{W}_d, \mathcal{R}_h)$ is an abelian model structure on $(\mathcal{A}, \mathcal{E}_h)$. Since $\mathcal{W}_d$ is already given to satisfy the 2 out of 3 property for $\mathcal{E}_d$ it certainly does for $\mathcal{E}_h$ too. By Lemma 11.3 we have $(\mathcal{Q}_m \cap \mathcal{W}_d, \mathcal{R}_h) = (\mathcal{Q}_h \cap \mathcal{W}_h, \mathcal{R}_h)$, which is already given to be a complete cotorsion pair on $\mathcal{E}_h$.

We now show that $(\mathcal{Q}_m, \mathcal{W}_d \cap \mathcal{R}_h)$ is a cotorsion pair. Of course $\mathcal{Q}_m := {}^{\perp_h}(\mathcal{W}_d \cap \mathcal{R}_h)$, so in order to prove we have a cotorsion pair it is left to show $\mathcal{Q}_m^{\perp_h} \subseteq \mathcal{W}_d \cap \mathcal{R}_h$. We know $\mathcal{Q}_d \subseteq \mathcal{Q}_m$ by Lemma 11.2. So we get $\mathcal{Q}_m^{\perp_h} \subseteq \mathcal{Q}_d^{\perp_h}$, and it follows from Lemma 11.5 that we have $\mathcal{Q}_d^{\perp_h} = \mathcal{Q}_d^{\perp_d} = \mathcal{W}_d \cap \mathcal{R}_d$. So now it is only left to show $\mathcal{Q}_m^{\perp_h} \subseteq \mathcal{R}_h$. For this we note $\mathcal{Q}_h \cap \mathcal{W}_h = \mathcal{Q}_m \cap \mathcal{W}_d \subseteq \mathcal{Q}_m$, so $\mathcal{Q}_m^{\perp_h} \subseteq (\mathcal{Q}_h \cap \mathcal{W}_h)^{\perp_h} = \mathcal{R}_h$. This proves that $(\mathcal{Q}_m, \mathcal{W}_d \cap \mathcal{R}_h)$ is a cotorsion pair.

Finally, it is left to show that the cotorsion pair $(\mathcal{Q}_m, \mathcal{W}_d \cap \mathcal{R}_h)$ is complete. But the middle row of the diagram in Lemma 11.4 proves that this cotorsion pair has enough projectives. To see that it has enough injectives we use a variation of Salce's Trick. Given any $A \in \mathcal{A}$, we start by using enough injectives of $\mathfrak{M}_h$ to write an h-exact sequence $A \rightarrowtail \tilde{R}_h \twoheadrightarrow Q_h$ where $Q_h \in \mathcal{Q}_h$ and $\tilde{R}_h \in \mathcal{W}_h \cap \mathcal{R}_h \subseteq \mathcal{W}_d \cap \mathcal{R}_h$. Then we use that $(\mathcal{Q}_m, \mathcal{W}_d \cap \mathcal{R}_h)$ has enough projectives to write another h-exact sequence $R_m \rightarrowtail Q_m \twoheadrightarrow Q_h$ where $Q_m \in \mathcal{Q}_m$ and $R_m \in \mathcal{W}_d \cap \mathcal{R}_h$. Using Lemma 1.10, form the pullback diagram in $(\mathcal{A}, \mathcal{E}_h)$ shown:

$$\begin{array}{ccccc} & & R_m & = & R_m \\ & & \downarrow & & \downarrow \\ A & \rightarrowtail & P & \twoheadrightarrow & Q_m \\ \| & & \downarrow & & \downarrow \\ A & \rightarrowtail & \tilde{R}_h & \twoheadrightarrow & Q_h. \end{array}$$

Since $\tilde{R}_h$ and R_m are each in $\mathcal{W}_d \cap \mathcal{R}_h$, we see that the h-extension P is also in $\mathcal{W}_d \cap \mathcal{R}_h$. Hence the middle row proves that $(\mathcal{Q}_m, \mathcal{W}_d \cap \mathcal{R}_h)$ has enough injectives and so it forms a complete cotorsion pair relative to $\mathcal{E}_h$. □

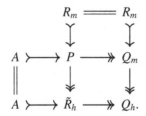

Exercise 11.1.1 Verify or prove each of the following statements for a mixed model structure $\mathfrak{M}_m$ and a morphism $f \in \mathcal{A}$.

(1) f is an m-fibration if and only if f is an h-fibration.
(2) f is an m-trivial cofibration if and only if f is an h-trivial cofibration.
(3) If f is a d-cofibration then f is an m-cofibration. And, if f is an m-cofibration then f is an h-cofibration.
(4) If f is an h-trivial fibration then f is an m-trivial fibration. And, if f is an m-trivial fibration then f is a d-trivial fibration.
(5) The class of m-weak equivalences coincides with the class of d-weak equivalences, and contains the class of all h-weak equivalences.

Exercise 11.1.2 Show that if $\mathfrak{M}_h = (\mathcal{Q}_h, \mathcal{W}_h, \mathcal{R}_h)$ is hereditary, then the mixed model structure $\mathfrak{M}_m = (\mathcal{Q}_m, \mathcal{W}_d, \mathcal{R}_h)$ is hereditary too.

11.2 Cofibrant Objects in Mixed Abelian Model Structures

We would like to better understand the class $\mathcal{Q}_m$ of cofibrant objects constructed in Cole's Theorem 11.7(1). So the goal here is to characterize the m-cofibrant objects. We continue to work exclusively with the hypotheses of part (1) of that theorem. We leave it to the reader to formulate the dual characterization of the class $\mathcal{R}_m$ of fibrant objects appearing in part (2) of Cole's Theorem.

Proposition 11.8 *[Cole, 2006, corollary 3.4] A d-weak equivalence between two m-cofibrant objects must be an h-weak equivalence.*

Proof Let A and B each be m-cofibrant and $f\colon A \to B$ a d-weak equivalence. Then f is a weak equivalence in $\mathfrak{M}_m$; see Exercise 11.1.1(5). So we have a factorization

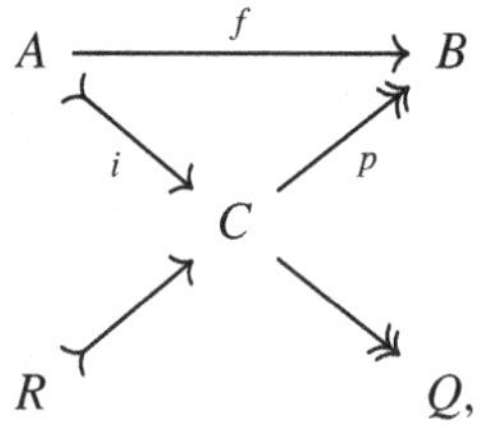

where p is an h-epic with kernel $R \in \mathcal{W}_d \cap \mathcal{R}_h$ and i is an h-monic with cokernel $Q \in \mathcal{Q}_m \cap \mathcal{W}_d = \mathcal{Q}_h \cap \mathcal{W}_h$. Note that C is an h-extension of A and Q, and so we have $C \in \mathcal{Q}_m$. Also $\mathrm{Ext}^1_h(B, R) = 0$, so p is a split epimorphism. Thus R is a direct summand of C, and so $R \in \mathcal{Q}_m \cap \mathcal{W}_d \cap \mathcal{R}_h = \mathcal{Q}_h \cap \mathcal{W}_h \cap \mathcal{R}_h$. Since i is an h-trivial monic and p is an h-trivial epic, it follows that the composition f is an h-weak equivalence. □

Proposition 11.9 (Cofibrant Objects in Mixed Model Structures) *An object Q is in $\mathcal{Q}_m$ if and only if $Q \in \mathcal{Q}_h$ and there is an h-weak equivalence $Q_d \to Q$ for some $Q_d \in \mathcal{Q}_d$. In fact, for any $Q \in \mathcal{Q}_m$, there exists an admissible d-epic morphism $p\colon Q_d \twoheadrightarrow Q$ that is an h-weak equivalence.*

Proof Suppose $Q \in \mathcal{Q}_m$. By Lemma 11.4, there is a bicartesian diagram

$$\begin{array}{ccccc}
\tilde{R}_d & \rightarrowtail & Q_d & \overset{p}{\twoheadrightarrow} & Q \\
\downarrow & & \downarrow{\scriptstyle j} & & \| \\
R_m & \rightarrowtail & Q_m & \overset{q}{\twoheadrightarrow} & Q \\
\downarrow & & \downarrow & & \\
\tilde{Q}_h & = & \tilde{Q}_h & &
\end{array}$$

with all rows and columns d-exact, in fact with middle row and middle column in $\mathcal{E}_h$, and with $Q_d \in \mathcal{Q}_d$, $\tilde{R}_d \in \mathcal{W}_d \cap \mathcal{R}_d$, $Q_m \in \mathcal{Q}_m$, $R_m \in \mathcal{W}_d \cap \mathcal{R}_h$, and $\tilde{Q}_h \in \mathcal{Q}_m \cap \mathcal{W}_d = \mathcal{Q}_h \cap \mathcal{W}_h$. But now since $Q \in \mathcal{Q}_m := {}^{\perp_h}(\mathcal{W}_d \cap \mathcal{R}_h)$, we see that the middle row must be a split exact sequence. This means R_m is a direct summand of Q_m and hence $R_m \in \mathcal{Q}_m$ too. Therefore, we have

$$R_m \in \mathcal{Q}_m \cap (\mathcal{W}_d \cap \mathcal{R}_h) = (\mathcal{Q}_m \cap \mathcal{W}_d) \cap \mathcal{R}_h = (\mathcal{Q}_h \cap \mathcal{W}_h) \cap \mathcal{R}_h.$$

In particular, q is an h-trivial fibration. Since i is an h-trivial cofibration, the composite p is an h-weak equivalence. Note too that p is an admissible d-epic, proving the second statement of the proposition.

Conversely, assume $Q \in \mathcal{Q}_h$ and let $w\colon Q_d \to Q$ be an h-weak equivalence for some $Q_d \in \mathcal{Q}_d$. Use enough projectives of $(\mathcal{Q}_m, \mathcal{W}_d \cap \mathcal{R}_h)$ to write an h-exact sequence $R \rightarrowtail Q_m \overset{g}{\twoheadrightarrow} Q$ where $R \in \mathcal{W}_d \cap \mathcal{R}_h$ and $Q_m \in \mathcal{Q}_m$. Since $\mathrm{Ext}^1_h(Q_d, R) = 0$, there exists a morphism $f\colon Q_d \to Q_m$ such that $w = gf$. We now apply the 2 out of 3 Axiom a couple of times to this last equation. On one hand, we know that f must be be a d-weak equivalence since both w and g are. However, w is actually the stronger h-weak equivalence, and we see f is too, by Proposition 11.8. So we now turn around and conclude g is an h-weak equivalence. Since g is already an h-epic, it follows from Proposition 4.7 that g is an h-trivial epic. That is, $R \in \mathcal{W}_h$. Since Q is assumed to be in $\mathcal{Q}_h$, this implies g is a split epimorphism and so Q is a direct summand of Q_m. Thus Q too is in $\mathcal{Q}_m$, proving (1). □

Example 11.10 We point out that the morphism p constructed in Proposition 11.9 need not be an h-trivial epic, just a d-epic that is also an h-weak equivalence. For instance, take the canonical example given at the beginning of Section 11.1: So $\mathfrak{M}_h$ is the standard Frobenius model for the homotopy category $K(R)$, and $\mathfrak{M}_d$ is the DG-projective model structure for the derived category $\mathcal{D}(R)$. Their mixed model structure is $\mathfrak{M}_m = \left(\text{K-}Proj, \widetilde{\mathcal{E}}, All\right)$, the K-projective model structure for $\mathcal{D}(R)$. Note that for any R-module M, there is an obvious short exact sequence of chain complexes $D^1(K) \rightarrowtail D^1(P) \xrightarrow{p} D^1(M)$ induced from any short exact sequence $K \rightarrowtail P \twoheadrightarrow M$ of R-modules with P projective. The short exact sequence of disks may be substituted for the top row of the diagram at the beginning of the proof to Proposition 11.9. Indeed $D^1(M)$ is K-projective, so $D^1(M) \in \mathcal{Q}_m$, and the morphism p is d-epic with $D^1(P)$ a DG-projective complex. The proof shows p to be an h-weak equivalence (homotopy equivalence), but p need not be an h-epic (i.e. a degreewise split epimorphism).

One can even mix abelian model structures having the same underlying exact structure; see Example 11.12. The following exercise gives a characterization of the m-cofibrant objects in this case.

Exercise 11.2.1 Assume $\mathfrak{M}_h$ and $\mathfrak{M}_d$ exist on the same exact structure $(\mathcal{A}, \mathcal{E})$, that is, $\mathcal{E}_h = \mathcal{E}_d = \mathcal{E}$. Show that $Q \in \mathcal{Q}_m$ if and only if $Q \in \mathcal{Q}_h$ and there exists a short exact sequence $R \rightarrowtail Q_d \twoheadrightarrow Q$ with $Q_d \in \mathcal{Q}_d$ and $R \in \mathcal{W}_h \cap \mathcal{R}_d$ (or just $R \in \mathcal{W}_h$).

Exercise 11.2.2 Let $Q \in \mathcal{Q}_m$ an m-cofibrant object in some mixed model structure. Show that there is an h-exact sequence $Q \rightarrowtail R \twoheadrightarrow C$ with $R \in \mathcal{R}_d$ and $C \in \mathcal{Q}_d \cap \mathcal{W}_h$.

11.2.1 Examples of Mixed Model Structures

Here we give some examples of mixed model structures on chain complexes of modules over a ring. This section is less self-contained in the sense that we will simply give references for the existence of some of the model structures. We let R be a ring and set the following notation throughout this section.

- $\text{Ch}(R)_{dw}$ denotes the Frobenius category of chain complexes along with the degreewise split exact structure, and $\mathfrak{M}_{K(R)} = (All, \mathcal{W}, All)$ denotes the associated Frobenius model structure; see Example 6 of the Introduction and Main Examples.
- $\text{Ch}(R)_{proj} = \left(dg\widetilde{\mathcal{P}}, \widetilde{\mathcal{E}}, All\right)$ and $\text{Ch}(R)_{inj} = \left(All, \widetilde{\mathcal{E}}, dg\widetilde{\mathcal{I}}\right)$ respectively denote

the standard projective and injective model structures on (unbounded) chain complexes; see Examples 3 and 4 of the Introduction and Main Examples.

- $\mathrm{Ch}(R)_{pur}$ denotes the exact category of chain complexes along with the short exact sequences that are pure exact in each degree.

Example 11.11 The canonical example of obtaining the K-projective model structure, $\mathfrak{M}_m = \left(\text{K-}Proj, \widetilde{\mathcal{E}}, All\right)$, as the mixed model structure of $\mathfrak{M}_{K(R)}$ and $\mathrm{Ch}(R)_{proj}$ has the following generalization: We may replace $\mathrm{Ch}(R)_{proj}$with any other projective cotorsion pair $\mathfrak{M}_d = (\mathcal{Q}, \mathcal{W}_d, All)$ on chain complexes, as long as $\mathcal{W}_d$ contains all the contractible complexes. For example, by Bravo et al. [2014, corollary 6.4] or Becker [2014, proposition 1.3.6], there is such an abelian model structure $\left(dw\widetilde{\mathcal{P}}, \mathcal{W}_{\mathrm{ctr}}, All\right)$ on $\mathrm{Ch}(R)$, where $dw\widetilde{\mathcal{P}}$ consists of all chain complexes with projective components. Taking this to be $\mathfrak{M}_d$ and mixing it with $\mathfrak{M}_h = \mathfrak{M}_{K(R)}$ results in a mixed model structure $\mathfrak{M}_m = (\mathrm{K}(Proj), \mathcal{W}_{\mathrm{ctr}}, All)$ on $\mathrm{Ch}(R)_{dw}$. Here $\mathrm{K}(Proj)$ denotes the class of all chain complexes that are homotopy equivalent to some chain complex of projectives; these are precisely the m-cofibrant objects by Proposition 11.9. The complexes in $\mathcal{W}_{\mathrm{ctr}}$ are sometimes called *contraacyclic (in the sense of Becker)*.

Note that the projective model structure $\mathfrak{M}_d$ may even be on some proper class of short exact sequences of complexes. For example, by [Gillespie, 2023b], there is a projective model structure, $\left(ex\widetilde{\mathcal{PP}}, \mathcal{V}, All\right)$ on $\mathrm{Ch}(R)_{pur}$, where $ex\widetilde{\mathcal{PP}}$ consists of all exact chain complexes with pure–projective components, and $\mathcal{V}$ denotes the class of all K-absolutely pure complexes from Emmanouil and Kaperonis [2024]. Taking this to be $\mathfrak{M}_d$ and mixing it with with $\mathfrak{M}_h = \mathfrak{M}_{K(R)}$ produces the mixed model structure $\mathfrak{M}_m = (\mathrm{K}_{ac}(\mathcal{PP}), \mathcal{V}, All)$ on $\mathrm{Ch}(R)_{dw}$. This time, the class $\mathrm{K}_{ac}(\mathcal{PP})$ of m-cofibrant objects is the class of all chain complexes that are homotopy equivalent to some exact (acyclic) complex of pure–projective R-modules.

Example 11.12 There is also an abelian model structure $\left(All, \mathcal{W}_{\mathrm{co}}, dw\widetilde{\mathcal{I}}\right)$ on $\mathrm{Ch}(R)$, where $dw\widetilde{\mathcal{I}}$ consists of all complexes of injective R-modules; see Example 8.8.1. The complexes in $\mathcal{W}_{\mathrm{co}}$ are called *coacyclic (in the sense of Becker)*. Taking $\mathfrak{M}_h = \left(All, \mathcal{W}_{\mathrm{co}}, dw\widetilde{\mathcal{I}}\right)$, and $\mathfrak{M}_d = \mathrm{Ch}(R)_{proj}$ to be the standard projective model structure, we obtain a mixed model structure $\mathfrak{M}_m = \left({}^{\perp}ex\widetilde{\mathcal{I}}, \widetilde{\mathcal{E}}, dw\widetilde{\mathcal{I}}\right)$, where $ex\widetilde{\mathcal{I}}$ is the class of all exact complexes of injective R-modules. By Exercise 11.2.1, a chain complex $X \in {}^{\perp}ex\widetilde{\mathcal{I}}$ if and only if there exists a short exact sequence $W \rightarrowtail P \twoheadrightarrow X$ where P is DG-projective and $W \in \mathcal{W}_{\mathrm{co}}$ (not just $\widetilde{\mathcal{E}}$).

Example 11.13 By the dual mixed model structure Theorem 11.7(2), and the dual of Proposition 11.9, the ideas in the previous examples dualize. For example, we have a model for $\mathcal{D}(R)$ whose fibrant objects are the K-injective complexes, etc. In addition to the references already given, see Bravo [2011], Šťovíček [2015, corollary 5.7], Gillespie [2016e], and Gillespie [2023a] for other examples of injective (and projective) model structures on chain complexes.

Let $\mathcal{A}_{pur}$ denote the class of all *pure acyclic* complexes in Ch(R). That is, $W \in \mathcal{A}_{pur}$ if and only if $M \otimes_R W$ is acyclic for any (right) R-module M. Then by Šťovíček [2015, corollary 5.7] there is both a projective and an injective model structure

$$\left(dw\widetilde{\mathcal{PP}}, \mathcal{A}_{pur}, All\right) \quad \text{and} \quad \left(All, \mathcal{A}_{pur}, dw\widetilde{\mathcal{PI}}\right)$$

on $\text{Ch}(R)_{pur}$, where $dw\widetilde{\mathcal{PP}}$ (resp. $dw\widetilde{\mathcal{PI}}$) consists of all chain complexes with pure–projective (resp. pure–injective) components. The homotopy category of each recovers $\mathcal{D}_{pur}(R)$, the pure derived category of R.

Exercise 11.2.3 Let $\mathfrak{M}_h = \left(dw\widetilde{\mathcal{PP}}, \mathcal{A}_{pur}, All\right)$ be the earlier pure–projective model structure on $\text{Ch}(R)_{pur}$.

(1) Taking $\mathfrak{M}_d = \text{Ch}(R)_{proj}$, describe the class of cofibrant objects in the resulting mixed model structure in terms of K-projective complexes.
(2) Taking $\mathfrak{M}_d = \left(dw\widetilde{\mathcal{P}}, \mathcal{W}_{\text{ctr}}, All\right)$, describe the class of cofibrant objects in the resulting mixed model structure.

Exercise 11.2.4 A chain complex X is called *K-flat* if the usual chain complex tensor product, $E \otimes_R X$, is exact whenever E is exact. It is shown in Gillespie [2023a, corollary 4.5] that X is K-flat if and only if there is a pure homology isomorphism $f\colon P \to X$ (so an isomorphism in $\mathcal{D}_{pur}(R)$) where P is some DG-projective chain complex. Use this to construct a (mixed) abelian model structure for the usual derived category, $\mathcal{D}(R)$, whose cofibrant objects are precisely the K-flat complexes.

Exercise 11.2.5 Show that there is a complete cotorsion pair $(\mathcal{X}, \mathcal{Y})$, in the exact category $\text{Ch}(R)_{pur}$, where $\mathcal{Y}$ is the class of all contraacylic chain complexes with pure–injective components. (See Example 11.11 for the class $\mathcal{W}_{\text{ctr}}$ of contraacyclic complexes.)

12

Cofibrant Generation and Well-Generated Homotopy Categories

A central idea in modern homological algebra is Neeman's notion of a well-generated triangulated category. In this chapter we prove an abelian version of a result of Rosický. He showed that the homotopy category of any (pointed) combinatorial model category, a concept introduced by Jeff Smith, is well generated. Following these ideas we show here that a cofibrantly generated abelian model structure on an accessible exact category has a well-generated homotopy category. We will take for granted some basic facts about accessible categories, giving reference to the standard source [Adámek and Rosický, 1994].

12.1 Accessible Categories and Exact Structures

In this section, and throughout the rest of the chapter, we utilize the convenient setting of *accessible categories* from Adámek and Rosický [1994]. In order to avoid an excursion into the fundamentals of accessible categories, we instead will briefly describe here the theory we need and refer the reader to Adámek and Rosický [1994] for details. Exact structures on accessible categories will provide a nice setting for us to develop an exact category version of Jeff Smith's notion of a combinatorial model category. This is also the setting we use to prove Jiri Rosický's result concerning well generated homotopy categories. As usual, we assume throughout this section that $\mathcal{A}$ is an additive category.

The basic idea is that an accessible category is one for which every object is built up, as a certain type of colimit, of a set of "small" objects. Let us make this precise. First, given any regular cardinal κ, a poset $(I, \leq)$ is said to be κ-*directed* if every subset $S \subseteq I$ with $|S| < \kappa$ has an upper bound in I. Considering $(I, \leq)$ as a category, it means that every such subcategory S has a cocone in I. By a κ-*direct limit* (or κ-*directed colimit*) we mean an existing colimit, $\varinjlim_I Y_i$, of some direct I-system $\{Y_i, f_{ij}\}$, indexed by some κ-directed poset $(I, \leq)$. An object $A \in \mathcal{A}$ is called **κ-presented** if $\mathrm{Hom}_{\mathcal{A}}(A, -)$ preserves κ-direct limits. It means that the canonical morphism $\xi\colon \varinjlim_I \mathrm{Hom}_{\mathcal{A}}(A, Y_i) \to \mathrm{Hom}_{\mathcal{A}}\left(A, \varinjlim_I Y_i\right)$ is an isomorphism for any given κ-direct limit $\varinjlim_I Y_i$.

Note that for regular cardinals $\lambda \geq \kappa$, any λ-directed poset is automatically κ-directed. So any κ-presented object is also λ-presented. In particular, any finitely presented object (i.e. $\aleph_0$-presented) is κ-presented for any regular κ.

Lemma 12.1 *Let κ be a regular cardinal.*

(1) *A retract of a κ-presented object is again κ-presented.*
(2) *Any coproduct $\bigoplus_{i\in I} A_i$ of κ-presented objects A_i is again κ-presented if $|I| < \kappa$. Similarly, any direct limit $\varinjlim_I A_i$ of κ-presented objects A_i, indexed by a poset $(I, \leq)$ of cardinality $|I| < \kappa$, is again κ-presented.*

Proof We leave the proofs as exercises. (The results follow from Adámek and Rosický [1994, proposition 1.16 and remark].) □

Definition 12.2 Let κ be a regular cardinal. $\mathcal{A}$ is said to be **κ-accessible** if all κ-direct limits exist in $\mathcal{A}$ and if $\mathcal{A}$ possesses a set $\mathcal{V}$ of κ-presented objects such that each object $A \in \mathcal{A}$ may be expressed as a κ-direct limit $A = \varinjlim_{i\in I} V_i$ of objects $V_i \in \mathcal{V}$. $\mathcal{A}$ is said to be **accessible** if it is κ-accessible for some regular cardinal κ.

A κ-accessible category $\mathcal{A}$ is called **locally κ-presentable** if it is cocomplete. So the class of accessible categories includes all locally presentable categories, which in turn is known to include all Grothendieck categories [Beke, 2000, proposition 3.10]. For example, the category R-Mod, of modules over a ring R, is locally $\aleph_0$-presentable (finitely presentable). The full subcategory of all flat R-modules is an $\aleph_0$-accessible category (finitely accessible). This follows from Lazard's Theorem: Each flat module is a direct limit of free R-modules of finite rank.

The following is a summary of the most crucial facts we need concerning accessible categories.

Proposition 12.3 (See Adámek and Rosický [1994]) *Let $\mathcal{A}$ be a κ-accessible (additive) category.*

(1) *Up to isomorphism, there exists a set (not just a proper class) of κ-presented objects in $\mathcal{A}$. We will let $\mathrm{Pres}_\kappa(\mathcal{A})$ denote a set of isomorphism representatives for all κ-presented objects of $\mathcal{A}$.*
(2) *Every object $A \in \mathcal{A}$ is λ-presented for some regular cardinal λ.*

(3) *There are arbitrarily large regular cardinals $\mu > \kappa$ such that $\mathcal{A}$ is also μ-accessible. Such cardinals μ may be specified as those with the following property: For any κ-directed poset $(I, \leq)$, the set*

$$\hat{I} = \{S \subseteq I \mid S \text{ is a } \kappa\text{-directed subposet of } I \text{ with cardinality } |S| < \mu\}$$

is a μ-directed poset under subset inclusion. That is, the poset $(\hat{I}, \subseteq)$ is always μ-directed. One says that κ is **sharply smaller** *than μ, written $\kappa \lhd \mu$. Then $\lhd$ is a transitive relation on regular cardinals.*

(4) *$\mathcal{A}$ is idempotent complete, so in particular weakly idempotent complete.*

(5) *If $\mathcal{A}$ is cocomplete (i.e. locally κ-presentable) then it is also complete.*

Proof We refer the reader to Adámek and Rosický [1994]. Specifically, see remark 2.2(1), (3), and (4) and theorem 2.11(iv). (The idempotent completeness is explained after Observation 2.4.) The previous definition of $\kappa \lhd \mu$ is slightly different than the one given in the cited theorem 2.11(iv). However, see the comments in remark 2.15(2). Conversely, theorem 2.11(iv) follows from the aforementioned definition (every singleton subset of a κ-directed poset is trivially κ-directed). Statement (5) is corollary 1.28 in Adámek and Rosický [1994]. □

It is helpful to see a few examples of the *sharply smaller* relation. We have $\aleph_0 \lhd \mu$ whenever μ is an uncountable regular cardinal. Also, for any regular cardinal κ, we have $\kappa \lhd \kappa^+$, where κ^+ is the cardinal successor of κ. However, the statement $\aleph_1 \lhd \aleph_{\omega+1}$ is *false*; see example 2.13 of Adámek and Rosický [1994] for details.

We note that Theorem 9.34 provides the following method for constructing complete cotorsion pairs relative to exact structures on accessible categories.

Corollary 12.4 (Accessible Categories and Complete Cotorsion Pairs) *Assume $\mathcal{A}$ is an accessible additive category with coproducts. Let $(\mathcal{A}, \mathcal{E})$ be an exact structure possessing a set of admissible generators. Then the cotorsion pair $({}^{\perp}(\mathcal{S}^{\perp}), \mathcal{S}^{\perp})$, functorially has enough injectives whenever $\mathcal{S}$ is an efficient set of objects. Moreover, if ${}^{\perp}(\mathcal{S}^{\perp})$ contains some set $\mathcal{U}$ of admissible generators, then this is a functorially complete cotorsion pair. In this case, we have ${}^{\perp}(\mathcal{S}^{\perp}) = \overline{\mathrm{Filt}(\mathcal{S} \cup \mathcal{U})}$, the closure of* $\mathrm{Filt}(\mathcal{S} \cup \mathcal{U})$ *under taking direct summands.*

Indeed a cotorsion pair $(\mathcal{X}, \mathcal{Y})$ in $(\mathcal{A}, \mathcal{E})$ is cofibrantly generated (see Definition 9.30) if and only if it is cogenerated by an efficient set $\mathcal{S}$, and $\mathcal{X}$ contains some set of admissible generators.

Proof By applying Proposition 9.29, the hypotheses of Theorem 9.34 are met. Indeed accessible categories are weakly idempotent complete and every object of $\mathcal{A}$ is λ-presented for some regular λ. Therefore, as alluded to in Remark 9.31, each object is certainly small relative to Filt($\mathcal{S}$)-cofibrations. □

The following lemma tells us that for any cotorsion pair arising by way of Corollary 12.4, the right-hand class, $\mathcal{S}^{\perp}$, is closed under λ-direct limits for sufficiently large cardinals λ.

Lemma 12.5 *Assume $\mathcal{A}$ is an accessible additive category with coproducts. Let $(\mathcal{A}, \mathcal{E})$ be an exact structure possessing a set of admissible generators. For any given efficient set $\mathcal{S}$, there exists a regular cardinal λ such that $\mathcal{S}^{\perp}$ is closed under λ-direct limits. Specifically, this is the case whenever $I_{\mathcal{S}}$ is a set of generating monics for $\mathcal{S}^{\perp}$, and L is λ-presented where $v\colon L \rightarrowtail V$ is the single generating monic obtained from $I_{\mathcal{S}}$ as in Lemma 9.32.*

Proof Let $I_{\mathcal{S}}$ be any given set of generating monics for $\mathcal{S}^{\perp}$ (one exists by Proposition 9.29), and let $v\colon L \rightarrowtail V$ be the single generating monic obtained by taking their coproduct as in Lemma 9.32. By Proposition 12.3, there exists a regular cardinal λ such that L is λ-presented. (Alternatively, we may take λ to be a regular cardinal such that the domain of each monic in the set $I_{\mathcal{S}}$ is λ-presented.) Now consider any morphism $f\colon L \to \varinjlim_{i\in I} Y_i$ into an existing direct limit of some λ-directed system of objects $\{Y_i\}_{i\in I}$ with each $Y_i \in \mathcal{S}^{\perp}$. Since L is λ-presented, it follows from the definition that f must factor as $f = \eta_i \circ f_i$, for some $i \in I$, where $\eta_i\colon Y_i \to \varinjlim_{i\in I} Y_i$ is the canonical map into the colimit. Since $Y_i \in \mathcal{S}^{\perp}$, the morphism $v\colon L \rightarrowtail U$ is Y_i-monic, meaning f_i extends over v. Consequently, f extends over v, proving $\varinjlim_{i\in I} Y_i \in \mathcal{S}^{\perp}$. □

Exercise 12.1.1 Let $(\mathcal{A}, \mathcal{E})$ be an exact structure on a locally presentable additive category $\mathcal{A}$. Argue that the standard projective model structure $\mathfrak{M} = \left(dg\widetilde{\mathcal{P}}, \widetilde{\mathcal{E}}, All\right)$ exists on $\mathrm{Ch}(\mathcal{A})_{\mathcal{E}}$, if $(\mathcal{A}, \mathcal{E})$ possesses a set of projective generators. (Use Proposition 12.3(5) and Corollary 10.42.) It is cofibrantly generated with homotopy category $\mathcal{D}(\mathcal{A}, \mathcal{E})$, the derived category of the exact structure.

12.2 Functorial Approximation Sequences in Accessible Exact Categories

The goal of the current section is to prove two results that will be vital to the proof that cofibrantly generated abelian model structures have well-generated homotopy categories. First, Proposition 12.10 implies the existence of functorial (co)fibrant approximation sequences that respect λ-direct limits, for sufficiently large λ. Second, Proposition 12.13 asserts further that these approximation functors also preserve μ-presented objects for sufficiently large regular cardinals μ.

To obtain these results we need to work with a nice enough exact structure on a nice enough accessible category. The following setup describes what we need. This is a vast class of exact structures; see the Remark 12.6 that follows.

Setup 12.2.1 Assume $\mathcal{A}$ is a κ-accessible additive category with coproducts. Assume $(\mathcal{A}, \mathcal{E})$ is an exact structure possessing a set of admissible generators and assume it satsfies the following *condition on κ-direct limits of admissible epics*: For any κ-directed $(I, \leq)$-system $\left\{K_i \overset{k_i}{\rightarrowtail} X_i \overset{e_i}{\twoheadrightarrow} Y_i\right\}$ of short exact sequences, if the induced map $\varinjlim_I e_i\colon \varinjlim_I X_i \to \varinjlim_I Y_i$ is an admissible epic, then its kernel coincides with the induced map $\varinjlim_I k_i\colon \varinjlim_I K_i \to \varinjlim_I X_i$. That is, we have an $\mathcal{E}$-short exact sequence

$$\varinjlim_I K_i \overset{\varinjlim_I k_i}{\rightarrowtail} \varinjlim_I X_i \overset{\varinjlim_I e_i}{\twoheadrightarrow} \varinjlim_I Y_i$$

whenever $\varinjlim_I e_i$ is an admissible epic.

Remark 12.6 Assume $\mathcal{A}$ is a κ-accessible additive category with coproducts, and that $(\mathcal{A}, \mathcal{E})$ is an exact structure with a set of admissible generators. Then the conditions of Setup 12.2.1 hold in ***either*** of the following scenarios.

- $\mathcal{A}$ is cocomplete, that is, $\mathcal{A}$ is locally κ-presentable. In this case, $\mathcal{A}$ is also complete and all finite limits (in particular kernels) commute with κ-direct limits; see Adámek and Rosický [1994, corollary 1.28 and proposition 1.59].
- $(\mathcal{A}, \mathcal{E})$ has exact κ-direct limits. That is, the class $\mathcal{E}$ of short exact sequences is closed under κ-direct limits (analogous to the definition given in Section 9.10, but for κ-directed systems). In this case we say that $(\mathcal{A}, \mathcal{E})$ is a **κ-accessible exact category**.

We say that $(\mathcal{A}, \mathcal{E})$ is an **accessible exact category** if it is a κ-accessible exact category for some regular cardinal κ. For example, let R be a ring. Then the category of all flat R-modules, with its naturally inherited exact structure, is a finitely accessible exact category ($\kappa = \aleph_0$).

Lemma 12.7 *Let $(\mathcal{A}, \mathcal{E})$ satisfy the conditions of Setup 12.2.1 for some regular cardinal κ. Then it again satisfies the conditions for any regular cardinal μ such that $\kappa \lhd \mu$. That is, $\mathcal{A}$ is μ-accessible and $(\mathcal{A}, \mathcal{E})$ satisfies the condition on μ-direct limits of admissible epics.*

Proof Assume $\kappa \lhd \mu$. First, the point of Proposition 12.3(3) is that it allows the accessibility of $\mathcal{A}$ to be raised from κ to μ. Second, $(\mathcal{A}, \mathcal{E})$ will satisfy the condition on μ-direct limits of admissible epics simply because any μ-directed poset is automatically κ-directed too. □

In order to prove Propositions 12.10 and Proposition 12.13, we need to revisit the construction of functorial approximation functors inside the proof of Theorem 9.37 of Section 9.8. We will need two lemmas. The first one clarifies how functors like Ff_*^L and Ff_*^V appearing in the commutative cube of Lemma 9.36 interact with κ-direct limits. Let us note first that for *any* given objects L and V, and any given endofunctor $F\colon \mathcal{A} \to \mathcal{A}$, the construction of Ff_*^V holds in general. That is, for any three given such L, V, and F, we have a functor taking an object A to $\bigoplus_{t\in H_{FA}^L} V_t$, and a morphism $f\colon A \to B$ to

$$Ff_*^V\colon \bigoplus_{t\in H_{FA}^L} V_t \to \bigoplus_{t\in H_{FB}^L} V_t.$$

Here, $V_t = V$ for all $t \in H_{FA}^L := \mathrm{Hom}_{\mathcal{A}}(L, FA)$, and Ff_*^V is the unique map out of the coproduct satisfying $Ff_*^V \circ \eta_t^V = \eta_{F(f)\circ t}^V$ for each $t \in H_{FA}^L$. The following lemma says that this functor preserves κ-direct limits, as long as F does and L is κ-presented.

Lemma 12.8 *Let L and V be objects of an additive category $\mathcal{A}$ with κ-direct limits, and let $F\colon \mathcal{A} \to \mathcal{A}$ be an endofunctor preserving κ-direct limits. If L is κ-presented then the functor taking a morphism $f\colon A \to B$ to*

$$Ff_*^V\colon \bigoplus_{t\in H_{FA}^L} V_t \to \bigoplus_{t\in H_{FB}^L} V_t$$

preserves κ-direct limits. That is, for each κ-directed $(I,\leq)$-system $\{A_i, f_{ij}\}_{i\in I}$ with colimit cone $\left(\varinjlim_{i\in I} A_i, \{a_i\}_{i\in I}\right)$, there is a canonical morphism

$$w\colon \varinjlim_{I} \bigoplus_{t\in H_{FA_i}^L} V_t \longrightarrow \bigoplus_{t\in H^L_{F(\varinjlim_I A_i)}} V_t$$

induced by $\{(Fa_i)_*^V\}_{i\in I}$, and it is an isomorphism.

Proof For any given κ-directed $(I,\leq)$-system $\{A_i, f_{ij}\}_{i\in I}$, with colimit cone $\left(\varinjlim_{i\in I} A_i, \{a_i\}_{i\in I}\right)$, applying the functor yields another κ-directed $(I,\leq)$-system, $\left\{\bigoplus_{t\in H_{FA_i}^L} V_t, (Ff_{ij})_*^V\right\}$, with a cocone

$$\left(\bigoplus_{t\in H^L_{F(\varinjlim_I A_i)}} V_t, \{(Fa_i)_*^V\}_{i\in I}\right).$$

We will show that this new cocone is in fact a colimit cone. So assume we are given another cocone $(Y, \{y_i\}_{i\in I})$ above $\left\{\bigoplus_{t\in H^L_{FA_i}} V_t, (Ff_{ij})^V_*\right\}$. Then for all $i \leq j$ we have $y_i = y_j \circ (Ff_{ij})^V_*$, equivalently, $y_i \circ \eta_{t_i} = y_j \circ (Ff_{ij})^V_* \circ \eta_{t_i}$ for all maps $t_i \colon L \to FA_i$. So we have the equations

$$y_i \circ \eta_{t_i} = y_j \circ \eta_{(Ff_{ij}\circ t_i)}. \tag{12.1}$$

But by hypothesis, $\left(F\left(\varinjlim_{i\in I} A_i\right), \{Fa_i\}_{i\in I}\right)$ is a colimit cone for the κ-directed system $\left\{FA_i, Ff_{ij}\right\}_{i\in I}$. Moreover, since L is κ-presented, the canonical map

$$\xi \colon \varinjlim_I \operatorname{Hom}_{\mathcal{A}}(L, FA_i) \to \operatorname{Hom}_{\mathcal{A}}\left(L, F\left(\varinjlim_I A_i\right)\right)$$

is an isomorphism; ξ being onto means that any given $t \in \operatorname{Hom}_{\mathcal{A}}\left(L, F\left(\varinjlim_I A_i\right)\right)$ factors as $Fa_i \circ t_i$ for some $t_i \colon L \to FA_i$. Because ξ is one-to-one we get that whenever $t = Fa_i \circ t_i = Fa_j \circ t_j$, there exists a $k \in I$ with $i \leq k$ and $j \leq k$ such that $(Ff_{ik})_*(t_i) = (Ff_{jk})_*(t_j)$, where here $(Ff_{ij})_* := \operatorname{Hom}_{\mathcal{A}}(L, Ff_{ij})$. is the usual abelian group homomorphism. This translates to the equations

$$Ff_{ik} \circ t_i = Ff_{jk} \circ t_j. \tag{12.2}$$

We are to show that there is a unique map $\psi \colon \bigoplus_{t\in H^L_{F(\varinjlim_I A_i)}} V_t \to Y$ such that $\psi \circ (Fa_i)^V_* = y_i$ for all $i \in I$. Note that such a ψ necessarily satisfies, for all $t_i \in \operatorname{Hom}_{\mathcal{A}}(L, FA_i)$, that $\psi \circ (Fa_i)^V_* \circ \eta_{t_i} = y_i \circ \eta_{t_i} \implies \psi \circ \eta_{Fa_i\circ t_i} = y_i \circ \eta_{t_i}$. In particular, whenever $t \colon L \to F\left(\varinjlim_I A_i\right)$ factors as $t = Fa_i \circ t_i$ then we must have $\psi \circ \eta_t = y_i \circ \eta_{t_i}$. So for any given such t we define ψ to be the unique map out of the coproduct satisfying $\psi \circ \eta_t = y_i \circ \eta_{t_i}$. Using Equations (12.1) and (12.2) one can see this is well defined. □

The next lemma is immediate from a standard categorical fact; the commutativity of taking colimits. It is also a straightforward exercise to check it directly.

Lemma 12.9 *Let $\mathcal{A}$ be an additive category with κ-direct limits. Assume $(I, \leq)$ is a κ-directed poset and let $\left\{C_i \xleftarrow{\alpha_i} A_i \xrightarrow{\beta_i} B_i\right\}_{i\in I}$ be a direct I-system of spans, each with an existing pushout diagram:*

$$\begin{array}{ccc} A_i & \xrightarrow{\beta_i} & B_i \\ {\scriptstyle \alpha_i}\downarrow & & \downarrow{\scriptstyle \bar{\alpha}_i} \\ C_i & \xrightarrow{\bar{\beta}_i} & P_i. \end{array}$$

Then the square obtained by taking direct limits pointwise is also a pushout:

$$\begin{array}{ccc} \varinjlim_I A_i & \xrightarrow{\varinjlim \beta_i} & \varinjlim_I B_i \\ {\scriptstyle \varinjlim \alpha_i}\downarrow & & \downarrow{\scriptstyle \varinjlim \bar{\alpha}_i} \\ \varinjlim_I C_i & \xrightarrow{\varinjlim \bar{\beta}_i} & \varinjlim_I P_i. \end{array}$$

Exercise 12.2.1 Prove Lemma 12.9.

Proposition 12.10 *Let $(\mathcal{X}, \mathcal{Y})$ be a cofibrantly generated cotorsion pair (see Corollary 12.4) on an exact category $(\mathcal{A}, \mathcal{E})$ as in Setup 12.2.1. Then there exists a triple $(\lambda, Y_\lambda, X_\lambda)$, where λ is a regular cardinal with $\kappa \lhd \lambda$, and $Y_\lambda : \mathcal{A} \to \mathcal{A}$ is a functorial $\mathcal{Y}$-approximation that preserves λ-direct limits, and $X_\lambda : \mathcal{A} \to \mathcal{A}$, is a functorial $\mathcal{X}$-approximation that preserves λ-direct limits.*

Remark 12.11 Note that if μ is another regular cardinal and $\lambda \lhd \mu$, then X_λ and Y_λ also preserve μ-direct limits.

Proof As per Corollary 12.4, let $\mathcal{S}$ be an efficient set of objects cogenerating a cotorsion pair $(\mathcal{X}, \mathcal{Y}) = ({}^\perp(\mathcal{S}^\perp), \mathcal{S}^\perp)$, for which $\mathcal{X}$ contains of a set of admissible generators. Also, let $I_\mathcal{S}$ be a corresponding set of generating monics for $\mathcal{S}^\perp = \mathcal{Y}$. Using Lemma 9.32, let $v : L \rightarrowtail V$ be the single generating monic for $\mathcal{S}^\perp = \mathcal{Y}$, obtained by taking the coproduct over all the maps in $I_\mathcal{S}$. By Proposition 12.3, we can find a regular cardinal λ such that (i) $\kappa \lhd \lambda$, (ii) the admissible generator $U = \bigoplus U_i$ is λ-presented, and (iii) the object L of the admissible generating monic $v : L \rightarrowtail V$ is λ-presented. Then turning again to the proof of Theorem 9.37, we check that the construction of P^v_λ, the functorial $\mathcal{Y}$-approximation, preserves λ-direct limits. Indeed suppose $\varinjlim_I A_i$ is the colimit of a direct I-system $\{A_i, f_{ij}\}$ indexed by some λ-directed poset $(I, \leq)$. There is a corresponding direct I-system of spans and pushouts, as indicated by the square shown on the left:

$$\begin{array}{ccc} \bigoplus_{t \in H^L_{A_i}} L_t & \xrightarrow{\oplus v} & \bigoplus_{t \in H^L_{A_i}} V_t \\ {\scriptstyle e_{A_i}}\downarrow & & \downarrow{\scriptstyle \bar{e}_{A_i}} \\ A_i & \xrightarrow{\bar{v}_{A_i}} & P^v A_i \end{array} \qquad \begin{array}{ccc} \bigoplus_{t \in H^L_{\varinjlim_I A_i}} L_t & \xrightarrow{\oplus v} & \bigoplus_{t \in H^L_{\varinjlim_I A_i}} V_t \\ {\scriptstyle e_{\varinjlim_I A_i}}\downarrow & & \downarrow{\scriptstyle \bar{e}_{\varinjlim_I A_i}} \\ \varinjlim_I A_i & \xrightarrow{\bar{v}_{\varinjlim A_i}} & P^v\left(\varinjlim_I A_i\right). \end{array}$$

By Lemma 12.8 (with $F = 1_\mathcal{A}$), the commutative square obtained by pointwise direct limits is canonically isomorphic to the square shown on the right.

Moreover, this must be a pushout square by Lemma 12.9. In particular, the admissible monic $\bar{v}_{\lim A_i} : \varinjlim_I A_i \rightarrowtail P^v\left(\varinjlim_I A_i\right)$ is canonically isomorphic to $\varinjlim \bar{v}_{A_i} : \varinjlim_I A_i \to \varinjlim_I (P^v A_i)$. So the natural admissible monic $\bar{v}_{0,1} : 1_{\mathcal{A}} \rightarrowtail P_1^v$ constructed in the first step of the induction respects λ-direct limits. Likewise, for the successor ordinal step of the induction, having already constructed $F = P_\alpha^v$ preserving λ-direct limits, the admissible monic $\bar{v}_{\alpha,\alpha+1} : P_\alpha^v \rightarrowtail P_{\alpha+1}^v$ respects λ-direct limits.

For limit ordinals γ we had set $P_\gamma^v := \varinjlim_{\alpha<\gamma} P_\alpha^v$. That is, for each $A \in \mathcal{A}$, $P_\gamma^v A$ is the transfinite extension $P_\gamma^v A := \varinjlim_{\alpha<\gamma} P_\alpha^v A$. Then P_γ^v also preserves λ-direct limits because we have natural isomorphisms

$$P_\gamma^v\left(\varinjlim_I A_i\right) := \varinjlim_{\alpha<\gamma} P_\alpha^v\left(\varinjlim_I A_i\right) \cong \varinjlim_{\alpha<\gamma}\left(\varinjlim_I P_\alpha^v A_i\right) \cong \varinjlim_I\left(\varinjlim_{\alpha<\gamma} P_\alpha^v A_i\right) = \varinjlim_I P_\gamma^v A_i,$$

where the last isomorphism is by the "interchange of colimits" theorem as in, for example, Mac Lane [1998, IX.2, equation (2)]. So the constructed functorial $\mathcal{Y}$-approximation, $Y_\lambda := P_\lambda^v$, preserves λ-direct limits whenever L is λ-presented.

As for X_λ, the functorial $\mathcal{X}$-approximation, we follow the argument in Theorem 9.37 regarding the functoriality of Salce's Trick. Since the admissible generator U is λ-presented, there is a morphism w as in Lemma 12.8 (taking $L = V = U$ and $F = 1_{\mathcal{A}}$) yielding a canonical isomorphism of short exact sequences (the top row is a short exact sequence by Lemma 12.7, since $\kappa \lhd \lambda$):

$$\begin{array}{ccccc}
\varinjlim \mathrm{Ker}\left(e^U_{A_i}\right) & \rightarrowtail & \varinjlim_I \bigoplus_{t\in H^U_{A_i}} U_t & \xrightarrow{\varinjlim e^U_{A_i}} & \varinjlim_I A_i \\
\downarrow & & \downarrow{\scriptstyle w} & & \| \\
\mathrm{Ker}\left(e^U_{\varinjlim A_i}\right) & \rightarrowtail & \bigoplus_{t\in H^U_{\varinjlim_I A_i}} U_t & \xrightarrow{e^U_{\varinjlim A_i}} & \varinjlim_I A_i.
\end{array}$$

Salce's Trick proceeds by applying the functor Y_λ, constructed earlier, to the kernels $\mathrm{Ker}\left(e^U_{A_i}\right)$, and then taking pushouts. Since Y_λ preserves λ-direct limits, it follows from Lemma 12.9 that the resulting functorial $\mathcal{X}$-approximation, X_λ, also preserves λ-direct limits. □

Remark 12.12 The proof of Proposition 12.10 shows that the functorial short exact sequences associated to $\mathcal{Y}_\lambda$ and $\mathcal{X}_\lambda$, are closed under λ-direct limits.

Proposition 12.13 *Continuing with the same hypotheses as Proposition 12.10, let $(\lambda, Y_\lambda, X_\lambda)$ be the triple constructed there. Then there exist arbitrarily large* *regular cardinals μ such that $\lambda \lhd \mu$ and such that Y_λ and X_λ each preserve μ-presented objects. That is, $Y_\lambda A$ and $X_\lambda A$ remain μ-presented whenever A is μ-presented.*

Proof By construction, λ is a regular cardinal such that $\kappa \lhd \lambda$. So by Proposition 12.3, there is a set $\mathrm{Pres}_\lambda(\mathcal{A})$ (that is not a proper class) of isomorphism representatives of all λ-presented objects in $\mathcal{A}$. Also, using parts (2) and (3) of Proposition 12.3, we may choose a regular cardinal μ to be so large that (i) $\lambda \lhd \mu$ and (ii) $Y_\lambda A$ and $X_\lambda A$ are each μ-presented for all $A \in \mathrm{Pres}_\lambda(\mathcal{A})$. Note that condition (ii) can be met by choosing a regular cardinal no smaller than $\sup\left\{\kappa_A, \kappa'_A \mid A \in \mathrm{Pres}_\lambda(\mathcal{A})\right\}$, where κ_A (resp. κ'_A) represents the least regular cardinal such that $Y_\lambda A$ is κ_A-presented (resp. $X_\lambda A$ is κ'_A-presented). We will show that $Y_\lambda A$ is μ-presented whenever A is μ-presented. The proof for X_λ works the same way.

So let A be a μ-presented object. The goal is to show that $Y_\lambda(A)$ is also μ-presented. First, because $\mathcal{A}$ is λ-accessible, we may write $A = \varinjlim_I A_i$ as a λ-direct limit of some direct I-system $\left\{A_i, f_{ij}\right\}$ of objects $A_i \in \mathrm{Pres}_\lambda(\mathcal{A})$, indexed by some λ-directed poset $(I, \leq)$. By our choice of μ, the corresponding poset $(\hat{I}, \subseteq)$ defined in Proposition 12.3(3) (now with $\kappa = \lambda$) is μ-directed under subset inclusion. For each $\mathcal{S} \in \hat{I}$, we set $A_\mathcal{S} := \varinjlim_{i \in \mathcal{S}} A_i$. Then for each $\mathcal{S} \subseteq \mathcal{S}'$, there is a canonical colimit morphism $f_{\mathcal{S},\mathcal{S}'} \colon A_\mathcal{S} \to A_{\mathcal{S}'}$. One may verify that $\{A_\mathcal{S}, f_{\mathcal{S},\mathcal{S}'}\}$ is an $(\hat{I}, \subseteq)$-directed system with colimit $\varinjlim_{\mathcal{S} \in \hat{I}} A_\mathcal{S} = \varinjlim_{i \in I} A_i = A$. For each $\mathcal{S} \in \hat{I}$, there is a canonical map $\eta_\mathcal{S} \colon A_\mathcal{S} \to A$ into the colimit, and $(A, \{\eta_\mathcal{S}\})$ constitutes a colimit cone of $\{A_\mathcal{S}, f_{\mathcal{S},\mathcal{S}'}\}$. (See also Adámek and Rosický [1994, p. 77, remark 2.15].)

But since A is μ-presented, and $A = \varinjlim_{\mathcal{S} \in \hat{I}} A_\mathcal{S}$ is a μ-direct limit, we have a canonical isomorphism

$$\varinjlim_{\mathcal{S} \in \hat{I}} \mathrm{Hom}_\mathcal{A}(A, A_\mathcal{S}) \cong \mathrm{Hom}_\mathcal{A}\left(A, \varinjlim_{\mathcal{S} \in \hat{I}} A_\mathcal{S}\right) = \mathrm{Hom}_\mathcal{A}(A, A).$$

In particular, it implies that the identity map $1_A \colon A \to A$ factors through some $\eta_\mathcal{S} \colon A_\mathcal{S} \to A$. So we may write $1_A = \eta_\mathcal{S} \circ \delta_\mathcal{S}$ for some morphism $\delta_\mathcal{S} \colon A \to A_\mathcal{S}$. Applying our functorial $\mathcal{Y}$-approximation, $Y_\lambda \colon \mathcal{A} \to \mathcal{A}$, we obtain a retraction $1_{Y_\lambda(A)} = Y_\lambda(\eta_\mathcal{S}) \circ Y_\lambda(\delta_\mathcal{S})$. That is, $Y_\lambda(A)$ is a direct summand of $Y_\lambda(A_\mathcal{S})$. Since μ-presented objects are closed under retracts, see Lemma 12.1, it suffices to show that $Y_\lambda(A_\mathcal{S})$ is μ-presented. But since $(\mathcal{S}, \leq)$ is a λ-directed poset, and Y_λ preserves any λ-direct limit, we have a canonical isomorphism

$$Y_\lambda(A_\mathcal{S}) = Y_\lambda\left(\varinjlim_{i \in \mathcal{S}} A_i\right) \cong \varinjlim_{i \in \mathcal{S}} Y_\lambda(A_i).$$

By our choice of μ, each object $Y_\lambda(A_i)$ is μ-presented. Moreover, since $|S| < \mu$, it follows from Lemma 12.1 that $\varinjlim_{i\in S} Y_\lambda(A_i)$ must also be a μ-presented object. Thus $Y_\lambda(A_S)$ is μ-presented and this completes the proof. □

12.3 Combinatorial Abelian Model Structures

We will now interpret the results of the previous section in the context of cofibrantly generated abelian model structures on $(\mathcal{A}, \mathcal{E})$. Following Jeff Smith's notion of a combinatorial model category, one might call these *combinatorial abelian model structures*. If $\mathcal{A}$ is bicomplete (equivalently cocomplete, equivalently locally presentable) then these are certainly combinatorial model categories in the sense of Smith.

Throughout this section, let $(\mathcal{A}, \mathcal{E})$ be an exact structure as in Setup 12.2.1. Because of what we observed in Corollary 12.4, an abelian model structure $\mathfrak{M} = (\mathcal{Q}, \mathcal{W}, \mathcal{R})$ on $(\mathcal{A}, \mathcal{E})$ is cofibrantly generated (see Definition 9.38) if both $(\mathcal{Q}_{\mathcal{W}}, \mathcal{R})$ and $(\mathcal{Q}, \mathcal{R}_{\mathcal{W}})$ are each cogenerated by some efficient set of objects, and if $\mathcal{Q}_{\mathcal{W}}$ contains a set $\mathcal{U} = \{U_i\}$ of admissible generators for $\mathcal{E}$. The following results now follow from Propositions 12.10 and 12.13.

Corollary 12.14 *Let $\mathfrak{M} = (\mathcal{Q}, \mathcal{W}, \mathcal{R})$ be a cofibrantly generated (i.e. combinatorial) abelian model structure on $(\mathcal{A}, \mathcal{E})$. Then the following hold.*

(1) *There exists a quintuple, $(\lambda, Q_\lambda, R_\lambda, \tilde{Q}_\lambda, \tilde{R}_\lambda)$, where λ is a regular cardinal with $\kappa \lhd \lambda$ and such that each of the following hold: Q_λ (resp. $\tilde{Q}_\lambda$) is a functorial cofibrant (resp. trivially cofibrant) approximation, R_λ (resp. $\tilde{R}_\lambda$) is a functorial fibrant (resp. trivially fibrant) approximation, and each of these four functors preserve λ-direct limits.*
(2) *For such λ, the class of fibrant objects, $\mathcal{R}$, and the class of trivially fibrant objects, $\mathcal{R}_{\mathcal{W}}$, are each closed under λ-direct limits.*
(3) *There exist arbitrarily large regular cardinals μ such that $\kappa \lhd \lambda \lhd \mu$, and such that each of the four functors Q_λ, R_λ, $\tilde{Q}_\lambda$, and $\tilde{R}_\lambda$ preserve μ-presented objects as well as μ-direct limits.*

Proof We refer to the notation set in Definition 9.38 for cofibrantly generated abelian model structures. Statement (1) follows from Proposition 12.10 and Remark 12.11. More directly, note that the construction in Proposition 12.10 can be applied to both cotorsion pairs at once. That is, by Proposition 12.3, we may choose a regular λ so that (i) $\kappa \lhd \lambda$, (ii) the single trivially cofibrant generator $U = \bigoplus U_i$ is λ-presented, and (iii) the domains of both $v\colon L \rightarrowtail V$ and $v'\colon L' \rightarrowtail V'$ from Definition 9.38 are λ-presented. Then the four resulting functorial approximations Q_λ, R_λ, $\tilde{Q}_\lambda$, and $\tilde{R}_\lambda$ will each preserve λ-direct limits.

With this choice of λ, statement (2) now follows at once from Lemma 12.5. Finally, statement (3) is immediate from Proposition 12.13 and Remark 12.11. Alternatively, we can again directly adjust the proof of Proposition 12.13 to require that $\mu \rhd \lambda$ satisfies that $Q_\lambda A$, $R_\lambda A$, $\tilde{Q}_\lambda A$, and $\tilde{R}_\lambda A$ are all μ-presented for all $A \in \mathrm{Pres}_\lambda(\mathcal{A})$. □

Proposition 12.15 *Continuing with the same hypotheses and the same regular cardinal λ as in Corollary 12.14, then the class $\mathcal{W}$ of trivial objects is closed under λ-direct limits.*

Proof Take λ as in the previous corollary, and let us be given a direct I-system $\{W_i\}$ indexed by some λ-directed poset $(I, \leq)$, and with each $W_i \in \mathcal{W}$. Using the functorial approximation functor $\tilde{Q}_\lambda$, we obtain a corresponding λ-directed system of short exact sequences $\left\{\Omega W_i \rightarrowtail \tilde{Q}_\lambda(W_i) \twoheadrightarrow W_i\right\}$ with each $\tilde{Q}_\lambda(W_i) \in \mathcal{Q}_\mathcal{W}$, and each $\Omega W_i \in \mathcal{R}$. Since each $W_i \in \mathcal{W}$, we have $\Omega W_i \in \mathcal{R}_\mathcal{W}$ too, by the 2 out of 3 property. Because $\tilde{Q}_\lambda$ preserves λ-direct limits, we have $\varinjlim_I \tilde{Q}_\lambda(W_i) \cong \tilde{Q}_\lambda\left(\varinjlim_I W_i\right) \in \mathcal{Q}_\mathcal{W}$. Since the λ-directed poset $(I, \leq)$ is also κ-directed, it then follows from Setup 12.2.1 that we have a short exact sequence

$$\varinjlim_I \Omega W_i \rightarrowtail \varinjlim_I \tilde{Q}_\lambda(W_i) \twoheadrightarrow \varinjlim_I W_i.$$

Moreover, it follows from Corollary 12.14(2), that $\varinjlim_I \Omega W_i \in \mathcal{R}_\mathcal{W}$. So again by the 2 out of 3 property we conclude $\varinjlim_I W_i \in \mathcal{W}$. □

Exercise 12.3.1 Take λ as in Corollary 12.14. Following up on Remark 12.12, verify that we have a functorial suspension functor Σ (and inverse $\Sigma^{-1} := \Omega$) that preserves λ-direct limits.

Exercise 12.3.2 Take λ as in Corollary 12.14. Show that we have functorial factorizations (see Exercise 9.8.3) of morphisms into trivial cofibrations followed by fibrations which preserve λ-direct limits. The same is true for the other factorizations of morphisms into cofibrations followed by trivial fibrations. Use this and Proposition 12.15 to show that the class of weak equivalences is closed under λ-direct limits.

12.4 Small Objects in Homotopy Categories

In Section 12.5 we will give the definition of a well-generated triangulated category. It relies on the central notion of *small* objects in triangulated categories with coproducts. The definition of small in this context is the same as it is in any additive category with coproducts; see Definition 10.39 to recall the notion of a *κ-small* object. The goal of this section is to give some ways to identify small objects in the homotopy category of an abelian model structure. Assume throughout that $\mathcal{A}$ is an additive category with coproducts and let κ denote a regular cardinal.

There is a convenient characterization of κ-smallness of an object S in terms of the functor $\mathrm{Hom}_{\mathcal{A}}(S,-)$ preserving a particular type of κ-direct limit. To set up the lemma describing this, suppose $\{A_i\}_{i\in I}$ is any given family of objects. Let I_κ denote the set of all subsets $J \subseteq I$ of cardinality $|J| < \kappa$. Note that I_κ is a κ-directed set when considered along with subset inclusion. That is, $(I_\kappa, \subseteq)$ is a κ-directed poset, because the union of $< \kappa$ many subsets of cardinality $< \kappa$ still has cardinality $< \kappa$. There is a κ-directed system

$$\left\{\bigoplus_{i\in J} A_i\,,\ \bigoplus_{i\in J} A_i \xrightarrow{\eta_J^{J'}} \bigoplus_{i\in J'} A_i\right\}$$

over $(I_\kappa, \subseteq)$, where $\eta_J^{J'} : \bigoplus_{i\in J} A_i \to \bigoplus_{i\in J'} A_i$ is the canonical inclusion whenever $J \subseteq J'$ is the inclusion of two sets from I_κ. (Here, $\eta_J^{J'}$ is the unique morphism satisfying $\eta_J^{J'} \circ \eta_i^J = \eta_i^{J'}$ for each $i \in J$.) One may check directly that $\left(\bigoplus_{i\in I} A_i, \{\eta_J\}\right)$ is a colimit cone for the κ-directed system $\left\{\bigoplus_{i\in J} A_i\,, \eta_J^{J'}\right\}$. In other words, any coproduct $\bigoplus_{i\in I} A_i$ may be expressed as the κ-direct limit

$$\varinjlim_{J\in I_\kappa}\left(\bigoplus_{i\in J} A_i\right) = \bigoplus_{i\in I} A_i.$$

Note too that this construction does not require all direct limits to exist in $\mathcal{A}$, merely that $\mathcal{A}$ has coproducts. Now by applying $\mathrm{Hom}_{\mathcal{A}}(S,-)$ to the entire colimit diagram we obtain another κ-directed system of abelian groups over $(I_\kappa, \subseteq)$, namely $\left\{\mathrm{Hom}_{\mathcal{A}}\left(S, \bigoplus_{i\in J} A_i\right)\,, \left(\eta_J^{J'}\right)_*\right\}$, and there is a canonical morphism

$$\xi_\kappa\colon \varinjlim_{J\in I_\kappa} \mathrm{Hom}_{\mathcal{A}}\left(S, \bigoplus_{i\in J} A_i\right) \to \mathrm{Hom}_{\mathcal{A}}\left(S, \bigoplus_{i\in I} A_i\right). \qquad (12.3)$$

Here, ξ_κ acts on a morphism $f_J\colon S \to \bigoplus_{i\in J} A_i$ representing an equivalence class for an element of $\varinjlim_{J\in I_\kappa} \mathrm{Hom}_{\mathcal{A}}\left(S, \bigoplus_{i\in J} A_i\right)$, by $\xi(f_J) = \eta_J \circ f_J$. We have the following lemma characterizing κ-small objects in terms of ξ_κ.

Lemma 12.16 *Let S be an object of an additive category $\mathcal{A}$ with coproducts and let κ be a regular cardinal. The canonical morphism ξ_κ satisfies the following.*

(1) *ξ_κ is one-to-one, for any given family $\{A_i\}_{i\in I}$.*
(2) *ξ_κ is onto (hence an isomorphism), for any given family $\{A_i\}_{i\in I}$, if and only if S is κ-small.*

It follows that any κ-presented object of $\mathcal{A}$ must in particular be κ-small.

Proof To see ξ_κ is always one-to-one, let $f_J \colon S \to \bigoplus_{i\in J} A_i$ be some morphism representing an equivalence class for an element of the colimit. If we have $\xi_\kappa(f_J) = \eta_J \circ f_J = 0$ in $\mathrm{Hom}_{\mathcal{A}}\left(A, \bigoplus_{i\in I} A_i\right)$, then we must have $f_J = 0$, because η_J can be seen to be a split monomorphism, and so it is left-cancellable.

Next, tracing through the definitions, we see that ξ_κ is onto if and only if for any morphism any $f \colon S \to \bigoplus_{i\in I} A_i$, there exists some $J \in I_\kappa$ and a morphism $f_J \colon S \to \bigoplus_{i\in J} A_i$ for which $f = \xi_\kappa(f_J) = \eta_J \circ f_J$. This exactly matches the definition of κ-small given in Definition 10.39.

Finally, it is immediate that any κ-presented object S must be κ-small. Indeed the definition of κ-presented means precisely that the canonical morphism induced by *any* (existing) κ-direct limit, analogous to how we defined ξ_κ, must be an isomorphism. The canonical morphisms ξ_κ are a special case. □

We just showed that if $S \in \mathcal{A}$ is κ-presented then it is κ-small. However, what we really want to know is this: When is S, viewed as an object in $\mathrm{Ho}(\mathfrak{M})$, where $\mathfrak{M}$ is some abelian model structure, a κ-small object? In this direction we have the following lemma and Proposition 12.18.

Lemma 12.17 *Assume $\mathcal{A}$ has coproducts, and let $\mathfrak{M} = (\mathcal{Q}, \mathcal{W}, \mathcal{R})$ be any abelian model structure with respect to an exact structure $(\mathcal{A}, \mathcal{E})$. The following are equivalent for any regular cardinal κ.*

1. *An object S is a κ-small object in the homotopy category,* $\mathrm{Ho}(\mathfrak{M})$.
2. *For any family $\{A_i\}_{i\in I}$ of $\mathcal{A}$-objects and for any $\mathcal{A}$-morphism $h \colon QS \to R\left(\bigoplus_{i\in I} QA_i\right)$, there exists $J \in I_\kappa$ and an $\mathcal{A}$-morphism $h_J \colon QS \to R\left(\bigoplus_{i\in J} QA_i\right)$ such that $[h] = [R(\eta_J)] \circ [h_J]$. Here, $\eta_J \colon \bigoplus_{i\in J} QA_i \to \bigoplus_{i\in I} QA_i$ is the canonical subcoproduct insertion in $\mathcal{A}$ that uniquely satisfies the factorization $\eta_i \colon QA_i \xrightarrow{\eta_i^J} \bigoplus_{i\in J} QA_i \xrightarrow{\eta_J} \bigoplus_{i\in I} QA_i$ for all $i \in J$.*

Proof Let $\{A_i\}_{i\in I}$ be any given family of $\mathcal{A}$-objects. Recall that its coproduct exists in $\mathrm{Ho}(\mathfrak{M})$, and is given by $\left(\bigoplus_{i\in I} QA_i, [R(\eta_i)]\right)$ where $R(\eta_i)$ represents a fibrant replacement of the canonical insersion $\eta_i \colon QA_i \to \bigoplus_{i\in I} QA_i$; see Proposition 5.28. Note that for any $J \in I_\kappa$ (recall it means $J \subseteq I$ has cardinality $|J| < \kappa$), the canonical subcoproduct insertion $\bigoplus_{i\in J} QA_i \to \bigoplus_{i\in I} QA_i$ in $\mathrm{Ho}(\mathfrak{M})$ is given by $[R(\eta_J)]$. Indeed since, in $\mathcal{A}$, the canonical insertion $\eta_J \colon \bigoplus_{i\in J} QA_i \to \bigoplus_{i\in I} QA_i$ uniquely satisfies $\eta_i = \eta_J \circ \eta_i^J$ for all $i \in J$, we get that $[R(\eta_i)] = [R(\eta_J)] \circ \left[R\left(\eta_i^J\right)\right]$ for all $i \in J$. This proves $[R(\eta_J)]$ is the canonical subcoproduct insertion $\bigoplus_{i\in J} QA_i \to \bigoplus_{i\in I} QA_i$ in $\mathrm{Ho}(\mathfrak{M})$. (Of course in the case that $R \colon \mathcal{A} \to \mathcal{A}$ is a functorial fibrant replacement functor we have actual equality $R(\eta_i) = R(\eta_J) \circ R\left(\eta_i^J\right)$.) Similarly, for elements $J \subseteq J'$ of I_κ, let $\eta_J^{J'} \colon \bigoplus_{i\in J} QA_i \to \bigoplus_{i\in J'} QA_i$ denote the transition morphism of the $(I_\kappa, \subseteq)$-directed system $\left\{\bigoplus_{i\in J} QA_i, \eta_J^{J'}\right\}$ associated to the coproduct of $\{QA_i\}_{i\in I}$, taken in $\mathcal{A}$. Then by the same reasoning as previously, the transition morphisms of the $(I_\kappa, \subseteq)$-directed system for the coproduct of $\{A_i\}_{i\in I}$, in $\mathrm{Ho}(\mathfrak{M})$, is given by $\left[R\left(\eta_J^{J'}\right)\right]$. By Lemma 12.16, an object S is κ-small in $\mathrm{Ho}(\mathfrak{M})$ if and only if the canonical monomorphism

$$\xi_\kappa\colon \varinjlim_{J\in I_\kappa} \mathrm{Ho}(\mathfrak{M})\left(S, \bigoplus_{i\in J} QA_i\right) \to \mathrm{Ho}(\mathfrak{M})\left(S, \bigoplus_{i\in I} QA_i\right)$$

is always an epimorphism. Note that ξ_κ acts by sending a morphism

$$[h_J] \in \mathrm{Ho}(\mathfrak{M})\left(S, \bigoplus_{i\in J} QA_i\right) := \underline{\mathrm{Hom}}\left(RQS, R\left(\bigoplus_{i\in J} QA_i\right)\right)$$

to $[R(\eta_J)] \circ [h_J]$. However, by Corollary 5.17 there is a natural equivalence $\mathrm{Ho}(\mathfrak{M})(X,Y) := \underline{\mathrm{Hom}}(QX, RY)$, and since in this case $Y = \bigoplus_{i\in I} QA_i$ is already cofibrant, the rule for this equivalence translates ξ_κ to a canonical monomorphism

$$\varinjlim_{J\in I_\kappa} \underline{\mathrm{Hom}}\left(QS, R\left(\bigoplus_{i\in J} QA_i\right)\right) \to \underline{\mathrm{Hom}}\left(QS, R\left(\bigoplus_{i\in I} QA_i\right)\right)$$

acting by $[h_J] \mapsto [R(\eta_J)] \circ [h_J]$ where now $h_J\colon QS \to R\left(\bigoplus_{i\in J} QA_i\right)$. Summarizing, we have that an object S is κ-small in $\mathrm{Ho}(\mathfrak{M})$ if and only if for any $\mathcal{A}$-morphism $h\colon QS \to R\left(\bigoplus_{i\in I} QA_i\right)$, there exists $J \in I_\kappa$ and an $\mathcal{A}$-morphism $h_J\colon QS \to R\left(\bigoplus_{i\in J} QA_i\right)$ such that $[h] = [R(\eta_J)] \circ [h_J]$. □

Proposition 12.18 *Let $\mathfrak{M} = (\mathcal{Q}, \mathcal{W}, \mathcal{R})$ be a cofibrantly generated (i.e. combinatorial) abelian model structure on an exact category $(\mathcal{A}, \mathcal{E})$ as in Setup 12.2.1. Let λ be a cardinal as guaranteed in Corollary 12.14(1). Then for any object S and regular cardinal $\mu \geq \lambda$, if S has a cofibrant approximation QS, which is μ-presented as an object of $\mathcal{A}$, then S is μ-small as an object in* $\mathrm{Ho}(\mathfrak{M})$.

In particular, let μ be a regular cardinal as guaranteed in Corollary 12.14(3). Then any μ-presented object of $\mathcal{A}$ is also a μ-small object of $\mathrm{Ho}(\mathfrak{M})$.

Proof Assume S is an object with a cofibrant approximation QS, which is μ-presented ($\mu \geq \lambda$) as an object of $\mathcal{A}$. We use Lemma 12.17, along with the functorial fibrant replacement $R_\lambda\colon \mathcal{A} \to \mathcal{A}$. So let $\{A_i\}_{i\in I}$ be a given family of $\mathcal{A}$-objects and let $h\colon QS \to R_\lambda\left(\bigoplus_{i\in I} QA_i\right)$ be a given $\mathcal{A}$-morphism. We want to show there exists $J \in I_\mu$ and an $\mathcal{A}$-morphism $h_J\colon QS \to R_\lambda\left(\bigoplus_{i\in J} QA_i\right)$ such that $[h] = [R_\lambda(\eta_J)] \circ [h_J]$.

Since $\mu \geq \lambda$, any μ-direct limit also exists and is preserved by R_λ. In particular, R_λ maps the colimit cone $\left(\bigoplus_{i\in I} QA_i, \{\eta_J\}\right)$ over the $(I_\mu, \subseteq)$-direct system $\left\{\bigoplus_{i\in J} QA_i, \eta_J^{J'}\right\}$ to a colimit cone $\left(R_\lambda\left(\bigoplus_{i\in I} QA_i\right), \{R_\lambda(\eta_J)\}\right)$ over the $(I_\mu, \subseteq)$-direct system $\left\{R_\lambda\left(\bigoplus_{i\in J} QA_i\right), R_\lambda\left(\eta_J^{J'}\right)\right\}$. Since QS is μ-presented, the functor

$\mathrm{Hom}_{\mathcal{A}}(QS, -)$ preserves μ-direct limits. In particular, it implies that any $\mathcal{A}$-morphism $h\colon QS \to R_\lambda\left(\bigoplus_{i\in I} QA_i\right)$ factors through some canonical insertion $R_\lambda(\eta_J)$ into the colimit. That is, there exists some $J \in I_\mu$ and an $\mathcal{A}$-morphism $h_J\colon QS \to R_\lambda\left(\bigoplus_{i\in J} QA_i\right)$ such that $h = R_\lambda(\eta_J) \circ h_J$. So the result follows from Lemma 12.17 which only asks for $[h] = [R_\lambda(\eta_J)] \circ [h_J]$; but the functoriality of the fibrant replacement R_λ gives an actual equality.

Now note that the final assertion of the proposition is immediate by taking $Q = Q_\lambda$, and where μ is any regular cardinal as in Corollary 12.14(3). □

12.5 Well-Generated Homotopy Categories

We now use the tools developed in this chapter to show that $\mathrm{Ho}(\mathfrak{M})$ is a well-generated triangulated category, whenever $\mathfrak{M}$ is a cofibrantly generated abelian model structure on an exact category $(\mathcal{A}, \mathcal{E})$ as in Setup 12.2.1 (i.e. a combinatorial abelian model structure). We start with the definition of well generated, which may actually be assigned to any additive category with coproducts.

Definition 12.19 Let $\mathcal{A}$ be an additive category with coproducts and let κ be a regular cardinal. We say that $\mathcal{A}$ is **κ-well generated (by $\mathcal{S}$)** if there exists a set $\mathcal{S}$ of objects of $\mathcal{A}$ such that the following hold.

(1) $\mathcal{S}$ is a set of weak generators for $\mathcal{A}$. Here we mean, if $\mathrm{Hom}_{\mathcal{A}}(S, A) = 0$ for all $S \in \mathcal{S}$ then $A = 0$.
(2) Each object $S \in \mathcal{S}$ is κ-small.
(3) For any morphism $f\colon S \to \bigoplus_{i\in I} A_i$ with $S \in \mathcal{S}$, there exist objects $S_i \in \mathcal{S}$ and morphisms $f_i\colon S_i \to A_i$ such that f factors as

$$S \xrightarrow{f'} \bigoplus_{i\in I} S_i \xrightarrow{\oplus f_i} \bigoplus_{i\in I} A_i.$$

An additive category with coproducts is called **well generated** if it is κ-well generated for some regular cardinal κ.

Note that if $\mathcal{U}$ is any set of (categorical) generators for an additive category $\mathcal{A}$, then it is automatically a set of weak generators as in (1). Indeed assume $\mathrm{Hom}_{\mathcal{A}}(U, A) = 0$ for all $U \in \mathcal{U}$ and by way of contradiction asssume $A \neq 0$, or equivalently, $1_A \neq 0_A$. Then by the standard definition of generator there would exist a $U \in \mathcal{U}$ and a morphism $f\colon U \to A$ such that $1_A \circ f \neq 0_A \circ f$, or, $f \neq 0$. This contradicts the assumption that $\mathrm{Hom}_{\mathcal{A}}(U, A) = 0$ for all $U \in \mathcal{U}$.

In fact, any κ-accessible additive category is κ-well generated by the (generating) "set" of all κ-presented objects. See Exercise 12.5.2. We note too that if $\mathcal{A} = \mathcal{T}$ has the structure of a triangulated category, then condition (1) in Definition 12.19 implies our previous definition of *weak generators*, given in Definition 8.20.

At this point, the biggest challenge left in showing Ho($\mathfrak{M}$) to be well generated is proving condition (3). The following lemma will be used for this.

Lemma 12.20 *Let $\mathcal{A}$ be an additive category with coproducts. Assume $\{A_i\}_{i\in I}$ is a family of objects for which each A_i is given to be a κ-direct limit $A_i = \varinjlim_{j_i \in J_i} S_{j_i}$ of a κ-directed $(J_i, \leq_i)$-system $\left\{S_{j_i}, f_{j_i j_i'}\right\}_{j_i \in J_i}$.*

(1) *The cartesian set product $\prod_{i\in I} J_i = \{(j_i)_{i\in I} \mid j_i \in J_i\}$, along with the usual pointwise ordering $(j_i)_{i\in I} \leq_\Pi (j_i')_{i\in I} \iff j_i \leq_i j_i'$ for each $i \in I$, is a κ-directed set.*

(2) *The functor (i.e. diagram) $\prod_{i\in I} J_i \to \mathcal{A}$ taking*

$$(j_i)_{i\in I} \mapsto \bigoplus_{i\in I} S_{j_i}$$

and

$$(j_i)_{i\in I} \leq_\Pi (j_i')_{i\in I} \mapsto \bigoplus_{i\in I} S_{j_i} \xrightarrow{\oplus f_{j_i j_i'}} \bigoplus_{i\in I} S_{j_i'},$$

where we are taking the coproducts over the transition morphisms $f_{j_i j_i'}$ coming from each direct system, produces a κ-direct system $\left\{\bigoplus_{i\in I} S_{j_i}, \oplus f_{j_i j_i'}\right\}$ over $(\prod_{i\in I} J_i, \leq_\Pi)$.

(3) *We have*

$$\bigoplus_{i\in I} A_i = \varinjlim_{(j_i)\in\prod J_i} \left(\bigoplus_{i\in I} S_{j_i}\right).$$

Given any "thread" $(j_i)_{i\in I} \in \prod_{i\in I} J_i$, and letting $\eta_{j_i} : S_{j_i} \to A_i$ denote the canonical map into A_i, then $\oplus_{i\in I}\eta_{j_i} : \bigoplus_{i\in I} S_{j_i} \to \bigoplus_{i\in I} A_i$ is the canonical map into the κ-direct limit $\bigoplus_{i\in I} A_i$.

Proof For (1), it is clear that $\leq_\Pi$ inherits the Reflexive and Transitive properties pointwise from the $\leq_i$ (and the Antisymmetric property if each $\leq_i$ is such). Now suppose that we have a subset $K \subseteq \prod_{i\in I} J_i$ of cardinality $|K| < \kappa$. Then for each coordinate $i \in I$, we have a subset $K_i \subseteq J_i$ defined by

$$K_i := \{j_i \in J_i \mid \exists\, (j_i)_{i\in I} \in K \text{ with } i\text{th-coordinate } j_i\}.$$

Clearly $|K_i| < \kappa$ for each $i \in I$. Since each $(J_i, \leq_i)$ is κ-directed we may choose, for each i, an upper bound $\ell_i \in J_i$, so satisfying $j_i \leq_i \ell_i$ for all $j_i \in K_i$. Then we have $(j_i)_{i\in I} \leq_\Pi (\ell_i)_{i\in I}$ for all $(j_i)_{i\in I} \in K$. So $(\ell_i)_{i\in I} \in \prod_{i\in I} J_i$ is an upper bound for K, proving $\prod_{i\in I} J_i$ is a κ-directed set.

The checking of statements (2) and (3) is straightforward and we leave it as Exercise 12.5.1. □

Exercise 12.5.1 Verify statements (2) and (3) of Lemma 12.20.

The next exercise may be helpful for understanding the definition of well generated. It also indicates how, in the case of triangulated categories where we typically do not have colimits beyond coproducts, well generated serves as a substitute for accessible (or locally presentable).

Exercise 12.5.2 Let $\mathcal{A}$ be an additive category with coproducts. Show that if $\mathcal{A}$ is κ-accessible, then it is κ-well generated by $\mathrm{Pres}_\kappa(\mathcal{A})$. Here again, $\mathrm{Pres}_\kappa(\mathcal{A})$ denotes a set of isomorphism representatives for all κ-presented objects in $\mathcal{A}$. (*Hint*: Use Lemma 12.20 to help with condition (3). The argument is an easier version of what is proved ahead in Theorem 12.21.)

We now can prove the main theorem of this chapter. See also the two corollaries that follow.

Theorem 12.21 (Rosický [2005]) *Assume $(\mathcal{A}, \mathcal{E})$ is an exact category satisfying the conditions of Setup 12.2.1. Let $\mathfrak{M} = (\mathcal{Q}, \mathcal{W}, \mathcal{R})$ be a cofibrantly generated (i.e. combinatorial) and hereditary model structure on $(\mathcal{A}, \mathcal{E})$. Then* $\mathrm{Ho}(\mathfrak{M})$ *is a well generated triangulated category.*

Specifically, if we let μ be any regular cardinal as guaranteed by Corollary 12.14(3), then $\mathrm{Ho}(\mathfrak{M})$ *is μ-well generated by the set* $\mathrm{Pres}_\mu(\mathcal{A})$ *of all (isomorphism representatives for all) μ-presented objects.*

Remark 12.22 The statement is even true without the hereditary hypothesis, but we add it to simplify our proof of condition (1) of Definition 12.19. In examples arising in applications, abelian model structures tend to be hereditary. Our proofs of conditions (2) and (3) do not use the hereditary condition and Hovey [1999, theorem 7.3.1] may be used to remedy condition (1) should the hereditary hypothesis ever be absent.

Proof By Corollary 12.14(3), we may choose a regular cardinal μ such that $\kappa \lhd \lambda \lhd \mu$ and each of the four functors Q_λ, R_λ, $\tilde{Q}_\lambda$, and $\tilde{R}_\lambda$ preserve μ-presented objects as well as μ-direct limits. In particular, $Q_\lambda A$ remains μ-presented whenever A is μ-presented. The goal is to show that $\mathrm{Ho}(\mathfrak{M})$ is μ-well generated by $\mathrm{Pres}_\mu(\mathcal{A})$. Firstly, condition (2) of Definition 12.19 has already been shown; it is the last assertion of Proposition 12.18.

Let us turn to condition (3) of Definition 12.19. For this we let S be a μ-presented object and we are to consider an arbitrary morphism $S \to \coprod_{i \in I} A_i$ into the coproduct of some set of objects $\{A_i\}_{i \in I}$. As in the proofs of Lemma

12.17 and Proposition 12.18, such a morphism in the homotopy category is represented by the homotopy class of an $\mathcal{A}$-morphism $h\colon Q_\lambda S \to R_\lambda\left(\bigoplus_{i\in I} Q_\lambda A_i\right)$. Now since $\mathcal{A}$ is μ-accessible, we may write each $A_i = \varinjlim_{j_i\in J_i} S_{j_i}$ as a μ-direct limit of μ-presented objects S_{j_i}. Since Q_λ preserves μ-direct limits we have $Q_\lambda A_i = \varinjlim_{j_i\in J_i} Q_\lambda S_{j_i}$ is again a μ-direct limit, and moreover each $Q_\lambda S_{j_i}$ remains μ-presented by our choice of μ. So now by Lemma 12.20(3), we may express the $\mathcal{A}$-coproduct as the μ-direct limit

$$\bigoplus_{i\in I} Q_\lambda A_i = \varinjlim_{(j_i)\in\prod J_i} \left(\bigoplus_{i\in I} Q_\lambda S_{j_i}\right),$$

and for a "thread" $(j_i)_{i\in I} \in \prod_{i\in I} J_i$, the canonical insertion map is given by $\oplus_{i\in I} Q_\lambda(\eta_{j_i})\colon \bigoplus_{i\in I} Q_\lambda S_{j_i} \to \bigoplus_{i\in I} Q_\lambda A_i$. Since $R_\lambda\colon \mathcal{A}\to\mathcal{A}$ preserves μ-direct limits, we have

$$R_\lambda\left(\bigoplus_{i\in I} Q_\lambda A_i\right) = \varinjlim_{(j_i)\in\prod J_i} R_\lambda\left(\bigoplus_{i\in I} Q_\lambda S_{j_i}\right),$$

and now with $R_\lambda\left(\oplus_{i\in I} Q_\lambda(\eta_{j_i})\right) : R_\lambda\left(\bigoplus_{i\in I} Q_\lambda S_{j_i}\right) \to R_\lambda\left(\bigoplus_{i\in I} Q_\lambda A_i\right)$ serving as canonical insertions. So since $Q_\lambda S$ is μ-presented, the morphism $h\colon Q_\lambda S \to R_\lambda\left(\bigoplus_{i\in I} Q_\lambda A_i\right)$ factors as

$$Q_\lambda S \xrightarrow{h_{(j_i)}} R_\lambda\left(\bigoplus_{i\in I} Q_\lambda S_{j_i}\right) \xrightarrow{R_\lambda(\oplus_{i\in I} Q_\lambda(\eta_{j_i}))} R_\lambda\left(\bigoplus_{i\in I} Q_\lambda A_i\right)$$

for some morphism $h_{(j_i)}$ associated to some thread $(j_i)_{i\in I} \in \prod_{i\in I} J_i$. We note that $R_\lambda\left(\oplus_{i\in I} Q_\lambda\left(\eta_{j_i}\right)\right)$ represents a coproduct of morphisms $\coprod_{i\in I} S_{j_i} \to \coprod_{i\in I} A_i$ in $\mathrm{Ho}(\mathfrak{M})$, as required. Indeed it coincides with the coproduct $\coprod_{i\in I}\left[\gamma\left(\eta_{j_i}\right)\right] : \coprod_{i\in I} S_{j_i} \to \coprod_{i\in I} A_i$ taken in $\mathrm{Ho}(\mathfrak{M})$. To see this, recall that the canonical insertion $S_{j_i} \to \coprod_{i\in I} S_{j_i}$ is represented by $R_\lambda\left(\eta_{Q_\lambda S_{j_i}}\right)\colon R_\lambda\left(Q_\lambda S_{j_i}\right)\to R_\lambda\left(\bigoplus_{i\in I} Q_\lambda S_{j_i}\right)$, and likewise the canonical insertion $A_i \to \coprod_{i\in I} A_i$ is represented by $R_\lambda(\eta_{Q_\lambda A_i})$: $R_\lambda(Q_\lambda A_i) \to R_\lambda\left(\bigoplus_{i\in I} Q_\lambda A_i\right)$. The map $\coprod_{i\in I}\left[\gamma\left(\eta_{j_i}\right)\right]$ is characterized as the unique morphism in $\mathrm{Ho}(\mathfrak{M})$ making these canonical insertions commute with each $\gamma\left(\eta_{j_i}\right) = \left[R_\lambda\left(Q_\lambda\left(\eta_{j_i}\right)\right)\right]$. So it amounts to showing

$$\left[R_\lambda\left(\oplus_{i\in I} Q_\lambda\left(\eta_{j_i}\right)\right)\right] \circ \left[R_\lambda\left(\eta_{Q_\lambda S_{j_i}}\right)\right] = \left[R_\lambda\left(\eta_{Q_\lambda A_i}\right)\right] \circ \left[R_\lambda\left(Q_\lambda\left(\eta_{j_i}\right)\right)\right].$$

But we certainly have the commutativity

$$\left(\oplus_{i\in I} Q_\lambda\left(\eta_{j_i}\right)\right) \circ \eta_{Q_\lambda S_{j_i}} = \eta_{Q_\lambda A_i} \circ Q_\lambda\left(\eta_{j_i}\right),$$

coming from the $\mathcal{A}$-coproduct $\oplus_{i\in I} Q_\lambda\left(\eta_{j_i}\right)\colon \bigoplus_{i\in I} Q_\lambda S_{j_i} \to \bigoplus_{i\in I} Q_\lambda A_i$. So the wanted equation follows by the functoriality of R_λ. This completes the proof of condition (3).

Finally, we turn to prove condition (1), assuming the hereditary hypothesis (see Remark 12.22). So assume $\mathrm{Ho}(\mathfrak{M})(X, Y) = 0$ for all $X \in \mathrm{Pres}_\mu(\mathcal{A})$. We will argue that $Y = 0$. Referring to our definition of cofibrantly generated given in Definition 9.38, let $\mathcal{S}$ be such a set that cogenerates $(Q, \mathcal{R}_W)$. Also, take $\mathcal{U} = \{U_i\}$ to be a set of trivially cofibrant generators for $\mathcal{E}$, and with this $\mathcal{U}$ apply Proposition 9.29 to obtain a set I_S of generating monics for $\mathcal{S}^\perp = \mathcal{R}_W$. So for each $S \in \mathcal{S}$, there is a short exact sequence $\Omega S \rightarrowtail \bigoplus U_i \twoheadrightarrow S$ where $k\colon \Omega S \rightarrowtail \bigoplus U_i$ is one of the admissible monics in the set I_S. We use the notation ΩS here because $\bigoplus U_i$ is trivial (indeed trivially cofibrant). By construction, k is a direct summand of the single admissible monic $v\colon L \rightarrowtail V$ described in Definition 9.38. The proof of Corollary 12.14 constructs the regular cardinal λ so that L is λ-presented. Hence the direct summand ΩS is also λ-presented, thus it is μ-presented. So if $\mathrm{Ho}(\mathfrak{M})(X, Y) = 0$ for all $X \in \mathrm{Pres}_\mu(\mathcal{A})$, then this includes $\mathrm{Ho}(\mathfrak{M})(\Omega S, Y) = 0$ for all $S \in \mathcal{S}$. Then by Theorem 8.18 (note also Exercises 8.5.1 and 8.5.2) we have, for each $S \in \mathcal{S}$,

$$0 = \mathrm{Ho}(\mathfrak{M})(\Omega S, Y) \cong \mathrm{Ext}^1_{\mathcal{E}}(S, RY).$$

Hence $RY \in \mathcal{S}^\perp = \mathcal{R}_W$. In particular, $RY \in \mathcal{W}$. By the 2 out of 3 property on $\mathcal{W}$, this happens if and only if $Y \in \mathcal{W}$. Since $\mathcal{W}$ is the class of 0 objects in $\mathrm{Ho}(\mathfrak{M})$, this proves $\mathrm{Pres}_\mu(\mathcal{A})$ is a set of weak generators for $\mathrm{Ho}(\mathfrak{M})$.

This completes the proof that $\mathrm{Ho}(\mathfrak{M})$ is μ-well generated by the set $\mathrm{Pres}_\mu(\mathcal{A})$.

□

In particular we have the following powerful methods for constructing well-generated triangulated categories; see Remark 12.6 following Setup 12.2.1.

Corollary 12.23 *Assume $(\mathcal{A}, \mathcal{E})$ is an accessible exact category with coproducts and a set of admissible generators. If $\mathfrak{M} = (Q, \mathcal{W}, \mathcal{R})$ is a cofibrantly generated and hereditary model structure on $(\mathcal{A}, \mathcal{E})$, then* $\mathrm{Ho}(\mathfrak{M})$ *is a well-generated triangulated category.*

Corollary 12.24 *Assume $(\mathcal{A}, \mathcal{E})$ is an exact structure with a set of admissible generators, on a locally presentable additive category $\mathcal{A}$. If $\mathfrak{M} = (Q, \mathcal{W}, \mathcal{R})$ is a cofibrantly generated and hereditary model structure on $(\mathcal{A}, \mathcal{E})$, then* $\mathrm{Ho}(\mathfrak{M})$ *is a well-generated triangulated category.*

Appendix A

Hovey's Correspondence for General Exact Categories

Here we let $(\mathcal{A}, \mathcal{E})$ denote an arbitrary exact category; we allow that it may not be weakly idempotent complete. The original definition of a model category given in Quillen [1967] did not require that it be *closed*, meaning at least one of the classes of cofibrations, fibrations, or weak equivalences may not be closed under retracts. Here we give a bijective correspondence between Hovey Ext-triples satisfying the 2 out of 3 Axiom on $\mathfrak{M}$-weak equivalences and not necessarily closed abelian model structures on $(\mathcal{A}, \mathcal{E})$. We start by giving Quillen's original definition of a model category. Actually, we list the axioms slightly differently but the reader can easily verify that the following definition is equivalent to the definition given in Quillen [1967].

Definition A.1 (Quillen [1967]) A **classical model category**, which we also might call a **(not necessarily closed) model category**, is a category C along with a triple $\mathcal{M} = (Cof, We, Fib)$ of classes of morphisms in C, respectively called *cofibrations*, *weak equivalences*, and *fibrations*, all satisfying axioms **(M0)–(M4)** from Definition 4.23 along with the Pullbacks and Pushouts Axiom **(QM5)** as follows. By definition, we call a map a *trivial cofibration* if it is both a cofibration and a weak equivalence. Similarly a *trivial fibration* is both a fibration and a weak equivalence.

(QM5) (Pullbacks and Pushouts Axiom) The base change map of any pullback along any fibration (resp. trivial fibration) is again a fibration (resp. trivial fibration). That is, the pullback of any fibration (resp. trivial fibration) along any morphism is again a fibration (resp. trivial fibration). Dually, the cobase change map of any pushout along any cofibration (resp. trivial cofibration) is again a cofibration (resp. trivial cofibration).

By a *classical model structure*, or a *(not necessarily closed) model structure*, we mean a triple $\mathcal{M} = (Cof, We, Fib)$ on C satisfying axioms (M1)–(M4) and (QM5).

The next exercise asks the reader to show that any closed model structure on a category C is necessarily a classical model structure. Conversely, a classical model structure is closed if and only if each of the classes *Cof*, *We*, and *Fib* are closed under retracts.

Exercise A.0.1 Show that any closed model structure satisfies the Pullbacks and Pushouts Axiom (QM5); see Exercise 4.6.1. Therefore a closed model structure is precisely a classical model structure satisfying the Retracts Axiom (M5).

The entire discussion following Definition 4.23 applies here too. In particular, we call $\mathcal{M} = (Cof, We, Fib)$ a *classical abelian model structure* if it satisfies Definition 4.24. Similarly, given such an $\mathcal{M} = (Cof, We, Fib)$, we define *trivial* objects, *(trivially) cofibrant* objects, and *(trivially) fibrant* objects in the same way we did there.

Now we give the generalization of Hovey's correspondence to classical abelian model structures on general exact categories.

Theorem A.2 (Hovey's Correspondence for Arbitrary Exact Categories) *Let* $(\mathcal{A}, \mathcal{E})$ *be an arbitrary exact category.*

(1) *There is a bijective correspondence between the class of all Hovey Ext-triples* $\mathfrak{M} = (\mathcal{Q}, \mathcal{W}, \mathcal{R})$ *on* $(\mathcal{A}, \mathcal{E})$ *satisfying the 2 out of 3 Axiom (M4) on* $\mathfrak{M}$*-weak equivalences, and the class of all classical abelian model structures on* $(\mathcal{A}, \mathcal{E})$*. The correspondence acts by*

$$\mathfrak{M} = (\mathcal{Q}, \mathcal{W}, \mathcal{R}) \hookrightarrow (\mathfrak{M}\textit{-cofibrations}, \mathfrak{M}\textit{-weak equivalences}, \mathfrak{M}\textit{-fibrations})$$

and the inverse is given by

$$\mathcal{M} = (Cof, We, Fib) \hookrightarrow (\textit{Cofibrant objects}, \textit{Trivial objects}, \textit{Fibrant objects}).$$

In the case that $\mathcal{A}$ *is a weakly idempotent complete category, then the 2 out of 3 Axiom on* $\mathfrak{M}$*-weak equivalences is automatic. Thus we have a bijective correspondence between Hovey Ext-triples and classical abelian model structures on* $(\mathcal{A}, \mathcal{E})$*.*

(2) *The correspondence restricts to a bijective correspondence between Hovey triples* $\mathfrak{M} = (\mathcal{Q}, \mathcal{W}, \mathcal{R})$ *on* $(\mathcal{A}, \mathcal{E})$ *satisfying both*
   (a) *the 2 out of 3 Axiom (M4) on* $\mathfrak{M}$*-weak equivalences, and*
   (b) *the Retracts Axiom (M5) on* $\mathfrak{M}$*-cofibrations,* $\mathfrak{M}$*-weak equivalences, and* $\mathfrak{M}$*-fibrations,*

   and (closed) abelian model structures on $(\mathcal{A}, \mathcal{E})$*. In the case that* $\mathcal{A}$ *is a weakly idempotent complete category, these two conditions are automatic and we recover Hovey's Correspondence from Theorem 4.25.*

Proof Much of the proof is the same or analogous to the proof of Theorem 4.25. So we will reference that proof, pointing out modifications.

Let us start by proving (1). Given a Hovey Ext-triple $\mathfrak{M} = (\mathcal{Q}, \mathcal{W}, \mathcal{R})$ satisfying the 2 out of 3 Axiom on $\mathfrak{M}$-weak equivalences, we will argue that

$$(\mathfrak{M}\text{-cofibrations}, \mathfrak{M}\text{-weak equivalences}, \mathfrak{M}\text{-fibrations})$$

determines a classical abelian model structure. First, the proofs of the Lifting Axiom (M2) and Factorization Axiom (M3) from Theorem 4.25 work the same way for Ext-pairs. The 2 out of 3 Axiom (M4) holds by hypothesis. The proof of the Subcategories Axiom (M1) also holds as it did in the proof of Theorem 4.25: Corollary 4.9 holds because the definition of Ext-pair requires that the classes $\mathcal{Q}$, $\mathcal{Q}_{\mathcal{W}}$, $\mathcal{R}$, and $\mathcal{R}_{\mathcal{W}}$ are each closed under extensions and contain 0.

It is left to prove the Pullbacks and Pushouts Axiom (QM5). We will show that any pullback of any (trivial) $\mathfrak{M}$-fibration is again a (trivial) $\mathfrak{M}$-fibration. (The dual argument will correspond to pushouts of (trivial) $\mathfrak{M}$-cofibrations.) So let $p\colon A \twoheadrightarrow B$ be any (trivial) $\mathfrak{M}$-fibration and suppose that $f\colon C \to B$ is an arbitrary morphism. The axioms of exact categories ensure that the pullback of p along f exists and yields another admissible epic p'. In fact, by Lemma 1.10, we get a commutative diagram with short exact rows as shown in the following:

$$\begin{array}{ccccc}
\operatorname{Ker} p & \rightarrowtail & P & \overset{p'}{\twoheadrightarrow} & C \\
\| & & \downarrow{\scriptstyle f'} & & \downarrow{\scriptstyle f} \\
\operatorname{Ker} p & \rightarrowtail & A & \overset{p}{\twoheadrightarrow} & B.
\end{array}$$

So clearly (trivial) $\mathfrak{M}$-fibrations are stable under taking pullbacks.

This completes the proof that

$$(\mathfrak{M}\text{-cofibrations}, \mathfrak{M}\text{-weak equivalences}, \mathfrak{M}\text{-fibrations})$$

is a classical abelian model structure whenever $\mathfrak{M} = (\mathcal{Q}, \mathcal{W}, \mathcal{R})$ is a Hovey Ext-triple.

The converse also follows by arguments similar to those in the proof of Theorem 4.25. Indeed assume we are given a classical model structure $\mathcal{M} = (\mathit{Cof}, \mathit{We}, \mathit{Fib})$ that is abelian in the sense of Definition 4.24, and let $(\mathcal{Q}, \mathcal{W}, \mathcal{R})$ be the associated triple of cofibrant objects, trivial objects, and fibrant objects. We must show that $(\mathcal{Q}, \mathcal{W}, \mathcal{R})$ is a Hovey Ext-triple. Set $\mathcal{Q}_{\mathcal{W}} := \mathcal{Q} \cap \mathcal{W}$ to be the class of trivially cofibrant objects, and $\mathcal{R}_{\mathcal{W}} := \mathcal{W} \cap \mathcal{R}$ to be the class of trivially fibrant objects. We first wish to see that $(\mathcal{Q}_{\mathcal{W}}, \mathcal{R})$ and $(\mathcal{Q}, \mathcal{R}_{\mathcal{W}})$ are each complete Ext-pairs. We only do this for $(\mathcal{Q}_{\mathcal{W}}, \mathcal{R})$ as the proof for $(\mathcal{Q}, \mathcal{R}_{\mathcal{W}})$ is similar. The fact that $\operatorname{Ext}^1_{\mathcal{E}}(\mathcal{Q}_{\mathcal{W}}, \mathcal{R}) = 0$ follows immediately from the Lifting Axiom (M2) and Proposition 2.11. To finish showing we have an Ext-pair, we need to show that $\mathcal{Q}_{\mathcal{W}}$ and $\mathcal{R}$ each contain 0 and are closed under extensions. Let us just show this for $\mathcal{Q}_{\mathcal{W}}$. First, the Subcategories Axioms (M1) implies that any isomorphism is a trivial cofibration. Since isomorphisms are admissible monics with cokernel 0, it means that $0 \in \mathcal{Q}_{\mathcal{W}}$. Axiom (M1) also implies that trivial cofibrations are closed under compositions. It follows that $\mathcal{Q}_{\mathcal{W}}$ is closed under extensions. Indeed if $Q' \overset{i}{\rightarrowtail} X \overset{p}{\twoheadrightarrow} Q$ is a short exact sequence with $Q, Q' \in \mathcal{Q}_{\mathcal{W}}$, then $0 \rightarrowtail Q' \overset{i}{\rightarrowtail} X$ is the composition of two trivial cofibrations and hence is a trivial cofibration. In other words, X is trivially cofibrant. This completes the proof that $(\mathcal{Q}_{\mathcal{W}}, \mathcal{R})$ is an Ext-pair. The fact that it is a complete Ext-pair follows from the Factorization Axiom (M3), along with Proposition 2.13.

The proof that $\mathcal{W}$ satisfies the 2 out of 3 property on short exact sequences follows as shown in the proof of Theorem 4.25. However, since $\mathcal{A}$ is not necessarily weakly idempotent complete we must revise the argument that the morphism j in Diagram ($\star$) in that proof must be a weak equivalence. We do this as follows: First, note that by Proposition 1.12 we get that the left square of that diagram is a pushout square. Letting $q = \operatorname{cok} i$, one can directly show that qk is the cokernel of j. (This is Exercise 1.8(2), first use the universal property of the pushout square followed by the universal property of the cokernel q.) Since i and f are admissible monics, it follows that the composition $kj = if$ is an admissible monic. Furthermore k is an admissible monic. Since j has a cokernel it follows from Obscure Axiom (Corollary 1.19) that j is also an admissible monic. Since the target of q is trivially cofibrant we see that the target of qk is also trivially cofibrant. Therefore j is a trivial cofibration. In particular j is a weak equivalence.

This completes the proof that $\mathfrak{M} = (\mathcal{Q}, \mathcal{W}, \mathcal{R})$ forms a Hovey Ext-triple. As in the proof of Theorem 4.25, the 2 out of 3 Axiom (M4) implies that the weak equivalences of $\mathcal{M}$ coincide with the $\mathfrak{M}$-weak equivalences.

Finally, in the case that $\mathcal{A}$ is weakly idempotent complete, then satisfaction of the 2 out of 3 Axiom on $\mathfrak{M}$-weak equivalences is automatic from the 2 out of 3 property on $\mathcal{W}$, by Theorem 4.21. Thus we have a bijective correspondence between Hovey Ext-triples and classical abelian model structures in this case. This completes the proof of (1).

To prove (2), one can similarly trace the proof of Theorem 4.25. However, we can employ the result of Exercise A.0.1. Indeed if we are handed a Hovey triple $\mathfrak{M} = (\mathcal{Q}, \mathcal{W}, \mathcal{R})$ satisfying property (a), then by what we just proved in (1) we get a classical abelian model structure. But property (b) is precisely the Rectracts Axiom (M5), so Exercise A.0.1 tells us the model structure is closed. Conversely, suppose we are given a closed model structure $\mathcal{M} = (\mathit{Cof}, \mathit{We}, \mathit{Fib})$

that is abelian in the sense of Definition 4.24. Then by what we just proved in (1), we obtain a Hovey Ext-triple $\mathfrak{M} = (Q, \mathcal{W}, \mathcal{R})$ on $(\mathcal{A}, \mathcal{E})$, satisfying properties (a) and (b), as they correspond to axioms (M4) and (M5). But it follows from Corollary 2.5 that $\mathfrak{M}$ must actually be a Hovey triple, since property (b) certainly implies that each of the classes Q, $Q_{\mathcal{W}}$, $\mathcal{R}$, and $\mathcal{R}_{\mathcal{W}}$ are closed under direct summands. □

Remark A.3 A subtle point the author has noticed is that to develop the theory of abelian model categories, the full statement of the Retracts Axiom (M5) is rarely, if ever, needed. It seems to suffice that $\mathfrak{M} = (Q, \mathcal{W}, \mathcal{R})$ be a classical abelian model structure for which only Q, $\mathcal{W}$ and $\mathcal{R}$ are each closed under retracts. In other words, that $\mathfrak{M}$ be a Hovey triple satisfying the 2 out of 3 Axiom on $\mathfrak{M}$-weak equivalences.

In light of Remark A.3, we point out the following corollary of Theorem A.2(2), and then provide a counterexample, related to the Retracts Axiom (M5) and Remark A.3.

Corollary A.4 *Let $(\mathcal{A}, \mathcal{E})$ be any exact category possessing a (closed) abelian model structure $\mathcal{M} = (Cof, We, Fib)$. Then the associated triple $\mathfrak{M} = (Q, \mathcal{W}, \mathcal{R})$ of cofibrant objects, trivial objects, and fibrant objects is a Hovey triple and satisfies the 2 out of 3 Axiom on $\mathfrak{M}$-weak equivalences.*

Examples and Counterexampes Involving Frobenius Categories Recall that a Frobenius category is an exact category $(\mathcal{A}, \mathcal{E})$ with enough injectives and enough projectives and such that the projective and injective objects coincide. The next proposition shows that any Frobenius category provides an example of a classical abelian model structure and it is closed if and only if $\mathcal{A}$ is weakly idempotent complete.

Proposition A.5 *Let $(\mathcal{A}, \mathcal{E})$ be an exact category. Then $(\mathcal{A}, \mathcal{E})$ is Frobenius with ω the class of projective–injective objects if and only if $\mathfrak{M} = (\mathcal{A}, \omega, \mathcal{A})$ is a Hovey triple. In this case, it automatically satisfies the 2 out of 3 Axiom (M4) on $\mathfrak{M}$-weak equivalences, which are precisely the ω-stable equivalences. Moreover, the Retracts Axiom (M5) holds if and only if the underlying category $\mathcal{A}$ is weakly idempotent complete.*

Proof It is clear that $(\mathcal{A}, \omega, \mathcal{A})$ is a Hovey triple if and only if it is a Hovey Ext-triple if and only if $(\mathcal{A}, \mathcal{E})$ is Frobenius with ω the class of projective-injectives. Recall that an ω-stable equivalence is, by definition, an isomorphism in $\mathrm{St}_{\omega}(\mathcal{A})$. Corollary 5.4 implies that a morphism is an $\mathfrak{M}$-weak equivalence if and only if it is an ω-stable equivalence. (As noted shortly in Remark A.7, the corollary holds without assuming weak idempotent completeness.) Since isomorphisms in $\mathrm{St}_{\omega}(\mathcal{A})$ certainly satisfy the 2 out of 3 Axiom, we see that $\mathfrak{M} = (\mathcal{A}, \omega, \mathcal{A})$ determines a classical abelian model structure by Theorem A.2. In the case that $\mathcal{A}$ is weakly idempotent complete then this is a closed abelian model structure by Theorem 4.25. On the other hand, if the classical model structure resulting from $\mathfrak{M} = (\mathcal{A}, \omega, \mathcal{A})$ is closed, then the class of all admissible monomorphisms (or admissible epimorphisms) is closed under retracts. By Proposition 1.31 this implies that $\mathcal{A}$ must be weakly idempotent complete. □

Note that Proposition A.5 confirms it is possible for $\mathfrak{M}$-weak equivalences to satisfy the 2 out of 3 Axiom even if $\mathcal{A}$ is not weakly idempotent complete, for there exist Frobenius categories that are not weakly idempotent complete. Moreover, a Hovey triple satisfying the 2 out of 3 Axiom on weak equivalences may not satisfy the Retracts Axiom (M5). We now give an explicit example of these facts based on Eilenberg's trick.

Example A.6 We wish to make the point that a Hovey triple satisfying the 2 out of 3 Axiom still may not determine a *closed* abelian model structure. In particular, the Retracts Axiom (M5) may fail if the underlying category is not weakly idempotent complete. For a particular example, let R be a ring admitting a nonfree projective module P. It must be a direct summand of a free module, so let $P \bigoplus Q$ be free. Define free R-modules by setting

$$F := \bigoplus_{n \in \mathbb{N}} \left(P_n \bigoplus Q_n\right) = \left(P \bigoplus Q\right) \bigoplus \left(P \bigoplus Q\right) \bigoplus \left(P \bigoplus Q\right) \bigoplus \cdots$$

and

$$F' := \bigoplus_{n \in \mathbb{N}} \left(Q_n \bigoplus P_n\right) = \left(Q \bigoplus P\right) \bigoplus \left(Q \bigoplus P\right) \bigoplus \left(Q \bigoplus P\right) \bigoplus \cdots .$$

There is an obvious isomorphism $\tau\colon F' \to F$ defined by

$$\Sigma_{n \in \mathbb{N}}(x_n, y_n) \mapsto \Sigma_{n \in \mathbb{N}}(y_n, x_n),$$

and there is an R-module epimorphism $[0 \quad \tau]\colon P \bigoplus F' \to F$ defined by mapping all elements of P to 0 and acting by τ on elements of F'. This map has kernel P. Adding a copy of F' to the domain, we see that $[0 \quad \tau]$ is a retract of the map $[0 \quad \tau \quad 0]$ as shown in the commutative diagram:

$$\begin{array}{ccccc}
P \bigoplus F' & \xrightarrow{i_{P \bigoplus F'}} & P \bigoplus F' \bigoplus F' & \xrightarrow{\pi_{P \bigoplus F'}} & P \bigoplus F' \\
\downarrow{\scriptstyle [0 \quad \tau]} & & \downarrow{\scriptstyle [0 \quad \tau \quad 0]} & & \downarrow{\scriptstyle [0 \quad \tau]} \\
F & \xrightarrow[1_F]{} & F & \xrightarrow[1_F]{} & F.
\end{array}$$

Note that $P \bigoplus F' \cong F$ is a free R-module. In particular, the diagram exists in the Frobenius category Free(R), of free R-modules. We conclude that $[0 \quad \tau \quad 0]$ is an admissible epic (fibration) in the Frobenius category Free(R), but the retract $[0 \quad \tau] \colon P \bigoplus F' \to F$ is not an admissible epic because it does not have a kernel. (Its kernel is P which is not free.) Letting $(\mathcal{A}, \mathcal{E}) = \text{Free}(R)$ with the split exact structure, we see that $\mathfrak{M} = (\mathcal{A}, \mathcal{A}, \mathcal{A})$ is trivially a Hovey triple, not just a Hovey Ext-triple in the exact category Free(R). However, $\mathfrak{M}$ does not determine a closed abelian model structure on Free(R) because we just showed that the $\mathfrak{M}$-fibrations are not closed under retracts.

Remark A.7 Chapter 5 was presented with the underlying assumption that $\mathcal{A}$ is weakly idempotent complete. However, most results hold for general exact categories, at the cost of explicitly requiring the 2 out of 3 Axiom from Section 4.5. Here we explicitly point out the minimum hypotheses needed (that the author is aware of) to obtain the results from each section of Chapter 5.

- *Section 5.1 – Homotopy Equivalences* All the results of Section 5.1 hold exactly as written for any Hovey triple $\mathfrak{M} = (\mathcal{Q}, \mathcal{W}, \mathcal{R})$ on any exact category $(\mathcal{A}, \mathcal{E})$; it need not be weakly idempotent complete. However, it is not enough that $\mathfrak{M}$ be just a Hovey Ext-triple – we use that $\mathcal{Q}_{\mathcal{W}}$ and $\mathcal{R}_{\mathcal{W}}$ are closed under retracts. On the other hand, the results of this section do not require the use of the 2 out of 3 Axiom of Section 4.5.
- *Section 5.2 – Bifibrant Approximations* Throughout all of Section 5.2, $\mathfrak{M} = (\mathcal{Q}, \mathcal{W}, \mathcal{R})$ only needs to be a Hovey Ext-triple on an arbitrary exact category $(\mathcal{A}, \mathcal{E})$, but $\mathfrak{M}$ must satisfy the 2 out of 3 Axiom of Section 4.5. Equivalently, $\mathfrak{M}$ should at least be a classical abelian model structure in the sense of Theorem A.2.
- *Sections 5.3–5.5* The results of these sections all hold with the exact same hypotheses on $\mathfrak{M} = (\mathcal{Q}, \mathcal{W}, \mathcal{R})$ and $(\mathcal{A}, \mathcal{E})$ as in the previous section: $\mathfrak{M}$ only needs to be a classical abelian model structure on any exact category $(\mathcal{A}, \mathcal{E})$. In particular, the Localization Theorem 5.21 holds for classical abelian model structures on general exact categories. However, for Section 5.4.1, in particular Proposition 5.23, we need $\mathfrak{M}$ to be Hovey triple (not just a Hovey Ext-triple) satisfying the 2 out of 3 Axiom.
- *Section 5.6 – Homotopy Coproducts and Products* The results of Section 5.6 hold for Hovey triples $\mathfrak{M} = (\mathcal{Q}, \mathcal{W}, \mathcal{R})$ satisfying the 2 out of 3 Axiom. $(\mathcal{A}, \mathcal{E})$ may be any exact category. The remarks at the end of the section on generalizations to Hovey Ext-triples hold, again assuming the 2 out of 3 Axiom.

Appendix B

Right and Left Homotopy Relations

In model category theory, the notion of right homotopic maps is defined in terms of path objects. The dual notion of left homotopic maps is defined in terms of cylinder objects. It turns out that the two notions coincide for abelian model structures on weakly idempotent complete exact categories. This stems from the fact that the biproduct $A \bigoplus B$ serves as both the coproduct $A \coprod B$ and the product $A \prod B$.

Throughout this appendix, we let $\mathfrak{M} = (Q, \mathcal{W}, \mathcal{R})$ be an abelian model structure on a weakly idempotent complete exact category $(\mathcal{A}, \mathcal{E})$. The goal here is to prove Theorem 4.14 from Section 4.3. It gives a comprehensive characterization of the formal left and right homotopy relations in terms of the simple equivalence relations $\sim$, $\sim^{\ell}$, $\sim^{r}$, and $\sim^{\omega}$. Recall that these equivalence relations are defined on morphism sets $\mathrm{Hom}_{\mathcal{A}}(A, B)$ as follows: The relation $\sim := \sim^{\mathcal{W}}$ is defined by $f \sim g$ if and only if $g - f$ factors through some object $W \in \mathcal{W}$. The relation $\sim^{r}$ (resp. $\sim^{\ell}$) is defined by $f \sim^{r} g$ (resp. $f \sim^{\ell} g$) if and only if $g - f$ factors through some object $Q \in Q_{\mathcal{W}}$ (resp. $R \in \mathcal{R}_{\mathcal{W}}$). Finally, $f \sim^{\omega} g$ if and only if $g - f$ factors through some object $W \in \omega := Q \cap \mathcal{W} \cap \mathcal{R}$.

B.1 Path Objects and Right Homotopy

The first goal is to characterize the formal right homotopy relation.

Let us recall a few basic things from Section 1.1. First, finite products, coproducts, and biproducts are all equivalent. By the universal property of $A \bigoplus B$ as a product, a morphism $X \to A \bigoplus B$ is equivalent to a column matrix $\begin{bmatrix} f \\ g \end{bmatrix}$, where $X \xrightarrow{f} A$ and $X \xrightarrow{g} B$. On the other hand, we represent a morphism $A \bigoplus B \to Y$ out of a copoduct as a row matrix $[\, s \;\; t \,]$. In this notation $\pi_A = [\, 1_A \;\; 0 \,]$

and $\pi_B = [\,0\ 1_B\,]$ are the canonical projections, and $\eta_A = \begin{bmatrix} 1_A \\ 0 \end{bmatrix}, \eta_B = \begin{bmatrix} 0 \\ 1_B \end{bmatrix}$ the canonical injections.

Definition B.1 A **path object** for $B \in \mathcal{A}$ is an object $\widetilde{B} \in \mathcal{A}$ along with a factorization of the diagonal map

$$\begin{bmatrix} 1_B \\ 1_B \end{bmatrix} : B \to B \bigoplus B$$

as

$$B \xrightarrow{w} \widetilde{B} \xrightarrow{p} B \bigoplus B$$

where w is a weak equivalence. Furthermore, we have the following.

(1) The factorization is called a **good path object** for B if p is a fibration.
(2) The factorization is called a **very good path object** for B if p is a fibration and w is a (trivial) cofibration.

Note that p must take the form $p = \begin{bmatrix} p_0 \\ p_1 \end{bmatrix}$ for some maps $p_0, p_1 : \widetilde{B} \to B$ satisfying $p_0 = \pi_0 p$ and $p_1 = \pi_1 p$ where π_i are the canonical projections.

By the Factorization Axiom, very good path objects exist for any B. In fact, the following shows how one may construct very good path objects.

Lemma B.2 *Let B be any given object. Then for any morphism $q\colon W \to B$ with domain $W \in \mathcal{W}$ we have a path object for B as shown:*

$$B \xrightarrow{\begin{bmatrix} 1_B \\ 0 \end{bmatrix}} B \bigoplus W \xrightarrow{\begin{bmatrix} 1_B & 0 \\ 1_B & q \end{bmatrix}} B \bigoplus B.$$

Moreover, we have the following.

(1) *This is a good path object if and only if q is a fibration.*
(2) *This is a very good path object if and only if q is a fibration and $W \in \mathcal{Q}_{\mathcal{W}}$.*

Note that we can obtain a fibration q as in (2), and thus a very good path object for B, by using enough projectives of the cotorsion pair $(\mathcal{Q}_{\mathcal{W}}, \mathcal{R})$.

Proof The morphism $\begin{bmatrix} 1_B \\ 0 \end{bmatrix} : B \rightarrowtail B \bigoplus W$ is a weak equivalence since it is an admissible monic with trivial cokernel. So it is clear that the diagram describes a path object. Now suppose that q is a fibration with kernel $i\colon R \rightarrowtail W$ for some $R \in \mathcal{R}$. Then we get a commutative diagram as follows, and its right square is bicartesian. Indeed, by extending the columns of the right square to split short exact sequences, it follows from Proposition 1.12(1) that the right square is bicartesian:

$$\begin{array}{ccccc} R & \overset{i}{\rightarrowtail} & W & \overset{q}{\twoheadrightarrow} & B \\ \| & & \downarrow{\scriptstyle \left[\begin{smallmatrix} 0 \\ 1_W \end{smallmatrix}\right]} & & \downarrow{\scriptstyle \left[\begin{smallmatrix} 0 \\ 1_B \end{smallmatrix}\right]} \\ R & \underset{\left[\begin{smallmatrix} 0 \\ i \end{smallmatrix}\right]}{\rightarrowtail} & B \bigoplus W & \underset{\left[\begin{smallmatrix} 1_B & 0 \\ 1_B & q \end{smallmatrix}\right]}{\twoheadrightarrow} & B \bigoplus B. \end{array}$$

It then follows from Corollary 1.14(2) that the second row is also a short exact sequence. Hence the morphism $\left[\begin{smallmatrix} 1_B & 0 \\ 1_B & q \end{smallmatrix}\right]: B \bigoplus W \to B \bigoplus B$ is also a fibration. Conversely, if $\left[\begin{smallmatrix} 1_B & 0 \\ 1_B & q \end{smallmatrix}\right]$ is a fibration, then so is q since it is the pullback. This proves statement (1). For (2), we just note that $\left[\begin{smallmatrix} 1_B \\ 0 \end{smallmatrix}\right]: B \rightarrowtail B \bigoplus W$ is a trivial cofibration if and only if $W \in \mathcal{Q}_W$. □

In fact, the following asserts that *any* path object for B takes the form of one as in Lemma B.2.

Proposition B.3 *A path object for B is equivalent to a diagram of the form*

$$B \xrightarrow{\left[\begin{smallmatrix} 1_B \\ 0 \end{smallmatrix}\right]} B \bigoplus W \xrightarrow{\left[\begin{smallmatrix} 1_B & 0 \\ 1_B & q \end{smallmatrix}\right]} B \bigoplus B,$$

where $q: W \to B$ is any morphism with domain W some trivial object. Moreover, we have the following.

(1) *This is a good path object if and only if q is a fibration.*
(2) *This is a very good path object if and only if q is a fibration and $W \in \mathcal{Q}_W$.*

Proof For a general path object

$$B \xrightarrow{w} \widetilde{B} \xrightarrow{p} B \bigoplus B,$$

where w is a weak equivalence, p must take the form $\left[\begin{smallmatrix} p_0 \\ p_1 \end{smallmatrix}\right]$ for some maps $p_i: \widetilde{B} \to B$. So then $pw = \left[\begin{smallmatrix} p_0 w \\ p_1 w \end{smallmatrix}\right] = \left[\begin{smallmatrix} 1_B \\ 1_B \end{smallmatrix}\right]$ which implies $p_0 w = 1_B$ and $p_1 w = 1_B$. Therefore w must be a split monic and in particular an admissible monic by the weakly idempotent complete hypothesis. Being a weak equivalence, w must have a trivial cokernel by Proposition 4.7. Setting $g = \operatorname{cok} w: \widetilde{B} \twoheadrightarrow W$, and $s = \ker p_0: W \rightarrowtail \widetilde{B}$ we have, by Proposition 1.4, the biproduct diagram

$$B \underset{w}{\overset{p_0}{\leftrightarrows}} \widetilde{B} \underset{g}{\overset{s}{\leftrightarrows}} W$$

satisfying (i) $p_0 w = 1_B$, (ii) $gs = 1_W$, and (iii) $wp_0 + sg = 1_{\widetilde{B}}$. By Lemma 1.3 the morphism $\left[\begin{smallmatrix} p_0 \\ g \end{smallmatrix}\right]: \widetilde{B} \to B \bigoplus W$ is an isomorphism with inverse $\left[\begin{smallmatrix} w & s \end{smallmatrix}\right]: B \bigoplus W \to \widetilde{B}$, and these isomorphisms are compatible with all maps in the two biproduct diagrams. Through this isomorphism, the path object (w, p) takes the form

$$B \xrightarrow{\left[\begin{smallmatrix} 1_B \\ 0 \end{smallmatrix}\right]} B \bigoplus W \xrightarrow{\left[\begin{smallmatrix} \alpha & \beta \\ \gamma & \delta \end{smallmatrix}\right]} B \bigoplus B$$

for some morphisms $\alpha, \gamma \colon B \to B$ and $\beta, \delta \colon W \to B$. So first,

$$\left[\begin{smallmatrix} 1_B \\ 1_B \end{smallmatrix}\right] = \left[\begin{smallmatrix} \alpha & \beta \\ \gamma & \delta \end{smallmatrix}\right]\left[\begin{smallmatrix} 1_B \\ 0 \end{smallmatrix}\right] \implies \left[\begin{smallmatrix} 1_B \\ 1_B \end{smallmatrix}\right] = \left[\begin{smallmatrix} \alpha \\ \gamma \end{smallmatrix}\right] \implies \alpha = 1_B, \gamma = 1_B.$$

But second, we need to have

$$\left[\begin{smallmatrix} \alpha & \beta \\ \gamma & \delta \end{smallmatrix}\right]\left[\begin{smallmatrix} p_0 \\ g \end{smallmatrix}\right] = \left[\begin{smallmatrix} p_0 \\ p_1 \end{smallmatrix}\right] \implies \left[\begin{smallmatrix} 1_B & \beta \\ 1_B & \delta \end{smallmatrix}\right]\left[\begin{smallmatrix} p_0 \\ g \end{smallmatrix}\right] = \left[\begin{smallmatrix} p_0 \\ p_1 \end{smallmatrix}\right] \implies \left[\begin{smallmatrix} p_0+\beta g \\ p_0+\delta g \end{smallmatrix}\right] = \left[\begin{smallmatrix} p_0 \\ p_1 \end{smallmatrix}\right]$$

$$\implies p_0 + \beta g = p_0 \implies \beta g = 0 \implies \beta g s = 0 \implies \beta = 0.$$

So now we have $\left[\begin{smallmatrix} \alpha & \beta \\ \gamma & \delta \end{smallmatrix}\right] = \left[\begin{smallmatrix} 1_B & 0 \\ 1_B & \delta \end{smallmatrix}\right]$, and we are done since $\delta \colon W \to B$ is the morphism q we have claimed. Now the rest was shown in Lemma B.2. □

Definition B.4 Two maps $f, g \colon A \to B$ are called **right homotopic** if there exists a path object for B,

$$B \xrightarrow{w} \widetilde{B} \xrightarrow{p} B \bigoplus B,$$

such that the product map

$$\left[\begin{smallmatrix} f \\ g \end{smallmatrix}\right] \colon A \to B \bigoplus B$$

lifts over p to a map $H \colon A \to \widetilde{B}$. So $\left[\begin{smallmatrix} f \\ g \end{smallmatrix}\right] = pH$, equivalently, $f = p_0 H$ and $g = p_1 H$. H is then said to be a **right homotopy** from f to g and said to be a **good right homotopy** (resp. **very good right homotopy**) whenever the aforementioned path object is good (resp. very good).

Example B.5 For any B, we may consider B to be a path object on itself by the trivial factorization

$$B \xrightarrow{1_B} B \xrightarrow{\left[\begin{smallmatrix} 1_B \\ 1_B \end{smallmatrix}\right]} B \bigoplus B$$

and one can easily see that $f, g \colon A \to B$ are right homotopic by way of this particular path object if and only if $f = g (= H)$.

The language of Definition B.4 is standard in model category theory. However, for abelian model structures on (weakly idempotent complete) exact categories we have the simpler characterization in Proposition B.6.

Proposition B.6 *Two maps $f, g \colon A \to B$ are right homotopic if and only if their difference factors through a trivial object. That is, f and g are right homotopic if and only if $f \sim g$.*

In fact, f is right homotopic to g by way of some right homotopy for a path object of the form

$$B \xrightarrow{\begin{bmatrix} 1_B \\ 0 \end{bmatrix}} B \bigoplus W \xrightarrow{\begin{bmatrix} 1_B & 0 \\ 1_B & q \end{bmatrix}} B \bigoplus B$$

if and only if $g - f$ factors through this same specific $W \in \mathcal{W}$.

Proof For the "if" part, let $f, g\colon A \to B$ be two morphisms whose difference factors through a trivial object. That is, assume there exists $W \in \mathcal{W}$ and maps $u\colon A \to W$ and $v\colon W \to B$ such that $g - f = vu$. Then taking $p = \begin{bmatrix} 1_B & 0 \\ 1_B & v \end{bmatrix}$ and $H = \begin{bmatrix} f \\ u \end{bmatrix}$, we obtain a path object

$$B \xrightarrow{\begin{bmatrix} 1_B \\ 0 \end{bmatrix}} B \bigoplus W \xrightarrow{\begin{bmatrix} 1_B & 0 \\ 1_B & v \end{bmatrix}} B \bigoplus B$$

and a right homotopy $H\colon A \xrightarrow{\begin{bmatrix} f \\ u \end{bmatrix}} B \bigoplus W$ from f to g. Indeed we have $pH = \begin{bmatrix} 1_B & 0 \\ 1_B & v \end{bmatrix} \begin{bmatrix} f \\ u \end{bmatrix} = \begin{bmatrix} f \\ f+vu \end{bmatrix} = \begin{bmatrix} f \\ g \end{bmatrix}$.

To prove the "only if" part, we use that by Proposition B.3, any path object $\widetilde{B}$ takes the form

$$B \xrightarrow{\begin{bmatrix} 1_B \\ 0 \end{bmatrix}} B \bigoplus W \xrightarrow{\begin{bmatrix} 1_B & o \\ 1_B & q \end{bmatrix}} B \bigoplus B$$

for some morphism $q\colon W \to B$ where W is some trivial object. So turning to Definition B.4, if $f, g\colon A \to B$ are right homotopic, then there exists such a path object and a map $H = \begin{bmatrix} h_0 \\ h_1 \end{bmatrix}\colon A \to B \bigoplus W$ such that

$$\begin{bmatrix} f \\ g \end{bmatrix} = \begin{bmatrix} 1_B & 0 \\ 1_B & q \end{bmatrix} \begin{bmatrix} h_0 \\ h_1 \end{bmatrix} = \begin{bmatrix} h_0 \\ h_0 + qh_1 \end{bmatrix}.$$

So $f = h_0$ and $g = h_0 + qh_1$, and this implies $g - f = qh_1$. In particular, $g - f$ factors through the trivial object W. □

Recall that in Section 4.3 we defined $f \sim^r g$ to mean $g - f$ factors through some object of $\mathcal{Q}_{\mathcal{W}}$. Let us now characterize this relation in terms of the formal right homotopy relation.

Proposition B.7 *Let $f, g\colon A \to B$. Then $f \sim^r g$ if and only if there is a very good right homotopy from f to g. In more detail, the following are equivalent.*

(1) *$f \sim^r g$, that is, $g - f$ factors through some $Q \in \mathcal{Q}_{\mathcal{W}}$.*
(2) *f is right homotopic to g by way of some path object of the form*

$$B \xrightarrow{\begin{bmatrix} 1_B \\ 0 \end{bmatrix}} B \bigoplus Q \xrightarrow{\begin{bmatrix} 1_B & 0 \\ 1_B & q \end{bmatrix}} B \bigoplus B$$

where $Q \in \mathcal{Q}_{\mathcal{W}}$ is the same specific Q as in (1).

(3) *There is a very good right homotopy from f to g.*

(4) *f is right homotopic to g by way of any given (very) good path object for B.*

Proof Statements (1) and (2) are equivalent by the second statement of Proposition B.6. Now we will prove (1) implies (4). So let us be given an arbitrary good path object for B. Then by Proposition B.3 it must take the form

$$B \xrightarrow{\left[\begin{smallmatrix} 1_B \\ 0 \end{smallmatrix}\right]} B \bigoplus W \xrightarrow{\left[\begin{smallmatrix} 1_B & 0 \\ 1_B & q \end{smallmatrix}\right]} B \bigoplus B,$$

where $q\colon W \twoheadrightarrow B$ is some fibration with $W \in \mathcal{W}$. By assumption, $g - f$ factors through an object of $\mathcal{Q}_{\mathcal{W}}$, so as $g - f = ts$. It is easy to see that the t in the factorization lifts over q, since $\mathrm{Ext}^1_{\mathcal{E}}(\mathcal{Q}_{\mathcal{W}}, \mathcal{R}) = 0$. Consequently we can write $g - f = qu$ for some map $u\colon A \to W$. So as in Proposition B.6 we see that $H = \left[\begin{smallmatrix} f \\ u \end{smallmatrix}\right]\colon A \to B \bigoplus W$ is a right homotopy from f to g, compatible with $p = \left[\begin{smallmatrix} 1_B & 0 \\ 1_B & q \end{smallmatrix}\right]$.

Now, (4) implies (3) is trivial since very good path objects for B always exist (by using enough projectives of the cotorsion pair $(\mathcal{Q}_{\mathcal{W}}, \mathcal{R})$ to get a fibration q with domain $W \in \mathcal{Q}_{\mathcal{W}}$). It is left to see (3) implies (1). So let us be given a very good right homotopy from f to g by way of some very good path object. Again, the path object must take the form described in Proposition B.3(2). So by the second statement of Proposition B.6 we conclude that $g - f$ factors through the specific $W \in \mathcal{Q}_{\mathcal{W}}$ appearing in that path object. □

We have justified why in Section 4.3 we call the stronger relation, $\sim^r$, the **very good right homotopy relation**. The "very good" terminology is standard, but unfortunately it leads here to the awkward terminology "very good right homotopic". Hence the alternative *r-homotopic* was also introduced in Section 4.3. Regardless of terminology, we have completed the proof of the following portion of the main Theorem 4.14.

Theorem B.8 *Let $f, g\colon A \to B$ be morphisms and assume $\mathcal{A}$ is weakly idempotent complete.*

(1) *$f \sim g$ if and only if f is formally right homotopic to g.*

(2) *$f \sim^r g$ if and only if there is a very good right homotopy from f to g.*

Thus formal right homotopy and very good right homotopy are each equivalence relations, even when A and B are neither fibrant nor cofibrant.

There are a couple of obvious questions regarding right homotopy that we have not yet addressed. In particular, if f is right homotopic to g, is there always a (very) good right homotopy from f to g? The following exercises provide answers.

Exercise B.1.1 Show that if f is right homotopic to g, then there exists a good right homotopy from f to g.

Notice that Exercise B.1.1 shows that if we define a relation called *good right homotopy*, it does not give anything new. It again coincides with the standard right homotopy relation. On the other hand, we have the following.

Exercise B.1.2 By way of example, show that the very good right homotopy relation $\sim^r$ is stronger than the general right homotopy relation $\sim$. (Of course you will need to look for morphisms whose domain is *not* cofibrant.)

B.2 Cylinder Objects and Left Homotopy

For reference, and to complete the proof of Theorem 4.14, we state the most notable dual results from Section B.1.

Definition B.9 A **cylinder object** for $A \in \mathcal{A}$ is an object $\widehat{A} \in \mathcal{A}$ along with a factorization of the codiagonal map

$$[\, 1_A \; 1_A \,] \colon A \bigoplus A \to A$$

as

$$A \bigoplus A \xrightarrow{j} \widehat{A} \xrightarrow{w} A,$$

where w is a weak equivalence. Furthermore we have the following.

(1) The factorization is called a **good cylinder object** for A if j is a cofibration.
(2) The factorization is called a **very good cylinder object** for A if j is a cofibration and w is a (trivial) fibration.

Note that j must take the form $j = [\, j_0 \; j_1 \,]$ for some maps $j_0, j_1 \colon A \to \widehat{A}$ satisfying $j_0 = j\eta_0$ and $j_1 = j\eta_1$ where η_i are the canonical injections.

Lemma B.10 *A cylinder object for A is equivalent to a diagram of the form*

$$A \bigoplus A \xrightarrow{\left[\begin{smallmatrix} 1_A & 1_A \\ 0 & i \end{smallmatrix}\right]} A \bigoplus W \xrightarrow{[\, 1_A \; 0 \,]} A,$$

where $i \colon A \to W$ is any morphism with codomain W some trivial object. Moreover, we have the following.

(1) *This is a good cyclinder object if and only if i is a cofibration.*

(2) *This is a very good cylinder object if and only if i is a cofibration and $W \in \mathcal{R}_W$.*

Definition B.11 Two maps $f, g\colon A \to B$ are called **left homotopic** if there exists a cylinder object for A

$$A \bigoplus A \xrightarrow{j} \widehat{A} \xrightarrow{w} A$$

such that the coproduct map

$$[\, f \; g \,]\colon A \bigoplus A \to B$$

extends over j to a map $H\colon \widehat{A} \to B$. So $[\, f \; g \,] = Hj$, equivalently, $f = Hj_0$ and $g = Hj_1$. H is then said to be a **left homotopy** from f to g and said to be a **good left homotopy** (resp. **very good left homotopy**) whenever the aforementioned cylinder object is good (resp. very good).

We now get to the dual of Proposition B.6. It implies that left and right homotopy coincide. Moreover, they are always an equivalence relation, specifically the equivalence relation $\sim := \sim^{\mathcal{W}}$.

Proposition B.12 *Two maps $f, g\colon A \to B$ are left homotopic if and only if their difference factors through a trivial object. That is, f and g are left homotopic if and only if $f \sim g$.*

In fact, f is left homotopic to g by way of some left homotopy for a cylinder object of the form

$$A \bigoplus A \xrightarrow{\left[\begin{smallmatrix} 1_A & 1_A \\ 0 & i \end{smallmatrix}\right]} A \bigoplus W \xrightarrow{[\, 1_A \; 0 \,]} A$$

if and only if $g - f$ factors through this same specific $W \in \mathcal{W}$.

Recall $f \sim^{\ell} g$ means $g - f$ factors through some object of $\mathcal{R}_W$. We can call this the **very good left homotopy relation**.

Proposition B.13 *Let $f, g\colon A \to B$. Then $f \sim^{\ell} g$ if and only if there is a very good left homotopy from f to g. In more detail, the following are equivalent.*

(1) *$f \sim^{\ell} g$, that is, $g - f$ factors through some $R \in \mathcal{R}_W$.*
(2) *f is left homotopic to g by way of some cylinder object of the form*

$$A \bigoplus A \xrightarrow{\left[\begin{smallmatrix} 1_A & 1_A \\ 0 & i \end{smallmatrix}\right]} A \bigoplus R \xrightarrow{[\, 1_A \; 0 \,]} A,$$

where $R \in \mathcal{R}_W$ is the same specific R as in (1).
(3) *There is a very good left homotopy from f to g.*

(4) *f is left homotopic to g by way of any given (very) good cylinder object for A.*

The dual of Theorem B.8 is the following characterization of the (very good) left homotopy relation.

Theorem B.14 *Let $f, g\colon A \to B$ be morphisms and assume $\mathcal{A}$ is weakly idempotent complete.*

(1) *$f \sim g$ if and only if f is formally left homotopic to g.*
(2) *$f \sim^{\ell} g$ if and only if there is a very good left homotopy from f to g.*

Thus formal left homotopy and very good left homotopy are each equivalence relations, even when A and B are neither fibrant nor cofibrant.

B.3 Proof of the Main Theorem

We have now proved the main Theorem 4.14. Indeed statements (1) and (2) are immediate from Theorems B.8 and B.14. Then statements (3) and (4) of Theorem 4.14 are immediate from Lemma 4.11.

Appendix C

Bibliographical Notes

Chapter 1 The axioms for an exact category as stated in Definition 1.9 are equivalent to Quillen's original definition from Quillen [1973]. We have followed Bühler's excellent survey [Bühler, 2010]. In fact, this book would not appear the way it does without Bühler's article and there is much overlap between this chapter and Bühler [2010]. Two relevant works of Bernhard Keller on the subject of exact categories are Keller [1990, 1996]; in particular, the redundancy of Quillen's Obscure Axiom was shown in Keller [1990, appendix A]. Bühler's article is also recommended for the more subtle history of exact categories that it gives. The equivalence between exact structures and Mac Lane's notion of a proper class in an abelian category was shown in Gillespie [2016c, appendix B]. Other results in this chapter are straightforward generalizations of results for abelian categories. For instance, the proof of Lemma 1.24 appeared in the context of abelian categories in Colpi and Fuller [2007, proposition 8.1].

Chapters 2 and 3 Cotorsion pairs (first called *cotorsion theories*) were introduced in the context of abelian groups by Luigi Salce [1979]. They became standard gadgets in homological algebra especially after their use in Eklof and Trlifaj [2001] and Bican et al. [2001] to settle Enochs' Flat Cover Conjecture. Some standard references for cotorsion pairs in categories of modules are Enochs and Jenda [2000] and Göbel and Trlifaj [2006].

Interactions between (co)torsion pairs and Quillen model structures on additive and abelian categories were also studied in Beligiannis and Reiten [2007]. The Lifting Extension Lemma 2.9 explicitly appears in Beligiannis and Reiten [2007, lemma 3.1, p. 139]. The Hereditary Test, Theorem 2.16, appeared in Šťovíček [2014, lemma 6.17] and previously in Saorín and Šťovíček [2011, lemma 4.25]. The Generalized Horseshoe Lemma and its proof given in Lemma 2.18 goes back at least to Auslander and Reiten [1991, proposition 3.6].

Chapters 4 and 5 Abelian model structures were defined, and Hovey's Correspondence was proved, in Hovey [2002]. After reading Bühler [2010], the author extended abelian model structures and Hovey's Correspondence to weakly idempotent complete exact categories in Gillespie [2011]. Šťovíček gave a more thorough and nicely readable account in Šťovíček [2014]. Our approach to the homotopy category in Chapter 5 is based on the approach in Dwyer and Spalinski [1995]. The proof of Proposition 5.1 is based on Li [2017, lemma 2.3].

Chapter 6 Triangulated categories go back to Verdier [1967] and have been studied by many people since then. A standard modern reference is Neeman [2001]. May showed in May [2001] that (Tr-3) is a redundant axiom, following from the others and the stronger (Tr-4). The material from Section 6.1 is based on Beligiannis and Marmaridis [1994] and Assem et al. [1998]. One-sided triangulations implicitly go back to Heller [1960], and they appear explicitly in Keller and Vossieck [1987]. The material from Section 6.3 was proved in Beligiannis and Marmaridis [1994] though in a slightly different setting and with slightly different language. We have also encorporated some of the ideas of the approach in Li [2015]. These constructions are rooted in the classical construction of the triangulated stable category of a Frobenius category [Happel 1988]. As for Sections 6.4 and 6.5, we learned that $\mathrm{Ho}(\mathfrak{M})$ is always triangulated, even for not necessarily hereditary model structures, from Nakaoka and Palu [2019]. In fact, they introduced the more general notion of an extriangulated category, and prove Hovey's correspondence in this context. See also Palu [2024].

Chapter 7 Most results in Sections 7.1, 7.2, and 7.4 are additive/abelian versions of standard facts concerning derived functors; the author especially consulted Dwyer and Spalinski [1995] and Hovey [1999]. The material on abelian monoidal model structures is based on Hovey [1999] and Hovey [2002].

Chapter 8 Much of Sections 8.1 and 8.2 comes from Gillespie [2011], Šťovíček [2014], or Becker [2015]. Again, the triangulatation on the stable category of a Frobenius category is from Happel [1988]. The main result of Section 8.4 was proved in Gillespie [2015b]. The approach taken in Section 8.4, in particular the generalization to nonhereditary cotorsion pairs given in Theorem 8.15, is from Nakaoka and Palu's work in Nakaoka and Palu [2019]. The results of Section 8.5 come from Gillespie [2021].

The standard reference for recollement is Beilinson et al. [1982], and the work in Krause [2005] has been an important influence. Model category intepretations of recollements were first given by Hanno Becker in Becker [2014] and Becker [2015], and later generalized in Gillespie [2016a], Gillespie [2016b], and Gillespie [2016d]. With only some modifications made here, Section 8.7 was first published in Gillespie [2016d]. We note that Proposition 8.29 is a generalization of results from Becker [2014].

Chapter 9 Efficient exact categories were introduced in Saorín and Šťovíček [2011], and many of the results of this chapter (in particular, Propositions 9.19, 9.21, 9.29, and Theorem 9.34) are inspired by this paper. Eklof's Lemma (and its dual) goes back to Eklof [1977], Eklof and Trlifaj [2001], and Trlifaj [2003, lemma 2.3], where it arose in the context of modules over a ring. It was also shown in Eklof and Trlifaj [2001] that any cotorsion pair of R-modules is complete whenever it is cogenerated by a set. Many authors have presented generalizations of these results over the years. It was pointed out in Positselski and Šťovíček [2023] that Eklof's Lemma holds for general exact categories.

What we call a *cofibrantly generated cotorsion pair* is very close to what Hovey called a *small* cotorsion pair in Hovey [2002]. Our approach avoids assuming that all admissible monics are closed under transfinite compositions, and provides an explicit notion of smallness for the domains of the generating monics. The proof of Theorem 9.34 is based on Quillen's small object argument and the proof of Enochs and Jenda [2000, theorem 7.4.1]. The proof there is for R-modules and uses enough projectives, but Proposition 9.29 allows us to avoid using enough projectives. The material from Section 9.10 brings together some results scattered about in Šťovíček [2015], Gillespie [2017b], Estrada et al. [2024], and Positselski and Šťovíček [2023].

Chapter 10 Since here we are working in general exact categories, much of the early sections of this chapter is again influenced by Bühler's survey [Bühler, 2010]. In particular, Proposition 10.15 is extracted from Bühler and is due to Keller. See Keller [1990] and Keller [1996]. There are several ways to approach the topic in Section 10.9. Besides Gillespie [2004] we have followed the approach taken in Yang and Liu [2011] and later advocated in Šťovíček [2014]. The result in Exercise 10.9.5 was shown in Bazzoni et al. [2020, theorem 5.3].

Chapter 11 The author first learned of mixed model structures from Timothée Moreau, when he shared his work on mixing abelian model structures [Moreau, 2020]. Chapter 11 is just the author's interpretation of Moreau's work. The general idea of mixed model structures comes from Cole's paper [2006].

Chapter 12 Jeff Smith's idea of a combinatorial model category was described and developed by Dan Dugger in Dugger [2001]. Well-generated triangulated categories were introduced in Neeman [2001], and a revised definition appeared in Krause [2001]. Jiri Rosický showed in Rosický [2005] that the homotopy category of any (pointed) combinatorial model category is well generated. We have followed Dugger [2001] and Rosický [2005] throughout the chapter to prove the main Theorem 12.21.

Appendix B The proof in Appendix B of Theorem 4.14 fixes a mistake published in Gillespie [2011, proposition 4.4]. This was brought to the author's attention by Sergio Estrada after a question was raised by a student of his. Fortunately the misstatement there is rather harmless because in practice one applies the result to morphisms with cofibrant domain and fibrant codomain.

Further References In addition to the works previously cited in the text, the reader may find the following further references to be helpful. This is by no means (even an attempt to give) an exhaustive list of the relevant literature.

Classical or influential works related to stable module theory and Gorenstein rings include Bass [1963], Auslander and Bridger [1969], Olaf and Buchweitz [1986], Enochs and Jenda [1995], Beligiannis [2001], Holm [2004], and Jorgensen [2007]. Iwanaga–Gorenstein rings originated in Iwanaga [1979] and Iwanaga [1980], and their coherent analog appeared in Ding and Chen [1993] and Ding and Chen [1996]. The Gorenstein flat model structure on modules was constructed in Gillespie [2017a].

Just a few additional works that study chain complexes include García-Rozas [1999], Enochs and Jenda [2011], Golasiński and Gromadzki [1982], Neeman [2008], and Enochs, Estrada, and Iacob [2014]. The first paper to study the chain homotopy category of projective modules was Jørgensen [2005], and another early study of compactly generated chain homotopy categories was done in Holm and Jørgensen [2007].

Many of the works previously cited in this book construct abelian model structures on (complexes of) sheaves or quasi-coherent sheaves. Beyond those already cited, further resources for this include Hovey [2001], Enochs and Oyonarte [2001], Gillespie [2006], Estrada, Gillespie, and Odabaşi [2017] and Estrada and Gillespie [2019].

This book did not consider abelian model structures on functor categories. We refer the interested reader to Holm and Jørgensen [2019a], Holm and Jørgensen [2019b], Holm and Jørgensen [2022], Holm and Jørgensen [2024a], and Holm and Jørgensen [2024b].

Finally, another standard text on the general theory of Quillen model categories is Hirschhorn [2003]. The book Pérez [2016] is another treatment of the special case of abelian model structures.

References

Adámek, J. and J. Rosický, *Locally presentable and accessible categories*, London Mathematical Society Lecture Note Series vol. 189, Cambridge University Press, Cambridge, 1994.

Assem, I., A. Beligiannis, and N. Marmaridis, *Right triangulated categories with right semi-equivalences*, Canadian Mathematical Society Conference Proceedings vol. 24, 1998, pp. 17–37.

Auslander, M. and M. Bridger, *Stable module theory*, Mem. Amer. Math. Soc. No. 94, American Mathematical Society, Providence, RI, 1969, 146pp.

Auslander, M. and I. Reiten, *Applications of contravariantly finite subcategories*, Adv. Math. vol. 86, 1991, pp. 111–152.

Balmer, P., *The spectrum of prime ideals in tensor triangulated categories*, J. Reine Angew. Math. no. 588, 2005, pp. 149–168.

Bass, H. *On the ubiquity of Gorenstein rings*, Math. Zeit. vol. 82, 1963, pp. 8–28.

Bazzoni, S., M. Cortés-Izurdiaga, and S. Estrada, *Periodic modules and acyclic complexes*, Algebr. Represent. Theory vol. 23, no. 5, 2020, pp. 1861–1883.

Becker, H. *Models for singularity categories*, Adv. Math. vol. 254, 2014, pp. 187–232.

Becker, H., *Homotopy-theoretic studies of Khovanov-Rozansky homology*, PhD thesis, University of Bonn, 2015 (online at www.math.uni-bonn.de/people/habecker/).

Beilinson, A. A., J. Bernstein, and P. Deligne, *Faisceaux pervers*, Astérisque vol. 100, 1982, pp. 5–171.

Beke, T., *Sheafifiable homotopy model categories*, Math. Proc. Camb. Phil. Soc. vol. 129, no. 3, 2000, pp. 447–475.

Beligiannis, A., *Homotopy theory of modules and Gorenstein rings*, Math. Scand. vol. 89, 2001, pp. 5–45.

Beligiannis, A. and N. Marmaridis, *Left triangulated categories arising from contravariantly finite subcategories*, Comm. Alg. vol. 22, no. 12, 1994, pp. 5021–5036.

Beligiannis, A. and I. Reiten, *Homological and homotopical aspects of torsion theories*, Mem. Amer. Math. Soc. vol. 188, no. 883, 2007.

Bican, L., R. El Bashir and E. Enochs, *All modules have flat covers*, Bull. London Math. Soc. vol. 33, no. 4, 2001, pp. 385–390.

Bravo, D., *The stable derived category of a ring via model categories*, PhD thesis, Wesleyan University, May 2011.

Bravo, D., J. Gillespie and M. Hovey, *The stable module category of a general ring*, 2014, arXiv:1405.5768.

Buchweitz, R.-O., *Maximal Cohen–Macaulay modules and Tate-cohomology over Gorenstein rings*, Unpublished manuscript, http://hdl.handle.net/1807/16682, 1986.

Bühler, T., *Exact Categories*, Expo. Math. vol. 28, no. 1, 2010, pp. 1–69.

Cole, M., *Mixing model structures*, Topology Appl. vol. 153, 2006, pp. 1016–1032.

Christensen, L. W., S. Estrada, and P. Thompson, *The stable category of Gorenstein flat sheaves on a Noetherian scheme*, Proc. Amer. Math. Soc. vol. 149, no. 2, 2021, pp. 525–538.

Colpi, R. and K. R. Fuller, *Tilting objects in abelian categories and quasitilted rings*, Trans. Amer. Math. Soc. vol. 359, no. 2, 2007, pp. 741–765.

Ding, N. and J. Chen, *The flat dimensions of injective modules*, Manuscripta Math. vol. 78, no. 2, 1993, pp. 165–177.

Ding, N. and J. Chen, *Coherent rings with finite self-FP-injective dimension*, Comm. Algebra vol. 24, no. 9, 1996, pp. 2963–2980.

Dugger, D., *Combinatorial model categories have presentations*, Adv. Math. vol. 164, 2001, pp. 177–201.

Dwyer, W. G. and J. Spalinski, *Homotopy theories and model categories*, Handbook of Algebraic Topology, North-Holland, Amsterdam, 1995, pp. 73–126.

Eklof, P. C., *Homological algebra and set theory*, Trans. Amer. Math. Soc. vol. 227, 1977, pp. 207–225.

Eklof, P. C. and J. Trlifaj, *How to make Ext vanish*, Bull. London Math. Soc. vol. 33, no. 1, 2001, pp. 41–51.

Emmanouil, I., *On pure acyclic complexes*, J. Algebra vol. 465, 2016, pp. 190–213.

Emmanouil, I. and I. Kaperonis, *On K-absolutely pure complexes*, J. Algebra vol. 640, 2024, pp. 274–299.

Enochs, E., and S. Estrada, *Relative homological algebra in the category of quasi-coherent sheaves*, Adv. Math. vol. 194, no. 2, 2005, pp. 284–295.

Enochs, E., S. Estrada, and A. Iacob, *Cotorsion pairs, model structures and adjoints in homotopy categories*, Houston J. Math. vol. 40, no. 1, 2014, pp. 43–61.

Enochs, E. E. and O. M. G. Jenda, *Gorenstein injective and projective modules*, Math. Zeit. vol. 220, 1995, pp. 611–633.

Enocks, E. and O. Jenda, *Relative homological algebra*, de Gruyter Expositions in Mathematics vol. 30, Walter de Gruyter, Berlin, 2000.

Enochs, E. and O. Jenda, *Relative homological algebra*, Volume 2, de Gruyter Expositions in Mathematics vol. 54, Walter de Gruyter, Berlin, 2011, xii+96 pp.

Enochs, E. and L. Oyonarte, *Flat covers and cotorsion envelopes of sheaves*, Proc. Amer. Maths. Soc. vol. 130, no. 5, 2001, pp. 1285–1292.

Estrada, S. and J. Gillespie, *The projective stable category of a coherent scheme*, Proc. Royal Soc. Edinburgh vol. 149, no. 1, 2019, pp. 15–43.

Estrada, S. and J. Gillespie, *Quillen equivalences inducing Grothendieck duality for unbounded chain complexes of sheaves*, Commun. Contemp. Math., 2024. https://doi.org/10.1142/S0219199724500342.

Estrada, S., J. Gillespie, and S. Odabaşı, *Pure exact structures and the pure derived category of a scheme*, Math. Proc. Cambridge Philos. Soc. vol. 163, no. 2, 2017, pp. 251–264.

Estrada, S., J. Gillespie, and S. Odabaşı, *K-flatness in Grothendieck categories: Application to quasi-coherent sheaves*, Collect. Math., 2024. https://doi.org/10.1007/s13348-024-00439-7

Frerik, L. and D. Sieg, *Exact categories in functional analysis*, 2010. Online lecture notes, available at: www.math.uni-trier.de/abteilung/analysis/HomAlg.pdf.

Freyd, P. *Splitting homotopy idempotents*, Proceeding of the Conference on Categorical Algebra, La Jolla 1965, Springer, New York, 1966, pp. 173–176.

García-Rozas, J. R., *Covers and envelopes in the category of complexes of modules*, Research Notes in Mathematics no. 407, Chapman & Hall/CRC, Boca Raton, FL, 1999.

Gillespie, J., *The flat model structure on Ch(R)*, Trans. Amer. Math. Soc. vol. 356, no. 8, 2004, pp. 3369–3390.

Gillespie, J., *The flat model structure on chain complexes of sheaves*, Trans. Amer. Math. Soc. vol. 358, no. 7, 2006, pp. 2855–2874.

Gillespie, J., *Kaplansky classes and derived categories*, Math. Zeit. vol. 257, no. 4, 2007, pp. 811–843.

Gillespie, J., *Model structures on modules over Ding–Chen rings*, Homology, Homotopy Appl. vol. 12, no. 1, 2010, pp. 61–73.

Gillespie, J., *Model structures on exact categories*, J. Pure. Appl. Algebra vol. 215, 2011, pp. 2892–2902.

Gillespie, J., *The homotopy category of N-complexes is a homotopy category*, J. Homotopy and Related Structures vol. 10, no. 1, 2015a, pp. 93–106.

Gillespie, J., *How to construct a Hovey triple from two cotorsion pairs*, Fundamenta Mathematicae vol. 230, no. 3, 2015b, pp. 281–289.

Gillespie, J., *Gorenstein complexes and recollements from cotorsion pairs*, Adv. Math. vol. 291, 2016a, pp. 859–911.

Gillespie, J., *Exact model structures and recollements*, J. Algebra vol. 458, 2016b, pp. 265–306.

Gillespie, J., *The derived category with respect to a generator*, Ann. Mat. Pura Appl. (4) vol. 195, no. 2, 2016c, pp. 371–402.

Gillespie, J., *Models for mock homotopy categories of projectives*, Homology, Homotopy Appl. vol. 18, no. 1, 2016d, pp. 247–263.

Gillespie, J., *Hereditary abelian model categories*, Bull. Lond. Math. Soc. vol. 48, no. 6, 2016e, pp. 895–922.

Gillespie, J., *The flat stable module category of a coherent ring*, J. Pure Appl. Algebra vol. 221, no. 8, 2017a, pp. 2025–2031.

Gillespie, J., *On Ding injective, Ding projective and Ding flat modules and complexes*, Rocky Mountain J. Math. vol. 47, no. 8, 2017b, pp. 2641–2673.

Gillespie, J., *Canonical resolutions in hereditary abelian model categories*, Pacific J. Math. vol. 313, no. 2, 2021, pp. 365–411.

Gillespie, J., *K-flat complexes and derived categories*, Bull. London Math. Soc. vol. 55, no. 1, 2023a, pp. 119–136.

Gillespie, J., *The homotopy category of acyclic complexes of pure–projective modules*, Forum Mathematicum vol. 35, no. 2, 2023b, pp. 507–521.

Göbel, R. and J. Trlifaj, *Approximations and endomorphism algebras of modules*, de Gruyter Expositions in Mathematics vol. 41, Walter de Gruyter & Co., Berlin, 2006.

Golasiński, M. and G. Gromadzki, *The homotopy category of chain complexes is a homotopy category*, Colloq. Math. vol. 47, no. 2, 1982, pp. 173–178.

Happel, D. *Triangulated categories in the representation theory of finite-dimensional algebras*, London Mathematical Society Lecture Note Series vol. 119, Cambridge University Press, Cambridge, 1988.

Heller, A., *The loop-space functor in homological algebra*, Trans. Amer. Math. Soc. vol. 96, no. 3, 1960, pp. 382–394.

Hirschhorn, P. S., *Model categories and their localizations*, Mathematical Surveys and Monographs vol. 99, American Mathematical Society, 2003.

Holm, H., *Gorenstein homological dimensions*, J. Pure Appl. Algebra vol. 189, 2004, pp. 167–193.

Holm, H. and P. Jørgensen, *Compactly generated homotopy categories*, Homology, Homotopy Appl. vol. 9, no. 1, 2007, pp. 257–274.

Holm, H. and P. Jørgensen, *Cotorsion pairs in categories of quiver representations*, Kyoto J. Math. vol. 59, no. 3, 2019, pp. 575–606.

Holm, H. and P. Jørgensen, *Model categories of quiver representations*, Adv. Math. vol. 357, 2019, Article no. 106826, 46 pp.

Holm, H. and P. Jørgensen, *The Q-shaped derived category of a ring*, J. London Math. Soc. (2) vol. 106, no. 4, 2022, pp. 3263–3316.

Holm, H. and P. Jørgensen, *The Q-shaped derived category of a ring – compact and perfect objects*, Trans. Amer. Math. Soc. vol. 377, 2024a, pp. 3095–3128. https://doi.org/10.1090/tran/8979.

Holm, H. and P. Jørgensen, *A brief introduction to the Q-shaped derived category*, Abel Symp. vol. 17, 2024b, pp. 141–167. Proceedings of the Abel Symposium 2022.

Hovey, M., *Model categories*, Mathematical Surveys and Monographs vol. 63, American Mathematical Society, 1999.

Hovey, M., *Model category structures on chain complexes of sheaves*, Trans. Amer. Math. Soc. vol. 353, no. 6, 2001, pp. 2441–2457.

Hovey, M., *Cotorsion pairs, model category structures, and representation theory*, Mathematische Zeitschrift vol. 241, 2002, pp. 553–592.

Hovey, M., J. Palmieri, and N. Strickland, *Axiomatic stable homotopy theory*, Mem. Amer. Math. Soc. vol. 128, no. 610, 1997.

Iverson, B., *Cohomology of sheaves*, Universitext, Springer-Verlag, Berlin, 1986.

Iwanaga, Y., *On rings with finite self-injective dimension*, Comm. Algebra vol. 7, no. 4, 1979, pp. 393–414.

Iwanaga, Y., *On rings with finite self-injective dimension II*, Tsukuba J. Math. vol. 4, no. 1, 1980, pp. 107–113.

Jørgensen, P., *The homotopy category of complexes of projective modules*, Adv. Math. vol. 193, no. 1, 2005, pp. 223–232.

Jørgensen, P., *Existence of Gorenstein projective resolutions and Tate cohomology*, J. Eur. Math. Soc. (JEMS) vol. 9, no. 1, 2007, pp. 59–76.

Keller, B., *Chain complexes and stable categories*, Manuscripta Math. vol. 67, no. 4, 1990, pp. 379–417.

Keller, B., *Derived categories and their uses*, in: Handbook of Algebra vol. 1, North-Holland, Amsterdam, 1996, pp. 671–701.

Keller, B. and D. Vossieck, *Sous les catégories dérivées*, C. R. Acad. Sci. Paris vol. 305, 1987, pp. 225–228.

Kelly, J., *The homotopy theory of convenient modules*, 2023, preprint.

Kelly, J., *Homotopy in exact categories*, Mem. Amer. Math. Soc. vol. 298, no. 1490, 2024. https://doi.org/10.1090/memo/1490.

Krause, H., *On Neeman's well generated triangulated categories*, Doc. Math. 6, 2001, pp. 121–126.

Krause, H., *The stable derived category of a Noetherian scheme*, Compos. Math. vol. 141, no. 5, 2005, pp. 1128–1162.

Li, Z.-W., *The left and right triangulated structures of stable categories*, Comm. Alg. vol. 43, no. 9, 2015, pp. 3725–3753.

Li, Z.-W., *A note on model structures on arbitrary Frobenius categories*, Czech. Math. J. vol. 67, no. 2, 2017, pp. 329–337.

Mac Lane, S., *Homology*, Die Grundlehren der mathematischen Wissenschaften vol.114, Springer-Verlag, Berlin, Heidelberg, 1963.

Mac Lane, S., *Categories for the working mathematician*, Graduate Texts in Mathematics vol. 5, Springer-Verlag, New York, 2nd ed., 1998.

May, P., *The additivity of traces in triangulated categories*, Adv. Math. vol. 163, no. 1, 2001, pp. 34–73.

Mitchell, B., *Theory of categories*, Pure and Applied Mathematics vol. 17, Academic Press, New York, London, 1965.

Moreau, T., *Stability of (mixed) exact model categories*, personal communication, 2020.

Nakaoka, H. and Y. Palu, *Extriangulated categories, Hovey twin cotorsion pairs and model structures*, Cah. Topol. Géom. Différ. Catég. LX, 2019.

Neeman, A., *Triangulated categories*, Annals of Mathematics Studies vol. 148, Princeton University Press, Princeton, 2001.

Neeman, A., *The homotopy category of flat modules, and Grothendieck duality*, Invent. Math. vol. 174, no. 2, 2008, pp. 255–308.

Palu, Y., *Some applications of extriangulated categories*, Abel Symp. 17, 2024, pp. 217–254. Proceedings of the Abel Symposium 2022.

Pérez, M., *Introduction to abelian model structures and Gorenstein homological dimensions*, Monographs and Research Notes in Mathematics, Chapman & Hall, 2016.

Positselski, L. and J. Šťovíček, *Flat quasi-coherent sheaves as direct limits, and quasi-coherent cotorsion periodicity*, 2023, arXiv:2212.09639v1.

Prosmans, F., *Derived categories for functional analysis*, Publ. Res. Inst. Math. Sci. vol. 36, no. 1, 2000, pp. 19–83.

Prosmans, F. and J.-P. Schneiders, *A homological study of bornological spaces*, Laboratoire analyse, géométrie et applications, Unité mixte de recherche, Institut Galileé, Université Paris 13, CNRS, 2000.

Quillen, D., *Homotopical algebra*, SLNM vol. 43, Springer-Verlag, Berlin, Heidelberg, 1967.

Quillen, D. *Rational homotopy theory*, Ann. of Math. vol. 90, no. 2, 1969, pp. 205–295.

Quillen, D., *Higher algebraic K-theory I*, SLNM vol. 341, Springer-Verlag, Berlin, Heidelberg, 1973, pp. 85–147.

Rosický, J., *Generalized Brown representability in homotopy categories*, Theory Appl. Categ. vol. 14, no. 19, 2005, pp. 451–479.

Salce, L., *Cotorsion theories for abelian groups*, Symposia Math. vol. 23, 1979, pp. 11–32.

Saorín, M. and J. Šťovíček, *On exact categories and applications to triangulated adjoints and model structures*, Adv. Math. vol. 228, no. 2, 2011, pp. 968–1007.

Šaroch, J. and J. Šťovíček, *Singular compactness and definability for Σ-cotorsion and Gorenstein modules*, Selecta Math. (N.S.) vol. 26, no. 2, 2020, Paper No. 23, 40 pp.

Schneiders, J.-P., *Quasi-abelian categories and sheaves*, Mém. Soc. Math. Fr. vol. 76, Société mathématique de France, 1999.

Spaltenstein, N., *Resolutions of unbounded complexes*, Compos. Math. vol. 65, no. 2, 1988, pp. 121–154.

Stenström, B., *Rings of quotients*, Die Grundlehren der mathematischen Wissenschaften in Einzeldarstellungen Band 217, Springer-Verlag, New York, 1975.

Šťovíček, J., *Exact model categories, approximation theory, and cohomology of quasi-coherent sheaves*, Advances in Representation Theory of Algebras (ICRA Bielefeld, Germany, 8–17 August, 2012), EMS Series of Congress Reports, European Mathematical Society Publishing House, Helsinki, 2014, pp. 297–367.

Šťovíček, J., *On purity and applications to coderived and singularity categories*, 2015, arXiv:1412.1615.

Trlifaj, J., *Ext and inverse limits*, Illinois J. Math. vol. 47, nos. 1–2, 2003, pp. 529–538.

Verdier, J.-L., *Des catégories dérivées des catégories abéliennes*, Astérisque vol. 239 [1967], Société Mathématique de France, 1996, pp. xii + 253.

Weibel, C. A., *An introduction to homological algebra*, Cambridge Studies in Advanced Mathematics vol. 38, Cambridge University Press, Cambridge, 1994.

Yang, G. and Z. Liu, *Cotorsion pairs and model structures on* Ch(R), Proc. Edinburgh Math. Soc. (2) vol. 54, no. 3, 2011, pp. 783–797.

Index

Printed in the United States
by Baker & Taylor Publisher Services